KU-777-942

Fluid Power Control

edited by

JOHN F. BLACKBURN
Research Associate in Mechanical Engineering
Massachusetts Institute of Technology

GERHARD REETHOF
General Electric Company
Evendale, Ohio

J. LOWEN SHEARER
Associate Professor of Mechanical Engineering
Massachusetts Institute of Technology

contributing
authors

JOHN F. BLACKBURN

JAMES L. COAKLEY

FREDDIE D. EZEKIEL

RICHARD H. FRAZIER

THOMAS E. HOFFMAN

JOHN A. HRONES

SHIH-YING LEE

HENRY M. PAYNTER

GERHARD REETHOF

J. LOWEN SHEARER

ALAN H. STENNING

*Staff Members of
the Dynamic Analysis and Control Laboratory
of the Massachusetts Institute of Technology*

Fluid Power Control

THE UNIVERSITY OF STRATHCLYDE
ANDERSONIAN
LIBRARY

The M.I.T. Press

Massachusetts Institute of Technology

Cambridge, Massachusetts

58278
5·5·65

Copyright © 1960 by The Massachusetts Institute of Technology ═══

All Rights Reserved
This book or any part thereof must not
be reproduced in any form without the
written permission of the publisher.

Library of Congress Catalog Card Number: 59–6759

Printed in the United States of America

D
621·2
FLU

621 - 522

HS

To Nathaniel McLean Sage for ever-staunch support, wise counsel, and deep understanding of people

Foreword

Fluid power has long been used by man. Sailboats were in use approximately 5000 years ago, and there is evidence that the first water wheels were built between 200 and 100 B.C. Indeed, before the development of the steam engine the only sources of mechanical power were human or animal muscle, and moving fluids—water or wind power. The steam boiler and engine made possible the conversion to mechanical power—via a fluid-power link—of the energy stored in fossil fuels, but this naturally increased man's interest in the problems and possibilities of fluid power.

Since the transmission of water power over relatively long distances is difficult and expensive, power-using operations were originally located on rivers near the sources of power. The development of the techniques of electric-power transmission has permitted power-using industries to locate in spots most convenient from other standpoints, and has also permitted the utilization of many water-power sources that would otherwise have been wasted. In connection with this utilization, the hydraulic engineer has had to deal with the problems of using water and steam power. He has developed the theoretical background, supported by extensive measurements, to deal with steady-flow situations. Only in recent years, however, has it become necessary to analyze the dynamics of systems within which the rate of flow of fluid power is rapidly varying.

The outbreak of World War II forced an enormous acceleration of research and development effort, aimed at the rapid evolution of armament and strategy. One important aspect of this work was the development of automatic fire-control systems, including mechanisms that would cause heavy guns to respond accurately and rapidly to positioning signals from a control center or director. Similar mechanisms with very high speeds of response were needed to permit the required rapid increase in the speeds of operation of military aircraft. In both of these fields, hydraulic systems proved to be particularly useful because of their very fast response and their great stiffness as seen by the load.

The Servomechanisms Laboratory of the Massachusetts Institute of Technology began work in this general area in 1939 with a very small staff. During the war it grew rapidly and made major contributions, not only in the development of new mechanisms for military uses but also in the application to the problems of high-performance control systems of many mathematical and analytical techniques, some already widely used in other areas and others entirely new. The close association of people trained in a wide variety of disciplines, including electrical, mechanical, and aeronautical engineering, mathematics, and physics, greatly facilitated the analysis of the dynamics of complicated systems and the application of the results of this analysis to the design of actual hardware whose performance far exceeded that previously attained.

As the required level of performance of these systems increased, limitations in our knowledge of the dynamic characteristics of the hydraulic components of the systems became apparent. Toward the end of the war, the gathering momentum of the missile program accentuated the need for further and more penetrating research into the dynamics of systems in which fluid flow occurs. In addition, the operational requirements of missiles placed tremendous emphasis on the saving of weight and volume but at the same time on much higher performance. In 1945, Dr. A. C. Hall and some of his associates at the Servomechanisms Laboratory formed the Dynamic Analysis and Control Laboratory. One of the early acts of this laboratory was the initiation of a program of basic research on fluid power. This program, continuing over the intervening years, has grown in size and has made very significant contributions, both in new knowledge and in the development of highly trained, competent people. Much of the material upon which this book is based has come from this program.

Under Dr. Hall's leadership and encouragement, Dr. J. F. Blackburn and Dr. S.-Y. Lee carried on the research and development work necessary for the construction of a high-speed flight table, an essential component of the M.I.T. Flight Simulator, which was a new analogue computer designed, developed, and constructed in the Laboratory for the study of the dynamics of large and complex systems. Both Dr. Blackburn and Dr. Lee have been members of the organization ever since.

Dr. Lee's creative approach to difficult problems, his ability to apply fundamental relationships to problems of design, and his patience and skill in the laboratory have been invaluable assets. Dr. Blackburn's wealth of knowledge and experience in many diverse fields of activity, his ability to make critical appraisals, and his talent for exposition have made possible the dissemination of much information to a considerably larger audience than would otherwise have been possible.

During recent years the Fluid Power Research Group in DACL has been supervised by Dr. J. Lowen Shearer. His able leadership of the research work, his own outstanding technical contributions, particularly in the area of high-pressure pneumatic systems, and his remarkable capacity for writing and editing have been indispensable in preparing the manuscript for much of this book.

A large part of the material was initially presented in special summer courses held at M.I.T. for representatives of industrial organizations. Dr. Shearer and Dr. Gerhard Reethof played major roles in organizing these courses and in preparing the notes which were essential to their success. The interchanges between our own staff and the representatives of industry who came to the courses were enormously profitable to us both in evaluating our own work and in supplementing it with their experience.

It is impossible to acknowledge individually the contributions made by many people, and only a few can be mentioned here. Beside his work in organizing the notes for the summer courses and in writing several important sections of the manuscript, Dr. Reethof ably led the earlier work on hot-gas servos. Other contributions were made by Dr. W. W. Seifert, Associate Director of DACL, by Mr. Emery St. George, Jr., and Dr. Thomas C. Searle, Assistant Directors, and by Prof. Richard H. Frazier, Prof. Robert W. Mann, Prof. J. B. Reswick, Prof. Freddie Ezekiel, Mr. James L. Coakley, and Mr. Herbert H. Richardson. Much of the work was made possible by the co-operation and support of the United States Navy and the United States Air Force.

The role of a Laboratory in an educational institution is not only to contribute new knowledge but also to provide an environment in which students can rapidly grow in maturity and in professional stature. The Dynamic Analysis and Control Laboratory of M.I.T. has been outstandingly successful in providing this environment for graduate students in mechanical and electrical engineering, and has provided a place for the exchange of ideas and information between people in these two disciplines. It has enabled them to carry on research and development in the very forefronts of the fields of their interests. Such men have contributed much to this book. Many of them have gone on to responsible positions in industry, with a maturity at an early age that can only result from the unusual opportunity which they had at DACL to assume major responsibility for challenging engineering work during their years of graduate study.

Full recognition is given to Dr. J. F. Blackburn for his work in editing and rearranging the material and for adding his own ideas in many spots.

Finally, I must express my own appreciation for having been fortunate enough to serve the Laboratory as its Director during the period 1950 through 1957, when this manuscript was prepared. It has been a stimulating and richly rewarding experience.

JOHN A. HRONES

═ *Preface*

The preface of a technical book is traditionally the editor's only opportunity to explain to such of his readers as see it (1) the book's history and *raison d'être*, (2) his indebtedness to those who have helped to produce it, and (3) his excuses for its shortcomings, in the hope that this may somewhat soften the hearts of the reviewers. In the present case part of this explanation is given in Dr. Hrones's Foreword and in Chapter 1, so this preface can be correspondingly brief.

The book is the successor of several sets of mimeographed notes for M.I.T. Summer Courses 2.781 and 2.789, refresher courses for engineers from industry. The courses have been given four times since 1951, and have been well received, as evidenced by the number of subsequent requests for reprints of the notes. Like the course notes, the book is based largely upon the experience of the fluid-power group of the Dynamic Analysis and Control Laboratory (referred to herein as DACL) and of the Department of Mechanical Engineering at M.I.T. For this reason it does not give a complete or balanced picture of the state of the art and science of fluid-power control, but it does reflect more or less accurately the collective experience and accomplishments of the DACL staff. The engineers who attended the courses seemed to think that this body of lore was useful, and we hope that the readers of the book will feel the same way.

Undoubtedly one complaint will be that of a lack of uniformity in treatment, or even of consistency in notation, from one chapter to another. We can only plead guilty to this accusation, pointing out in extenuation that the material was taken from many different sources, that our knowledge of the subject is constantly growing—several of the chapters were completely rewritten in the light of new knowledge when the manuscript was almost ready for the publisher—and that we felt that the limited time which the editors could spare from their other duties would be more profitably expended filling in gaps in the story than on polishing a text that said much less. Current research and development work in the regrouped DACL, now part of the Department of Mechanical Engineering at M.I.T., is concentrated on filling some of these gaps, thus providing an undiminished flow of graduate theses and professional papers of the type that led to the writing of this book.

In a co-operative effort such as the present one, it is obviously out of the question to enumerate all the individuals and organizations to whom we are indebted for help, and we can only thank them as a group. We feel, however, that in addition to the persons mentioned in the Foreword, we should at least acknowledge our special indebtedness to the Bureau of Ordnance, United States Navy, the Air Research and Development Command, United States Air Force, the Moog Valve Company of East Aurora, New York, and the Pantex Manufacturing Company of Pawtucket, Rhode Island, for financial support of much of the work reported herein, and to the Raytheon Company, Waltham, Massachusetts, for contributing secretarial help and Blackburn's time during the later phases of editing, and to Vickers, Inc., division of Sperry Rand Corporation, Detroit, Michigan, for a similar contribution for Dr. Reethof. Finally we must mention the efficient assistance of Miss Constance D. Boyd of The Technology Press, formerly of the DACL.

JOHN F. BLACKBURN

GERHARD REETHOF

J. LOWEN SHEARER

Belmont, Massachusetts, and
Green Hills, Ohio
October 1959

Contents

1

J. F. Blackburn

Fluid Power

1.1. INTRODUCTION

The subject of fluid power control offers an excellent example of the development of a branch of engineering. With the invention of economically useful prime movers such as Watt's steam engine and the development of the factory system, there arose a need for a method of transmitting power from the point of generation to a more or less distant power-using machine. This transmission could be accomplished mechanically, as by lineshaft and belting, but mechanical transmission was often expensive and inconvenient. Probably its greatest inconvenience was the difficulty of controlling the flow of power accurately and cheaply, and one solution to the problem lay in transmission by a fluid under pressure.

This development was pushed hard for a time, and in some industrial cities, particularly in England,[1] regular networks of high-pressure water pipes were laid between central generating stations, which contained steam-driven pumps, and the mills which used the power. In order to make such systems practical, it was necessary to invent a large number of auxiliary devices, such as hydraulic accumulators and various control and regulating valves, and to amass a considerable body of information about how the commoner problems of operation could be met. The demand produced the supply; numerous hydraulic devices, many of

[1] Ian McNeil, *Hydraulic Operation and Control of Machines*, Ronald Press, New York, 1955, pp. 137–140.

1

them most ingenious, appeared on the market or in the pages of the newly founded engineering magazines, and there arose a class of fluid-power specialists who were skilled in the art of generating, transmitting, and controlling power by means of high-pressure water.

Unfortunately, as is usually the case with a rapidly developing subject, this new art remained largely an art; practice far outran theory, and much of the lore remained qualitative rather than quantitative. The situation no doubt would have been remedied in time except that the economic foundation for the whole art was swept away by the phenomenal growth of the electric-power industry. Electrical power can be transmitted much more economically and efficiently and to vastly greater distances than the power produced by any hydraulic or pneumatic technique ever can, and as this was obvious from the beginning, the fluid-power art withered and nearly died for some generations.

For several decades now it has been undergoing a rebirth, however, and at the present time it is a subject of great interest and importance. This is due not to any recently discovered deficiencies in the electrical techniques of power transmission, but to the rapid rise in the demand for types of performance that are difficult or impossible to obtain from electromagnetic devices alone. In particular, the need for servomechanisms of large power and exceedingly rapid response for both military and industrial applications requires motors with torque-to-inertia ratios several orders of magnitude higher than can be obtained from an electric motor. In spite of its relatively great cost and its numerous inconveniences, the use of fluid power has been forced upon us, and we must learn how best to use it.

Now, as in the nineteenth century, much of the demand is being satisfied on an empirical basis; many current hydraulic devices are being designed but not engineered. The designer of today has innumerable advantages over his predecessor of a century ago, but like him he suffers from the lack of a unified and organized presentation of the basis of his work. This statement does not imply that the knowledge of the physical phenomena is lacking; after all, hydrodynamics is one of the older branches of physics, and within the last generation it has put on working clothes and changed its name to fluid mechanics, and there are excellent texts on the subject. The trouble with most of these texts is that they cover all of hydrodynamics, and many of them cover aerodynamics also. With such a vast field it is impossible to dig very deeply at any one spot, and current texts make no attempt to do so.

In this book only as much theory is given as is necessary for the exposition of the subject. It is assumed that other books are available to remedy any deficiencies in this respect. Comparatively few formulas are derived and there is little attempt at mathematical rigor or elegance. This is a book for the practicing engineer, and we hope that the information is presented in a way which will make it most useful to him, and be least likely to lead him astray through concealed assumptions or logical booby traps.

Throughout the book the theory is supplemented as far as is practicable by present practice in the field of fluid power. Because this field is developing very rapidly, its center of gravity is constantly shifting and it is impossible to give a truly balanced presentation of the subject. For the same reason the part of the book that applies to practice is somewhat ephemeral, but the fundamental part is of permanent value, and it is this fundamental part that makes present-day fluid-power control an engineering science rather than an art.

1.2. WHY FLUID POWER?

It has already been stated that there is a rapidly growing interest in fluid power, and since this is the primary concern of this book, it is in order to explain this interest and to compare fluid power with electrical and mechanical power.

In most respects fluid power occupies a position somewhere between electrical and mechanical power. It is easier to transmit over appreciable distances than mechanical power, but far more difficult to transmit than electrical power. Both fluid power and electrical power are much more easily and accurately controllable than mechanical power; electrical power is the best in this respect at low power levels, but its superiority is much less marked at higher power levels. From the standpoint of safety the three are about equal, except that it is somewhat harder to guard mechanical power-transmitting elements from accidental human contact. From the standpoint of efficiency when long-distance transmission is not involved, mechanical power is usually superior and the other two about equal. Cost comparisons are difficult to make since they depend so much upon each individual case. At low power levels electrical devices are much cheaper than those of the other two classes; at high power levels there is probably not much difference on the whole. From the standpoint of availability of a wide variety of components, electricity is far better than the others; furthermore, this abundance of components and techniques is reflected in a flexibility of application that they cannot claim. The same is true of speed of response, except for the all-important limitation that applies to electromechanical devices. This limitation and its consequences will be discussed later.

On the basis of the foregoing paragraph alone it might be concluded that there is little difference among the three types of power but that perhaps electricity is better on the whole. In the average case—if there is an average case—this is probably true, and it is for this reason that electrical techniques and equipment have been so much further developed than the others. Within a comparatively few years, however, advancing technology has forced the electrical designer against an apparently impenetrable barrier in his efforts to improve his equipment still further, and since to date he has been unable to penetrate the barrier, it is necessary to seek a path around it. The easiest and most direct path appears to be the use of fluid power.

The barrier just referred to is a property of matter, the fact that any known ferromagnetic material saturates at an inconveniently low flux density. This means that no more than a certain torque can be obtained from a pound of iron in a motor armature. This fact in turn puts a direct limit on the ratio of torque to inertia that can be obtained with an electromagnetic device such as a torquer, a motor, or a tractive magnet.

A few decades ago this limitation was of little interest. Recently, however, there has been a great and ever-increasing demand for systems of very high response speeds, which demand fast components, that is, components with high ratios of torque to inertia. Electromagnetic devices, at least in the higher power ranges, simply are not fast enough to satisfy these requirements.

A rough numerical calculation will support this statement. Suppose we compare two similar devices, a ram and a tractive electromagnet. The saturation density of good electrical steels is about 135 kilomaxwells per square inch, which gives a tractive force of about 250 pounds per square inch for the magnet. This cannot be greatly increased by a change to other materials; even if the very costly Permendur (cobalt-iron alloy) is used, the attainable force increases only to about 315 psi. These are extreme cases; practical limits would be probably one-half to two-thirds of these values.

With a hydraulic ram the attainable pressure would usually be given by the safe working pressures of other parts of the system rather than that of the ram itself unless it were very large. At present, hydraulic systems operating at 5000 psi are not uncommon, and considerably higher pressures can be used where necessary. Thus it can be seen that forces can be obtained from fluid-power devices that are at least ten to twenty times larger than those obtained from electromagnets.

This calculation is ultraconservative also since the designer of fluid-powered equipment does not have to contend with the rather serious design limitations that are imposed by magnetic design requirements. He has a far wider range of materials to choose from and he can nearly always use his materials more effectively, thus greatly decreasing his moving mass from that which the electrical designer has to use. Therefore there is a very large increase in torque-to-inertia ratio in going from an electrical to a hydraulic motor; in practical cases this increase might be several thousand times in the larger-horsepower range. With small powers the increase would be much less because it is practical to build electric motors of only a few watts' output, but the smallest economically feasible hydraulic motor would have a rating of, say, half a horsepower if it were at all efficiently designed.

We may now consider the advantages and disadvantages of fluid power in somewhat greater detail. To consider the disadvantages first, probably the greatest one is that power-transmitting liquids are messy. Good design, good workmanship, and good maintenance can keep leakage and spillage to a minimum, but a completely clean and leakproof system probably does not exist. With reasonably good housekeeping the hazard is probably more psychological than real, but it exists. Of course this limitation does not apply to gas-powered systems, but at present these are relatively in their infancy. They will be considered in Chapters 8, 16, 17, 19, and 20 of this book.

A second and serious disadvantage of both hydraulic and pneumatic systems is their great vulnerability to dirt or other contamination of the fluid medium. Again the elimination of this trouble demands good design, good workmanship, and good maintenance, particularly the last, but at best a hydraulic system will not stand neglect or abuse as will a mechanical or electrical system.

A third disadvantage is the danger of bursts. This is more often given as an argument against high-pressure pneumatic than against hydraulic systems, though actually one seems to have little advantage over the other; most practical systems normally have a good deal of energy stored either in accumulators or in pressure bottles, and the effects of a burst will depend a great deal on where the burst occurs in the system. In most cases the danger is greatly exaggerated, but it is certainly real.

The most serious danger of hydraulic systems is that of fire and explosion. Even without an actual burst, a pinhole leak of liquid under high pressure will disperse the liquid into a fog of very minute droplets, and if the liquid is even slightly combustible this fog can explode violently. Explosions and fires from this cause have been sufficiently common and sufficiently serious to produce a great demand for a satisfactory nonflammable hydraulic fluid. At the time of writing there seems to be no such fluid; except for its flammability, petroleum-based fluid still seems to be our nearest approach to an all-around pressure medium. Progress is being made, however, and it is very likely that satisfactory nonflammable fluids will become available in the future.

To turn now to the advantages, one of the most important is the fact that a material medium, unlike electricity, serves to carry off the heat produced by power losses from the point where it is produced. This permits a great reduction in the size of a component for a given power, or conversely it permits a much higher power density in the system. Obviously, the losses must be dissipated somewhere, but in a hydraulic or pneumatic system this dissipation can be done with a heat exchanger located at any convenient point. The minimum size for an electrical component is usually determined by the maximum usable magnetic flux density and by heating; the minimum size of a hydraulic or pneumatic component is normally governed only by structural considerations. Currently produced aircraft hydraulic motors, for example, weigh well under 1 lb per hp, which is a figure that electrical machines cannot even approach, and the hydraulic motors are correspondingly compact.

A second major advantage of hydraulic systems is the fact that as seen from the load they are mechanically stiff. Ideally (though certainly not in practice) a cylinder full of oil looks infinitely stiff to the piston, while either a cylinder of gas (except at very high pressures) or a magnetic field looks very soft and springy. If it is necessary, as it usually is, to hold the load fixed in position until it is desired to move it, far less loop gain will be required with a hydraulic servo than with either a low-pressure pneumatic or an electric one. Also, since a major limitation in speed of response is the resonant frequency of the load acting against the equivalent spring of the driver, it is desirable that this spring be as stiff as possible, which is another point in favor of hydraulics.

The third and usually the conclusive advantage is the high attainable speed of response; stated concisely, in the present state of the art there are many important jobs that can be done only by fluid power, and many others where it is greatly to be preferred. Thus the system designer is often stuck with hydraulics whether he likes it or not. In the recent past most of the lore of the hydraulic designer was empirical, but that situation is rapidly changing; hydraulic design is becoming an engineering science rather than an art, and pneumatics is just beginning to follow suit. This book is an attempt to accelerate this process.

1.3. PLAN AND SCOPE OF THIS BOOK

Since in certain respects this volume is the first in its field, it seems worth while to define the limits of the field and to describe briefly the way in which it will be covered. Out of the vastly larger field of fluid mechanics this field concerns itself only with the use of fluids in the transmission of power at elevated pressures—pressures great enough so that gravitational effects are negligible—and with the control and, to some extent, the utilization of the power thus transmitted. This restriction of field permits us to ignore the flow of fluids in open channels, and also practically all of the subject of aerodynamics. Furthermore, for much of the book we can neglect the effects of compressibility on flow, though these effects are very important with gaseous media and even with liquid media when the compliance of trapped oil volumes affects the dynamic characteristics of the system. We also omit the consideration of non-Newtonian flow, in which the effective viscosity depends upon the rate of shear. This type of flow is very important in some applications, as in petroleum refineries and in many chemical processes, but the fluids used as power-transmitting media are at least approximately Newtonian.

At the present time there are very few books on fluid power and these few are largely or entirely descriptive. Our experience at DACL, particularly some years ago when even less information was available than at present, made us feel that there was a need for a book in which the emphasis would be not on the description of hydraulic or pneumatic components and systems, but upon the principles of operation, design, and application, and particularly upon the quantitative factors that are all-important to the engineer who wishes to put fluid-power devices and techniques to work. This book is very far from being a complete answer to this requirement—indeed, no one could satisfy it completely in the present state of the art—but we believe that it is at least a step in that direction.

The book divides itself naturally into three more-or-less distinct parts. The first is introductory, the second discusses in considerable detail the design of hydraulic and pneumatic drives and particularly valve-controlled drives, and the third deals with the applications of such drives to actual systems, with special emphasis on system dynamics.

The first part consists of five chapters. After the present explanation and apologia, Chapter 2 gives a brief summary of the properties of commercial fluid media, and some comments as to the choice of a medium. It is difficult at the present time to make this choice intelligently since hydraulic fluids are being very actively developed, the available information is abundant but rather uninformative, and the claims for the various proprietary products are confusing and sometimes controversial. In these circumstances, it is impossible to do more than to list a number of properties on the basis of which a choice should be made and (at some risk of being drawn into the fray) to give a rough indication of the relative ranks of the several products according to these criteria.

Chaper 3 is a brief review of those parts of the field of fluid mechanics which are of most interest to the fluid-power engineer. It has been included in spite of the general availability of excellent texts on the subject because we have found such a review to be very useful in refresher courses on the subject of fluid power and to be generally approved by the practicing engineers who attended the courses.

Chapter 4 deals with the generation and the utilization of fluid power. After a very brief descriptive section, it reviews briefly the W. E. Wilson theory of the operation of positive-displacement pumps and motors and then describes the effects of the "thermal wedge" on pump operation. This chapter is not intended to be a complete or balanced presentation of its subject, which can be found elsewhere, but to summarize these two more specialized and less well-known topics.

Chapter 5, which concludes the first part of the book, is even less exhaustive. It is intended primarily to suggest further work in the field of fluid-power transmission and is supplemented by a short bibliography.

The second part, consisting of eight chapters on the general subject of drive and valve design, is in some respects the heart of the book. It starts with Chapter 6, which discusses the way in which it is necessary to specify to a drive designer the character of a mechanical load and the motion which that load must be forced to execute. This information is applicable to all kinds of drives, but primarily to valve-controlled rams or motors.

Chapter 7 discusses the performance characteristics of hydraulic valves, various methods of expressing and presenting these characteristics, and the choice of a type of valve for various applications. The following chapter, Chapter 8, is a closely parallel exposition of the characteristics of pneumatic valves, which differ significantly from hydraulic valves because of the occurrence of the critical-flow phenomenon with gases.

Chapter 9 describes the commoner types of control valves and the simpler circuits in which they are used, outlines methods of spool-valve design, and gives some comments on techniques of manufacture.

Chapter 10 discusses the steady-state forces that are exerted on the moving members of valves by the working fluid. Its coverage of spool-, plate-, and flapper-type valve forces is fairly complete; very little is known at present about the forces on poppet or jet-pipe valves.

Chapter 11 presents a very complete and detailed theory of the moving-iron electromagnetic valve actuator, or "torque motor" as it is inappropriately known in the trade.

Chapter 12 deals with a number of types of valve instability, a phenomenon that has plagued every user of valves and that has been very difficult to diagnose or cure. We believe that the specific examples given in this chapter will be useful in themselves and that the general method of analysis will be widely applicable.

Finally, Chapter 13 closes the second part of the book with a short description of several two-stage valves, followed by a detailed account of the design of a miniature hydraulic servo which was intended for use as an actuator for larger valves.

The third and last part of the book consists of seven chapters in the general field of fluid-power applications. For the same reasons that led to the inclusion of Chapter 3, this part leads off with Chapter 14, a review of system dynamics, which presents certain tools that are indispensable to the worker in the field of system analysis and design.

Chapters 15 and 16 are companion chapters, the first on hydraulic and the second on pneumatic drives, and were included to fill gaps in the story. Chapter 15 is the only one in the book which does more than refer to the very important case of a variable-stroke pump driving a rotary motor, usually known as the hydraulic transmission. Omission of this chapter would have given the impression that the only feasible drive is the valve-controlled drive and would have made the book more one-sided than it is. Chapter 16 describes the analysis and design of a high-performance drive using high-pressure compressed air as a fluid medium. Gases have certain major advantages over liquids as fluid-power media, as well as great disadvantages, and the demonstration that good servo performance is obtainable with a gas is important and convincing.

In all the aforementioned chapters of the book, the possibility of closing a feedback loop around the drive is touched upon very lightly. Chapter 17 presents four examples of the analysis of closed-loop fluid-power systems, with particular emphasis on the dynamic performance and the stability of these systems. Its companion chapter, 18, describes the analysis, the paper design, and the performance of an automobile fluid-power steering mechanism, an excellent example of the application of the techniques of this book to a practical problem.

Chapter 19 gives a brief comparison of hydraulic and pneumatic power and discusses the factors which tend to make one or the other preferable in a particular application. With the rapid increase of interest in pneumatic systems, particularly in airborne applications, this chapter is significant and, we hope, useful.

The final chapter, Chapter 20, describes a particular application in which the pneumatic rather than the hydraulic approach was chosen, in contrast to previous practice in the field. This application was a control-surface actuator for a (hypothetical) guided missile, which derived its power from a pyrotechnic cartridge, the power medium being the hot, dirty products of combustion of the propellant. The high temperature, the dirt content, and the compressibility of the medium, the uncontrollable nature of the rate of power generation, and the excellent dynamic response required all added up to a difficult design problem, which was successfully solved in prototype form.

2

Gerhard Reethof

Properties of Fluids

2.1. INTRODUCTION

Since a fluid-power device requires a physical fluid for its operation, knowledge of the properties of the fluid in considerable detail is important. This chapter discusses those properties which are of greatest importance to the fluid-power engineer and the effects upon them of variations in temperature, pressure, and so on. Since both gaseous and liquid media are useful, the discussion will cover both, endeavoring to bring out similarities as well as contrasts.

The trend of the last few years toward operation over a much wider range of temperature than was formerly required has emphasized the shortcomings of the usual hydrocarbon-base "hydraulic fluid" and the importance of finding acceptable substitutes. At the present time (1959) there seems to be no universally applicable fluid. There are, however, a great many liquids which have properties or combinations of properties that are useful for particular types of service, and much effort is being expended on the evaluation and improvement of these liquids and on the development of new ones. This situation means that any compilation of numerical data will inevitably become obsolete within a few years, while the sheer abundance of the available data makes it impracticable to present them all here. It has been necessary therefore to confine this chapter to a discussion of the more important properties of fluid-power media, plus a brief discussion of the basis for choosing one fluid rather than another for a particular application. Typical values for some of the properties are given for illustrative purposes, but the chapter makes no attempt to give complete information on any one topic.

2.2. DENSITY

We start by defining the quantities of interest:

Density is the mass, or quantity of matter, contained in unit volume of the substance under consideration. It will be denoted by the symbol ρ. In this book we use force rather than mass as one of the fundamental quantities, so in our system of units the dimensions of density are $[FL^2T^{-4}]$, or lb sec^2/in.^{4}

Specific weight, w, is the weight (*not* the mass) of unit volume of a substance. Its dimensions are $[FL^{-3}]$, or lb/in.^{3}

Specific volume, which is the reciprocal of specific weight, is often used in discussing gases. It is denoted by v.

Compressibility is the decrease in volume of unit volume of the substance when the ambient pressure is increased by a unit amount:

$$\kappa = \frac{1}{V}\frac{dV}{dP} \tag{2.1}$$

Its dimensions are in.^{2}/lb.

It is necessary to specify the conditions of the compression, especially with gases. If the temperature is kept constant by allowing heat to flow into or out of the volume of matter being compressed, the resulting quantity is the *isothermal* compressibility, κ_T. If on the other hand the heat flow is prevented, a somewhat different value will be obtained, the *adiabatic* compressibility, κ_s.

Bulk modulus, β, is the reciprocal of the compressibility. Naturally it will have two values, the isothermal and the adiabatic bulk moduli. The former is the one usually given in the tables; the latter should be used in calculating the velocity of sound in the fluid. In a liquid the two values are nearly the same, but they differ considerably for a gas.

Thermal expansion coefficient (or cubical expansion coefficient) is the change in volume of unit volume of the substance when its temperature is changed by unit amount:

$$\alpha = \frac{1}{V}\frac{dV}{dT} \tag{2.2}$$

It has the dimensions 1/temperature. It is usually evaluated at constant pressure.

2.21. Density of Gases; Equation of State

For any quantitatively definable property of a physical substance, an equation can be written which expresses the relationship between that property and the ambient pressure, temperature, or other conditions to which the substance is exposed. When the given property is the density (or some equivalent quantity) the equation is called the *equation of state*. For many substances this equation is too complicated for convenient use, or its constants have never been accurately determined, but for gases and vapors it is a very useful and concise way of expressing a large amount of information.

For a "perfect" gas the equation of state is

$$\frac{P}{\rho} = RT \tag{2.3}$$

where P = pressure, $lb/in.^2$

ρ = density, $lb\ sec^2/in.^4$

R = a constant, characteristic for the gas, $in.^2/sec^2\ °R$

T = absolute temperature, $°R$ (Rankine; $°F + 458.6$)

Frequently the specific volume is used instead of the density and the equation becomes the *perfect-gas law* in its usual form:

$$Pv = \frac{RT}{g} \tag{2.4}$$

where g is the acceleration of gravity.

If we differentiate Eq. 2.4 and combine it with Eq. 2.1, we get the isothermal compressibility

$$\kappa_T = \frac{1}{P} \tag{2.5}$$

If the compression is carried out isentropically (adiabatically) it can be shown that

$$\kappa_s = \frac{1}{kP} \tag{2.6}$$

where k = the ratio $\dfrac{\text{specific heat at constant pressure, } C_p}{\text{specific heat at constant volume, } C_v}$

For the isentropic process the quantity $(P/\rho)^k$ remains constant.

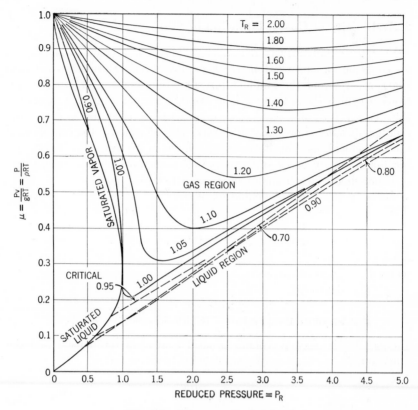

Fig. 2.1. The generalized μ chart.

For most pneumatic processes, especially those using "permanent" gases such as air or nitrogen, the perfect-gas law is followed fairly accurately. This is the case if:

1. The pressures are not excessive (up to perhaps 3000 psi) and the temperatures are above the critical temperatures; or

2. The temperatures are subcritical but the pressures do not exceed about 100 psi.

Instead of using separate charts for each gas, it is much more economical of space to use a generalized chart such as those of Figs. 2.1 and 2.2.[1,2] The variables used in these charts are the reduced temperature, $T_R = T/T_c$; the reduced pressure, $P_R = P/P_c$; and the volume factor

---

[1] J. H. Keenan, *Thermodynamics*, John Wiley and Sons, New York, 1941.

[2] H. C. Weber, *Thermodynamics for Chemical Engineers*, John Wiley and Sons, New York, 1939, p. 109.

$\mu = P/\rho RT$; where T_c and P_c are the critical temperature and pressure for the gas in question. The chart can be used with fair accuracy not only for the "permanent" gases but also for easily condensable ones such as organic vapors.

As an example of the accuracy of these charts, consider the case of nitrogen, for which $T_c = 227°R$ and $P_c = 493$ psig (gage pressure) $= 508$ psia (absolute pressure). At $70°F$ ($530°R$) $T_R = 2.34$, and at 5000 psia $P_R = 9.84$. From the chart we see that $\mu = 1.15$. An error

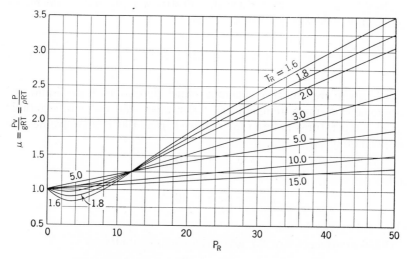

Fig. 2.2. μ chart for high pressures and temperatures.

of 15 per cent is not bad for the perfect-gas law under these rather extreme conditions. If we double the pressure to 10,000 psia, however, $\mu = 1.65$ and the actual density is 65 per cent higher than that calculated from the perfect-gas law.

For many gases the charts give excellent accuracy. Thus for steam at $1200°F$ and 5500 psia, the error from the chart is 1.7 per cent while that from the perfect-gas law is 17 per cent. For some gases, particularly hydrogen, helium, and neon, the chart is inaccurate if the actual critical constants are used in computing the reduced parameters but is fairly accurate if these constants are suitably altered. Even with mixtures of gases, pseudocritical constants can be calculated that will still permit the use of the charts.[3,4]

[3] Lionel S. Marks, *Mechanical Engineer's Handbook*, 5th ed., McGraw-Hill Book Co., New York, 1951, pp. 291–293.

[4] A. J. Rutgers, *Physical Chemistry*, Interscience Publishers, New York, 1954.

2.22. Density of Liquids

For liquids the equations of state are considerably more complicated than for gases, but fortunately the small values of the coefficients involved make simplified linear expressions sufficiently accurate for most purposes. To a first approximation, liquids are incompressible, and are so considered in many of the expressions in this book. Actually, over the usual range of operating pressures a quadratic expression is required for good accuracy since the effective compressibility decreases appreciably at higher pressures. Thus

$$\rho = \rho_0(1 + aP + bP^2) \tag{2.7}$$

where a and b are empirical constants which depend upon the liquid and also somewhat upon temperature. For most mineral and vegetable oils at temperatures not too far removed from 100°F, their values are approximately

$$a = 4.38 \times 10^{-6}$$

and

$$b = 5.65 \times 10^{-11}$$

when P is given in psi.

Differentiation of Eq. 2.7 gives the compressibility as

$$\kappa = \frac{a + 2bP}{1 + aP + bP^2} \tag{2.8}$$

At zero pressure the compressibility from Eq. 2.8 would be simply a, or 4.38×10^{-6} in.^{2}/lb. Increasing temperature increases the values of both a and b, as can be seen from Table 2.1.

Table 2.1. COMPRESSIBILITY OF MINERAL OILS
(IN UNITS OF 10^{-6} IN.^{2}/LB)

Temperature, °F	20	220	440
$P = 0$ psig	3.96	4.70	4.80
$P = 5000$ psig	3.17	4.15	4.45

In general the compressibilities of heavy oils are somewhat greater than those of light oils of the same types. Most solids are much less compressible than oils; on the other hand, some liquids are considerably more so.[5]   These relatively compressible liquids are of little interest as hydraulic fluids, but they might make very good liquid springs.[6]

The values of compressibility given in tables such as Table 2.1 or calculated from the various formulas given in the literature are not necessarily the values which are effective in a particular application. They make no allowance for the effect of stretching of the containing walls, which is small for heavy-walled steel tubing but may be large for high-pressure hose, and they neglect the effect of entrained air. Almost every hydraulic system draws in at least a little air and mixes it thoroughly with the oil, which becomes saturated with the dissolved air and may also carry appreciable volumes of bubbles. This effect is greater if the liquid has a tendency to foam. The air that is in true solution apparently has little effect in most cases, but air bubbles may cause trouble. If the entrainment problem is serious it may be necessary to provide for deaeration of the oil.

At low pressures even a small bubble of air will greatly increase the effective compressibility of a considerable volume of oil. Since as the pressure increases much of the air will dissolve in the oil and since according to Eq. 2.5 the compressibility of a gas varies inversely as the pressure, the effect of bubbles is much less at the higher operating pressures. Since the effects of air bubbles and of stretch of the walls are somewhat indeterminate, it is common practice to assume that the effective compressibility is 5×10^{-6} in.^{2}/lb. The logical basis of the practice is open to question but it seems to work.

In hydraulic work the thermal-expansion coefficient of a liquid is occasionally important, particularly in cases where a volume of liquid is trapped between two shutoff valves. Here a rise in temperature can cause a dangerous pressure rise if there is no leakage. Another pertinent case is that of the parallel-plate hydrodynamic bearing, where a significant effect on the load-carrying capacity is produced by the "thermal wedge." This is further discussed in Sec. 4.31.

Thermal expansion is usually expressed as a power series:

$$\rho = \rho_1(1 - \alpha \, \Delta T - \beta \, \Delta T^2 + \cdots) \tag{2.9}$$

where $\Delta T = T - T_1$

For most liquids, powers of ΔT higher than the second can be neglected, and for engineering purposes it is usually sufficient to neglect the term in T^2 also. For MIL-O-5606 aircraft hydraulic fluid, $a = 4.26 \times 10^{-4}/°$F.

[5] Graham W. Marks, "Variation of Acoustic Velocity with Temperature in Some Low Velocity Liquids and Solutions," *J. Acoustical Soc. of America*, Vol. 27 (July 1955), pp. 680–688. One liquid reported was 3½ times as compressible as water.

[6] A. E. Bingham, "Liquid Springs: Progress in Design and Application," *Proc. Inst. Mech. Engs. (London)*, Vol. 169, No. 43 (1955), pp. 881–893.

It is frequently useful to express physical properties as dimensionless ratios of the actual value to some standard value of the same property. Perhaps the commonest example is the specific gravity, σ, which is the ratio of the density of the substance to that of water, usually taken at 60°F in engineering work. Relative densities of petroleum products are often given in API degrees.[7] According to an agreement among the

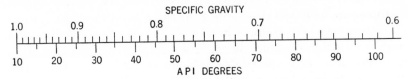

Fig. 2.3. Conversion of specific gravity to API degrees.

American Petroleum Institute, the National Bureau of Standards, and other organizations, the API degree is defined as:

$$\text{Degrees API} = \frac{141.5}{\text{Specific gravity } 60°/60°\text{F} - 131.5} \qquad (2.10)$$

and conversely

$$\text{Specific gravity} = \frac{141.5}{\text{API degrees}} + 131.5 \qquad (2.11)$$

For engineering calculations the API degrees must always be translated into specific gravity or density. This translation will be facilitated by Fig. 2.3.

2.3. VISCOSITY

A true fluid is a substance which cannot support a shear stress applied infinitely slowly. A finite rate of shear in a physical fluid, however, requires a finite stress for its maintenance, and if the stress is proportional to the rate, the fluid is said to be Newtonian and the factor of proportionality is called the absolute viscosity.

In Fig. 2.4, consider the elementary cube of fluid $dx\,dy\,dz$ in the space between the two parallel surfaces, the upper one of which is moving

[7] ASTM Standard D287-55, *API Gravity of Petroleum and Its Products. Hydrometer Method*. This and the other ASTM publications referred to below can be obtained from the American Society for Testing Materials, 1916 Race Street, Philadelphia 3, Pennsylvania. Most technical libraries keep a file of ASTM standards.

to the right with the velocity U. The force dF on the upper face of the cube will be $\tau\, dx\, dz$, where τ is the shear stress. The rate of shear is the same as the velocity gradient dU/dy. If the fluid is Newtonian,

$$\tau\, dx\, dz = \mu\, dx\, dz\, \frac{dU}{dy} \tag{2.12}$$

Canceling the area term from each side and rearranging, we have

$$\mu = \frac{\tau}{dU/dy} \tag{2.13}$$

The absolute viscosity therefore has the dimensions $[FT/L^2]$, or in

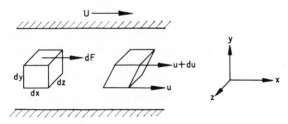

Fig. 2.4. Fluid shear diagram.

English units, lb sec/in.[2] This unit is frequently called the Reyn, after Osborne Reynolds.

The Reyn is a very large unit; a liquid with a viscosity of 1 Reyn would be almost a solid, like tar. A much more convenient unit is the centipoise (1 cp = 0.01 poise; 1 poise = 1 dyne sec/cm^2), which is approximately the viscosity of water at 68°F. 1 cp = 1.45×10^{-7} Reyn; 1 Reyn = 6.9×10^6 cp.

For a series of liquids of the same chemical type the absolute viscosity increases (somewhat irregularly) with the molecular weight. For hydrocarbon mixtures at about 70°F, the viscosity ranges are about as follows:

Gasolines	0.35 to 1 cp
Kerosenes	1.5 to 2
Light lubricating oils	10 to 100
Heavy oils	100 to 1000

Most "hydraulic fluids" fall in the 10-to-100 cp range. At 100°F the standard MIL-O-5606 aircraft hydraulic fluid has a viscosity of about 1.7×10^{-6} Reyn or 12 cp.

Most liquids which are used as hydraulic fluids are at least approximately Newtonian, but many other common substances are not. Greases, for example, act as plastic solids for small strains but as liquids for higher strain rates. Gases are almost perfectly Newtonian, with viscosities at room temperature in the range 0.1 to 0.25 millipoise.

In many of the equations of fluid mechanics the viscosity occurs as the ratio of absolute viscosity to fluid density, and this ratio is usually called the kinematic viscosity. (The absolute viscosity is sometimes called the dynamic viscosity, though it is not immediately obvious just what is dynamic or kinematic about either of the quantities.) The usual unit of kinematic viscosity is the stoke, which is equal to 1 cm^2/sec. It is numerically equal to the absolute viscosity in poises divided by the specific gravity. The corresponding English unit (in.^{2}/sec) is sometimes used but has never been named. The viscosities of petroleum products were formerly determined by the now-obsolete Saybolt efflux viscometer,[8] and the industry still reports viscosities in seconds Saybolt Universal (SSU). The British use the very similar Redwood instrument, and the Germans and most other Europeans the Engler. None of these gives meaningful results for liquids with viscosities under a few centistokes, and up to perhaps 30 cs the conversion is a nuisance to use. For high efflux times the viscosity is proportional to the time. The usual formulas for the Saybolt instrument are:[9]

For efflux times from 32 to 100 SSU

$$\nu = 0.226t - \frac{195}{t} \qquad (2.14)$$

and for efflux times greater than 100 SSU

$$\nu = 0.220t - \frac{135}{t} \qquad (2.15)$$

where t is the efflux time in seconds and ν is the kinematic viscosity in centistokes. Obviously for viscous liquids the efflux times are long and the second term becomes negligible.

Since the Saybolt second is physically meaningless, it will not be used in this book. As in the case of the API degree, the formulas are included as a convenience, and the translation from the arbitrary to the physical language will be aided by Fig. 2.5. Of the meaningful units the centistoke has the most convenient size, and it and the centi-

---

[8] ASTM Standard D88-53, *Viscosity by Means of the Saybolt Viscosimeter.*

[9] ASTM Standard D446-53, *Conversion of Kinematic Viscosity to Saybolt Universal Viscosity.*

poise will be frequently used hereafter even in expressions involving English units. A great deal of time and confusion would be saved if this practice were universal.

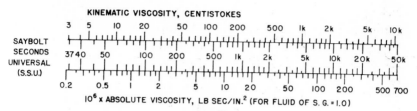

Fig. 2.5. Viscosity conversion diagram.

2.31. Effect of Pressure

According to the kinetic theory of gases, the viscosity of a perfect gas would be independent of pressure. Real gases obey this rule fairly accurately over the ranges of temperature and pressure where they also obey the equation of state of a perfect gas. The viscosity does increase for very high pressures at which the molecular mean free path is no longer large compared to the diameter of the molecules. For most gases the viscosity can be considered to be independent of pressure from about 1.4 to about 1000 psia. Above this pressure the viscosity increases very rapidly. Thus carbon dioxide at 20°C has a viscosity of 0.16 millipoise ± 10 per cent from 15 to 750 psia, but at 1245 psia the viscosity rises to 0.823 mp.[10]

The viscosities of liquids increase considerably with increasing pressure, usually approximately according to the expression [11]

$$\log_{10} \frac{\mu}{\mu_0} = cP \tag{2.16}$$

Most petroleum products at room temperature increase in viscosity by a factor of about 2.25 when the pressure is raised to 5000 psi. This gives a value of $c = 7 \times 10^{-4}/$psi.

2.32. Effect of Temperature

The viscosities of both gases and liquids are considerably affected by changes of temperature. In the case of gases the change is described

[10] *International Critical Tables*, Vol. 5, McGraw-Hill Book Co., New York, 1929, p. 4.

[11] "Viscosity and Density of Over 40 Lubricating Fluids of Known Composition at Pressures to 150,000 Psi and Temperatures to 425°F," Vols. I and II, *ASME Research Report*, 1953.

fairly accurately by Sutherland's formula:

$$\mu = \mu_0 \left(\frac{T_0 + C}{T + C} \right) \left(\frac{T}{T_0} \right)^{\frac{3}{2}} \tag{2.17}$$

where μ = viscosity at (absolute) temperature, T
     μ_0 = viscosity at reference temperature, T_0
     C = a constant, characteristic for the gas in question

Sutherland's equation can also be written

$$\mu = \frac{A T^{\frac{3}{2}}}{T + C} \tag{2.18}$$

where

$$A = \frac{\mu_0}{T_0} \left(1 + \frac{C}{T_0} \right)$$

This form of the equation shows more clearly than the other that the viscosity of a gas rises rapidly with temperature.

All liquids show the opposite effect, except for molten sulfur and a few other materials where chemical changes such as association or reversible polymerization are caused by increasing temperature. Many empirical formulas have been proposed to describe the variation of viscosity with temperature, but no fully satisfactory one exists, even for a pure chemical compound. In view of the inherent complexity of the phenomenon it seems unlikely that one will ever be found, especially for liquids as complex and variable in composition as commercial oils and hydraulic fluids.

When an empirical expression is used, it is very desirable that it be in a form convenient for mathematical analysis. One which is useful over a restricted temperature range is

$$\mu_T = \mu_0 e^{-\lambda(T - T_0)} \tag{2.19}$$

where μ_T = viscosity at temperature T
     μ_0 = viscosity at reference temperature T_0
     λ = a constant characteristic of the particular liquid

This expression will be used in Sec. 4.32, where the actual and the assumed viscosity-temperature curves for seven typical hydraulic fluids are plotted in Fig. 4.18.

With actual liquids no reasonable form of equation seems to fit over more than an inconveniently narrow temperature range.[12] The one

---

[12] R. Houwink, *Elasticity, Plasticity, and the Structure of Matter*, Cambridge University Press, 1937.

which fits best for certain types of petroleum-base fluids is the Walther formula:

$$\log_{10} \log_{10} (\nu + c) = n \log_{10} T + C \qquad (2.20)$$

where ν is given in centistokes, $c = 0.8$ (usually), and n and C are characteristic of the given oil. Beside being phenomenally poorly adapted to mathematical manipulation, this equation and the ASTM charts [13] based on it have the following disadvantages: [14, 15]

1. It is completely devoid of physical meaning; if any other unit, such as stokes or Reyns, were used the resultant curve would no longer be even approximately linear, nor would the curves for several liquids of the same family intersect in a "pole."

2. It applies fairly accurately over the range of temperatures encountered in automobile engines but comparatively poorly outside this range.

3. It is reasonably accurate for "straight" petroleum-base lubricating oils, but not for solvent-refined oils, nor for those with appreciable percentages of certain common types of additives, especially the so-called viscosity-index improvers.

4. It is fairly accurate for some types of synthetic liquids but not for other equally important types.

5. Because of the nature of the log-log ordinate scale it tends to exaggerate the proportional change of viscosity with temperature at low viscosities but to minimize it for viscous liquids.

In spite of these and other disadvantages the Walther equation and its various descendants have been and still are the most common methods of representing viscosity-temperature data. The commonest of these, at least in the United States, is the ASTM chart. Figure 2.6 shows on an approximation of such a chart the viscosity-temperature curves for several different grades of the Univis series of lubricating oils of the Esso Standard Oil Company. Figure 2.7 shows the characteristics of hydraulic fluids of several different chemical classes.[16] It can be seen that some of these deviate from linearity much more than others but that none is truly straight at extreme temperatures.

[13] ASTM D341-43, *Standard Viscosity-Temperature Charts for Liquid Petroleum Products.*

[14] H. H. Zuidema, *The Performance of Lubricating Oils*, Reinhold Publishing Corp., New York, 1952, pp. 29–37.

[15] A. Bondi, *Physical Chemistry of Lubricating Oils*, Reinhold Publishing Corp., New York, 1951, pp. 50–59.

[16] C. M. Murphy, J. B. Romans, and W. A. Zisman, "Viscosities and Densities of Lubricating Fluids from −40 to 700°F," *Trans. ASME*, Vol. 71, No. 5 (July 1949), pp. 561–574.

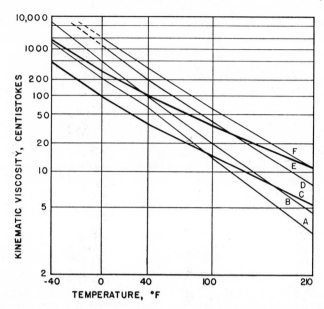

Fig. 2.6. Viscosity-temperature characteristics of six petroleum-base hydraulic fluids.

Typical Univis Fluids, Made by Esso Standard Oil Co.

Curve	Grade of Fluid
A	P-38
B	40
C	J-43
D	P-48
E	PJ-59
F	60

(The J prefix indicates the use of a viscosity-improving additive. J-43 is one type of MIL-O-5606 aircraft hydraulic fluid.)

It would be very convenient if the effect of temperature on viscosity could be described by a single number, and many attempts have been made to devise such a quantity. In Europe the Walther-Ubbelohde pole-height has been used, and in the United States the Dean and Davis Viscosity Index [17] is the most popular. Both are derived from the Walther equation and inherit its defects, and the viscosity index in particular greatly exaggerates small changes in the temperature effect in certain ranges while minimizing those in other ranges. Unfortunately for our purposes, although it has been useful in the past in the field of automotive lubricants, it is almost completely useless when applied to

[17] ASTM Standard D567-53, *Calculating Viscosity Index.*

most practical hydraulic fluids, whether petroleum-based or synthetic. It is still quoted, but the fluid-power engineer should analyze the quotation with some care to determine if the proposed fluid is actually suitable for his application.

Perhaps the simplest criterion of viscosity change would be the temperature drop required to double the viscosity from that at a suitable

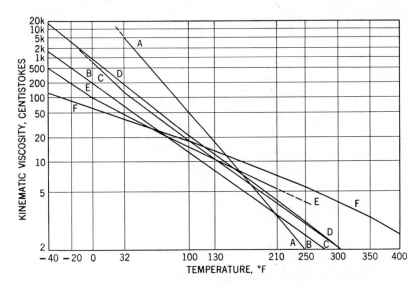

Fig. 2.7. Viscosity-temperature curves for various aircraft hydraulic fluids.

Curve	Material
A	Fluoro lubricant FCD-331
B	Di(2-ethylhexyl) sebacate
C	Poly(methyl-phenyl) siloxane DC-702
D	Polypropylene glycol derivative LB-100-X
E	Petroleum-base Univis J-43
F	Polymethyl siloxane DC-500-A

reference temperature, say 210°F for usual conditions. For the Univis fluids of Fig. 2.6 this temperature drop runs from 50 to 60°F, being less for the more viscous fluids. Viscosity-index improvers increase this by about 40 per cent.

2.33. Effect of Shear

For a Newtonian fluid, the viscosity is independent of the rate of shear, and many liquids are almost perfectly Newtonian. Many others, however, are not, and among these are some of our most useful hydraulic fluids, especially those which contain appreciable amounts of compounds of high molecular weight. These compounds may be natural components of the fluid, or they may be additives, especially the viscosity-index improvers.

When a fluid of this type is subjected to shear, the effective viscosity decreases with increasing rate of shear. This increase appears to be instantaneously reversible for moderate shear rates and for most fluids. As the shear rate increases, however, the viscosity change also increases, and persists for a time after the flow ceases. At extremely high rates of shear the decrease of viscosity may be as much as 40 per cent, and part of this decrease is permanent.

One plausible explanation of these effects is that these three types of viscosity decrease are due respectively to increasing orientation of elongated molecules parallel to the flow lines, to uncoiling and orientation of coiled large molecules, and to actual fracturing of large molecules, accompanied by oxidation and other chemical reactions. This last explanation is supported by the fact that for some oils antioxidants help to increase the shear stability.

The rate of shear necessary to cause appreciable changes in viscosity, either transient or permanent, naturally varies with the liquid. Temporary changes may occur with many commercial hydraulic fluids and with many types of machines, such as positive-displacement pumps and motors, in piston or high-speed-bearing clearances.[18-22] The higher shear stresses necessary to cause permanent viscosity changes are ordinarily attained only under conditions of extreme turbulence, as in throttling valves and orifices under high pressure drops. The effects are greater for oils containing high-polymer thickeners; this should be kept in mind when making detailed analyses of viscous-flow phenomena for such oils. The effective viscosities may be very different from those measured at low rates of shear in conventional viscometers.

The considerable changes in effective viscosity produced by shear, as well as by changes in temperature and pressure, the variability in

[18] E. E. Klaus and M. R. Fenske, "Some Viscosity-Shear Characteristics of Lubricants," *Lubrication Eng.*, Vol. 11, No. 2 (March–April 1955), pp. 101–108.

[19] N. D. Lawson, "Determination of the Shear Stability of Non-Newtonian Liquids," *ASTM Spec. Tech. Publ. No. 182.*

[20] "Symposium on Methods of Measuring Viscosity at High Rates of Shear," *ASTM Spec. Tech. Publ. No. 111.*

[21] R. N. Weltmann, "Thixotropic Behavior of Oils," *Ind. Eng. Chem., Anal. Ed.,* Vol. 15 (1943), p. 424.

[22] N. W. Furby, R. L. Peeler, and R. I. Stirton, "Oronite High Temperature Hydraulic Fluids 8200 and 8515," *Trans. ASME*, Vol. 79, No. 5 (July 1957), pp. 1029–1038.

the properties of commercial fluids as received by the user, and the effects of aging and contamination, suggest that there is really little point in trying to predict viscosity (or other properties) with high accuracy. This is true from a purely empirical standpoint, but there are two arguments against this point of view. One is that eventually we shall know more about all of these effects, and this greater knowledge will permit better control of the performance of our hydraulic systems. The other argument is that the constant improvement in the fluids themselves, and still more in the quality of maintenance of the average hydraulic system, will greatly decrease the variability of the fluid properties. In any given case it is necessary for the engineer to decide for himself how much importance he must attach to the considerations that have been outlined here.

2.34. Dilatational Viscosity

One fluid property that is not now known to be of importance to the hydraulic engineer, but which may eventually prove to be so, is dilatational viscosity, or "second" viscosity. Ordinary viscosity represents the loss of mechanical energy consequent upon deformation of the shape of a volume of fluid by shearing. If the volume is expanded or compressed without change of shape there is no shear, but energy may be dissipated in the change of volume, and this dissipation is represented by the dilatational viscosity.[23, 24] At the present time this phenomenon appears to be somewhat obscure, at least in its relationship to hydraulic machinery, but it may be significant both for the transmission of very rapidly varying pressure waves through conduits and for certain types of valve instability.

2.35. Pour Point

One rough but practical measure of the low-temperature limit of usefulness of a liquid is its pour point.[25] As the name indicates, this is the temperature at which it will no longer pour from a beaker when tested according to a standard procedure. Actually it will be too viscous at considerably higher temperatures for use in a hydraulic system, but the

[23] S. M. Karim and L. Rosenhead, "Second Coefficient of Viscosity of Liquids and Gases," *Revs. Modern Phys.*, Vol. 24, No. 2 (April 1952), pp. 108–116. This article includes an excellent bibliography.

[24] L. M. Liebermann, "The Second Viscosity of Liquids," *Phys. Rev.*, Vol. 75 (May 1949), pp. 1415–1422.

[25] ASTM Standard D97-47, *Cloud and Pour Points*.

pour point is a fairly good measure of the comparative low-temperature properties of various liquids.

A related measurement is that of the cloud point, which is the temperature at which a cloudy precipitate begins to form in the liquid. For many liquids this is greatly affected by the rate of cooling, and the actual cloud point is somewhat indeterminate, but at least it is a crude indication of the likelihood of non-Newtonian behavior.

2.4. CHEMICAL PROPERTIES

Ideally a fluid-power medium would be completely inert to the materials with which it comes in contact. With the hardly significant exception of the noble gases, no actual fluid satisfies this requirement completely, and for most of them chemical reactivity is a real problem. Because it is less serious for gases than for liquids, in the following section we shall consider only the latter.

2.41. Thermal Stability

Even if confined in containers whose surfaces are completely inert to them, many liquids will change chemically when heated to relatively moderate temperatures. The change may be cracking, with a decrease in the average molecular weight and the generation of more volatile materials, or it may be polymerization, with the formation of resins, tars, and even cokes, or both processes may occur together. As the temperature increases, the rate of these reactions usually goes up very rapidly, and a limiting temperature is soon reached above which the use of the material as a working fluid becomes impractical.

In an actual system this process of degradation may be greatly accelerated by contact with metals or other materials which act as catalysts for the process. Most of our useful liquids are mixtures of organic compounds, and the resulting products of degradation, together with other foreign material such as water and dirt, form a sludge which is partly suspended in the liquid and partly settled out at various points in the system. Much of it ends up in the filter, which must be frequently changed to prevent plugging.

At normal operating temperatures this sludge formation usually takes place fairly slowly, causing trouble only if proper maintenance, particularly filter changing or cleaning, is neglected. At high temperatures, however, the process may be so rapid that the useful life of a charge of fluid may be only a few hours, or even minutes in extreme cases. Demands for higher-temperature operation are becoming more and more insistent; there is therefore a very active program for developing high-temperature fluids. Some of these will be very briefly discussed later in this chapter.

2.42. Oxidative Stability

More serious than pyrolysis, in general, is the reaction of the fluid with atmospheric air or other oxidizing agents. The products of this reaction are usually more or less acidic in nature and often cause serious corrosion of some of the metals of which the hydraulic system is made. In addition, oxidized compounds are usually more reactive than the products of pyrolysis, and therefore are likely to form sludge more rapidly.

Much more serious than the slow oxidation of fluid at normal temperatures is the possibility of fire or explosion. This is actually the principal weakness of hydrocarbon-base fluids; if they were not oxidizable the incentive for the development of synthetic hydraulic fluids would be much less. The occasional spectacular and disastrous event which can be charged to flammable fluids is sufficient to keep this development active.

The relative likelihood of trouble from the ignition of petroleum products has for many years been evaluated by three somewhat arbitrary tests.[26] The oil is heated in a cup and a small open pilot flame is periodically lowered to the liquid surface. The temperature at which enough vapor is evolved to cause a transient flame is called the *flash point*. As the temperature rises, the evolution of vapor becomes more rapid until finally the vapor will continue to burn after removal of the pilot flame. This temperature is the *fire point*. At a considerably higher temperature (determined by a different test procedure), called the *autogenous ignition* (or autoignition) *temperature*, the liquid or vapor will catch fire upon contact with air, without a pilot flame or other means of ignition. These three temperatures cannot be related to the actual probability of fire or explosion in a given situation, but they are indications of the relative hazards of various fluids.

The autoignition temperature, in particular, is greatly affected by the conditions of the test, such as the nature of the heated surface with which the liquid or vapor comes in contact, the oxygen content and pressure of the air, the pressure with which the liquid is sprayed against the surface, and the degree of atomization. Perhaps the greatest hazard

---

[26] ASTM Standard D92-52, *Flash and Fire Points by Means of Cleveland Open Cup.*

in the operation of a hydraulic system arises not from the evolution of inflammable vapors but from the possibility of creating an explosive mixture of finely dispersed droplets of oxidizable liquid in air. Such a mixture can explode with great violence even if the bulk liquid is comparatively inert. The danger is particularly great in vehicles, where hot exhaust pipes serve as ignitors for the vapor or fog, and in industrial applications such as furnace controls or diecasting machines, where the hydraulic fluid must be brought into the immediate proximity of highly heated metal surfaces if not of actual flames. In either case a burst or even a pinhole leak may cause disaster.

2.43. Hydrolytic Stability

Except in carefully sealed hydraulic systems, water of condensation will usually accumulate more or less rapidly in the fluid. Much of it will remain in the bottom of the sump, at least with petroleum fluids, but a certain amount will circulate as an emulsion, or will become one of the constituents of the sludge. The water will also cause a considerable increase in the corrosive power of the acidic oxidation products of the oil.

With nonhydrocarbon fluids, particularly with some types of esters such as the silicates, the water will react with the fluid itself, hydrolyzing it into alcoholic and acidic fractions. These fractions may dissolve in the fluid or they may form separate liquid or solid phases. Because the organic silicates seem to be particularly susceptible to hydrolysis, considerable care should be employed to exclude water from systems in which they are used.

2.44. Compatibility

Besides being affected by other materials in a hydraulic system, the hydraulic fluid itself affects them. These effects are of various kinds. The fluid or its degradation products, particularly the acids produced by oxidation, will corrode many metals, sometimes very rapidly. This corrosion can be minimized by careful choice of the metals used, by the use of anticorrosion devices such as sacrificial anodes and bichromate cartridges, and most of all by good maintenance of the system. Filters should be changed before sludge has a chance to build up, condensed water and sludge should be removed from the sump, and in extreme cases the system should be completely drained and flushed at suitable intervals.

Nearly all hydraulic systems use rubber-like substances extensively in seals, hoses, accumulator bladders, and so on. Several types of elastomers have been developed that are compatible with hydrocarbons and give very good service. When synthetic fluids are used, however, trouble is frequently encountered since some of these fluids seriously affect the usual synthetic rubbers. If the base compound of the rubber is even slightly soluble in the fluid, the part will swell and soften and may even dissolve. (This has been one fruitful cause of sludge formation.) In other cases the fluid will dissolve plasticizers from the rubber part, which will shrink and become horny or brittle. In either case the part becomes useless and the system usually becomes inoperative.

Special elastomers have been developed for use with the synthetic fluids, and if a system is changed from one fluid to another it will often be necessary to change all of the seals and hoses and the accumulator bladder. The maker of the particular fluid can usually supply information about sources of molded parts with which his fluid will be compatible.

When changing fluids, great care often must be taken to clean the system thoroughly of the last traces of the old fluid before putting in the new since some pairs of fluids are mutually incompatible. The glycols, for example, are alleged to precipitate out the viscosity-index-improving additive from petroleum-base aircraft oil, and there is some evidence that something in the oil catalyzes the polymerization of the polyglycol. In either case the result of mixing even small amounts of one fluid with the other is the formation of large quantities of a material much like a very soft chewing gum, which can effectively disable the system and is very hard to remove.

Other organic materials which are used in hydraulic systems include paints, varnishes, and enamels of various types, and electrical insulation. Again there is the problem of compatibility of the fluid with these materials, which is serious because of their great diversity and because of the difficulty of obtaining information about them. In some cases it is necessary to run compatibility tests with the proposed fluid on the paints and plastics which are to be used in or in close proximity to the system.

2.45. Toxicity

Since it is almost impossible to avoid at least occasional personal contact with it, the hydraulic fluid should not be poisonous. Some hydraulic fluids are somewhat toxic themselves, and their degradation products may be more so. For most fluids and in most circumstances the danger is vanishingly small, but in exceptional cases it may be sig-

nificant. For example, tricresyl phosphate is a very common additive
to petroleum oils. It is highly toxic when taken internally, and there
have been a few cases of poisoning from the vapor. It is not known to
be ingested through the skin, but enough possibility of poisoning by
this route exists to make one somewhat suspicious of the material.
Again the actual danger is small, but should be considered if extensive
contact of the fluid with the skin is likely.

One not uncommon accident with high-pressure systems is the piercing
of the skin by a jet of liquid. Even with a biologically inert material
this may cause extensive mechanical damage to the tissues, with consequent necrosis. If the liquid contains appreciable percentages of
toxic material, whether additives or degradation products, the danger
is naturally greater.

2.5. SURFACE PROPERTIES

The study of the surface properties of hydraulic fluids has just begun,
and the problems involved are so complicated that firm conclusions are
difficult to draw even in the simplest situations.[27] A good deal is known
about surface phenomena with carefully cleaned surfaces and highly
purified single chemical compounds, but many important surface phenomena are greatly affected by even very small concentrations of foreign
molecules, and conclusions drawn from experiments made under simplified conditions cannot safely be extended to cover phenomena in the
operation of practical hydraulic systems. About all that can be done is
to use the results of the idealized experiments to suggest remedies for
actual operating troubles.

When two different phases are in contact—say a liquid and a gas, or
two different liquids—a certain amount of energy will be associated with
each unit area of the surface which separates the two. This energy per
unit area is dimensionally, though not physically, the same as a force per
unit distance, and it is convenient at times to speak of the latter. When
the interface separates a liquid and a gas, it is called a free surface, and
the force is called surface tension. It is usually given in dynes/cm.

Free surfaces are unimportant in pressure hydraulic systems except
for the phenomenon of foaming.[28, 29] A foam is an emulsion of gas bubbles

[27] N. K. Adam, *The Physics and Chemistry of Surfaces*, Oxford University Press,
1941. See also Bondi, Reference 15 above, Chapter 3.

[28] A. M. Schwartz and J. W. Perry, *Surface Active Agents*, Interscience Publishers,
New York, 1949.

[29] S. Ross and J. W. McBain, "Inhibition of Foaming in Solvents Containing
Known Foamers," *Ind. Eng. Chem.*, Vol. 36 (1944), p. 570.

in a liquid, and under some circumstances it may be very stable and persistent. Naturally it is far more compressible than the liquid itself, and if it is drawn into the pump and injected into the hydraulic system, it may seriously disturb system operation. Also, because of the large liquid-gas interfacial areas in a foam, oxidation and similar reactions may be considerably accelerated by its presence.

The mechanism of foam formation and stabilization is little understood, and that of foam destabilization by defoaming agents is even less so. It is known that for a given liquid there may be several classes of compounds that aid in foam formation and preservation and that other compounds, sometimes even in very small concentrations, act as effective agents to prevent foaming. The same agent may even act as a foamer in one concentration and as a defoamer in another. One plausible explanation for foam formation is that the additive concentrates preferentially at the gas-liquid interface but is sufficiently mobile to increase the surface tension at thin points of the film and to decrease it at thick ones so as to keep the film thickness uniform and the film stable. A defoamer is less mobile under the same conditions and forms weak patches, thus rupturing the film and destroying the foam.

Several classes of defoaming agents have been used with varying success on the various hydraulic fluids. Among the more successful of these agents are dimethyl silicones, dialkyl orthophosphates, trialkyl thiophosphates, fluorinated oils, sulfonated fatty acids and their amino salts, and certain soaps. Some hydraulic fluids are much more subject to foaming than others, and some (unfortunately not necessarily the same ones) are relatively easy to defoam by the use of the proper additives. In cases of trouble from foaming, if there is no obvious mechanical reason for the excessive foam, the manufacturer of the fluid should be consulted for recommendations.

Generally speaking, the surfaces between fluid and solid in a hydraulic system are much more important than the free surfaces. This is especially true where neighboring solid surfaces are in relative motion, with a thin film of liquid between. In such a situation the molecular structure of the interfaces will largely determine the nature and amount of friction and wear and thus the performance and useful life of the system.

It would be very convenient if true hydrodynamic lubrication could be provided at all times and at all wearing areas in a hydraulic system. Unfortunately this is usually impractical and there is always a certain amount of boundary-layer friction.[30] The base materials of most hy-

---

[30] M. C. Shaw and E. F. Macks, "Analysis and Lubrication of Bearings," McGraw-Hill Book Co., New York, 1949. There is also a brief discussion of boundary-layer friction and related phenomena in Sec. 10.22.

draulic fluids are poor boundary-layer formers; for this reason, so-called lubricity additives are often provided to assist in the layer formation. These are usually long-chain polar compounds, such as long-chain fatty acids or their esters, which migrate to the metal surfaces and stick tightly. If the surface finish is poor enough or the normal forces are great enough, under load the boundary layer on the high spots will be torn off and destroyed. In such cases it is customary, and usually fairly effectual, to add one of the so-called extra-pressure (EP) additives, such as tricresyl phosphate or an organic halogen compound. Under the high instantaneous temperatures which occur at points of metal-to-metal friction, these compounds decompose and attack the metal, forming soft low-melting phosphides or halides, which are smeared down into the valleys of the metal surface. In this way the higher asperities are rounded and removed, the surface finish is improved, and the boundary layer can renew itself with a better chance of survival. For petroleum oils and at normal operating temperatures both lubricity and EP additives are effective and useful. For some other fluids, particularly the fluorocarbons and silicones, and at high temperatures, suitable additives are difficult to find. Unfortunately it is just these classes of fluids that are most in need of improvement.

2.6. THERMAL AND MISCELLANEOUS PROPERTIES

Two of the thermal properties of fluids are of considerable importance to the fluid-power engineer, namely the specific heat and the thermal conductivity.

The specific heat of a material is the derivative with respect to temperature of its enthalpy (if the pressure is kept constant) or of its internal energy (for constant volume). Less rigorously but more commonly, specific heat is defined as the amount of heat which must be supplied to unit weight of a substance to raise its temperature by unit amount. Water is taken as the standard substance, and 1 British thermal unit (Btu) is the heat required to raise the temperature of 1 lb of water by 1°F, at the freezing point of 32°F.

The defining equations of specific heat are

$$C_p = \left(\frac{\partial h}{\partial T}\right)_p \tag{2.21}$$

and

$$C_v = \left(\frac{\partial u}{\partial T}\right)_v \tag{2.22}$$

where C_p = specific heat at constant pressure, Btu/lb mole °F
   C_v = specific heat at constant volume, Btu/lb mole °F
   h = enthalpy, Btu/lb mole
   u = internal energy, Btu/lb mole

The subscript denotes the condition held constant. 1 Btu = 9339 in. lb.

For real fluids the specific heats are not constants but vary with temperature, as indicated in Tables 2.2 and 2.3. For most fluids the specific heats increase considerably with the temperature.

For gases the two specific heats are considerably different, and their ratio (usually designated k in engineering work), C_p/C_v, is an important quantity in most gas-dynamics equations. Because of the high bulk modulus of a normal liquid, its two specific heats are nearly the same.

The thermal conductivity of a gas increases with temperature but is

Table 2.2. C_p AND k VERSUS TEMPERATURE FOR NITROGEN *

Absolute Temperature, °R	C_p, Btu/lb mole °F	k
400	6.950	1.400
600	6.961	1.399
800	7.012	1.395
1000	7.132	1.386
2000	7.962	1.332

* J. H. Keenan and J. Kaye, *Gas Tables*, John Wiley and Sons, New York, 1948, p. 102.

Table 2.3.   SPECIFIC HEAT OF PETROLEUM OILS * (IN BTU/LB °F)

Absolute Temperature, °R	Specific Gravity at 60/60°F			
	1.000	0.934	0.876	0.825
460	0.388	0.410	0.416	0.427
560	0.432	0.448	0.463	0.477
660	0.478	0.495	0.512	0.526
760	0.523	0.542	0.558	0.576
860	0.568	0.588	0.608	0.626
960	0.612	0.633	0.656	0.676
1060	0.659	0.682	0.703	0.724
1160	0.704	0.726	0.798	...

* Adapted from J. C. Hunsaker and B. G. Rightmire, *Engineering Applications of Fluid Mechanics*, McGraw-Hill Book Co., New York, 1947, p. 285.

almost independent of pressure. It is related to viscosity by the equation

$$k_t = \tfrac{1}{4}(9k - 5)\mu C_v \qquad (2.23)$$

This equation makes it possible to calculate thermal conductivity from Sutherland's equation or from tabulated values of the viscosity of a gas. Since the viscosity increases with temperature, the conductivity does also. The thermal conductivity of air at 32°F is 0.314×10^{-6} Btu in./in.^{2} sec °F.

Under most conditions the transfer of heat takes place much more rapidly through fluids than through solids of the same conductivities because of the additional transport by convection.

The thermal conductivities of liquids are greater than that of air and decrease with increasing temperature. For most petroleum products the thermal conductivity is about 1.5 to 1.8×10^{-6} Btu in./in.^{2} sec °F. Other nonpolar organic liquids show similar values; strongly polar ones such as the glycols may be twice as conductive

The electrical properties of fluids are of some interest. It is often convenient and sometimes necessary to have electrical elements such as solenoid and torque-motor coils and motor windings within the spaces normally filled with the working fluid. If the fluid is a sufficiently good insulator, the coils and their leads need not be sealed or specially insulated, which fact is a great convenience. Most commercial hydraulic fluids are good to excellent insulators when pure—the hydrolubes are naturally a conspicuous exception—and it is often permissible to omit the extra insulation. This omission is impractical if the system maintenance is not good, in which case the coils and terminals will become covered with a sludge of relatively high conductivity. This may lead to flashovers, and perhaps to fire or explosion. Also, if the coils are not carefully impregnated, they will absorb water from the fluid, if any is present; and this will lead to electrolytic corrosion and open circuiting, or to insulation breakdown. The characteristics of interest are the insulation resistance and the breakdown voltage gradient of the fluid, and in some cases the dielectric constant also. Conductivities are given for six of the fluids in Table 2.5.

2.7. THE CHOICE OF A FLUID-POWER MEDIUM

As in any design problem, the choice of the fluid-power medium for a particular system depends upon a large number of factors which must be weighted according to their relative importance for the particular application. One choice must be made at the start, that of a liquid versus a gaseous medium. Chapters 8, 16, and 20 of this book, and parts of some of the other chapters, are concerned directly with pneumatic operation, and Chapter 19 is devoted to a comparison of hydraulics and pneumatics; so for the purposes of this section we may omit gaseous media and consider only liquids.

Probably no two experienced hydraulic engineers would draw up exactly the same list of requirements for a good hydraulic fluid, but all of the lists would be more or less similar. The list in Table 2.4 is presented because it is fairly all-inclusive and represents a great deal of operating experience.

Table 2.4.   Recommended Characteristics of Hydraulic Fluids *

1. Good lubricity with presently available materials for bearings and sealing surfaces.

2. Small change in viscosity over a wide temperature range (hopefully −70 to 500°F, at present).

3. Inertness to presently used materials for hydraulic equipment, including metals, paints, plastics, and elastomers.

4. A viscosity value compatible with present hydraulic fits and clearances.

5. Long service life; stability against heat, water, oxidation, and shear.

6. Very low toxicity of the liquid and its vapor, even after decomposition in service.

7. High bulk modulus.

8. Low tendency to foam.

9. Low specific gravity.

10. Low cost and high availability.

11. Nonflammability (high but not necessarily complete).

12. Low air absorption or solubility.

13. High heat-transfer coefficient.

14. Low vapor pressure and high boiling point, with oily residual film.

15. Low thermal-expansion coefficient.

16. Good dielectric properties, even if contaminated.

17. Nonhydroscopicity and very low mutual solubility with water.

18. Compatibility with present hydraulic fluids up to at least 10 per cent dilution.

19. High specific heat.

20. Lack of odor, or at least a pleasant odor.

21. Small change of viscosity with change of pressure and rate of shear.

22. Transparency and distinctive color.

* This list was originally compiled at Vickers, Inc., and is quoted with slight modification from Reference 22.

Table 2.5. PROPERTIES OF COMMERCIAL

Fluid Designation Manufacturer Base Material	Univis J-43 Esso Standard Oil Co. Petroleum	DC-200 Dow Chemical Co. Dimethyl Siloxane	Oronite 8515 Calif. Research Corp. Disiloxane
Specific gravity, 60/60°F	0.848	0.955	0.93
$10^{-3} \times$ bulk modulus, psi	270	170	218
$10^3 \times$ thermal expansion co- efficient/°F	0.50	0.57	0.445
Viscosity, centipoises, at			
−65 °F	2130	...	2357
−40	466	125	600
0	96	60	150
60	25	25	40
100	14.3	15.5	24.3
210	5.1	6.6	8.1
400	1.9	2.5 *	2.6
Pour point, °F	−90	−76	<-100
Flash point, °F	225	520	410
Fire point, °F	...	...	450
Autogenous ignition temp, °F	700	890	760
$10^6 \times$ thermal conductivity, Btu in./in.^{2} sec °F	1.85	1.87	2.1
Specific heat, Btu/lb °F, at (T) °F	0.50 (100)	0.35 (68)	0.47 (68)
Surface tension, dynes/cm	31.5	20.5	...

* Values doubtful; obtained by extrapolation on ASTM chart.
† Values for the very similar material Pydraul F-9.

There are many "hydraulic fluids" available on the U.S. market, and it is a matter of considerable difficulty to obtain adequate and accurate information on more than a few of them, or to evaluate the claims that are made for them. Table 2.5 lists some of the properties of fluids of seven different chemical types in an attempt to assist in the selection of a fluid for a particular application. The fluids listed are all "aircraft type," and in many cases other fluids of the same class are available with properties varying somewhat from the values given.

Inspection of the table will show that many of the properties are nearly the same for all of the fluids listed. Other properties, however, vary considerably from one class to another; it is these properties that must form the criteria of selection. Without having a specific application in mind it is impossible to go into detail concerning the selection process, but a rough comparison can be made with the aid of Table 2.6, which compares eight fluids with respect to eight factors. There seems to be approximate agreement on the relative ratings of the several fluids with respect to each of these factors, though there are occasional

AIRCRAFT-TYPE HYDRAULIC FLUIDS

Koolanol 45 Monsanto Chemical Co. Orthosilicate Ester	Skydrol Monsanto Chemical Co. Phosphate Ester	Aroclor 1242 Monsanto Chemical Co. Chlorinated Biphenyl	Ucon U-4 Union Carbide Water
0.892	1.086	1.41	1.054
253	387	...	...
...	...	...	...
2230	...	...	...
...	...	...	1980
125 *	240	...	262
25 *	37	150	39
12.1	15.5	18	16.6
3.95	3.85	2	(Boils)
1.2	...	...	...
−85	−70	+2	−60
365	430 †	348	...
425	675 †	"None"	...
705	1100 †	...	...
1.84	1.1	1.6	...
0.45 (68)	0.42 (68)	0.285 (68)	...
...	...	50.3	...

Table 2.6.   CHARACTERISTICS OF HYDRAULIC-FLUID BASE STOCKS *

Class	Viscosity Temperature	Viscosity Volatility	Thermal Stability	Hydrolytic Stability	Fire Resistance	Lubricity	Additive Response	Cost, dollars/gal
Silicate esters	E	G	E	F	F	G	G	10–20
Diesters	G	G	F	G	F	G	G	5–10
Silicones	E	E	G	E	F	P	P	30–40
Phosphate esters	G	F	F	F	E	E	G	4–12
Polyglycol esters	G	G	G	G	F	G	F	2–6
Chlorinated hydrocarbons	P	F	E	G	E	F	G	2–10
Fluorolubes	P	P	E	G	E	F	P	100–250
Petroleum	G	P	G	E	P	F	E	0.5–2

E = Excellent
G = Good
F = Fair
P = Poor
* Reprinted with the permission of Monsanto Chemical Co.

cases of disagreement.[31] This disagreement is only to be expected in a field as new and as rapidly developing as this one. Time will undoubtedly eliminate some of the present contenders and bring forward new ones, but at present the only sure test for a satisfactory fluid is successful operation in the system in question.

Demands for hydraulic fluids to cover wider and wider ranges of operating temperatures are rising rapidly, and will undoubtedly continue to rise. At the present time aircraft operation requires a temperature range of −65 to 275°F, and with the newer planes now in production the upper limit must be raised still higher. Future planes and guided missiles, and industrial applications such as furnace controls and diecasting machines, demand an upper limit of 400°, and 700 or even 1000°F will be required soon. Fortunately the low-temperature limit for industrial operation is much more reasonable; 32°F, with no damage on storage at 0°F, is fairly realistic. This relaxation of the lower limit permits the use of fluids that would solidify completely at the lower aircraft temperatures.

Parallel with the demand for wide temperature ranges is that for complete nonflammability, or at least a much reduced likelihood of fire or explosion in case of an accidental burst. This demand has been met in two ways: by the use of base materials that are inherently difficult to oxidize, such as chlorinated or perfluorinated compounds or phosphate esters, and by the use of a double-base liquid, where one principal component acts as a "snuffer" and thus prevents the other from burning. The snuffer liquid is usually water, and when it evaporates the fluid may no longer be fireproof. The high volatility of water naturally limits the operating temperature of these fluids to about 150°F, but in spite of this handicap they have been used to some extent in aircraft and extensively in industry.

Perhaps one of the easiest ways to select a fluid is by a process of elimination. To this end we may list the major disadvantages of the various classes:

1. Petroleum-base fluids have rather poor viscosity indices and are comparatively volatile and highly flammable. It should be noted, however, that when used in a pressurized system at very high temperatures, good mineral oils will outlast any other fluid now available.[32]

[31] Anon., "Synthetic Oils Start Moving," *Chem. Eng. News*, Vol. 34, No. 36 (Sept. 3, 1956), pp. 4244–4248.

[32] E. Erwin Klaus and Merrell R. Fenske, "Mineral Oils as High Temperature Fluids and Lubricants," *WADC Tech. Rept. 56–254* (Wright Air Development Center), August 1956.

2. Esters of dibasic organic acids have poor low-temperature viscosity characteristics, poor thermal stability, and poor compatibility with elastomers and other organic materials. (Some of them make excellent paint removers.)

3. Phosphate esters are good in many respects and have been much used in both aircraft and industrial applications, but their thermal stability is very poor above 250°F.

4. Polyglycols have poor thermal stability above 400°F and their decomposition is catalyzed by many metals. They are incompatible with even trace amounts of hydrocarbon oils, and this incompatibility, which results in the formation of a voluminous gummy precipitate, has caused a great deal of trouble in the field.

5. Fluorocarbons are fantastically expensive (though the price would fall rapidly if there were enough demand to warrant quantity manufacture), their viscosity indices are exceedingly poor, it is almost impossible to find satisfactory additives, and they are heavy. Chlorinated hydrocarbons are generally similar, but much cheaper. Neither class seems promising for most applications.

6. Silicones are very expensive and seem likely to remain so, their lubricity is poor, and they are comparatively unresponsive to additives.

Two of the most promising classes of fluids for wide-temperature-range operation are the polysiloxanes developed by the California Research Corporation and the orthosilicate esters [33] developed by the Monsanto Chemical Company. Both of these seem to be fairly satisfactory over the range of −65 to 400°F. Their biggest disadvantage is poor hydrolytic stability,[34] which requires that water be rigorously excluded from systems in which they are used. They are about as flammable as petroleum (indeed most organic compounds are unless they contain built-in snuffers) and their thermal and oxidation stabilities and their lubricities should be improved. Their relatively good operating temperature ranges compared to those of other commercial fluids can be seen in Fig. 2.8.

There is one other class of fluid that has received little attention until recently but now is much in the news. This class is that of the liquid

[33] H. S. Litzsinger and R. E. Hatton, "The Technical Appraisal of the Performance of OS-45 and/or OS-45-1 High Temperature Hydraulic Fluids," presented at the Vickers Turbojet Engine Hydraulics Symposium, Detroit, Michigan, February 2, 1956.

[34] B. Rubin and H. M. Schiefer, "Hydraulic Fluids for High Temperature Systems," presented at the Vickers Turbojet Engine Hydraulics Symposium, Detroit, Michigan, February 2, 1956.

metals, which are already in use as heat-transfer media in nuclear reactors. Besides mercury, sodium or the liquid sodium-potassium eutectic and various alloys of lead, bismuth, and other low-melting metals have been used, at least on an experimental basis. The use of metals, whose

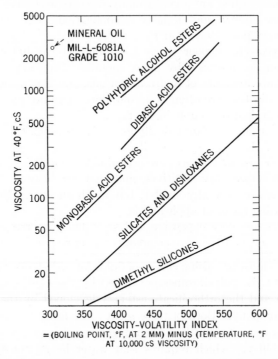

Fig. 2.8. Viscosity-volatility characteristics of synthetic hydraulic fluids. (Courtesy of California Research Corp.)

properties differ radically from those of the organic liquids ordinarily used as hydraulic fluids, brings a whole series of new problems in the choice of materials and the design of components for the systems in which they will be used, but at least at extreme temperatures or in other very unfavorable environments there seem to be few substitutes in sight. One good alternative in many respects is the use of gases, which introduces a similar set of problems. Both offer a major challenge to the fluid-power engineer.

3

A. H. Stenning
J. L. Shearer

Fundamentals of Fluid Flow

3.1. INTRODUCTION

To design fluid machinery, or to predict its performance, it is necessary to understand and to be able to apply the fundamental laws of fluid mechanics. Frequently a machine which will do its job fairly well can be produced by trial and error, and with no analysis. Indeed, much of present-day engineering achievement would not exist if engineers always waited until a satisfactory theory were available to explain the operation of a new device, idea, or system. Nevertheless, an optimum design can be achieved economically only by intelligent application of existing theoretical and empirical knowledge. Few systems are pure enough or simple enough to conform very closely to basic theoretical concepts. Most fluid-flow systems are sufficiently complex to require a great deal of judgment in applying the theoretical concepts (often called fundamental laws or basic principles) to them. Experience in the form of empirical data taken by others and in the form of personal observations from everyday life and engineering practice is a vital factor in most successful engineering analyses. This chapter presents the basic principles of fluid flow, together with some examples of how they may be applied to simple systems.

43

3.11. The Variables

The condition or state of a fluid at any point in a region is defined in terms of certain properties such as pressure, temperature, density, viscosity, etc. In addition, if the fluid is not at rest, it is necessary to know its velocity and direction of motion as a function of time. Application of the laws of fluid mechanics, together with the relationships which exist between interdependent properties of the fluid, makes it possible to describe a system with a number of equations equal to the number of variables. For the general case of a compressible fluid in three dimensions with friction and heat transfer, it is difficult to obtain an exact solution in many important problems. It then becomes necessary to simplify the system of equations by making intelligent approximations to the system so that a simple solution sufficiently accurate for the purpose is obtainable. The limitations of this approach must always be kept in mind lest erroneous conclusions be drawn from an analysis so simplified that some of the important factors have been left out.

3.2. DERIVATION OF FUNDAMENTAL RELATIONSHIPS

3.21. The "Control Volume" Method

Fluid mechanics employs the same basic laws which Newton and later mathematicians applied to systems of particles and rigid bodies of fixed identity.[1,2] However, many problems in fluid mechanics can be more readily handled by use of the "control volume concept."[3,4] The reason for this is that when studying fluid motion we are not so often interested in following particles of fixed identity as in observing how the flow varies at one particular point in space, or within a given volume— as for example the space within the blades of a fan, or within a heat-exchanger tube. Accordingly, we use the basic laws to develop equations which apply to flow through a fixed space called a control volume, whose surface is called a control surface.

[1] F. B. Seeley and N. E. Ensign, *Analytical Mechanics for Engineers*, John Wiley and Sons, New York, 1952.

[2] J. P. Den Hartog, *Mechanics*, McGraw-Hill Book Co., New York, 1948.

[3] J. C. Hunsaker and B. G. Rightmire, *Engineering Applications of Fluid Mechanics*, McGraw-Hill Book Co., New York, 1947.

[4] R. C. Binder, *Advanced Fluid Mechanics and Fluid Machinery*, Prentice-Hall, New York, 1951.

3.22. Conservation of Mass

If, in the flow which we consider, fluid is neither created nor destroyed, then material flowing into the control volume must either flow out again or stay inside; that is, *rate of mass flow into the volume equals rate of mass flow out plus rate at which mass accumulates inside.*

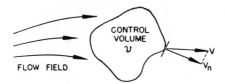

Fig. 3.1. Control volume in a flow field.

Expressed mathematically, the law of the conservation of mass applied to a control volume such as that shown in Fig. 3.1 becomes the continuity equation

$$\int_{A_s} \rho V_n \, dA_s = -\frac{dm_{cv}}{dt} = -\frac{d}{dt}\int_{\upsilon} \rho \, d\upsilon \qquad (3.1)$$

where ρ = density at a point, lb sec^2/in.^{4}

V_n = velocity normal to surface ($+$ when directed outward), in./sec

dA_s = element of area of control surface, in.^{2}

$d\upsilon$ = element of control volume, in.^{3}

m_{cv} = mass in control volume, lb sec^2/in.

3.221. Applications.

(*a*) Steady flow of liquid into and out of a tank as shown in Fig. 3.2.

A_1 = area of inlet flow passage, in.^{2}

A_2 = area of exit flow passage, in.^{2}

V_1 = inlet velocity, in./sec

V_2 = exit velocity, in./sec

L = height of water in tank, in.

A_t = cross-sectional area of tank, in.^{2}

Since the flow is steady, L does not vary and the mass in the tank is constant. Hence, rate of mass flow in is equal to rate of mass flow out; that is,

$$\rho V_1 A_1 = \rho V_2 A_2$$

or
$$V_1 A_1 = V_2 A_2 \qquad (3.2)$$

(*b*) Unsteady flow into a tank (*L* may now vary with time). If the liquid level is rising at the rate dL/dt, then the mass inside the tank increases at the rate $\rho A_t(dL/dt)$. Hence, the rate at which mass of liquid

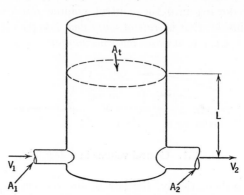

Fig. 3.2. Storage of liquid in a tank.

flows out must be $\rho A_t(dL/dt)$ less than the rate at which mass of liquid enters.

$$\rho V_1 A_1 - \rho V_2 A_2 = \rho A_t \frac{dL}{dt}$$

or   (3.3)

$$V_1 A_1 - V_2 A_2 = A_t \frac{dL}{dt}$$

3.23. Conservation of Momentum

Newton's second law states that rate of change of momentum of a system of fixed identity is equal to the net force acting on it. This may be applied as follows when dealing with a control volume as shown in Fig. 3.3.

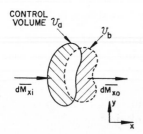

Fig. 3.3. Conservation of momentum applied to a control volume.

Consider a system of fixed identity comprising a quantity of fluid which at time t_a occupies the space in a control volume $\mathcal{U}_a$. At time $t_b = t_a + dt$, a little later, this system of fluid particles will occupy the space $\mathcal{U}_b$. Part of the fluid will have moved out of the control volume $\mathcal{U}_a$. Other fluid will have moved in to replace it. The momentum in the x-direction of that part of the system lying outside $\mathcal{U}_a$ at t_b we shall call $d\overline{M}_{xo}$ and the momentum in the x-direction of the new matter inside $\mathcal{U}_a$

at t_b but not included in the system we shall call $d\overline{M}_{xi}$. Representing the x-momentum of all the matter inside the control volume of $\mathcal{V}_a$ at any time by $\overline{M}_x{}'$ and representing the x-momentum of the system of fixed identity at any time by $\overline{M}_x$, we see that the initial and final momenta of the system of fixed identity are, respectively,

$$\overline{M}_{xa} = \overline{M}_{xa}{}' \qquad \overline{M}_{xb} = \overline{M}_{xb}{}' - d\overline{M}_{xo} - d\overline{M}_{xi}$$

whence

$$\frac{d\overline{M}_x}{dt} = \frac{\overline{M}_{xb} - \overline{M}_{xa}}{dt} = \frac{\overline{M}_{xb}{}' - \overline{M}_{xa}{}'}{dt} + \frac{d\overline{M}_{xo} - d\overline{M}_{xi}}{dt} =$$

$$\frac{d\overline{M}_x{}'}{dt} + \frac{d\overline{M}_{xo}}{dt} - \frac{d\overline{M}_{xi}}{dt} \qquad (3.4)$$

The rate of change of x-momentum of the system is thus expressed as the sum of the rate of change of x-momentum of the material inside the control volume and the net rate of outflow of x-momentum through the control surface. The instantaneous force acting in the x-direction on the matter momentarily occupying the fixed volume $\mathcal{V}_a$ equals the rate of change of x-momentum inside plus the net rate of outflow of x-momentum through the control surface. This may be expressed mathematically as

$$F_x = \frac{d}{dt} \int_{\mathcal{V}_a} \rho V_x \, d\mathcal{V}_a + \int_{A_s} \rho V_x V_n \, dA_s \qquad (3.5)$$

Similarly

$$F_y = \frac{d}{dt} \int_{\mathcal{V}_a} \rho V_y \, d\mathcal{V}_a + \int_{A_s} \rho V_y V_n \, dA_s \qquad (3.6)$$

$$F_z = \frac{d}{dt} \int_{\mathcal{V}_a} \rho V_z \, d\mathcal{V}_a + \int_{A_s} \rho V_z V_n \, dA_s \qquad (3.7)$$

For steady-state conditions, the first term on the right-hand side of each of the above equations is zero, and the force equals the net outflow of momentum from the control volume.

3.231. Steady-Flow Example. What is the external force, F_x, required for equilibrium of the container shown in Fig. 3.4 under steady-flow conditions if $P_1 = P_2$?

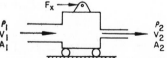

Fig. 3.4. Propulsion thrust developed as the result of a steady flow.

$$\text{Inflow of } x\text{-momentum} = V_1 \cdot \rho_1 A_1 V_1$$

$$= \rho_1 A_1 V_1{}^2 \tag{3.8}$$

$$\text{Outflow of } x\text{-momentum} = \rho_2 A_2 V_2{}^2 \tag{3.9}$$

$$\text{Net outflow} = \rho_2 A_2 V_2{}^2 - \rho_1 A_1 V_1{}^2 \tag{3.10}$$

If $P_1 = P_2$, F_x = net outflow of x-momentum:

$$F_x = \rho_2 A_2 V_2{}^2 - \rho_1 A_1 V_1{}^2 \tag{3.11}$$

From continuity considerations

$$\rho_1 A_1 V_1 = \rho_2 A_2 V_2 \tag{3.12}$$

$$F_x = \rho_1 A_1 V_1 (V_2 - V_1) \tag{3.13}$$

3.232. Unsteady-Flow Example. Find how the pressure difference $(P_1 - P_2)$ is related to the rate of change of flow rate of a *frictionless,*

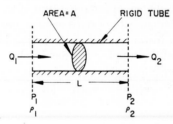

Fig. 3.5. Unsteady flow of a frictionless, incompressible fluid through a constant-area tube.

incompressible fluid in a uniform tube of length L such as that shown in Fig. 3.5.

Applying the continuity equation to the volume bounded by the tube of length L, we obtain

$$\rho Q_1 - \rho Q_2 = 0 \tag{3.14}$$

$$Q_1 = Q_2 = Q \tag{3.15}$$

The momentum equation gives

$$P_1 A - P_2 A = \rho Q V - \rho Q V + \frac{d}{dt}(\rho A L V) \tag{3.16}$$

$$(P_1 - P_2)A = \rho A L \frac{d}{dt}\left(\frac{Q}{A}\right) = \rho L \frac{dQ}{dt} \tag{3.17}$$

$$P_1 - P_2 = \frac{\rho L}{A}\frac{dQ}{dt} \tag{3.18}$$

3.24. Conservation of Angular Momentum

Angular momentum of particle m about an axis through P is defined as $mVr\cos\alpha$, as shown in Fig. 3.6. An analysis similar to that used in the preceding section will show that under steady-state conditions the net rate of outflow of angular momentum from a control volume equals the net torque acting on the control volume. When unsteady flow exists, the net torque is equal to the net rate of outflow of angular momentum plus the rate of change of angular momentum within the control volume.

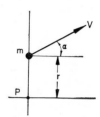

Fig. 3.6. Conservation of angular momentum of a particle, m.

3.241. Application of the Law of Conservation of Angular Momentum to a Pump (Steady Flow).

Consider a centrifugal pump in which the fluid enters at a radius r_1 with a tangential velocity V_{t1} and leaves at a radius r_2 with a tangential velocity V_{t2}, as shown in Fig. 3.7. Then, entering angular momentum per unit mass of fluid about C is given by r_1V_{t1}, and leaving angular momentum per unit mass of fluid is given by r_2V_{t2}. Change in angular momentum per unit mass of fluid flowing is $(r_2V_{t2} - r_1V_{t1})$. But torque is equal to net rate of outflow of angular momentum:

$$T = (r_2V_{t2} - r_1V_{t1})\rho Q \qquad (3.19)$$

where T = torque exerted by impeller on the fluid, lb-in.

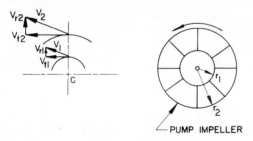

Fig. 3.7. Velocity diagram for a centrifugal pump.

3.25. Conservation of Energy

The first law of thermodynamics states that the increase in internal energy of a system of fixed identity is equal to the work done on the system plus the heat added to the system. Extension of this law to flow through a control volume by the method used previously for the law of conservation of momentum gives for the system shown in Fig. 3.8

$$\frac{dQ_h}{dt} - \frac{dW_x}{dt} = \frac{dE}{dt} + \int_{A_1} \left(\frac{P}{\rho} + e\right) \rho V_n \, dA \tag{3.20}$$

where W_x = shaft and shear work done by the system on its surroundings, in. lb

Q_h = heat flow to the control volume, in. lb (1 Btu = 9336 in. lb)

E = total internal energy of the fluid inside the control volume, in. lb

$e = u + gz + V^2/2$ = total internal energy per unit mass of fluid, in.^{2}/sec^2

u = intrinsic internal energy per unit mass of fluid, in.^{2}/sec^2 (a function of the state of the fluid)

g = acceleration due to gravity, 386 in./sec^2

z = height above datum, in. (gz = potential energy per unit mass of fluid, in.^{2}/sec^2)

V_n = velocity normal to the surface of the control volume, in./sec

P = pressure on an element of area at the surface, lb/in.^{2}

This is often called the energy equation.

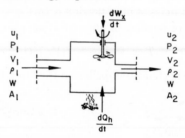

Fig. 3.8. Conservation of energy applied to a simple control volume.

For the case of steady flow in which changes in height are unimportant and in which the fluid enters the volume at only one place and leaves at only one other place, the equation simplifies to

$$\frac{d}{dt}Q_h - \frac{d}{dt}W_x = \left(\frac{P_2}{\rho_2} - \frac{P_1}{\rho_1} + u_2 - u_1 + \frac{V_2{}^2 - V_1{}^2}{2}\right)\frac{W}{g} \tag{3.21}$$

where $\dfrac{W}{g} = \rho_1 V_1 A_1 = \rho_2 V_2 A_2$ \tag{3.22}

W = weight flow rate of fluid, lb/sec (a pound of fluid is the mass of that fluid which is held in equilibrium by a 1-lb force when it is in a gravitational field of 386 in./sec^2)

Or

$$\frac{P_1}{\rho_1} + u_1 + \frac{V_1^2}{2} = \frac{P_2}{\rho_2} + u_2 + \frac{V_2^2}{2} + \frac{(dW_x/dt) - (dQ_h/dt)}{W/g} \qquad (3.23)$$

It is important to note that the rates of flow of both heat and work, as well as the rate of flow of fluid, must be constant for the flow to be truly steady.

The relationships that have just been derived for a control volume which is fixed in space may also be applied without modification to a control volume that is moving along a straight line at constant velocity. These relationships *do not* apply to a control volume that is undergoing any kind of acceleration.

3.3. STEADY FLOW OF FLUIDS IN CLOSED PASSAGES

In the previous paragraphs dealing with the equations of conservation of mass, momentum, and energy, it has been shown that if conditions within the control volume do not change with time, the equations are greatly simplified.

Of course, in most practical cases conditions do not remain the same indefinitely, and frequently the flow changes continuously. However, if the changes occur slowly, then the terms representing time rates of change of mass, energy, or momentum within the control volume considered may be so small as to be negligible compared with the other terms. Such a flow is said to be quasi-steady.

STEADY FLOW OF AN INCOMPRESSIBLE FLUID. It is often possible to consider the fluid density as constant throughout the region under examination. This is almost always true for steady flow of liquids and frequently for low-speed gas flows. In these circumstances the equations are greatly simplified and satisfactory solutions may be obtained with little effort.

Special cases are discussed later for which even small density changes must be included.

One of the most important fundamental examples of a steady-state incompressible flow is that of frictionless flow through a passage of changing area, in which fluid properties are approximately constant over any cross section and vary only in the direction of flow.

3.31. One-Dimensional, Frictionless, Incompressible, Streamline Flow with Area Change

Consider flow through a passage in which property changes occur only in the direction of motion and in which the curvature is small, as shown in Fig. 3.9.

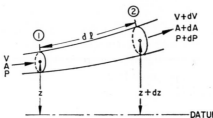

Fig. 3.9. One-dimensional, frictionless, steady, streamline flow of an incompressible fluid with area change.

At section 1:
Let the area be A, velocity V, pressure P, height above datum z, and density ρ.

At section 2:
Let the area be $A + dA$, velocity $V + dV$, pressure $P + dP$, height above datum $z + dz$, and density ρ.

Let the distance between 1 and 2 be dl.

Let the control volume be bounded by section 1, section 2, and the wall of the passage.

From continuity considerations

$$\rho V A = \rho(V + dV)(A + dA) \tag{3.24}$$

$$V \, dA + A \, dV = 0 \tag{3.25}$$

If there is no friction, then the only forces acting on the fluid inside the control volume are pressure forces and gravity forces. Pressure forces in the direction of flow are

$$F_p = PA + \left(P + \frac{dP}{2}\right) dA - (P + dP)(A + dA) \approx -A \, dP \tag{3.26}$$

Gravity forces in the direction of flow are

$$F_g = -g\rho A \, dl \, \frac{dz}{dl} = -g\rho A \, dz \tag{3.27}$$

Net momentum outflow from the control volume is

$$\frac{d}{dt}\overline{M}_o - \frac{d}{dt}\overline{M}_i$$

$$= \rho(V + dV)^2(A + dA) - \rho(A)V^2 \approx \rho V^2\,dA + 2A\rho V\,dV \quad (3.28)$$

Therefore

$$-g\rho A\,dz - A\,dP = \rho V^2\,dA + 2A\rho V\,dV \quad (3.29)$$

$$-g\rho\,dz - dP = \rho V^2\frac{dA}{A} + \rho V\,dV \quad (3.30)$$

But from continuity

$$(dA/A) = -(dV/V) \quad (3.31)$$

Therefore

$$-g\rho\,dz - dP = -\rho V\,dV + 2\rho V\,dV \quad (3.32)$$

$$-g\rho\,dz - dP = \rho V\,dV \quad (3.33)$$

Dividing the above equation by ρ gives Euler's equation for steady flow,

$$V\,dV + \frac{dP}{\rho} + g\,dz = 0 \quad (3.34)$$

Since ρ is constant, Euler's equation can be integrated to give

$$\frac{V^2}{2g} + \frac{P}{\rho g} + z = \text{constant} \quad (3.35)$$

The quantity $\left(\dfrac{V^2}{2g} + \dfrac{P}{\rho g} + z\right)$ is called the total head of the fluid stream.

Equation 3.35 is known as Bernoulli's equation for an incompressible fluid. Note that it was possible to obtain a solution relating velocity, pressure, and height by using only the continuity and momentum equations. By integrating between limits for flow between sections 1 and 2 along the passage, we obtain

$$\frac{P_1}{\rho g} + \frac{V_1{}^2}{2g} + z_1 = \frac{P_2}{\rho g} + \frac{V_2{}^2}{2g} + z_2 \quad (3.36)$$

Introducing the energy equation and using as the control volume the space bounded by sections 1 and 2 and the walls of the passage, we have

$$\frac{P_1}{\rho} + u_1 + \frac{V_1{}^2}{2} + gz_1 = \frac{P_2}{\rho} + u_2 + \frac{V_2{}^2}{2} + gz_2 + \frac{(dW_x/dt) - (dQ_h/dt)}{W/g}$$

$$(3.37)$$

But

$$\frac{P_1}{\rho} + \frac{V_1{}^2}{2} + gz_1 = \frac{P_2}{\rho} + \frac{V_2{}^2}{2} + gz_2 \quad \text{(Bernoulli's equation)} \quad (3.38)$$

And in developing Bernoulli's equation we stated that the only forces acting were pressure and gravity forces. Therefore

$$\frac{dW_x}{dt} = 0 \tag{3.39}$$

And hence

$$\frac{dQ_h}{dt} = \frac{W}{g}(u_2 - u_1) = \frac{W}{g} C_h(T_2 - T_1) \tag{3.40}$$

where C_h is the specific heat of the fluid. This equation enables us to solve for temperature changes due to heat transfer. If there is no heat transfer, $T_1 = T_2 = $ constant. Note that in this case temperature changes are independent of pressure, velocity, and height changes. This is generally true for frictionless, incompressible flow because temperature changes have small effect on the density and hence little effect on the flow (except for free-convection flows).

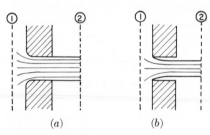

Fig. 3.10. Frictionless flow through nozzles and orifices.

(a) Nozzle or round-edged orifice.
(b) Sharp-edged orifice.

3.311. Frictionless Flow through Nozzles and Orifices. Consider the flow from a reservoir as shown in Fig. 3.10, where the velocity is negligible, through an orifice or nozzle into a chamber. Let the conditions in the reservoir be represented by suffix 1. Let the conditions at exhaust be represented by suffix 2. At 2 the fluid properties are uniform across the stream and fluid pressure in the stream is equal to the pressure in the downstream chamber. The velocity at 1 is negligible.

Then from Bernoulli's equation

$$\frac{P_1}{\rho} = \frac{P_2}{\rho} + \frac{V_2{}^2}{2} \tag{3.41}$$

$$V_2{}^2 = \frac{2}{\rho}(P_1 - P_2) \tag{3.42}$$

$$V_2 = \sqrt{(2/\rho)(P_1 - P_2)} \tag{3.43}$$

$$\rho Q = \text{mass flow rate} = \rho A_2 V_2 = \frac{W}{g} \tag{3.44}$$

$$\rho Q = \rho A_2 \sqrt{(2/\rho)(P_1 - P_2)} = A_2 \sqrt{2\rho(P_1 - P_2)} \tag{3.45}$$

For nozzles and rounded-edge orifices A_2 is very nearly equal to the nozzle exit area, A_o. Hence

$$\rho Q = A_o \sqrt{2\rho(P_1 - P_2)} \tag{3.46}$$

For a sharp-edged orifice, A_2 is less than A_0, and

$$\rho Q = C_d A_o \sqrt{2\rho(P_1 - P_2)} \tag{3.47}$$

where the discharge coefficient, C_d, varies from 0.6 to 1.0, depending on the geometry of the upstream passage.

Although Eqs. 3.46 and 3.47 indicate that the flow rate, Q, is instantaneously related to the orifice area, A_0, and pressure drop, $(P_2 - P_1)$, it does take some time to establish a new flow regime in the orifice if either A_o or $(P_2 - P_1)$ is changed suddenly. This time-lag effect is very small and is usually negligible compared with the other lags inherent in fluid-power-control systems. The reader who may be concerned with the lag effects within an orifice configuration should refer to the paper by Daily, Hankey, Olive, and Jordan.[5] In this paper Daily and his associates discuss the fluid accelerations in unsteady flow and show how the lag effect is caused by changes in the fluid kinetic energy within the orifice passages as the pressure drop across the orifice changes.

3.32. Effects of Friction in a One-Dimensional Flow

If we consider, in addition, a friction stress, τ, acting on the boundaries

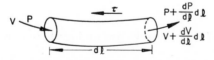

Fig. 3.11. Effects of friction in a one-dimensional incompressible flow.

of the fluid at the wall of the passage as shown in Fig. 3.11, we have

[5] J. W. Daily, W. L. Hankey, Jr., R. W. Olive, and J. M. Jordan, Jr., "Resistance Coefficients for Accelerated and Decelerated Flows Through Smooth Tubes and Orifices," *Trans. ASME*, Vol. 78, No. 5 (July 1956), pp. 1071–1077.

from the momentum equation

$$-A \, dP - \tau S \, dl - A g \rho \, dz = \rho V^2 \, dA + 2A \rho V \, dV \tag{3.48}$$

where S = average perimeter of tube wall, inches

And since the continuity equation gives

$$(dA/A) = -(dV/V) \tag{3.49}$$

then

$$-g\rho \, dz - dP - \tau \frac{S}{A} \, dl = \rho V \, dV \tag{3.50}$$

or

$$g\rho \, dz + dP + \tau \frac{S}{A} \, dl + \rho V \, dV = 0 \tag{3.51}$$

If $\tau(S/A)$ is constant along a tube of length l we may integrate and obtain

$$gz + \frac{P}{\rho} + \frac{\tau S l}{\rho A} + \frac{V^2}{2} = \text{constant} \tag{3.52}$$

Or between sections 1 and 2 a distance of $(l_2 - l_1)$ apart

$$\left(\frac{P_1}{\rho} + \frac{V_1{}^2}{2} + gz_1 \right) = \left(\frac{P_2}{\rho} + \frac{V_2{}^2}{2} + gz_2 \right) + \frac{\tau S}{\rho A} (l_2 - l_1) \tag{3.53}$$

Thus the total head, $(P/\rho) + (V^2/2) + gz$, decreases in the direction of flow.

The energy equation may now be applied to the control volume shown in Fig. 3.12 in order to determine how the temperature varies along the passage.

$$\frac{P_1}{\rho} + u_1 + \frac{V_1{}^2}{2} + gz_1 = \frac{P_2}{\rho} + u_2 + \frac{V_2{}^2}{2} + gz_2 + \frac{(dW_x/dt) - (dQ_h/dt)}{W/g} \tag{3.54}$$

As before, $dW_x/dt = 0$ since we are taking the control surface outside the channel. But

$$\left(\frac{P_1}{\rho} + \frac{V_1{}^2}{2} + gz_1 \right) = \left(\frac{P_2}{\rho} + \frac{V_2{}^2}{2} + gz_2 \right) + \frac{\tau S}{\rho A} (l_2 - l_1) \tag{3.55}$$

$$\frac{dQ_h}{dt} + \frac{\tau S(l_2 - l_1)}{\rho A} \frac{W}{g} = \frac{W}{g} (u_2 - u_1) \tag{3.56}$$

and even if $dQ_h/dt = 0$, u_2 is greater than u_1. *Thus friction in a steady incompressible flow causes an internal energy rise in the direction of flow,* an explanation of the temperature rise observed with many liquids.

In the preceding analysis it was assumed that the flow was one-dimensional, that is, that the fluid properties were constant throughout any cross section of the flow. Such a one-dimensional flow with friction can exist only when friction effects are concentrated in a very thin layer at the wall of the passage. This condition is encountered only with very short passages in hydraulic systems. However, the one-dimensional analysis does give a qualitative picture of the effects of friction even if it occurs throughout the fluid stream, and Eq. 3.53 may still be employed if appropriate average velocities are used for V_1^2 and V_2^2.

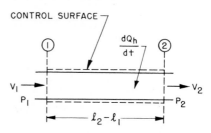

Fig. 3.12. Incompressible flow through a passage with friction and heat transfer.

3.321. Friction and Viscosity. When a real fluid flows steadily past a stationary wall, the particles next to the wall adhere to it, and there is in the fluid a velocity gradient normal to the direction of flow. The retarding force exerted by the boundary on the fluid depends on both the velocity gradient at the boundary and the viscosity of the fluid. For Reynolds numbers less than 2000, $R_e = \rho V D/\mu$, where D is the passage diameter, V is the average velocity, ρ is the density, and μ is the viscosity of the fluid—the flow in a passage is not turbulent, so the path followed by each particle is a smooth curve called a streamline. This type of flow is designated as laminar, or viscous, flow, and for the type of unidirectional flow shown in Fig. 3.13, the friction force per unit area at any point of the flow is $\tau = \mu \, \partial V/\partial y$.

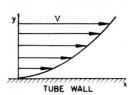

Fig. 3.13. Laminar flow past a stationary wall.

In many cases of leakage flow past pistons, glands, etc., the flow is laminar and may be readily analyzed in this way. Results for several such cases are shown in Table 3.1. For larger Reynolds numbers, the flow may consist of random lateral motions superimposed on the main forward motion. The flow no longer consists of steady streamlines. Under these conditions the relationship between shear stress and velocity gradient is very complex, and no exact theory exists to describe it. It is most convenient to define $\tau = f\rho(V^2/2)$ where f is called the friction

Table 3.1. EQUATIONS OF FULLY DEVELOPED LAMINAR FLOW FOR
SEVERAL TYPICAL CASES (THE TEMPERATURE AND THE
VISCOSITY OF THE FLUID ARE ASSUMED CONSTANT)

A) STEADY FLOW THROUGH CIRCULAR PIPE

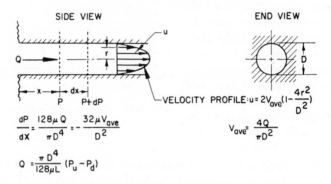

$$\frac{dP}{dX} = \frac{128\mu Q}{\pi D^4} = -\frac{32\mu V_{ave}}{D^2}$$

$$Q = \frac{\pi D^4}{128\mu L}(P_u - P_d)$$

VELOCITY PROFILE: $u = 2V_{ave}\left(1 - \frac{4r^2}{D^2}\right)$

$$V_{ave} = \frac{4Q}{\pi D^2}$$

B) STEADY FLOW BETWEEN STATIONARY FLAT PLATES

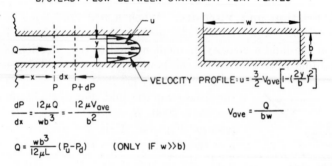

$$\frac{dP}{dx} = \frac{12\mu Q}{wb^3} = -\frac{12\mu V_{ave}}{b^2}$$

$$Q = \frac{wb^3}{12\mu L}(P_u - P_d) \quad \text{(ONLY IF } w \gg b\text{)}$$

VELOCITY PROFILE: $u = \frac{3}{2}V_{ave}\left[1 - \left(\frac{2y}{b}\right)^2\right]$

$$V_{ave} = \frac{Q}{bw}$$

C) STEADY FLOW IN ANNULUS BETWEEN CIRCULAR SHAFT AND CYLINDER

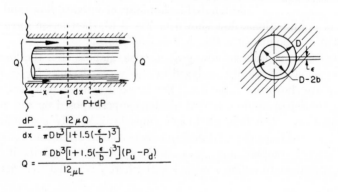

$$\frac{dP}{dx} = \frac{12\mu Q}{\pi Db^3\left[1 + 1.5\left(\frac{\epsilon}{b}\right)^3\right]}$$

$$Q = \frac{\pi Db^3\left[1 + 1.5\left(\frac{\epsilon}{b}\right)^3\right](P_u - P_d)}{12\mu L}$$

Table 3.1. (CONTINUED)

D) STEADY FLOW BETWEEN STATIONARY AND MOVING FLAT, PARALLEL PLATES

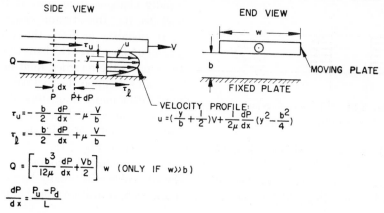

$$\tau_u = -\frac{b}{2}\frac{dP}{dx} - \mu\frac{V}{b}$$

$$\tau_\ell = -\frac{b}{2}\frac{dP}{dx} + \mu\frac{V}{b}$$

VELOCITY PROFILE:

$$u = \left(\frac{y}{b} + \frac{1}{2}\right)V + \frac{1}{2\mu}\frac{dP}{dx}\left(y^2 - \frac{b^2}{4}\right)$$

$$Q = \left[-\frac{b^3}{12\mu}\frac{dP}{dx} + \frac{Vb}{2}\right]w \quad \text{(ONLY IF } w \gg b)$$

$$\frac{dP}{dx} = \frac{P_u - P_d}{L}$$

Q IS MEASURED RELATIVE TO A POINT ON THE FIXED PLATE

NOMENCLATURE

μ = FLUID VISCOSITY, LB SEC/SQ IN.
Q = VOLUME FLOW RATE, CU IN/SEC
P = PRESSURE, LB/SQ IN.
x = DISTANCE ALONG PASSAGE, IN.
D = DIAMETER OF PASSAGE, IN.
V = VELOCITY, IN. SEC
w = PASSAGE WIDTH, IN.
b = PASSAGE HEIGHT, IN.
τ_u = SHEAR STRESS AT UPPER WALL, LB/SQ IN.
τ_ℓ = SHEAR STRESS AT LOWER WALL, LB/SQ IN.
P_u = UPSTREAM PRESSURE, LB/SQ IN.
P_d = DOWNSTREAM PRESSURE, LB/SQ IN.
L = PASSAGE LENGTH (IN DIRECTION OF FLOW), IN.

factor. For a pipe, the friction factor, f, is a function of Reynolds number, surface roughness, and distance from the pipe inlet. A considerable body of information on turbulent friction factors has been amassed experimentally. This friction factor f usually is between 0.002 and 0.01. For greater detail, the reader is referred to a good fluid mechanics text (fns. 3 and 4).

3.322. Friction at the Entrance to a Pipe. If a fluid flows into a pipe from a large chamber, as shown in Fig. 3.14, the velocity distribution over the cross section will be very nearly uniform at the entrance. At the entrance there will be a very strong shearing action between the pipe wall and the fluid next to it, but the effect of this shearing action will be felt only a short distance from the wall, in a region called the boundary layer, where the velocity rapidly changes from zero at the wall to the free-stream value a short distance in from the wall. The boundary layer thickens along the pipe as the retarded fluid near the wall acts upon the free stream, until, after many diameters along the pipe, the whole flow is affected and a velocity profile develops which does not change along the pipe, as shown in Fig. 3.15.

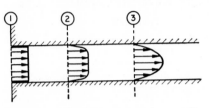

Fig. 3.14. Change of velocity profile near the entrance of a pipe.

For Reynolds numbers less than 2000, the flow development will usually be laminar and the fully developed profile will usually be parabolic. For Reynolds numbers greater than 2000, transition from laminar

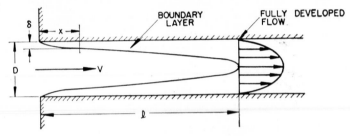

Fig. 3.15. Growth of laminar boundary layer at the entrance of a pipe.

to turbulent flow may occur at some point before the boundary layer fills the pipe, as shown in Fig. 3.16, and the final velocity distribution will be that of turbulent flow. If the Reynolds number is very high,

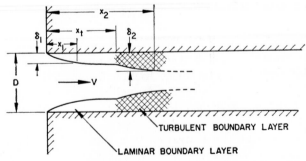

Fig. 3.16. Transition from laminar to turbulent boundary-layer flow near the entrance of a pipe.

transition may occur very close to the entrance of the pipe, so for purposes of calculation the boundary layer may be assumed turbulent along the entire pipe.

3.323. Development of Laminar Flow in a Pipe.

$$\frac{\delta}{x} = \frac{5}{\sqrt{\rho V x/\mu}} \qquad \text{(Holds only for } \delta < D/20\text{)} \qquad (3.57)$$

$$\frac{l}{D} = 0.058 R_e \qquad \left(R_e = \frac{\rho V D}{\mu} \right) \qquad (3.58)$$

where l = distance required to attain fully developed flow (parabolic profile)

3.324. Development of Turbulent Flow in a Pipe.

$$\frac{\delta_1}{x_1} = \frac{5}{\sqrt{\rho V x/\mu}} \qquad \begin{array}{l}\text{(Holds only for laminar bound-}\\ \text{ary layer with } \delta < D/20\text{)}\end{array} \qquad (3.59)$$

For high Reynolds numbers ($R_e = \rho V D/\mu$) greater than 2000 and very turbulent chamber conditions, x_t is negligible and δ_2 may be obtained from the following equation:

$$\frac{\delta_2}{x_2} = \frac{0.376}{(\rho V x/\mu)^{\frac{1}{5}}} \qquad (3.60)$$

3.33. Steady Flow of a Compressible Fluid

From the previous section, the general characteristics of incompressible, one-dimensional, streamline flows are as follows:

1. Decreasing the flow area increases the velocity.
2. Decreasing pressure accompanies increasing velocity if the height changes are negligible.
3. In flow with friction, the total head decreases in the direction of motion and the intrinsic internal energy increases.
4. Heat transfer does not affect the velocity distribution along the pipe.

The main objective of the next section is to find the extent to which these effects are also true of flows with considerable density changes, and in what circumstances they are no longer true. It will be shown that when the stream velocities are small, there are many similarities between incompressible and compressible flows and that when the velocity is greater than the speed of sound (supersonic), most of the effects are reversed. For example, in supersonic flow, increasing area is accompanied by increasing velocity.

Before a mathematical treatment of compressible flow is begun, it is profitable to consider this subject from a qualitative point of view.

In any flow, increasing velocity means increasing momentum, and in consequence a force must act in the direction of motion. Where there are no changes in height, this force must be provided by a pressure drop. We would therefore expect a negative pressure gradient to accompany a positive velocity gradient no matter how great the density changes. The continuity equation for flow through a passage requires that $\rho A V$ be a constant. If the density is constant, then velocity is inversely proportional to area. However, if density changes are considerable, it is no longer possible to tell immediately how the velocity will respond to varying flow area.

It will be shown for a compressible fluid when velocities are low, that density changes are small and the velocity-area relation is nearly the same as for incompressible flow. The greater the velocity, the more the area-velocity relation diverges from that for incompressible flow until at sonic speed [6-8] the nature of the flow changes completely.

Since density and temperature are closely related for a compressible fluid, heat transfer can no longer be neglected in solving for the velocity distribution in even a one-dimensional flow problem.

In the next section are considered the pressure-velocity relation and the velocity-area relation for frictionless, compressible flow without heat transfer and also for flow with the added effects of friction and heat transfer.

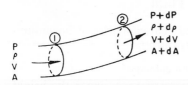

Fig. 3.17. One-dimensional, adiabatic, frictionless flow of a compressible fluid with negligible changes in height.

The mathematical analyses are considerably simplified by assuming that the fluid is a perfect gas, that is, a gas obeying the perfect-gas law, $p = \rho R T$, and having constant specific heat at constant pressure, C_p, and constant specific heat at constant volume C_v so that $R = C_p - C_v$. Use of the perfect gas equations is justified for many gas flows, and the analytical results give a good qualitative

[6] A. H. Shapiro, *The Dynamics and Thermodynamics of Compressible Fluid Flow*, Vols. I and II, Ronald Press, New York, 1953.

[7] H. W. Liepmann and A. E. Puckett, *Introduction to Aerodynamics of a Compressible Fluid*, John Wiley and Sons, New York, 1947.

[8] N. A. Hall, *Thermodynamics of Fluid Flow*, Prentice-Hall, New York, 1951.

picture of important effects that can be expected with nonperfect gas flows.

3.331. One-Dimensional, Adiabatic, Frictionless Flow of a Perfect Gas with Negligible Changes in Height. The only forces acting on an element of fluid (Fig. 3.17) are pressure forces. The net force on the control volume in direction of motion $= -dPA$. The net efflux of momentum is given by

$$\frac{d\overline{M}}{dt} = (\rho + d\rho)(A + dA)(V + dV)^2 - \rho A V^2 \qquad (3.61)$$

$$\frac{d\overline{M}}{dt} = A V^2 \, d\rho + 2A V \rho \, dV + \rho V^2 \, dA \qquad (3.62)$$

$$\frac{d\overline{M}}{dt} = d(\rho A V^2) = V d(\rho A V) + \rho A V \, dV \qquad (3.63)$$

From continuity

$$\rho A V = (A + dA)(\rho + d\rho)(V + dV) \qquad (3.64)$$

$$d(\rho A V) = 0 \qquad (3.65)$$

Substituting in the equation for efflux of momentum gives $d\overline{M}/dt = \rho A V \, dV$. Therefore

$$-A \, dP = \frac{d\overline{M}}{dt} = \rho A V \, dV \qquad (3.66)$$

$$\frac{dP}{\rho} + V \, dV = 0 \qquad (3.67)$$

This equation (3.67), known as Euler's equation, is the same as the one obtained for an incompressible fluid, but now ρ is not constant. Hence

$$\frac{1}{2} V^2 + \int \frac{dP}{\rho} = \text{constant} \qquad (3.68)$$

and it is necessary to know the pressure-density relationship to find the integral of dP/ρ. For an adiabatic, frictionless process in a perfect gas, we know that $P = C\rho^k$, where C is a constant and k is the ratio of the specific heat at constant pressure to the specific heat at constant volume. Therefore

$$\rho = C^{-1/k} P^{1/k} \qquad (3.69)$$

Substituting in Eq. 3.68 gives

$$\frac{1}{2} V^2 + \int C^{1/k} \frac{dP}{P^{1/k}} = \text{constant} \tag{3.70}$$

Or, using limits representing two points in such a flow, we have

$$\frac{V_2{}^2 - V_1{}^2}{2} = \frac{P_1{}^{1/k}}{\rho_1} \left(\frac{k}{k-1}\right) [P_1{}^{(k-1)/k} - P_2{}^{(k-1)/k}] \tag{3.71}$$

This is the expression relating velocity and pressure for *frictionless* flow of a compressible fluid with no heat transfer.

It should be noted that an *increase in velocity* is accompanied by a *decrease in pressure* as for an incompressible flow.

In order to find how velocity is related to area it is also necessary to include the continuity equation

$$d(\rho V A) = 0 \tag{3.72}$$

or

$$\frac{d\rho}{\rho} + \frac{dA}{A} + \frac{dV}{V} = 0 \tag{3.73}$$

Similarly, the logarithmic differential form of Eq. 3.69 is

$$\frac{dP}{P} = k \frac{d\rho}{\rho} \tag{3.74}$$

Combining Eqs. 3.67, 3.73, and 3.74 gives

$$-\frac{\rho V \, dV}{kP} + \frac{dA}{A} + \frac{dV}{V} = 0 \tag{3.75}$$

Or, since $P = \rho R T$,

$$-\frac{V \, dV}{kRT} + \frac{dA}{A} + \frac{dV}{V} = 0 \tag{3.76}$$

$$\frac{dA}{A} = \frac{dV}{V}\left(\frac{V^2}{kRT} - 1\right) \tag{3.77}$$

It is convenient to work with a dimensionless parameter called the Mach number, named after the Austrian physicist. The Mach number is defined as (fluid velocity)/(local speed of sound). The speed of sound, c, is also the speed at which small disturbances are propagated through a fluid. The speed of sound in a perfect gas is given by

$$c = \sqrt{\frac{\partial P}{\partial \rho}}\Bigg|_{\text{const entropy}} = \sqrt{kRT} \tag{3.78}$$

where k = ratio of specific heats
  R = gas constant, in.^{2}/sec^2 °R
  T = absolute local temperature, °R

And the Mach number is then given by

$$M = \frac{V}{\sqrt{kRT}} \tag{3.79}$$

The speed of sound is discussed in detail in a later section dealing with unsteady flows.

Combining Eqs. 3.77 and 3.79 gives

$$\frac{dA}{A} = \frac{dV}{V}(M^2 - 1) \tag{3.80}$$

or

$$\frac{dA}{A} = -\frac{dV}{V}(1 - M^2) \tag{3.81}$$

This relationship shows very clearly the influence of the Mach number. If $M < 1$, dA/A is negative for positive values of dV/V; that is, *increasing velocity* corresponds to *decreasing area*. If $M > 1$, dA/A is positive for positive values of dV/V; that is, we must *increase* the *area* of flow to *increase velocity*.

If $M = 1$, $dA/A = 0$ for dV/V positive or negative. This indicates that to obtain a high Mach number in a nozzle, the area must first decrease until $M = 1$ is reached and then increase again as shown in Fig. 3.18. Furthermore, if M^2 is small compared with unity ($M = 0.2$ say), then $dA/A \approx -(dV/V)$ as for frictionless, incompressible flow. Hence, a compressible fluid behaves very much like an incompressible fluid for low Mach numbers, that is, when the density change is small. The velocity of sound in air at room temperatures is approximately 13,000 in./sec. $M = 0.2$ corresponds then to a flow velocity of about 2640 in./sec. Thus, for velocities less than 200 ft/sec, incompressible theory may often be employed. Consequently, most pipe-flow problems in pneumatic systems may be treated by the methods used for incompressible flow for computations of velocity, pressure, and area changes. On the other hand, where high velocities accompanying large pressure changes occur, as in valves and orifices, the errors associated with

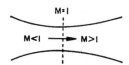

Fig. 3.18. Change of Mach number with area in a converging-diverging nozzle.

neglecting density changes are too great for this practice to be acceptable.

3.332. Flow through Orifices and Converging Nozzles. It has been observed experimentally that if air is supplied at constant inlet conditions to a converging nozzle and the exhaust pressure is decreased gradually, the weight rate of flow * of air through the nozzle increases until the pressure ratio, P_u/P_d, reaches the critical value 1.89. Thereafter, further decrease in the exhaust pressure has no effect on the flow, and the flow is said to be "choked," as shown in Fig. 3.19. Thus for

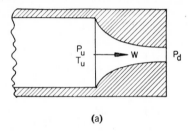

(a)

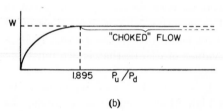

(b)

Fig. 3.19. (a) Converging nozzle.
(b) Weight rate of flow, W, versus P_u/P_d when P_u and T_u are held constant.

$P_u/P_d > 1.89$ or $P_d/P_u < 0.528$, the weight rate of flow of air is given by

$$\left(\frac{W}{A}\right)_{\text{max}} = 0.532 \frac{P_u}{\sqrt{T_u}} \tag{3.82}$$

where W = weight rate of flow, lb/sec
  A = area at throat of nozzle, in.^{2}
  P_u = upstream stagnation pressure, lb/in.^{2}
  T_u = upstream stagnation temperature, °R

---

* The phrase "weight rate of flow" is synonymous with the phrase "mass rate of flow" when fluid mass is reckoned in terms of fluid weight in a gravitational field of 386 in./sec^2.

The explanation for this phenomenon is readily obtained from Eq. 3.81. For the nozzle, dA is negative in the direction of flow. As the exhaust pressure decreases, the exit velocity increases until at the critical-pressure ratio the Mach number at the exit becomes equal to one. For high Mach numbers at the exit it would be necessary to have $M = 1$ at some intermediate point between 1 and 2. But from the relation $dA/A = -(dV/V)(1 - M^2)$, it is obvious that for $M = 1$ and $dA/A = 0$, the area must be a minimum at the point where $M = 1$; and since no throat exists in the nozzle passage, between 1 and 2, the throat is at or very near the exit, and its area is very nearly the same as the exit area.

An analytical expression for the weight rate of flow through a nozzle may be derived from Eq. 3.71. In the case of nozzle and orifice flows, the velocity at a point (1) well upstream from the throat is negligible if the throat area is small compared to the approaching passage area. Thus, $V_1 = V_u \doteq 0$, and the velocity at the throat, $V_t = V_2$, is given by

$$V_t = \sqrt{\frac{2k}{k-1} \frac{P_u^{1/k}}{\rho_u} (P_u^{(k-1)/k} - P_d^{(k-1)/k})} \tag{3.83}$$

The weight rate of flow through the nozzle is then found to be

$$W = \rho_t g A V_t \tag{3.84}$$

where $\rho_t = \rho_u \left(\dfrac{P_d}{P_u}\right)^{1/k}$ = density of the fluid at the throat of the nozzle $\tag{3.85}$

Thus the following equation holds whenever the flow is not "choked":

$$W = gA \sqrt{\frac{2k}{R(k-1)}} \frac{P_u}{\sqrt{T_u}} \left(\frac{P_d}{P_u}\right)^{1/k} \sqrt{1 - \left(\frac{P_d}{P_u}\right)^{(k-1)/k}} \tag{3.86}$$

Choked flow occurs when this expression reaches a maximum, that is, when

$$\frac{P_d}{P_u} = \left(\frac{2}{k+1}\right)^{k/(k-1)} \tag{3.87}$$

and the flow is said to be choked whenever

$$\frac{P_d}{P_u} < \left(\frac{2}{k+1}\right)^{k/(k-1)} \tag{3.88}$$

Let the critical weight rate of flow per unit area occurring at the throat be $(W/A)_{max}$ and the weight rate of flow per unit area at the throat for a pressure drop less than the critical be W/A. Then W/A is smaller than $(W/A)_{max}$ and the ratio $(W/A)/(W/A)_{max}$ will be a function of the pressure ratio. In the graph of Fig. 3.20 this ratio, together with the exit Mach number, is plotted against the nozzle-pressure ratio for air. For any nozzle the weight flow rate for a given pressure ratio may be computed by calculating $(W/A)_{max} = 0.532P_u/\sqrt{T_u}$ and multiplying it by the value of $(W/A)/(W/A)_{max}$ obtained from the graph. Corresponding information on many other gases, as well as air, is given in Reference 9. It is interesting to note that the pressure in the air stream as it leaves the nozzle with $P_d/P_u < 0.528$ is always equal to $0.528P_u$.

3.333. Example. Find the weight flow rate of air through a nozzle of throat area 1 sq in. when the supply pressure is 100 psia, the temperature is 500°R, and the exhaust pressure is 80 psia.

$$\frac{P_d}{P_u} = \frac{80}{100} = 0.8$$

From the graph, for $P_d/P_u = 0.8$

$$\frac{(W/A)}{(W/A)_{max}} = 0.82$$

and

$$\left(\frac{W}{A}\right)_{max} = 0.532\,\frac{P_u}{\sqrt{T_u}}$$

$$\left(\frac{W}{A}\right)_{max} = \frac{0.532 \times 100}{\sqrt{500}}$$

$$\left(\frac{W}{A}\right)_{max} = 2.38 \text{ lb/sec/sq in.}$$

$$\frac{W}{A} = 0.82\left(\frac{W}{A}\right)_{max} = 1.95 \text{ lb/sec/sq in.}$$

If $A = 1$ sq in.
$W = 1.95$ lb/sec

For an orifice the preceding results and methods of calculation also hold, except that a discharge coefficient must be used in calculating weight flow rate because the throat is well downstream from the orifice edge and has less area than the orifice.

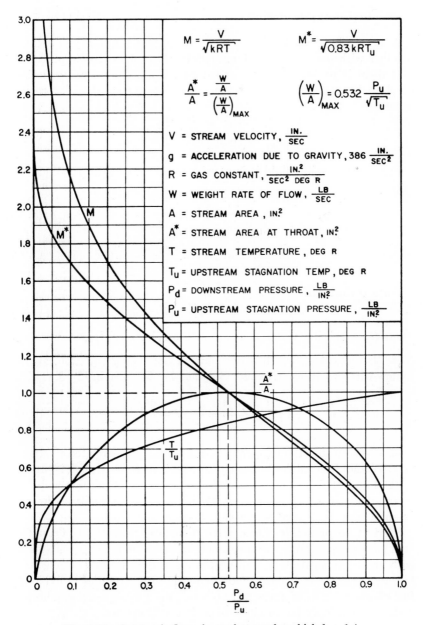

Fig. 3.20. Isentropic flow of a perfect gas for which $k = 1.4$.

3.34. The Energy Equation

So far we have used only the momentum and continuity equations, together with the pressure-volume relationship for isentropic processes.

If there is no heat transfer or external work, the energy equation, when used to study changes between two points in a steady-flow process, becomes

$$u_1 + \frac{P_1}{\rho_1} + \frac{V_1^2}{2} = u_2 + \frac{P_2}{\rho_2} + \frac{V_2^2}{2} \tag{3.89}$$

For a compressible fluid it is more convenient to write the equation in the form

$$h_1 + \frac{V_1^2}{2} = h_2 + \frac{V_2^2}{2} \tag{3.90}$$

where $h = u + \dfrac{P}{\rho}$ = enthalpy, in.^{2}/sec^2 $\qquad$ (3.91)

For a perfect gas

$$h = C_p T \tag{3.92}$$

where C_p = specific heat at constant pressure, in.^{2}/sec^2 °R

Note that Eq. 3.89, which is one form of the energy equation, applies to flow with or without friction. Substituting Eq. 3.90 in Eq. 3.92 gives

$$C_p T_1 + \frac{V_1^2}{2} = C_p T_2 + \frac{V_2^2}{2} \tag{3.93}$$

Thus it is seen that the temperature of a perfect gas is related to stream velocity and does *not* depend solely on whether or not friction is present in the flow. This means that the temperatures are the same at points in an adiabatic, compressible flow having the same velocity as long as no shaft work is done at any place along the flow path.

3.341. Compressible Flow with Friction and Heat Transfer. It is possible to analyze the one-dimensional flow of a compressible fluid with area change, friction, and heat transfer occurring simultaneously, but the analysis is lengthy and involved. By studying individually the effects of each of these variables on the flow, a good qualitative understanding may be obtained of their combined actions.

Frictionless flow with area change and no heat transfer has already been considered, and the same method of approach will be applied to flow with no area change, but with either friction or heat transfer.

3.342. Compressible One-Dimensional Flow through a Constant-Area Pipe with Wall Friction Only. Experience has shown that most of the effects of friction in turbulent flow occur very near the wall, so it is reasonable to assume a one-dimensional flow to exist in the pipe, as in Fig. 3.21. The momentum equation gives

$$PA - (P + dP)A - \tau S\, dx = (\rho + d\rho)A(V + dV)^2 - \rho A V^2 \quad (3.94)$$

or

$$-dP - \frac{\tau S}{A}\, dx = 2\rho V\, dV + V^2\, d\rho \quad (3.95)$$

From the continuity equation

$$V\, d\rho + \rho\, dV = 0 \quad (3.96)$$

or

$$\rho\, dV = -V\, d\rho \quad (3.97)$$

Substituting for $\rho\, dV$ in Eq. 3.95 gives

$$-dP - \frac{\tau S}{A}\, dx = -2V^2\, d\rho + V^2\, d\rho \quad (3.98)$$

$$dP + \frac{\tau S}{A}\, dx = V^2\, d\rho \quad (3.99)$$

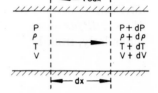

Fig. 3.21. One-dimensional flow of a compressible fluid through a constant-area pipe with wall friction only.

In order to find how $d\rho$ is related to dP, we may use the perfect-gas equation, the energy equation, and the continuity equations as follows

Perfect gas:

$$dP = \frac{P}{\rho}\, d\rho + \frac{P}{T}\, dT \quad (3.100)$$

Energy:

$$C_p\, dT = -V\, dV \quad (3.101)$$

Combining Eqs. 3.97 and 3.101 gives

$$dT = \frac{V^2}{C_p \rho}\, d\rho \quad (3.102)$$

Now Eqs. 3.100 and 3.102 yield

$$dP = \left(\frac{P}{\rho} + \frac{P}{\rho}\frac{V^2}{C_p T}\right) d\rho \quad (3.103)$$

Noting that $C_p = kR/(k-1)$ and $P = \rho RT$, we find that rearrangement gives

$$dp = \frac{dP}{RT[1 + (k - 1)M^2]} \tag{3.104}$$

Substituting for $d\rho$ in Eq. 3.99, we have

$$\left\{ \frac{V^2}{RT[1 + (k - 1)M^2]} - 1 \right\} dP = \frac{\tau S}{A} dx \tag{3.105}$$

or

$$\frac{dP}{dx} = \left[\frac{1 + (k - 1)M^2}{M^2 - 1} \right] \frac{\tau S}{A} \tag{3.106}$$

Hence when $M < 1$, dP/dx will be negative (pressure decreasing in the direction of motion); when $M > 1$, dP/dx will be positive; and when $M \to 1$, $dP/dx \to \infty$.

Similarly, from Eqs. 3.95 and 3.97,

$$-dP - \frac{\tau S}{A} dx = \rho V \, dV \tag{3.107}$$

And combining this equation with Eq. 3.106, we find that

$$\left[\frac{1 + (k - 1)M^2}{1 - M^2} - 1 \right] \frac{\tau S}{A} dx = \rho V \, dV \tag{3.108}$$

or

$$\frac{dV}{dx} = \left(\frac{kM^2}{1 - M^2} \right) \frac{\tau S}{W/g} \tag{3.109}$$

Hence when $M < 1$, dV/dx will be positive (velocity increasing in the direction of motion); when $M > 1$, dV/dx will be negative; when $M \to 1$, $dV/dx \to \infty$; and when $M \to 0$, $dV/dx \to 0$.

Combining Eqs. 3.101 and 3.109 gives

$$\frac{dT}{dx} = \frac{V}{C_p} \left(\frac{kM^2}{M^2 - 1} \right) \frac{\tau S}{W/g} \tag{3.110}$$

Hence when $M < 1$, dT/dx will be negative, and when $M > 1$, dT/dx will be positive.

In order to see how the Mach number itself changes along the pipe we may differentiate the identity for the Mach number after taking its logarithm.

$$M^2 = \frac{V^2}{kRT} \tag{3.111}$$

$$\frac{dT}{T} + 2\frac{dM}{M} = 2\frac{dV}{V} \tag{3.112}$$

or

$$\frac{dM}{dx} = \frac{M}{V}\frac{dV}{dx} - \frac{M}{2T}\frac{dT}{dx} \tag{3.113}$$

This, together with Eq. 3.101, gives

$$\frac{dM}{dx} = \left[\frac{M}{V} - \frac{M}{2T}\left(\frac{-V}{C_p}\right)\right]\frac{dV}{dx} \tag{3.114}$$

$$\frac{dM}{dx} = \frac{M}{V}\left(1 + \frac{V^2}{2C_pT}\right)\frac{dV}{dx} \tag{3.115}$$

$$\frac{dM}{dx} = \frac{M}{V}\left[1 + \left(\frac{k-1}{2}\right)M^2\right]\frac{dV}{dx} \tag{3.116}$$

Combining Eqs. 3.109 and 3.116 gives

$$\frac{dM}{dx} = \frac{M}{V}\left[1 + \left(\frac{k-1}{2}\right)M^2\right]\left(\frac{kM^2}{1-M^2}\right)\frac{\tau S}{W/g} \tag{3.117}$$

Hence when $M < 1$, dM/dx is positive, and when $M > 1$, dM/dx is negative.

To summarize these results for compressible flow with friction but with no area change and no heat transfer: for subsonic flow, the velocity and Mach number gradients are positive, and the temperature and pressure gradients are negative; for supersonic flow, the velocity and Mach number gradients are negative, and the temperature and pressure gradients are positive. The tendency then is for the Mach number to become unity, regardless of its initial value.

Equations 3.106, 3.109, 3.110, and 3.117 may be readily integrated when the friction factor is constant along a pipe. These equations have been systematically integrated for a wide range of boundary conditions, and the results are tabulated in Tables 42 through 47 in the *Gas Tables*.[9]

Thus, the results for a compressible fluid are radically different from those for an incompressible fluid, if M is large; however, if M is small compared to unity, the compressible fluid behaves very much like an incompressible fluid.

[9] J. H. Keenan and J. Kaye, *Gas Tables*, John Wiley and Sons, New York, 1948.

Thus, if M^2 is small compared with 1,

$$\frac{dV}{dx} = 0 \tag{3.118}$$

and

$$\frac{dP}{dx} \approx -\frac{\tau S}{A} \tag{3.119}$$

These results are the same as those for a fluid with constant density.

Experimental data on friction coefficients for fully developed compressible flow in smooth tubes show that the von Karman-Nikuradse formula for incompressible flows may be used for subsonic compressible flows at points located farther than 50 diameters from the inlet:

$$\frac{1}{\sqrt{4f}} = -0.8 + 2\log_{10}\left(R_e\sqrt{4f}\right) \tag{3.120}$$

where R_e is the Reynolds number and f is the friction factor required in the following equation for the wall shear stress, τ:

$$\tau = \frac{1}{2}f\rho V^2 \tag{3.121}$$

For a perfect gas

$$\tau = \frac{1}{2}f\rho kRT\frac{V^2}{c^2} = \frac{1}{2}fkPM^2 \tag{3.122}$$

and

$$R_e = \frac{\rho V D}{\mu} = \left(\frac{PMD}{\mu}\right)\left(\frac{\rho V}{P}\right)\left(\frac{\sqrt{kRT}}{V}\right) = \frac{PMD}{\mu}\sqrt{\frac{k}{RT}} = \frac{WD}{g\mu A} \tag{3.123}$$

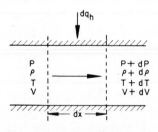

Fig. 3.22. One-dimensional flow of a compressible fluid through a constant-area pipe with heat transfer only through the wall.

For a Mach number larger than unity and a Reynolds number varying from 25,000 to 7,000,000, the friction factor varies between 0.002 and 0.003. For fully developed incompressible flow in the same range of Reynolds number, the friction coefficients vary from 0.003 to 0.0065. For supersonic flow, therefore, the friction coefficients are only about half as great as the coefficients for incompressible flow.

3.343. Frictionless One-Dimensional Flow through a Constant-Area Pipe with Heat Transfer Only. Although it is likely that the temperature may vary considerably across any given section of the flow when heat is being added to the stream, it is possible to use an average temperature for the entire section as long as the flow is one-dimensional. The qualitative results obtained from an analysis based on uniform temperature across each section of the flow (that is, perfect transverse mixing) are very useful in gaining an insight into the actual behavior of a stream which receives heat through the pipe wall, as in Fig. 3.22.

The momentum equation gives

$$-A \, dP = A d(\rho V^2) = A V^2 \, d\rho + 2\rho A V \, dV \qquad (3.124)$$

$$-dP = V^2 \, d\rho + 2\rho V \, dV \qquad (3.125)$$

And the continuity equation (3.97) still holds.

$$V \, d\rho = -\rho \, dV \qquad (3.97)$$

The energy equation gives

$$d\left(h + \frac{V^2}{2}\right) = dh + V \, dV = dq_h \qquad (3.126)$$

where dq_h = increment of heat flow per unit mass of fluid flowing,

$$(\text{in. lb})/(\text{lb sec}^2/\text{in.}) = \text{in.}^2/\text{sec}^2$$

If it is more convenient to use the heat-flow rate per unit surface area, we may write

$$dq_h = q_a \frac{S \, dx}{\rho A V} \qquad (3.127)$$

where q_a = heat flow per unit surface area, in. lb/sec in.^{2} = lb/sec in.

Combining Eqs. 3.126 and 3.127 gives

$$dh + V \, dV = q_a \frac{S \, dx}{\rho A V} \qquad (3.128)$$

When the fluid is a perfect gas,

$$dh = C_p \, dT \qquad (3.129)$$

and P, ρ, and T are related by Eq. 3.100:

$$dP = \frac{P}{\rho} \, d\rho + \frac{P}{T} \, dT \qquad (3.100)$$

Combining Eqs. 3.128 and 3.129 gives

$$C_p \, dT + V \, dV = q_a \frac{S \, dx}{\rho A V} \qquad (3.130)$$

Combining Eqs. 3.100 and 3.125, we have

$$\frac{P}{\rho} d\rho + \frac{P}{T} dT = -V^2 d\rho - 2\rho V \, dV \qquad (3.131)$$

or

$$dT = -\frac{T}{P}\left[\left(\frac{P}{\rho} + V^2\right) d\rho + 2\rho V \, dV\right] \qquad (3.132)$$

Using Eq. 3.97 to substitute for $d\rho$ gives

$$dT = -\frac{T}{P}\left[\left(\frac{P}{\rho} + V^2\right)\left(-\frac{\rho}{V} dV\right) + 2\rho V \, dV\right] \qquad (3.133)$$

or

$$dT = \left(\frac{T}{V} - \frac{T\rho V}{P}\right) dV \qquad (3.134)$$

Substituting for dT in Eq. 3.130 yields

$$\left[C_p\left(\frac{T}{V^2} - \frac{T\rho}{P}\right) + 1\right] V \, dV = \frac{q_a S \, dx}{\rho A V} \qquad (3.135)$$

$$\left[C_p\left(\frac{1}{kRM^2} - \frac{1}{R}\right) + 1\right] V^2 \, dV = q_a \frac{S \, dx}{\rho A} \qquad (3.136)$$

$$\left[\frac{C_p}{R}\left(\frac{1}{kM^2} - 1\right) + 1\right] kM^2 \, dV = q_a \frac{S \, dx}{PA} \qquad (3.137)$$

Noting that

$$\frac{C_p}{R} = \frac{k}{k-1} \qquad (3.138)$$

we see that

$$\left[\frac{k}{k-1}\left(\frac{1}{kM^2} - 1\right) + 1\right] kM^2 \, dV = q_a \frac{S \, dx}{PA} \qquad (3.139)$$

$$\frac{k}{k-1}(1 - M^2) \, dV = \frac{q_a S \, dx}{PA} \qquad (3.140)$$

Rearranging gives

$$\frac{dV}{dx} = \frac{k-1}{k(1-M^2)} \frac{q_a S}{PA} \qquad (3.141)$$

Thus when $M < 1$, dV/dx will be positive; when $M > 1$, dV/dx will be negative; and when $M \to 1$, $dV/dx \to \infty$.

In order to find how the pressure gradient, dP/dx, is related to the Mach number, M, it is convenient to use the momentum equation

(3.125) and the continuity equation (3.97) to obtain

$$-dP = \rho V \, dV \tag{3.142}$$

or

$$\frac{dP}{dx} = -\rho V \frac{dV}{dx} \tag{3.143}$$

Combining Eqs. 3.141 and 3.143 gives

$$\frac{dP}{dx} = \frac{-\rho V(k-1)q_a S}{k(1-M^2)PA} \tag{3.144}$$

Thus when $M < 1$, dP/dx will be negative; when $M > 1$, dP/dx will be positive; and when $M \to 1$, $dP/dx \to \infty$.

By combining Eqs. 3.134 and 3.141, we find that the temperature gradient is given by

$$\frac{dT}{dx} = \frac{(k-1)(1-kM^2)}{k^2 M^2 (1-M^2)} \frac{V q_a S}{RPA} \tag{3.145}$$

Thus dT/dx will always be positive except when $\sqrt{1/k} < M < 1$; $dT/dx \to 0$ when $M = \sqrt{1/k}$; and $dT/dx \to \infty$ when $M \to 1$ or when $M \to 0$.

It may also be shown that when $M < 1$, the Mach number gradient, dM/dx, will be positive, while when $M > 1$, dM/dx will be negative. When cooling takes place instead of heating, the effects are reversed.

3.344. Comparison of Flow with Friction and Flow with Heat Transfer. Consider the flow of air through a long pipe of constant area. If there is wall friction but no heat transfer to the fluid, an initially subsonic flow will have greater velocity and a greater Mach number in the downstream direction, and the pressure and temperature will be lower in the downstream direction. If the pipe is long enough, the exit velocity may equal the velocity of sound but will never exceed it. On the other hand, an initially supersonic flow will exhibit decreased velocity and Mach number in the downstream direction, accompanied by increased temperature and pressure.

If heat is added along the pipe but the flow is frictionless, the velocity, Mach number, and pressure gradients will be the same as with wall friction, but the temperature will always increase except when $1/k < M < 1$.

The results of the preceding discussion of one-dimensional compressible flows with area change, friction, and heat transfer are given qualitatively in Table 3.2. It is important to note that no mention has been made of sudden discontinuities (shock waves of various kinds) that occur in some supersonic flows. For a more detailed discussion of compressible flow with area change, friction, and heating, the reader is referred to *The Dynamics and Thermodynamics of Compressible Fluid Flow*, by A. H. Shapiro (fn. 6).

3.345. Flow of Gases in Narrow Passages. In the preceding section it was assumed that the effects of friction were concentrated

Table 3.2. Qualitative Results for Flow of a Compressible Fluid

	Heating		Cooling		Area Increase		Area Decrease		Friction	
	$M < 1$	$M > 1$	$M < 1$	$M > 1$	$M < 1$	$M > 1$	$M < 1$	$M > 1$	$M < 1$	$M > 1$	
M	+	−	−	+	−	+	+	−	+	−	
V	+	−	−	+	−	+	+	−	+	−	
P	−	+	+	−	+	−	−	+	−	+	
	$M < \frac{1}{k}$	$\frac{1}{k} < M < 1$	$M > 1$	$M < \frac{1}{k}$	$\frac{1}{k} < M < 1$	$M > 1$					
T	+	−	+	−	+	−	−	+	+	−	−	+

+ means increase in downstream direction.
− means decrease in downstream direction.

in a relatively small portion of the fluid at the wall of the pipe. There are instances when the effects of friction are important throughout the cross section of the flow passage. This is especially true in the consideration of the flow in small clearance spaces of the type encountered between moving and fixed parts of a valve, around rotating and sliding shafts, past pistons, and so on. This type of flow is often called capillary flow but is not to be confused with the type of flow associated with surface-tension effects (capillary attraction).

In certain circumstances, the fluid velocities in the passage are small enough to permit neglecting the momentum effects in analyzing the flow (that is, the kinetic energy of the fluid stream is negligible). Neglecting momentum effects leads to less than 13 per cent error when the Mach number M is less than 0.3.

When momentum effects may be neglected, only pressure and friction effects remain; so the analysis of fully developed flow in a small elemental length of duct or passage results in the same equation for volume rate of flow that is used for an incompressible flow. For example, in a circular passage of diameter D (see Fig. 3.23)

$$Q = \frac{-\pi D^4}{128\mu} \frac{dP}{dx} \qquad (3.146)$$

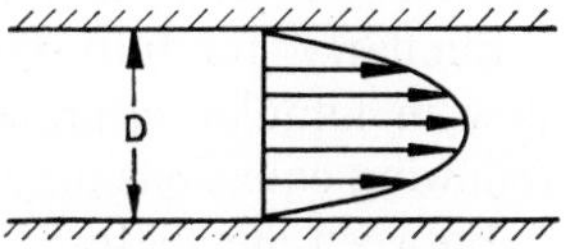

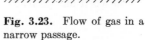

and the velocity profile is parabolic.

However, the volume rate of flow, Q, is not constant along the passage because the fluid expands as its pressure decreases.

Fig. 3.23. Flow of gas in a narrow passage.

Because of continuity, though, the weight rate of flow is constant along the passage. In fact, the weight rate of flow is often a better measure of flow rate than volume rate of flow when one is working with pneumatic systems. The equation for weight rate of flow is

$$W = g\rho Q = \frac{-\pi D^4 g\rho}{128\mu} \frac{dP}{dx} \qquad (3.147)$$

Noting that $\rho = P/RT$, we find that

$$W = \frac{-\pi D^4 g}{128\mu RT} P \frac{dP}{dx} \qquad (3.148)$$

From the energy equation, it is readily seen that the stream temperature, T, will not change along the passage unless heat is added to the fluid along the passage because $C_p\, dT = dq_h$. Thus, a gas which flows in a passage with wall temperatures the same as the entering gas will not undergo a temperature change along the passage.

If T is constant, Eq. 3.148 may be integrated over a passage of length L to give

$$W(X_2 - X_1) = WL = \frac{\pi D^4 g}{256\mu RT} (P_1{}^2 - P_2{}^2) \qquad (3.149)$$

or

$$W = \frac{\pi D^4 g}{256\mu RTL} (P_u{}^2 - P_d{}^2) \qquad (3.150)$$

where P_u = upstream pressure, lb/in.^{2}
$\quad P_d$ = downstream pressure, lb/in.^{2}
$\quad L$ = length of passage, in.
$\quad D$ = passage diameter, in.
$\quad g$ = acceleration of gravity, 386 in./sec^2
$\quad \mu$ = fluid viscosity, lb sec/in.^{2}
$\quad R$ = gas constant, in.^{2}/sec^2 °R
For air at room temperature: $\mu = 2.65 \times 10^{-9}$ lb sec/in.^{2}
$\qquad\qquad\qquad\qquad\qquad R = 247{,}000$ in.^{2}/sec^2 °R

It is important to remember that this equation holds only for fully developed laminar flow with negligible changes of fluid velocities.

Similarly, for fully developed laminar flow between flat plates, and flow in annular clearance spaces (see illustrations in Table 3.1), the following equations may be used:

Between flat plates:

$$W = \frac{wb^3 g}{24\mu RTL} (P_u{}^2 - P_d{}^2) \tag{3.151}$$

Through annular clearance space:

$$W = \frac{\pi Db^3 [1 + 1.5(\epsilon/b)]^2}{12\mu L} (P_u{}^2 - P_d{}^2) \tag{3.152}$$

Between flat plates with one plate moving at velocity V (flow is relative to nonmoving plate):

$$W = \left[\frac{wbVP}{2} + \frac{wb^3}{24\mu L} (P_u{}^2 - P_d{}^2) \right] \frac{g}{RT} \tag{3.153}$$

where P = fluid pressure at point of observation

3.346. Limitations of Capillary-Flow Equations.   In order to determine whether Eqs. 3.150 through 3.153 may be successfully applied, it is important to have a Reynolds number, $R_e = PVD/RT\mu$, less than 2000 as well as to have a negligible velocity gradient along the passage under consideration.   It is possible to have a low Reynolds number but a high velocity gradient in the direction of flow.   When this happens, the momentum equation must be included in the analysis.   A qualitative insight into such a problem often may be gained by assuming the effects of friction to be concentrated at the wall, by working with the average stream velocity, and by using laminar-flow friction coefficients in equations of the type developed in Sec. 3.342.

Experimental and analytical work by Egli,[10] Grinnell,[11] and Tao [12] should be studied by anyone interested in this work.

[10] A. Egli, "The Leakage of Gases Through Narrow Channels," *J. Appl. Mechanics,* Vol. 4, No. 2 (June 1937), p. A-63.

[11] S. K. Grinnell, "Flow of a Compressible Fluid in a Thin Passage," *Trans. ASME,* Vol. 78, No. 4 (May 1956), pp. 765–771.

[12] L. N. Tao and W. F. Donavan, "Through Flow in Concentric and Eccentric Annuli of Fine Clearance With and Without Relative Motion of the Boundaries," *Trans. ASME,* Vol. 77, No. 8 (Nov. 1955), pp. 1291–1301.

3.35. Unsteady Frictionless Compressible Flows

3.351. Unsteady Flow with Negligible Momentum Effects.
Many of the dynamic problems encountered in the design, development, and operation of fluid-power control systems are associated with the increase of energy stored in the working fluid when its pressure increases. Although this elastic effect is more predominant in pneumatic systems, it is also important in many hydraulic systems where fast dynamic response is required. Lines, passages, and working chambers of various kinds are places where fluid compliance can often be considered to the

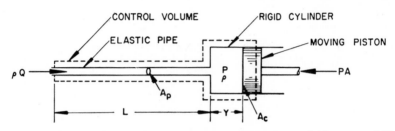

Fig. 3.24. Unsteady flow in a system having compliance due to fluid compressibility and elastic pipe walls.

exclusion of momentum effects. In some cases, the boundaries of the system containing the fluid are also elastic, or movable in some other way; therefore the following analysis includes the effects of moving walls.

Consider the system composed of an elastic pipe, a cylinder with rigid walls, and a moving piston, shown in Fig. 3.24, which receives a fluid flow at rate ρQ at the left end of the pipe. If momentum and friction effects can be neglected, the fluid pressure, P, will be uniform throughout the system. The volume of the elastic pipe may be given by the equation

$$\mathcal{V}_p = A_p L = V_{pi} + k_e P \tag{3.154}$$

where $\mathcal{V}_p$ = volume of fluid in pipe, in.^{3}
  A_p = cross-sectional area of pipe, in.^{2}
  L = length of pipe, in.
  $\mathcal{V}_{pi}$ = volume of pipe at zero pressure, in.^{3}
  k_e = elasticity coefficient of pipe, in.^{5}/lb
  P = system pressure, lb/in.^{2}

The volume of the chamber in the cylinder is

$$\mathcal{V}_c = A_c Y \tag{3.155}$$

where $\mathcal{V}_c$ = volume of fluid in cylinder, in.[3]
   Y = distance from end of cylinder to piston, in.

Applying the continuity equation to the control volume which includes the pipe, cylinder, and piston gives

$$\rho Q = \frac{d}{dt}(\rho \mathcal{V}_t) = \frac{d}{dt}[\rho(\mathcal{V}_p + \mathcal{V}_c)] \tag{3.156}$$

where $\mathcal{V}_t$ = total volume of fluid, in.[3]

Differentiating gives

$$\rho Q = \rho\left(\frac{d\mathcal{V}_p}{dt} + \frac{d\mathcal{V}_c}{dt}\right) + (\mathcal{V}_p + \mathcal{V}_c)\frac{d\rho}{dt} \tag{3.157}$$

When the fluid is a pure liquid, at constant temperature, the fluid density is related to fluid pressure by

$$\rho = \rho_i + \frac{\rho_i}{\beta}P \tag{3.158}$$

where ρ_i = fluid density at zero pressure, lb sec^2/in.[4]
   β = bulk modulus of elasticity of liquid, lb/in.[2]

When the fluid is a perfect gas,

$$\rho = \frac{P}{RT} \tag{3.159}$$

The two types of fluid may be compared at constant temperature by differentiating Eqs. 3.158 and 3.159 with respect to time.
   Liquid:

$$\frac{d\rho}{dt} = \frac{\rho_i}{\beta}\frac{dP}{dt} \tag{3.160}$$

   Gas:

$$\frac{d\rho}{dt} = \frac{1}{RT}\frac{dP}{dt} = \frac{\rho}{P}\frac{dP}{dt} \tag{3.161}$$

Thus, the bulk modulus of elasticity of a perfect gas at constant temperature is equal to its pressure whereas the bulk modulus of elasticity is independent of pressure in a liquid at constant temperature. The gas behaves like a nonlinear "spring" compared to the linear "spring" characteristic of a pure liquid. On the other hand, the bulk modulus of a perfect gas is independent of temperature while the bulk modulus of most liquids is very dependent upon temperature.

Substituting $\dfrac{d\rho}{dt}$, $\mathcal{V}_p$, and $\mathcal{V}_c$ in Eq. 3.157 gives

$$\rho Q = \rho \left(k_e \frac{dP}{dt} + A_c \frac{dY}{dt} \right) + (\mathcal{V}_p + \mathcal{V}_c) \frac{\rho}{\beta} \frac{dP}{dt} \tag{3.162}$$

or

$$Q = \left(k_e + \frac{\mathcal{V}t}{\beta} \right) \frac{dP}{dt} + A_c \frac{dY}{dt} \tag{3.163}$$

In some dynamic studies, $\mathcal{V}_t$ varies by only small amounts, so an average value may be substituted for $\mathcal{V}_t$ in Eq. 3.163. Integrating Eq. 3.163 with respect to time with $\mathcal{V}_t/\beta =$ constant gives

$$P = \left(\frac{1}{k_e + \mathcal{V}_t/\beta} \right) \left(A_c Y + \int_0^t Q\, dt \right) \tag{3.164}$$

Thus, the pressure, P, varies directly with ram position, Y, and the time integral of volume rate of flow, Q.

The effects of temperature variations in the working fluid, which are sometimes important in pneumatic systems, are discussed in a later chapter dealing with pneumatic processes. Whenever the system is subjected to disturbances varying at frequencies approaching the lowest natural frequency of the pipe itself, the above analysis cannot be used. (See Chapter 5, Fluid-Power Transmission.)

3.352. Unsteady, Frictionless, Compressible Flow in a Uniform Elastic Pipe. In some unsteady flows in lines and passages the distributed mass of the working fluid is as important as the fluid compressibility because of momentum effects that cannot be neglected. When

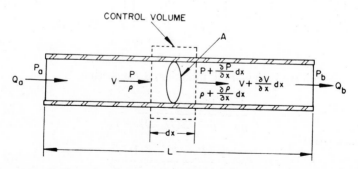

Fig. 3.25. Unsteady, frictionless, compressible flow in a uniform elastic pipe.

such a situation occurs, the pressure is not the same throughout the passage, and both the fluid mass and the fluid compressibility must be considered simultaneously as distributed characteristics. Similarly, the density, velocity, and flow rate differ from one point to another in the system as well as varying from one time to another at each point. For generality, the ensuing analysis includes the effects of elasticity in the passage wall.

Consider first the events that occur in a small control volume of length dx in the uniform pipe of length L shown in Fig. 3.25. Using the continuity equation gives

$$\frac{-\partial(\rho V A)\, dx}{\partial x} = \frac{\partial(\rho A)\, dx}{\partial t} \tag{3.165}$$

$$-\left(V A \frac{\partial \rho}{\partial x} + \rho A \frac{\partial V}{\partial x} + \rho V \frac{\partial A}{\partial x}\right) = A \frac{\partial \rho}{\partial t} + \rho \frac{\partial A}{\partial t} \tag{3.166}$$

From the equation of state of a compressible fluid

$$d\rho = \frac{\rho}{\beta} dP \tag{3.167}$$

and for a thin-walled elastic pipe

$$dA = \frac{A}{E(D_o/D_i - 1)} dP \tag{3.168}$$

where D_o = outside diameter of pipe, in.
   D_i = inside diameter of pipe, in.
   E = Young's modulus of elasticity of pipe material, lb/in.[2]

Combining Eqs. 3.166, 3.167, and 3.168, we have

$$-\left[\frac{1}{\beta} + \frac{1}{E(D_o/D_i - 1)}\right] V \frac{\partial P}{\partial x} - \frac{\partial V}{\partial x} = \left[\frac{1}{\beta} + \frac{1}{E(D_o/D_i - 1)}\right] \frac{\partial P}{\partial t} \tag{3.169}$$

Defining an equivalent bulk modulus, β_e, such that

$$\frac{1}{\beta_e} = \frac{1}{\beta} + \frac{1}{E(D_o/D_i - 1)} \tag{3.170}$$

we may write

$$-\frac{V}{\beta_e} \frac{\partial P}{\partial x} - \frac{\partial V}{\partial x} = \frac{1}{\beta_e} \frac{\partial P}{\partial t} \tag{3.171}$$

Now the momentum equation may be applied.

$$-A\frac{\partial P}{dx}dx = \frac{\partial(\rho V^2 A)}{\partial x}dx + \frac{\partial(\rho V A)}{\partial t}dx \qquad (3.172)$$

Combining Eqs. 3.172, 3.167, and 3.168 gives

$$-\left(1 + \frac{\rho V^2}{\beta_e}\right)\frac{\partial P}{\partial x} = \rho\frac{\partial V}{\partial t} + 2\rho V\frac{\partial V}{\partial x} + \frac{\rho V}{\beta_e}\frac{\partial P}{\partial t} \qquad (3.173)$$

In many problems the velocity, V, is always small enough to be negligible, so the two basic equations, 3.171 and 3.173, may be simplified to give

Continuity:

$$-\frac{\partial V}{\partial x} = \frac{1}{\beta_e}\frac{\partial P}{\partial t} \qquad (3.174)$$

and

Momentum:

$$-\frac{\partial P}{\partial x} = \rho\frac{\partial V}{\partial t} \qquad (3.175)$$

This pair of simultaneous partial-differential equations is often known as the wave equations because their solutions have been found to describe the traveling waves that have been observed in a wide variety of distributed-parameter systems. In working with systems of this kind it is often sufficient to consider the pressure and flow effects at each end and to determine the functional relationships between P_a, Q_a, P_b, and Q_b (see Fig. 3.25). Thus, the distributed-parameter line is considered simply as a four-terminal network or device as shown in Fig. 3.26. Two of the four variables, P_a, Q_a, P_b, and Q_b, may be considered as independent while the other two are dependent. Experience has shown that any two of these variables may be independent as long as both do not occur at the same end of the system. In other words, it is impossible, for instance, to vary P_a and Q_a independently by external means, but it is possible to vary P_a and either P_b or Q_b independently as inputs to the system.

Fig. 3.26.   Block representation of a four-terminal network.

There are two different forms in which the solutions of the wave equations may be expressed. The first, but perhaps less useful, form is a set of differential equations with hyperbolic sine and cosine functions. The second form is a set of time-difference equations.

The set of four differential equations is as follows:

$$P_b = [\cosh \Gamma]P_a - [Z_s \sinh \Gamma]Q_a \qquad (3.176)$$

$$Q_b = -[Y_s \sinh \Gamma]P_a + [\cosh \Gamma]Q_a \qquad (3.177)$$

$$P_a = [\cosh \Gamma]P_b + [Z_s \sinh \Gamma]Q_b \qquad (3.178)$$

$$Q_a = [Y_s \sinh \Gamma]P_b + [\cosh \Gamma]Q_b \qquad (3.179)$$

where $\Gamma = L\sqrt{\rho/\beta_e}\ D$ \qquad (3.180)

D = differential operator, d/dt

Y_s = characteristic admittance, $\sqrt{A^2/\rho\beta_e}$ \qquad (3.181)

Z_s = characteristic impedance, $\sqrt{\rho\beta_e/A^2}$ \qquad (3.182)

These equations are basically the same equations that have been used to characterize lossless electric transmission lines and lossless pneumatic transmission lines.[13–16]

The set of time-difference equations is as follows:

$$\sqrt{A^2/\rho\beta_e}\ [P_b(t - 2T_e) + P_b(t) - 2P_a(t - T_e)]$$

$$= Q_b(t - 2T_e) - Q_b(t) \quad (3.183)$$

$$\sqrt{\rho\beta_e/A^2}\ [Q_b(t - 2T_e) + Q_b(t) - 2Q_a(t - T_e)]$$

$$= P_b(t - 2T_e) - P_b(t) \quad (3.184)$$

$$\sqrt{A^2/\rho\beta_e}\ [P_a(t - 2T_e) + P_a(t) - 2P_b(t - T_e)]$$

$$= Q_a(t) - Q_a(t - 2T_e) \quad (3.185)$$

$$\sqrt{\rho\beta_e/A^2}\ [Q_a(t - 2T_e) + Q_a(t) - 2Q_b(t - T_e)]$$

$$= P_a(t) - P_a(t - 2T_e) \quad (3.186)$$

[13] L. F. Woodruff, *Principles of Electric Power Transmission*, 2nd ed., John Wiley and Sons, New York, 1938.

[14] A. E. Knowlton, *Standard Handbook for Electrical Engineers*, McGraw-Hill Book Co., New York, 1949, pp. 1177–1200.

[15] W. L. Emery, *Ultra-High Frequency Radio Engineering*, The Macmillan Co., New York, 1944, p. 140.

[16] J. C. Moise, "Pneumatic Transmission Lines," *J. Instr. Soc. Am.*, Vol. 1, No. 4 (April 1954), pp. 35–40.

where $T_e = L \sqrt{\rho/\beta_e}$ = time for pressure disturbance to travel length
of tube, sec   (3.187)

$\sqrt{\beta_e/\rho} = C$ = velocity of sound, in./sec   (3.188)

When the fluid is a perfect gas and the flow is frictionless and adiabatic,

$$\frac{\beta}{\rho} = \frac{\partial P}{\partial \rho}\bigg\| = kRT = \text{(speed of sound)}^2, \text{ constant entropy}$$
$$\text{(for small pressure variations)}$$

When large pressure variations occur with gases, the flow is usually
not frictionless, and in addition V is not negligible in Eqs. 3.174 and
3.175, so the solutions obtained above do not hold. For work with
large pressure disturbances, the reader is referred to discussions by
Shapiro and others on shock waves (fns. 6, 7, and 8).

**3.353. Example of Flow through a Hydraulic Transmission
Line.** Consider the problem of finding how Q_a and Q_b vary with time

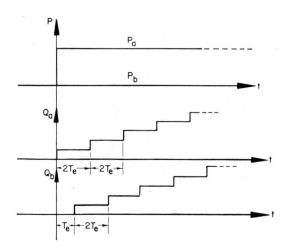

Fig. 3.27. Time variation of flow rates in a pipe for a step change in pressure at end a.

if a step change in P_a occurs while P_b is held constant. Using the first
and third time-difference equations and solving for Q_a and Q_b give the
results shown in Fig. 3.27. It may be seen that the flow Q_a changes
immediately when the step change in P_a occurs, while Q_b does not change
until T_e seconds later, and moreover, the pressure wave which required
T_e seconds to travel down the pipe in order to bring about a change in
Q_b requires another T_e seconds to travel back and again change Q_a, and

so on. The time-difference equations have proved very useful in graphical solutions of hydraulic transient problems.[17-21]

NOMENCLATURE

Symbol	Definition	Units
A	area	in.^{2}
b	depth	in.
C	constant	
C_d	discharge coefficient	
C_h	specific heat of liquid	in.^{2}/sec^2 °R
C_p	specific heat of a gas at constant pressure	in.^{2}/sec^2 °R
C_v	specific heat of a gas at constant volume	in.^{2}/sec^2 °R
D	differential operator, d/dt	1/sec
D	diameter	in.
D_i	inside diameter of pipe	in.
D_o	outside diameter of pipe	in.
e	total specific internal energy	in.^{2}/sec^2
E	Young's modulus of elasticity	lb/in.^{2}
E	total internal energy of fluid	in. lb
f	friction factor	
F	force	lb
g	standard acceleration of gravity	386 in./sec^2
h	specific enthalpy	in./sec^2
k	ratio of specific heats of a perfect gas, C_p/C_v	
k_e	elasticity coefficient of pipe	in.^{5}/lb
l	length	in.
L	length	in.
m	mass = weight/acceleration of gravity	lb sec^2/in.

[17] H. M. Paynter, "Methods and Results from M.I.T. Studies in Unsteady Flow," *J. Boston Soc. Civil Engrs.*, Vol. XXXIX, No. 2 (April 1952), pp. 135–148.

[18] K. DeJuhasz, "Graphical Analysis of Transient Phenomena in Linear Flow," *J. Franklin Inst.*, Vol. 223, No. 4 (April 1937), pp. 463–494; Vol. 223, No. 5 (May 1937), pp. 643–654; Vol. 223, No. 6 (June 1937), pp. 751–778.

[19] K. E. Sorensen, "Graphical Solution of Hydraulic Problems," *Proc. Am. Soc. Civil Engrs.*, Vol. 78, No. 116 (1952), pp. 1–17.

[20] C. A. M. Gray, "Analysis of Water Hammer by Characteristics," *Trans. Am. Soc. Civil Engrs.*, Vol. 119 (1954), pp. 1176–1194.

[21] G. Evangelisti, *La Regolazione delle turbine idrauliche*, Zanichelli, Bologna, Italy, 1947.

Symbol	Definition	Units
M	Mach number	
$\bar{M}$	momentum	lb sec
P	pressure	lb/in.^{2}
P_d	downstream pressure	lb/in.^{2}
P_u	upstream pressure	lb/in.^{2}
Q	volume rate of flow	in.^{3}/sec
Q_h	heat	in. lb
r	radius	in.
R	gas constant	in.^{2}/sec^2 °R
	(For air $R = 2.47 \times 10^5$ in.^{2}/sec^2—°R.)	
R_e	Reynolds number	
S	perimeter	in.
t	time	sec
T	absolute temperature	°R
T	torque	lb in.
T_e	transmission time	sec
T_u	upstream stagnation temperature	°R
u	specific internal energy	in.^{2}/sec^2
V	velocity	in./sec
V_r	relative velocity	in./sec
$\mho$	volume	in.^{3}
$\mho_c$	volume of fluid in cylinder	in.^{3}
$\mho_p$	volume of fluid in pipe	in.^{3}
$\mho_{pi}$	volume of pipe at zero pressure	in.^{3}
$\mho_t$	total volume	in.^{3}
w	width	in.
W	weight rate of flow	lb/sec
W_x	work	in. lb
x, y, z	fundamental linear dimensions	in.
Y_s	characteristic admittance	in.^{5}/lb sec
z	height above datum	in.
Z_s	characteristic impedance	lb sec/in.^{5}
α	angle	radians
α	angular acceleration	radians/sec^2
β	bulk modulus of elasticity	lb/in.^{2}
β_e	equivalent modulus of elasticity	lb/in.^{2}
Γ	propagation characteristic, $L\sqrt{\rho/\beta_e}D$	
δ	boundary-layer thickness	in.
ϵ	eccentricity	in.
μ	absolute viscosity	lb sec/in.^{2}
	(For air at 53°R, $\mu = 2.65 \times 10^{-9}$ lb sec/in.^{2})	
ρ	fluid density	lb sec^2/in.^{4}
ρ_i	fluid density at zero pressure	lb sec^2/in.^{4}
τ	friction force per unit area	lb/in.^{2}

4

Gerhard Reethof

Characteristics of Positive-Displacement Pumps and Motors

4.1. INTRODUCTION

In the rapidly expanding field of hydraulic power, positive-displacement pumps and motors of widely varying designs have been developed to meet the demands of high-pressure, high-speed applications. These designs are based upon three main types, which may be designated as vane, gear, and piston machines, illustrated in Figs. 4.1 through 4.4. Because of strength limitations and unbalanced pressure effects, the gear-type units have been largely confined in operating range to relatively low pressures (under 1500 psi). Vane pumps and motors are subject to much the same limitations as the gear types, but with proper balancing of pressure forces they have achieved higher pressure ranges (1500 to 2500 psi). The piston-type units, radial or axial, are the most versatile of the three basic types and are being used in most high-pressure applications (3000 psi and up). The piston designs are also the most readily available for variable-displacement applications at present, although some vane units are available. (Figs. 4.5 and 4.6.)

90

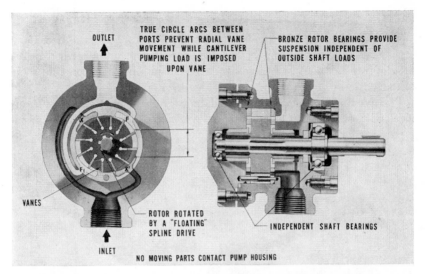

Fig. 4.1. Vane pump. (Courtesy of Vickers, Inc., Detroit, Mich.)

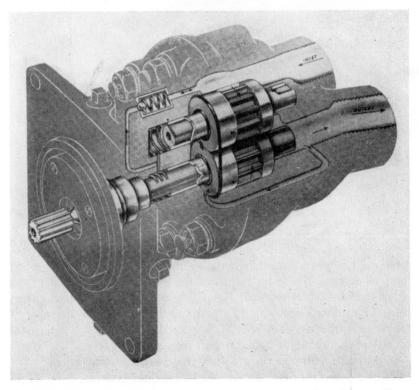

Fig. 4.2. Gear pump. (Courtesy of Pesco Products Div., Borg-Warner Corp., Bedford, Ohio.)

91

The combination of a fluid pump, a fluid motor, and the necessary accessory equipment forms a variable-speed drive, often termed a hydraulic transmission. Variable-displacement pumps or motors find their primary use in such applications. (Fig. 4.7.)

Positive-displacement pumps and motors are energy-conversion devices with the motion of a shaft or lever acting against a load as one

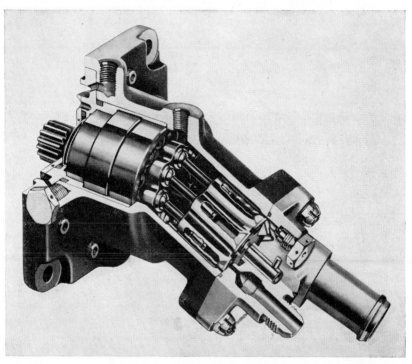

Fig. 4.3. Axial-piston pump. (Courtesy of Vickers, Inc., Detroit, Mich.)

form of energy, and displacement of hydraulic fluid under high pressures (and large pressure differentials) and low velocities as the other form of energy. In many instances the change in kinetic energy of either medium (shaft or fluid) is small in comparison with the energy being transmitted and can be safely neglected. Thus the energy-conversion process is not affected appreciably by fluid velocities, the density of the hydraulic fluid, or shaft inertia.

In contrast, hydrokinetic or hydrodynamic energy-conversion devices such as fans, centrifugal pumps, turbines, fluid clutches, and torque converters utilize the motion of the hydraulic fluid at high velocities, but under very low pressures (and low pressure differentials). Here the kinetic energy of the fluid is of the same order of magnitude as the

Fig. 4.4. Radial-piston pump. (Courtesy of The Oilgear Co., Milwaukee, Wis.)

energy being transmitted. Thus the process of conversion is very sensitive to velocity changes and the density of the fluid, but insensitive to the viscosity of the fluid. Hydrokinetic pumps and motors operate at low pressures compared with positive-displacement machines. Centrifugal pumps operating at exceptionally high speed can develop up to 1000 psi per stage, whereas positive-displacement pumps are used primarily for higher-pressure applications.

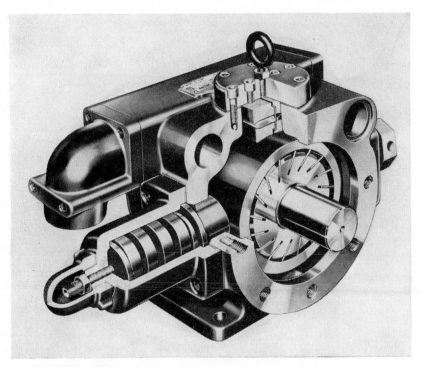

Fig. 4.5. Variable-displacement vane pump. (Courtesy of Racine Hydraulics and Machinery, Inc., Racine, Wis.)

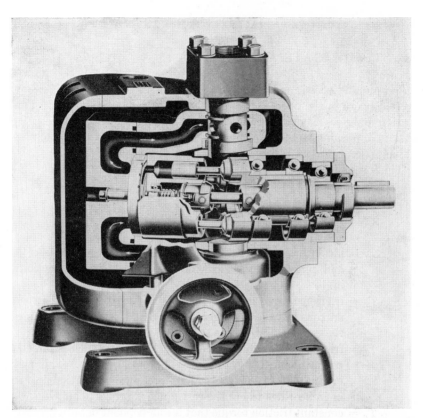

Fig. 4.6. Variable-displacement piston pump. (Courtesy of Vickers, Inc., Detroit, Mich.)

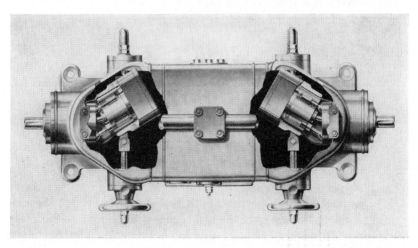

Fig. 4.7. Hydraulic transmission. (Courtesy of Vickers, Inc., Detroit, Mich.)

4.2. STEADY-STATE ANALYSIS

Torque and speed, as well as flow and pressure, are the variables of most concern in the operation of positive-displacement machines. This section presents a general analysis of the steady-state characteristics that is applicable to all three types of positive-displacement machines having one port at zero pressure.

The torque required to drive a positive-displacement pump at constant speed can be divided into four factors:

$$T_a = T_t + T_r + T_f + T_c \qquad (4.1)$$

where T_a = actual torque required, in. lb

T_t = ideal torque due to the pressure differential, $(P_1 - P_2)$, and physical dimensions of unit only (without friction), in. lb

T_r = friction torque due to viscous shearing of the fluid in the narrow passages between moving and stationary parts of the pump, in. lb

T_f = friction torque due to mechanical friction which is directly proportional to the pressure differential, in. lb. This may originate in seals if the sealing forces are proportional to P, or in bearings where the resistance is proportional to P.

T_c = constant friction torque that is independent of both pressure differential and speed, in. lb

The delivery of a pump can be expressed in a similar manner as:

$$Q_a = Q_t - Q_l - Q_r \qquad (4.2)$$

where Q_a = actual delivery, in.^{3}/sec

Q_t = ideal delivery (volume rate of flow) of the pump, which is a function of its physical dimensions and its shaft speed (assuming zero leakage and cavitation), in.^{3}/sec

Q_l = viscous-leakage flow, proportional to the pressure differential between inlet and outlet, in.^{3}/sec

Q_r = loss in delivery due to cavitation or entrained gases or vapor, in.^{3}/sec

When both port pressures are considerably greater than zero, it is necessary to account for leakage from both pressure chambers to the case (drain) as well as leakage from the higher-pressure chamber to the lower. At the present time very little information is available on leakage under these conditions.

4.21. Definition of Efficiencies

The various efficiencies are defined as follows:

1. *Volumetric efficiency.*

$$\eta_{vp} = \frac{\text{actual flow}}{\text{ideal flow}} \text{ for the pump}$$

$$\eta_{vm} = \frac{\text{ideal flow}}{\text{actual flow}} \text{ for the motor}$$

$$\eta_{vp} = \frac{Q_t - Q_l - Q_r}{Q_t} \tag{4.3}$$

In well-designed units cavitation losses are at a minimum, and the term Q_r is neglected. Thus

$$\eta_{vp} = \frac{Q_t - Q_l}{Q_t} = 1 - \frac{Q_l}{Q_t} \tag{4.4}$$

2. *Torque efficiency or mechanical efficiency.*

$$\eta_{mp} = \frac{\text{ideal torque required}}{\text{actual torque required}} \text{ for the pump}$$

and

$$\eta_{mm} = \frac{\text{actual torque output}}{\text{ideal torque output}} \text{ for the motor}$$

Therefore, for the pump

$$\eta_{mp} = \frac{T_t}{T_t + T_v + T_f + T_c} \tag{4.5}$$

3. *Over-all efficiency.* This is defined as

$$\eta_p = \frac{\text{power out}}{\text{power in}} \text{ for both pump and motor}$$

An expression for the torque of positive-displacement units can be obtained from a somewhat idealized consideration.

Consider an ideal pump consisting of a piston moving in a cylinder, as shown in Fig. 4.8. If F is the force required to move the piston, in traveling a length L, the pump displacement $D_p = AL$, and

$$F = A(P_1 - P_2) = \frac{(P_1 - P_2)D_p}{L} \tag{4.6}$$

The work, W, required to give this displacement is

$$W = FL = D_p(P_1 - P_2) \tag{4.7}$$

This is the work delivered by any pump in displacing a quantity of

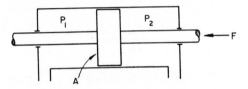

Fig. 4.8. Idealized pump.

liquid, D_p, at a pressure differential $(P_1 - P_2)$ regardless of pump configuration.

If D_p is defined as the displacement per radian in an ideal rotary pump, the work per radian, $D_p(P_1 - P_2) = T_t$, where T_t is the torque applied to the shaft, is

$$T_t = D_p(P_1 - P_2) \tag{4.8}$$

The power required to drive the shaft is

$$H_t = 2\pi N T_t = 2\pi N D_p(P_1 - P_2) \tag{4.9}$$

where N is the angular speed in rev/sec.

This is the expression for the ideal power. It is apparent that the ideal flow will be

$$Q_t = 2\pi N D_p \tag{4.10}$$

and the ideal power delivered to the fluid is

$$H_t = Q_t(P_1 - P_2) \tag{4.11}$$

The actual usable power delivered to the fluid is

$$H_a = Q_a(P_1 - P_2) \tag{4.12}$$

Then, returning to the expression for the over-all efficiency, we find that

$$\eta_p = \frac{Q_a(P_1 - P_2)}{2\pi N T_{ap}} \tag{4.13}$$

where Q_a is the actual flow out of the pump and T_{ap} is the actual torque required to drive the pump.

Since

$$T_a = \frac{T_t}{\eta_{mp}} = \frac{D_p(P_1 - P_2)}{\eta_{mp}} \qquad (4.14)$$

and

$$Q_a = \eta_{vp}Q_t = 2\pi ND_p\eta_{vp} \qquad (4.15)$$

it follows that

$$\eta_p = \frac{2\pi ND_p(P_1 - P_2)\eta_{mp}\eta_{vp}}{2\pi ND_p(P_1 - P_2)} = \eta_{vp}\eta_{mp} \qquad (4.16)$$

4.22. Derivation of Flow and Torque Equations

We shall next express each loss term in the flow and torque equations for the pump in terms of the physical characteristics of the unit, using nondimensional coefficients wherever possible.[1-7]

The delivery (the actual flow) from a positive-displacement hydraulic pump is

$$Q_a = Q_t - Q_l - Q_r \qquad (4.2)$$

The leakage that takes place in positive-displacement hydraulic components is caused primarily by the flow through the small clearance spaces between the various parts that separate high-pressure and low-pressure regions. The small spaces are often referred to as capillary passages. Most of these leakage paths are essentially two flat parallel plates with the flow occurring through the clearance space. It is, therefore, reasonable to apply the fundamental relationships for flow between flat plates. It was shown in Chapter 3 that, if the force balance is written

[1] W. E. Wilson, "Rotary Pump Theory," *Trans. ASME*, Vol. 68 (May 1946), pp. 371–384.

[2] W. E. Wilson, "Performance Criteria for Positive-Displacement Pumps and Fluid Motors," *Trans. ASME*, Vol. 71 (February 1949), pp. 115–120.

[3] W. E. Wilson, "Method of Evaluating Test Data Aids Design of Rotary Pumps," *Prod. Eng.*, Vol. 16 (October 1945), pp. 653–656.

[4] W. E. Wilson, "Design Analysis of Rotary Pumps to Obtain Maximum Efficiency," *Prod. Eng.*, Vol. 17 (February 1946), pp. 138–141.

[5] W. E. Wilson, "Hydraulic Pumps and Motors," *Machine Design*, Vol. 21 (January 1949), pp. 133–138.

[6] W. E. Wilson, "Positive Displacement Pumps and Fluid Motors," Pitman Publishing Corp., New York, 1950.

[7] W. E. Wilson, "Clearance Design in Positive-Displacement Pumps," *Machine Design*, Vol. 25 (February 1953), pp. 127–130.

for a fluid element in a two-dimensional flow where the relative motion of the surfaces is neglected, the pressure-induced flow becomes * (see Fig. 4.9):

$$Q_l = \frac{h^3 w (P_1 - P_2)}{12 L \mu} \tag{4.17}$$

where Q_l = leakage flow between flat plates for zero pressure gradient in the w direction, in.^{3}/sec

w = width of the flow, in.

L = length of the leakage path, in.

h = thickness of the clearance space, in.

μ = average viscosity of the fluid, lb sec/in.^{2}

If a series of geometrically similar units is considered, the physical dimensions of any one unit are by definition proportional to a charac-

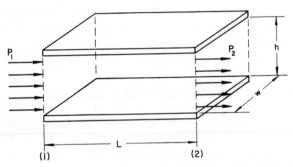

Fig. 4.9. Capillary flow between flat parallel plates.

teristic dimension, Λ. The introduction of a slip coefficient, C_s, has been found convenient. It is defined in such a way that the characteristic expression for the capillary flow becomes

$$Q_l = C_s \frac{2 \pi D_p (P_1 - P_2)}{\mu} \tag{4.18}$$

The slip coefficient, C_s, is given by

$$C_s = \frac{k_1 h^3}{D_p} \tag{4.19}$$

where D_p is the pump displacement per radian and k_1 is a constant. This procedure is permissible because of the assumed similarity of the series of pumps under consideration. Thus the displacement, D_p, is

* See Table 3.1 of Chapter 3, Fundamentals of Fluid Flow.

proportional to the cube of the characteristic dimension. The slip coefficient is therefore

$$C_s = k_2 \left(\frac{h}{\Lambda}\right)^3 \tag{4.20}$$

Thus the slip coefficient varies as the cube of the clearance ratio and can be expected to be very sensitive to manufacturing tolerances.

It should be noted that the derivation of the leakage flow and the subsequent definition of the slip coefficient are based on one leakage path. In actual units several different paths exist, each having different dimensions but basically similar characteristics. Therefore all these paths can be combined into one composite or equivalent path without detracting from the generality of the discussion. Variation of all clearances in a pump by the same percentage would thus vary the composite by the same percentage.

A question may be raised as to the effect of pump or motor size on the slip coefficient. For a series of geometrically similar units, it appears somewhat farfetched to assume that clearances will be directly proportional to a characteristic dimension. Since similar machine operations are used to produce the series of pumps, it would be far more realistic to predict that the clearances will not change so much as size on a percentage basis. It would, therefore, seem reasonable that the larger pumps should exhibit a consistently lower slip coefficient than the smaller pumps of a series.

The expression for the delivery of a pump then becomes, from Eq. 4.2,

$$Q_{ap} = 2\pi N D_p - \frac{2\pi C_s D_p (P_1 - P_2)}{\mu} - Q_r \tag{4.21}$$

The equivalent expression for the hydraulic motor is

$$Q_{am} = 2\pi N D_m + \frac{2\pi C_s D_m (P_1 - P_2)}{\mu} \tag{4.22}$$

It should be noted that these expressions are based on the assumption that clearances between bearing surfaces, and so on, remain fixed; but in some units, such as those employing pressure-loaded bearing surfaces, and in piston-type units where the pistons are neither constrained from moving normal to their axis nor prevented from cocking in the clearance space, the clearances may vary with pressure and in some cases with speed. These variations are often significant because C_s is extremely sensitive to clearances. A further deviation from the ideal leakage equations results from the occurrence of some orifice-type leakage in certain gear and vane units. The leakage flow across the vane tips and the gear teeth can be expected to vary roughly as the square root of the pressure drop because the length-to-thickness ratio of these clearance spaces is small. A later section will deal with some of these deviations from the ideal case.

As shown previously, the torque required to drive a pump can be expressed in terms of the theoretical torque and the loss terms. Thus

$$T_{ap} = T_t + T_r + T_f + T_c \qquad (4.1)$$

The first term on the right-hand side of the equation is the ideal torque required to obtain the desired pressure rise. The second term is the

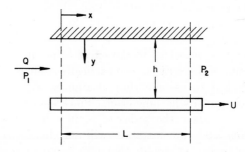

Fig. 4.10. Forces in the capillary flow between parallel moving plates.

viscous-friction term which is the result of the shearing of the oil in clearance passages between moving and stationary surfaces.

The shearing stress in the fluid between two flat parallel moving plates which are separated by a small distance, h, (as in Fig. 4.10) is given by

$$\tau = \frac{\mu U}{h} - \frac{dP}{dx}\left(\frac{h}{2} - y\right) \qquad (4.23)$$

At the surface where $y = h$, the unit force due to viscous friction which retards the moving plate is equal to

$$\tau_U = \frac{\mu U}{h} + \frac{h}{2}\frac{dP}{dx} \qquad (4.24)$$

The total force on the moving plate then becomes

$$F = \left(-\frac{P_1 - P_2}{L} \cdot \frac{h}{2} + \frac{\mu U}{h}\right) Lw \qquad (4.25)$$

and the torque at some radius, r, becomes

$$T_r = \left(-\frac{P_1 - P_2}{L} \cdot \frac{h}{2} + \frac{\mu U}{h} \right) Lwr \qquad (4.26)$$

Because the clearance spaces in which this viscous friction is developed are generally much more extensive than the regions which allow leakage, and also because the leakage is often normal to the velocity in these regions, the pressure-drop term in Eq. 4.26 usually is negligible. The torque due to viscous shear, therefore, simplifies to

$$T_r = \frac{\mu N r^2 Lw}{h} \qquad (4.27)$$

Rearrangement of terms permits us to say that

$$T_r \propto \frac{\mu N \Lambda^4}{h} \qquad (4.28)$$

A drag coefficient is introduced which is such that

$$T_r = C_d D_p N \mu \qquad (4.29)$$

Because D_p varies directly as D^3, the drag coefficient C_d must satisfy the relationship

$$C_d \propto \frac{\Lambda}{h} \qquad (4.30)$$

C_d is therefore inversely proportional to the clearance ratio, which has been defined as b/D.

The existence of sealing elements and bearings, both of which exert retarding torques that are related to working pressures, makes necessary the inclusion of a pressure-dependent friction-torque term. Experimental work on pumps and motors with one port at zero pressure has shown that this friction torque, to a first approximation, is proportional to the pressure drop across the unit.

By selecting a dimensionless parameter containing $(P_1 - P_2)$, we obtain

$$T_f = C_f (P_1 - P_2) D_p \qquad (4.31)$$

where C_f is a dimensionless coefficient relating the friction-torque parameter to the ideal torque. It should be noted that C_f is not related to the clearances of the pumping members and differs from one to another of a series of geometrically similar units, and that large values of C_f

are an indication of considerable metal-to-metal contact and probably of serious wear.

The resultant expression for the torque required to drive a positive-displacement pump is

$$T_{ap} = (P_1 - P_2)D_p + C_d D_p \mu N + C_f(P_1 - P_2)D_p + T_c \quad (4.32)$$

For the motor the equivalent torque output is

$$T_{am} = (P_1 - P_2)D_m - C_d D_m \mu N - C_f(P_1 - P_2)D_m - T_c \quad (4.33)$$

Equation 4.33 is based on the assumption that one of the ports is at zero pressure. In many applications, both ports of the fluid motor are at high pressures and the pressure differential exists between these high port pressures. The breakaway torque, T_c, increases rapidly with increasing mean port pressure for most piston and some vane motors.

Because these expressions also are based on the assumption that clearances remain fixed, the same reservations apply as were encountered in the leakage analysis. However, in the present case the clearance ratio enters only as a first power, and therefore any clearance variation as a function of pressure or speed results in only minor variations in the torque coefficients. Of more importance in applications to high-performance control systems is the fact that Eqs. 4.32 and 4.33 apply only to pumps and motors having one port at zero pressure.

4.23. Derivation of Efficiency Equations

The efficiencies, previously defined, can now be derived with the aid of the pertinent dimensionless coefficients.

4.231. Volumetric Efficiency. Because

$$\eta_{vp} = \frac{Q_t - Q_l - Q_r}{Q_t} \quad (4.3)$$

the volumetric efficiency of a pump becomes

$$\eta_{vp} = 1 - C_s \left(\frac{P_1 - P_2}{\mu N} \right) - \frac{Q_r}{2\pi D_p N} \quad (4.34)$$

For well-designed pumps the losses in flow due to cavitation are small and the last term in the expression can be neglected. Therefore

$$\eta_{vp} = 1 - C_s \left(\frac{P_1 - P_2}{\mu N} \right) \quad (4.35)$$

As already noted, the slip coefficient is proportional to the cube of the clearance ratio, so the volumetric efficiency should be expected to be the same in various-sized units of a geometrically similar series of pumps, operating with fluids of the same viscosity.

4.232. Mechanical Efficiency. Because

$$\eta_{mp} = \frac{T_t}{T_t + T_r + T_f + T_c} \tag{4.5}$$

the mechanical efficiency of a pump becomes

$$\eta_{mp} = \frac{(P_1 - P_2)D_p}{(P_1 - P_2)D_p + C_d D_p \mu N + C_f(P_1 - P_2)D_p + T_c} \tag{4.36}$$

$$\eta_{mp} = \frac{1}{1 + C_d \dfrac{\mu N}{(P_1 - P_2)} + C_f + \dfrac{T_c}{(P_1 - P_2)D_p}} \tag{4.37}$$

The drag coefficient, C_d, is inversely proportional to the first power of the clearance ratio and therefore is independent of the characteristic dimension for units in a geometrically similar series of pumps.

Similar conclusions cannot be reached regarding either C_f or T_c, which are more likely to depend on size alone.

4.233. Over-all Pump Efficiency. The output power of a pump is equal to

$$\text{Power out} = Q_{ap}(P_1 - P_2) \tag{4.38}$$

The power required to drive the pump is equal to

$$\text{Power in} = 2\pi N T_{ap} \tag{4.39}$$

Substituting the previously obtained expressions for Q_{ap} and T_{ap} gives

$$\text{Power out} = (P_1 - P_2)\left[2\pi N D_p - C_s \frac{2\pi D_p(P_1 - P_2)}{\mu}\right] \tag{4.40}$$

$$\text{Power in} = 2\pi N[(P_1 - P_2)D_p + C_d D_p \mu N + C_f(P_1 - P_2)D_p + T_c] \tag{4.41}$$

The ideal output power of the pump is given by:

$$H_{tp} = 2\pi N D_p(P_1 - P_2) \tag{4.42}$$

Thus

$$\text{Power out} = 2\pi N D_p (P_1 - P_2) \left(1 - C_s \frac{P_1 - P_2}{\mu N}\right) \tag{4.43}$$

$$\text{Power in} = 2\pi N D_p (P_1 - P_2) \left[1 + C_d \frac{\mu N}{P_1 - P_2} + C_f \right.$$

$$\left. + T_c \frac{1}{(P_1 - P_2)D_p}\right] \tag{4.44}$$

Then, since $\eta_p \overset{\Delta}{=}$ (power out)/(power in), the over-all pump efficiency is $\eta_p = \eta_{mp}\eta_{vp}$, in agreement with Eq. 4.16. If T_c can also be neglected, the efficiencies become

$$\eta_{vp} = 1 - C_s \left(\frac{P_1 - P_2}{\mu N}\right) \text{ (as before)} \tag{4.35}$$

$$\eta_{tp} = \left[1 + C_d \frac{\mu N}{(P_1 - P_2)} + C_f\right]^{-1} \tag{4.45}$$

and

$$\eta_p = \frac{1 - C_s \dfrac{(P_1 - P_2)}{\mu N}}{1 + C_d \left(\dfrac{\mu N}{P_1 - P_2}\right) + C_f} \tag{4.46}$$

Thus the efficiencies of a series of geometrically similar pumps are largely determined by four parameters: C_s, C_d, C_f, and $\mu N/(P_1 - P_2)$.

It should be mentioned that when clearances do not remain constant during operation, the efficiency expression is more complicated.

4.24. Dimensional Analysis

The concepts of dimensional analysis can be employed to substantiate the results just stated. First, consider all the variables that will affect the efficiency of a hydraulic pump:

$$\eta_p = f_2[N, (P_1 - P_2), \mu, c, \Lambda] \tag{4.47}$$

where c is now a composite measure of clearances.

It is to be noted that the density of the hydraulic fluid is omitted from this expression. The reason for this omission is that neither leakage nor viscous shear depends upon the density of the fluid, since laminar flow is assumed to exist in the clearance spaces.

In this array of variables we have six parameters and only three prime dimensions: namely, force, time, and length; therefore it should be possible to form three nondimensional coefficients, according to Buckingham's pi theorem.[8] A convenient set is

$$\Pi_1 = \frac{\mu N}{(P_1 - P_2)} \tag{4.48}$$

$$\Pi_2 = \frac{c}{\Lambda} \tag{4.49}$$

$$\Pi_3 = \eta_p \tag{4.50}$$

therefore

$$\Pi_3 = f_3(\Pi_1, \Pi_2)$$

or

$$\tag{4.51}$$

$$\eta_p = f_3\left[\left(\frac{\mu N}{P_1 - P_2}\right), \ \left(\frac{c}{\Lambda}\right)\right]$$

This method has been found to be extremely useful in experimental work, as the amount of data which must be taken can be reduced considerably by judicious choice of the coefficients.

4.25. Optimization of Efficiencies

The pump efficiency can be maximized with respect to the dimensionless parameter Π_1. Differentiating Eq. 4.46 with respect to Π_1 and setting it equal to zero gives

$$\frac{d\eta_p}{d\Pi_1} = \frac{C_s/\Pi_1{}^2}{1 + C_d\Pi_1 + C_f} - \frac{[1 - (C_s/\Pi_1)]C_d}{(1 + C_d\Pi_1 + C_f)^2} = 0 \tag{4.52}$$

The critical value of Π_1 is found to be

$$\Pi_{1\ \text{crit}} = C_s\left(1 + \sqrt{1 + \frac{1 + C_f}{C_s C_d}}\right) \tag{4.53}$$

With this result the maximum efficiency becomes

$$\eta_{p\ \text{max}} = \frac{1}{1 + C_f + 2C_s C_d\left(1 + \sqrt{1 + \dfrac{1 + C_f}{C_s C_d}}\right)}$$

$$= \frac{1}{1 + C_f + 2C_d\Pi_{1\ \text{crit}}} \tag{4.54}$$

---

[8] J. C. Hunsaker and B. G. Rightmire, *Engineering Applications of Fluid Mechanics*, McGraw-Hill Book Co., New York, 1947.

The important result obtained from these calculations is that the maximum efficiency is a function of only the product $C_s C_d$ and the coefficient C_f.

The effect of varying the clearance ratio will next be studied. If we neglect C_f, the maximum efficiency becomes

$$\eta_{p\ max} = \frac{1}{1 + 2C_s C_d [1 + \sqrt{1 + (1/C_s C_d)}\ ]} \tag{4.55}$$

If C_{so} and C_{do} are reference values, then as the clearance, c, is varied from C_o, from Eq. 4.20,

$$C_s = \left(\frac{c}{c_o}\right)^3 C_{so} = R^3 C_{so} \tag{4.56}$$

and similarly, from Eq. 4.30,

$$C_d = \left(\frac{c_o}{c}\right) C_{do} = \frac{C_{do}}{R} \tag{4.57}$$

where $R = c/c_o$. The maximum efficiency, $\eta_{p\ max}$, then becomes

$$\eta_{p\ max} = \frac{1}{1 + 2C_{so}C_{do}R^2[1 + \sqrt{1 + (1/C_{so}C_{do}R^2)}\ ]} \tag{4.58}$$

The effect of changes of the clearance ratio on either side of the reference value c_o now can be studied conveniently. Typical values of C_{so} and C_{do} are $C_{so} = 5 \times 10^{-7}$ and $C_{do} = 10^5$. For this case

$$\eta_{p\ max} = \frac{1}{1 + 0.10R^2[1 + \sqrt{1 + (20/R^2)}\ ]} \tag{4.59}$$

and the value of Π_1 for which maximum efficiency is attained is

$$\Pi_{1\ crit} = C_{so}R^3[1 + \sqrt{1 + (20/R^2)}\ ] \tag{4.60}$$

A warning should be given concerning the range of validity of Eqs. 4.59 and 4.60, which is limited by the nature of the assumptions on which they are based. It would be erroneous to extrapolate them to extremely large or extremely small clearance ratios. A realistic study will not, of course, consider unattainably small clearances. The minimum clearance is limited by manufacturing tolerances and also by the particle sizes of contaminants in the hydraulic fluid.

The plots of pump efficiency versus Π_1, the coefficient $\mu N/(P_1 - P_2)$, for various clearance ratios (Fig. 4.11), show the sensitivity of the peak efficiency to variations in the parameters. Thus the clearance, for example, can be increased by 40 per cent, but if the same operating output is required, a higher-viscosity oil should be used to allow operation near the peak efficiency.

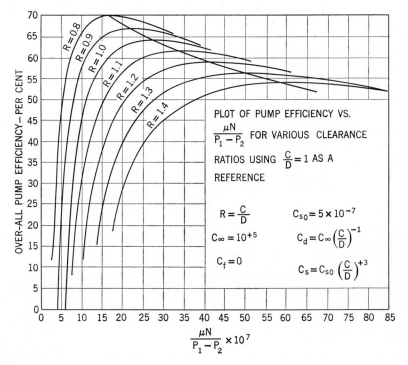

Fig. 4.11. Plot of over-all pump efficiency versus $\mu N/(P_1 - P_2)$ for various clearance ratios.

The effect of the omission of C_f from the analysis is shown in Fig. 4.12 for both a pump and a motor. Characteristic values of C_f for either pump or motor range from 0.04 to about 0.10 with the majority at approximately 0.07. As previously pointed out, a high value of C_f is an indication of excessive wear and should aid in the evaluation of the merit of any particular pump or motor. Similarly, a very low value of the parameter $\mu N/(P_1 - P_2)$ due to low viscosity tends to indicate poor lubrication, which causes high friction torque, T_c. The cavitation losses may become large (Q_r increases) for high values of flow. Thus, considerable judgment must be exercised in using this type of analysis.

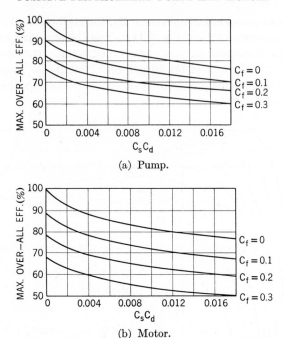

Fig. 4.12. Plot of maximum over-all efficiency of positive-displacement machines as a function of $C_s C_d$ for various values of C_f.

4.26. Experimental Determination of Performance Coefficients [9]

The experimental evaluation of the analysis just given requires both care and ingenuity. The quantities to be measured are:

Q_a = the actual flow of hydraulic fluid through the pump or motor, in.³/sec

$(P_1 - P_2)$ = the pressure drop between outlet and inlet ports of the pump or motor, lb/in.²

T_a = the torque in the pump or motor shaft, in. lb

N = the speed of rotation of pump or motor, rps

μ = the mean viscosity of the hydraulic fluid, lb sec/in.²

The actual delivery of a pump is

$$Q_{ap} = 2\pi N D_p - \frac{2\pi C_s D_p (P_1 - P_2)}{\mu} - Q_r \qquad (4.61)$$

A series of runs must be made, measuring the output flow as a function

⁹ Wilson, fn. 6.

of speed while holding the pressure drop and the viscosity constant. A theoretical delivery curve can be plotted for zero pressure drop across the pump; it will be a straight line with a slope equal to the displacement of the pump per revolution. The cavitation loss, Q_r, is assumed to be

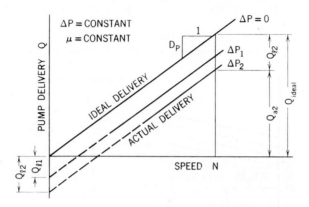

Fig. 4.13. Plot of pump delivery versus pump speed.

zero. The runs for different pressure drops are plotted as shown in Fig. 4.13. Oil temperature must be carefully controlled in order to keep the viscosity constant. Very high flows should be avoided as cavitation and turbulence losses will cause large errors.

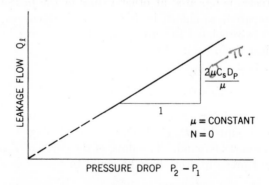

Fig. 4.14. Plot of leakage flow versus pressure drop.

If the constant-pressure-drop curves are extended to the zero-speed ordinate, the deviation from the ideal delivery must be the leakage flow due to the capillary passages. The leakage flows are then plotted against the pressure drops which cause them, as in Fig. 4.14. According to Eq. 4.18, the slope of this plot will be $2\Pi C_s D_p/\mu$, and from the measured value of this slope the value of the slip coefficient, C_s, can be calculated.

From Eq. 4.32 the torque input to the pump is:

$$T_{ap} = (P_1 - P_2)D_p + C_d D_p \mu N + C_f(P_1 - P_2)D_p + T_c \quad (4.32)$$

(For the motor, from Eq. 4.33, the signs of all terms after the first are reversed.) The torque required to drive the pump at various shaft

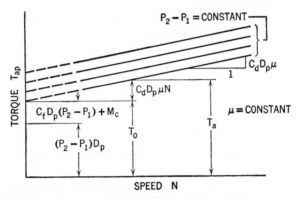

Fig. 4.15. Plot of torque to drive pump versus pump speed.

speeds for a constant pressure drop and a constant viscosity is measured, the process is repeated at other values of the pressure drop, and a plot is prepared, as shown in Fig. 4.15. The curves for constant $P_2 - P_1$ are extrapolated to zero speed. The zero-speed torque, T_0, is plotted against the pressure drop.

From Eq. 4.32

$$T_0 = (P_1 - P_2)D_p(C_f + 1) + T_c \quad (4.62)$$

The zero-pressure intercept of the curve of T_0 versus $(P_1 - P_2)$ is T_c, and the slope is equal to $D_p(C_f + 1)$, as shown in Fig. 4.16. The value of C_f can therefore be found. The slope of the curve of T_{ap} versus speed for a constant pressure drop must be equal to

$$\frac{T_{ap}}{N} = C_d D_p \mu \quad (4.63)$$

The drag coefficient, C_d, can now be obtained.

All the pertinent coefficients have now been evaluated and it has become possible to plot the efficiencies on a nondimensional basis to

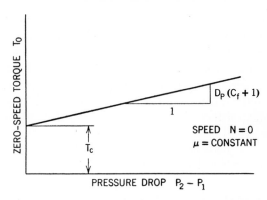

Fig. 4.16. Plot of zero-speed torque versus pressure drop.

assist the understanding of the steady-state operation of positive-displacement pumps and motors.

4.3.   FLUID FRICTION IN PUMPS AND MOTORS— THE THERMAL WEDGE

Like most theoretical analyses, the one just presented is based on simplifying assumptions that may introduce important errors. For example, the omission of the Coulomb friction, T_c, from Eq. 4.1, of the cavitation loss, Q_r, from Eq. 4.2, and of the friction factor, C_f, from Eq. 4.55 is hard to justify except on the grounds of mathematical expediency. Even more serious are the assumptions that the clearances are invariant with speed, pressure, and scale factor, that the fluid properties are constant and are the same throughout the clearance spaces, and that the momentum effects associated with orifice-type leakage can be neglected. An experimental study of axial-piston machines [10] showed that their actual characteristics correspond very poorly to the idealized expressions such as Eq. 4.21, and that a considerably more sophisticated analysis is badly needed. As yet, such an analysis has not appeared, but some progress has been made, and the remainder of this chapter will report an attempt to apply to hydraulic machines some of the concepts that have been developed for the study of lubricated bearings.

One of the most useful of these concepts is that of the "thermal wedge." This is a very simple idea, and obviously correct: the shear work that is

[10] D. K. Crockett, *A Study of Performance Coefficients of Positive-Displacement Hydraulic Pumps and Motors*, S. M. Thesis, Department of Mechanical Engineering, Massachusetts Institute of Technology, 1952.

done on the thin film of fluid which fills the clearance space between the two surfaces in relative motion heats the fluid and causes it to expand; this expansion is accommodated by a flow outward through the clearance, and the flow causes a pressure which, under conditions of normal hydrodynamic lubrication, sustains the load on the bearing. The work done on the fluid as it flows along the clearance space produces a temperature gradient, a thermal wedge, which is analogous to the physical wedge of the tilted-pad hydrodynamic bearing. The load-carrying capacity of parallel-plate bearings has been studied in England by Fogg [11] and in this country by Shaw and others.[12-15]

4.31. Theory of the Thermal Wedge

In the remainder of this chapter we shall apply some of the ideas developed in the lubrication studies just cited to the problem of the

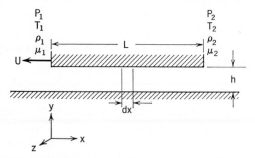

Fig. 4.17. Clearance space between two surfaces in relative motion.

motions of typical elements of a hydraulic pump or motor, such as pistons, vanes, and so on. We assume the following:

a. The geometry is ideal, as sketched in Fig. 4.17. The adjacent surfaces are smooth and parallel, and the slider moves parallel to the plane of the clearance with a velocity U.

[11] A. Fogg, "Fluid film Lubrication of Parallel Thrust Surfaces," *Proc. Inst. Mech. Engrs.* (*London*), Vol. 155, No. 14 (1946), pp. 49–53.

[12] M. C. Shaw and E. F. Macks, *Analysis and Lubrication of Bearings*, McGraw-Hill Book Co., New York, 1949.

[13] M. C. Shaw, "An Analysis of the Parallel-Surface Thrust Bearing," *Trans. ASME*, Vol. 69 (May 1947), pp. 381–387.

[14] F. Osterle, A. Charnes, and E. Saibel, "On the Solution of the Reynolds Equation for Slider-Bearing Lubrication—VI, The Parallel-Surface Slider Bearing Without Side Leakage," *Trans. ASME*, Vol. 75 (August 1953), pp. 1133–1136.

[15] G. Reethof, C. Goth, and H. Kord, "Thermal Effects in the Flow of Fluids Between Two Parallel Plates in Relative Motion," Am. Soc. Lubrication Engrs., Paper 58 AM 4A-1.

b. There are no edge or end effects. This is very nearly true because the height, h, of the clearance is very small compared with its other dimensions.

c. The fluid is Newtonian, and its properties are independent of y and z. This may not be strictly true, but for all practical purposes we may assume an average value of the property, which is what is ordinarily measured anyway.

d. The steady state has been established. This assumption is reasonable since in most cases the flow changes much more rapidly than does the motion of the slider.

e. The flow through the clearance space is everywhere laminar and parallel to x. This also is true or nearly so because of the small value of h.

The assumption of laminar flow permits the use of the principle of superposition, according to which we can consider the actual flow as being the sum of three separate and independent flows, and can study each one separately. These flows are:

1. The pure shear flow caused by the velocity, U, of the slider.

2. The pressure flow caused by the static-pressure difference between the two ends of the slider.

3. The thermal flow caused by the thermal expansion of the fluid.

These three flows and their sum are indicated symbolically in Fig. 4.18.

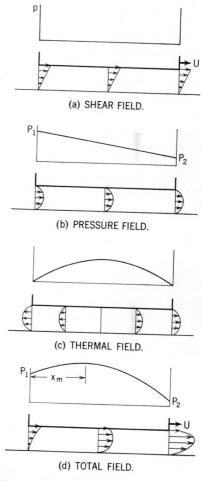

(a) SHEAR FIELD.

(b) PRESSURE FIELD.

(c) THERMAL FIELD.

(d) TOTAL FIELD.

Fig. 4.18. Component-flow fields for case of Fig. **4.17.**

It should be noted that this figure must be used with some care for two reasons. First, the superposition theorem is valid only to the extent that it is valid in any other physical situation. If we have both a d-c and an a-c current flowing through a physical resistor, for example, the two currents will be independent only if both are too small to cause appreciable heating of the resistor. If the resistance changes because of this heating, one current does affect the other; this is the principle of the bolometer. In the present case the same thing is true; the several flows and pressures are independent only to the extent that they do not affect the properties of the fluid.

In the second place, in many practical cases such as that of a pump or motor piston, the static-pressure difference is so large that the pressure flow is much larger than either of the other two. The derivations that follow would be made considerably more complicated by including the effects of this pressure flow. Accordingly, at some cost of generality, we shall assume (f) that the pressures at the ends of the slider, P_1 and P_2, are both zero. This implies that the pressure gradient dp/dx is only that due to the thermal flow. The effects of the static-pressure difference can be included in the theory, but so doing would invalidate several convenient approximations and would greatly complicate the mathematics with little added understanding of the phenomena.

We consider an elementary slice of the flow of Fig. 4.17, with height h, width w normal to the plane of the figure, and length δx. The power input to this slice will be

$$\frac{\delta E}{\delta t} = \tau U w \, \delta x - q \, \delta x \, \frac{dp}{dx} \tag{4.64}$$

where E = energy content of slice, in. lb
$\quad \tau$ = shear stress on top surface, lb/in.^{2}
$\quad q$ = rate of flow through slice, in.^{3}/sec
$\quad p$ = pressure, psi

For the conditions given,

$$\tau = \frac{\mu U}{h} + \frac{h}{2} \frac{dp}{dx} \tag{4.65}$$

and

$$q = \frac{hwU}{2} - \frac{h^3 w}{12} \frac{dp}{dx} \tag{4.66}$$

where μ = viscosity of fluid, lb sec/in.^{2}
$\quad U$ = velocity of slider, in./sec

Substituting Eqs. 4.65 and 4.66 into Eq. 4.64, we get

$$\frac{\delta E}{\delta t} = \frac{\mu U^2 w \, \delta x}{h} + \frac{h^3 w \, \delta x}{12\mu} \left(\frac{dp}{dx}\right)^2 \tag{4.67}$$

We now assume (g) that the thermal flow is negligible in comparison to the drag flow. This is true for a real liquid with practical ratios of clearance to length. In this case we can write for the average fluid velocity

$$\frac{\delta x}{\delta t} = \frac{U}{2} \tag{4.68}$$

Dividing Eq. 4.67 by Eq. 4.68, we obtain

$$\frac{\delta E}{\delta x} = \frac{\delta E}{\delta t} \cdot \frac{\delta t}{\delta x} = \frac{2\mu w \, \delta x}{hU} \left[U^2 + \frac{h^4}{12} \left(\frac{1}{\mu} \frac{dp}{dx} \right)^2 \right] \tag{4.69}$$

According to the principle of continuity, $(\rho q) = $ constant. Therefore

$$\frac{d}{dx} (\rho q) = 0 \tag{4.70}$$

where $\rho = $ fluid density, lb sec^2/in.^{4}

Multiplying Eq. 4.66 by $12\rho/h^3\mu$ and differentiating, we get

$$\frac{6U}{h^2} \frac{d\rho}{dx} - \frac{d}{dx} \left(\frac{\rho}{\mu} \frac{dp}{dx} \right) \tag{4.71}$$

Integration of Eq. 4.71 yields

$$\frac{6U\rho}{h^2} - \frac{\rho}{\mu} \frac{dp}{dx} + C_1 = 0 \tag{4.72}$$

Now consider the point at which the pressure is a maximum. Designate values of the variables at this point by the subscript m. Then when $x = x_m$, $dp/dx = 0$ and $C_1 = -6U\rho_m/h^2$. Substituting this value for C_1 in Eq. 4.72 and rearranging, we get

$$\frac{1}{\mu} \frac{dp}{dx} = \frac{6U}{h^2} \left(1 - \frac{\rho_m}{\rho} \right) \tag{4.73}$$

Squaring Eq. 4.73, substituting in Eq. 4.69, and rearranging, we finally obtain

$$\frac{\delta E}{\delta x} = \frac{2\mu w U \, \delta x}{h} \left[1 + 3 \left(1 - \frac{\rho_m}{\rho} \right)^2 \right] \tag{4.74}$$

We must now make some assumptions concerning the effects of pressure and temperature on the properties of the fluid. Again, in order to simplify the derivation but at the cost of generality, we assume (h) that pressure has no effect. This is false, and would introduce appreciable errors when the pressure change is large. In the present case, however, it is small, and the error is negligible. With respect to temperature, we assume (i) that the following three expressions hold:

$$\rho = \rho_1(1 - \alpha \, \Delta T) \tag{4.75}$$

$$E = E_1 + C_v \, \Delta T \tag{4.76}$$

$$\mu = \mu_1 e^{-\lambda \, \Delta T} \tag{4.77}$$

where $\Delta T = T - T_1$ = temperature change from temperature at $x = 0$, °F; C_v = specific heat per unit volume, in. lb/in.^{3} °F.

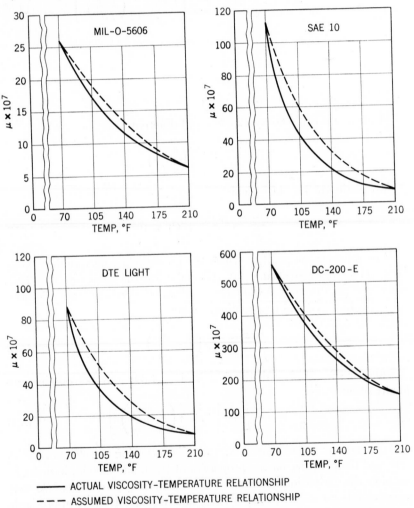

ACTUAL VISCOSITY-TEMPERATURE RELATIONSHIP
- - - ASSUMED VISCOSITY-TEMPERATURE RELATIONSHIP

Fig. 4.19. Viscosity-temperature characteristics of seven hydraulic fluids.

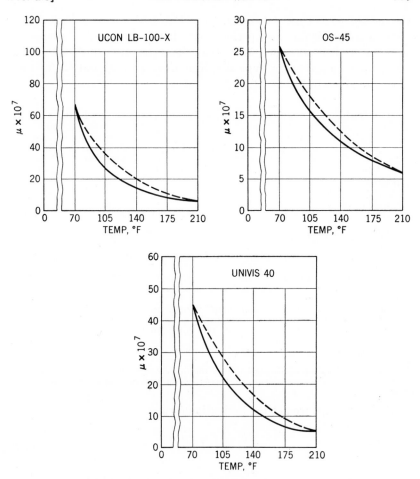

Fig. 4.19. (*Continued*). (Os-45 now known as Koolanol 45.)

Of these three equations, the first is fairly accurate, since in most practical cases the total temperature change is not large. The second, which implies that C_v is constant, is also reasonably well satisfied. It should be noted that for many liquids the specific heat *per unit weight* increases with temperature more rapidly than the density decreases, so that the net change in C_v is positive, but it is not large. The third equation is less accurate than the other two, but it is much better than a linear approximation, and it *is* amenable to mathematical manipulation, which the Walther formula (given in Eq. 2.20, Sec. 2.32) decidedly is *not*. Figure 4.19 shows the temperature-viscosity characteristics of seven typical fluids; the solid curves give the actual characteristics, and the dashed curves are calculated from Eq. 4.77. As can be seen, the correspondence is far from perfect, but at least for moderate ranges of temperature it is not much worse than the errors introduced by some of our other assumptions. Typical values of ρg, μ, α, λ, and C_v for these fluids are given in Table 4.1.

Table 4.1. PROPERTIES OF SEVEN HYDRAULIC FLUIDS

Fluid	Specific Weight, ρg, at 70 °F, lb/in.^{3}	Viscosity, μ, at 70 °F, lb sec/in.^{2}	Expansion Coefficient, α, 1/°F	Viscosity-Temperature Coefficient, λ, 1/°F	Specific Heat per Unit Volume, C_v, in. lb/°F in.^{3}
Petroleum-base fluids;					
Aircraft hydraulic oil, MIL-H-					
5606 *	0.0312	2.6×10^{-6}	4.26×10^{-4}	1.037×10^{-2}	149
Light motor oil,					
SAE 10	0.0316	11.5×10^{-6}	4.21×10^{-4}	1.937×10^{-2}	144
Light turbine oil,					
DTE light	0.0316	9.0×10^{-6}	4.21×10^{-4}	1.85×10^{-2}	144
Hydraulic oil,					
Univis 40 *	0.0310	4.56×10^{-6}	4.29×10^{-4}	1.59×10^{-2}	142
Synthetic fluids:					
Dimethyl silicone,					
DC-200-E	0.0351	56.7×10^{-6}	4.67×10^{-4}	0.972×10^{-2}	112
Polyalkylene glycol,					
Ucon LB-100-X	0.0353	6.4×10^{-6}	4.45×10^{-4}	1.74×10^{-2}	151
Orthosilicate ester,					
Koolanol 45 *	0.0321	2.59×10^{-6}	3.77×10^{-4}	1.058×10^{-2}	140

* *Note:* These fluids are non-Newtonian; their effective viscosities decrease with increasing shear rate. The decrease may be as much as 50 per cent for shear rates of 5×10^5/sec.

From Eq. 4.76 we get the derivative form

$$\frac{dE}{dx} = C_v h w \, \delta x \, \frac{dT}{dx} \tag{4.78}$$

Combining this with Eq. 4.74, we finally obtain for the temperature gradient

$$\frac{dT}{dx} = \frac{2\mu U}{h^2 C_v}\left[1 + 3\left(1 - \frac{\rho_m}{\rho}\right)\right]^2 \tag{4.79}$$

We now express ρ_m/ρ as a binomial-series expansion:

$$\frac{\rho_m}{\rho} = 1 - \alpha \, \Delta T + \alpha \, \Delta T_m + \alpha^2(\Delta T \, \Delta T_m) + \cdots \tag{4.80}$$

Since α is small, ΔT moderate, and ΔT_m less than ΔT, the products $\alpha \, \Delta T$ and $\alpha \, \Delta T_m$, as well as the higher-order terms, can be neglected without introducing an error of more than a few per cent in most cases.

Accordingly, we can write

$$\frac{dT}{dx} = \frac{2\mu U}{h^2 C_v} \tag{4.81}$$

Substituting Eq. 4.77 into Eq. 4.81 and integrating, we get

$$e^{\lambda \, \Delta T} = \frac{2U\mu_1\lambda}{h^2 C_v} x + C_2 \tag{4.82}$$

But when $x = 0$, $T = T_1$ and $\Delta T = 0$, whence $C_2 = 1$.   Therefore

$$\Delta T = \frac{1}{\lambda} \ln (1 + K_2 x) \tag{4.83}$$

where

$$K_2 \overset{\Delta}{=} \frac{2U\lambda\mu_1}{h^2 C_v} \tag{4.84}$$

Differentiating Eq. 4.83, we get for the temperature gradient

$$\frac{dT}{dx} = \frac{1}{\lambda}\left(\frac{K_2}{1 + K_2 x}\right) \tag{4.85}$$

This gradient is greatest when $x = 0$; that is, at the leading edge of the slider where the fluid is the coolest and most viscous. At this point

$$\left.\frac{dT}{dx}\right|_{x=0} = \frac{K_2}{\lambda} = \frac{2U\mu_1}{h^2 C_v} \tag{4.86}$$

In order to get a feel for the problem we insert some typical values into Eq. 4.86. Suppose we use MIL-H-5606 oil at 70 °F, for which $\mu_1 = 26 \times 10^{-7}$ lb sec/in.^{2} and $C_v = 149$ in. lb/in.^{3} °F, and that $U = 100$ in./sec and $h = 10^{-3}$ in. Then $dT/dx = 3.5$ °F/in. If the clearance had been ten times smaller, or 10^{-4} in., however, the gradient would have been 350 °F/in., and the temperature rise would have been excessive in even a very short bearing.

We now wish to determine the pressure distribution. From Eqs. 4.73 and 4.81, discarding higher powers of $\alpha \, \Delta T$, we obtain

$$\frac{dp}{dx} = \frac{6\mu U\alpha}{h^2} (\Delta T_m - \Delta T) \tag{4.87}$$

Substituting Eq. 4.83 into Eq. 4.87, we get

$$\frac{dp}{dx} = -\frac{6\mu U\alpha}{h^2\lambda} [\ln (1 + K_2 x) - \ln (1 + K_2 x_m)] \tag{4.88}$$

But from Eqs. 4.77 and 4.85,

$$\mu = \mu_1 e^{-\lambda \, \Delta T} = \mu_1 e^{-\ln \, (1 + K_2 x)} = \frac{\mu_1}{1 + K_2 x} \qquad (4.89)$$

From combining Eqs. 4.88 and 4.89, we have

$$\frac{dp}{dx} = -\frac{6\mu_1 U \alpha}{h^2 \lambda} \frac{1}{(1 + K_2 x)} [\ln \, (1 + K_2 x) - \ln \, (1 + K_2 x_m)] \qquad (4.90)$$

Integrating Eq. 4.90 from x_m to x, we get

$$p = P_m - \frac{3\alpha C_v}{2\lambda^2} \left[\ln \left(\frac{1 + K_2 x}{1 + K_2 x_m} \right) \right]^2 \qquad (4.91)$$

The total upward force on a strip of width w and length L will be

$$F = w \int_0^L p \, dx = P_m w L - \frac{3\alpha C_v w}{2\lambda^2} I \qquad (4.92)$$

where

$$I = \int_0^L \left[\ln \left(\frac{1 + K_2 x}{1 + K_2 x_m} \right) \right]^2 dx \qquad (4.93)$$

The evaluation of I is straightforward but somewhat complicated. The resulting expression may take several forms. One of these is

$$F = K_3 \left[\frac{2\lambda \, \Delta T_L}{e^{\lambda \, \Delta T_L} - 1} + \lambda \, \Delta T_L - 2 \right] \qquad (4.94)$$

where

$$K_3 \stackrel{\Delta}{=} \frac{3\alpha C_v L w}{2\lambda^2} \qquad (4.95)$$

If we use the relation

$$\lambda \, \Delta T_L = \ln \, (1 + K_2 L) \qquad (4.96)$$

Eq. 4.94 can be transformed to

$$F = K_3 \left[\frac{(2 + K_2 L)}{K_2 L} \ln \, (1 + K_2 L) - 2 \right] \qquad (4.97)$$

which in some respects is more useful than Eq. 4.94 since it does not involve the temperature. For small values of K_2L a more convenient form is

$$F = \frac{K_3}{6} [(K_2L)^2 - (K_2L)^3]$$ (4.98)

This approximation introduces less than 1 per cent error for values of K_2L up to about 0.1.

From the definition of K_2 (Eq. 4.84) we obtain

$$K_2L = \frac{2\lambda\mu_1 L}{C_v} \frac{U}{h^2}$$ (4.99)

From this equation and Eq. 4.95 we see that

a. F is proportional to α and to w.

b. It is a complicated function of λ, μ, L, C_v, and U/h^2.

c. At constant load, h will be proportional to the square root of the velocity, U.

It is instructive to plot clearance, temperature rise, and K_2L against the force. These plots will be most useful if we normalize the quantities involved. We define them by the equations

$$F^* = \frac{F}{F_1} = \frac{F}{K_3}$$ (4.100)

$$h^* = \frac{1}{\sqrt{K_2L}}$$ (4.101)

and

$$\Delta T^* = \lambda \Delta T_L = \ln (1 + K_2L)$$ (4.102)

Here F_1 and h_1 are the force and clearance for which $K_2L = 1$. In Fig. 4.20, h^*, K_2L, and ΔT^* are plotted against F^*, using logarithmic scales to cover the wide range of values involved.

The logarithmic scale tends to obscure the relationship between F^* and h^*. Accordingly, this relationship is plotted in Fig. 4.21, using two linear scales. This figure shows that as the clearance decreases, the initially small sustaining force increases enormously and rather suddenly.

It is instructive to calculate the force and the temperature rise for a typical case. Suppose that we have a slider 1 in. square and moving at

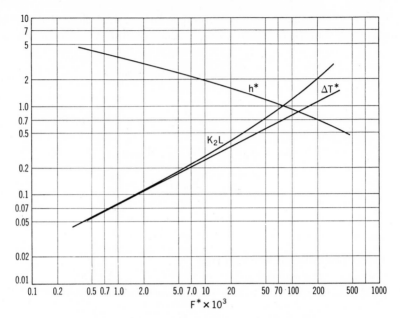

Fig. 4.20. Log-log plots of ΔT^*, K_2L, and h^* versus F^*.

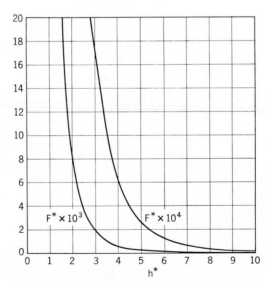

Fig. 4.21. Variation of load with clearance.

100 in./sec, lubricated with MIL-H-5606 oil at 70 °F. Then, using Table 4.1, we have

$$\mu_1 = 2.6 \times 10^{-6} \text{ lb sec/in.}^2$$
$$\lambda = 0.0104/°F$$
$$\alpha = 4.2 \times 10^{-4}/°F$$
$$C_v = 149 \text{ in. lb/°F in.}^3$$
$$L = 1 \text{ in.}$$
$$w = 1 \text{ in.}$$
$$U = 100 \text{ in./sec}$$

Using these values we find that $K_3 = 940$ lb and that $K_2L = 3.5 \times 10^{-8}/h^2$. If we assume that $h = 0.001$ in., then $K_2L = 0.035$, $\Delta T_L = 3$ °F, and $F = 0.185$ lb. If we decrease the clearance to 0.0002 in., then $K_2L = 0.875$, $\Delta T_L = 60$ °F, and $F = 61$ lb.

4.32. Experimental Verification

The theory just outlined was tested experimentally [16] using the test fixture shown in Fig. 4.22. A polished disc 4 in. in diameter and flat within 5 μin. was rotated at various controlled speeds. Bearing against the disc were three pads in the form of segments of a ring concentric with the disc, each $\frac{1}{2}$ in. in radial width and 1 in. long. The pads were attached to the three arms of a spider at a radius of 1.5 in. The spider was restrained from rotating by three piano wires attached to the spider arms and to the housing. The pads were loaded by cylindrical weights stacked on the spider stem. A fine probe extended through an axial hole in the stem and rested upon the disc. An extension of the probe carried a fine scribed line, and another was placed upon a window in the spider stem. The relative separation of these two lines gave the vertical displacement of the pads with respect to the disc. The separation was measured with a 750-power microscope furnished with a vertical illuminator and a calibrated micrometer eyepiece. The repeatability of a measurement was about ±4 μin.; the accuracy is believed to be about 10 μin.

An attempt was made to measure the difference in temperature between the leading and trailing edges of the pads. Small thermocouples were placed in pockets machined in the pads, and temperature differences were observed, but the measurements appeared to be untrustworthy because of the extreme thinness of the oil film.

The test results are plotted in Fig. 4.23. The generally good agreement between experiment and theory is gratifying and shows that the

---

[16] Reethof, Goth, and Kord, fn. 15.

(*a*) Test fixture assembled.

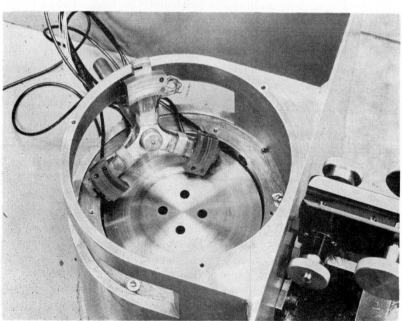

(*b*) Pads and disc.

Fig. 4.22.

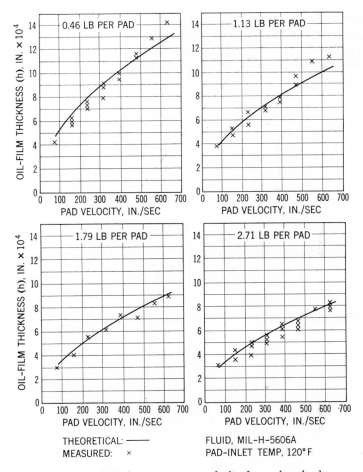

Fig. 4.23. Pad clearance versus velocity for various loads.

effects of side leakage were small. The measured clearance at lower speeds was slightly less than the calculated value. The high values measured at high speeds and light loads are believed to be due to vortex motion of the fluid, although many attempts were made to reduce this effect by shielding.

4.33. Application to Pumps

The theory outlined in Sec. 4.31 can be applied to several of the clearances of various types of hydraulic pumps and motors. Two examples will be described briefly: the valving surface of a rotating-cylinder axial-piston pump, and the rotor end surfaces of a vane or gear pump.

Consider the pump of Fig. 4.24. The cylinder block rotates against the stationary valve plate, separated from it by a thin film of the fluid being pumped. It is desired to design the bearing surfaces of the outer rim, the scallops, in such a way as to ensure a minimum safe separation

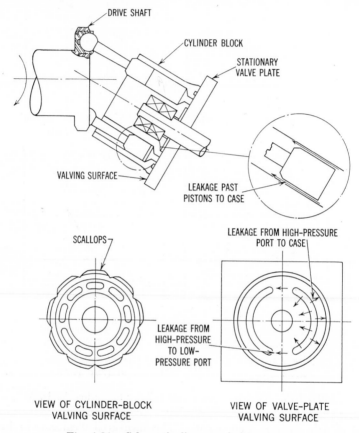

Fig. 4.24. Schematic diagram of piston pump.

between cylinder and valve plate. This minimum separation is particularly important when pumping fluids of poor lubricity, such as some of the newer synthetic hydraulic fluids, or propellants such as hydrogen peroxide, hydrazine, or ethylene oxide. If the minimum separation is not maintained, the cylinder and valve-plate faces will make contact, gall, and wear excessively.

With a reasonable allowance for side leakage, the operating clearance can be calculated from the known scallop dimensions, loading, and fluid properties, and either Eq. 4.97, or Eqs. 4.94 and 4.96, or Fig. 4.21. Temperature rise can be checked against Eq. 4.96.

The theory of Sec. 4.31 can be used to correct that given in earlier sections of this chapter. For example, since the speed does not occur in K_3, Eq. 4.94 shows that (within the accuracy of our theory) the temperature rise depends upon the loading but not on the speed. Also since h and U occur only as the ratio U/h^2, it becomes possible to correct expressions such as Eq. 4.21 for the increase of clearance with speed. If the loading, F, remains constant, as it will in the case of Fig. 4.24 so long as the input, output, and case pressures remain constant, h will be

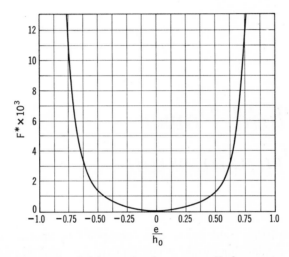

Fig. 4.25. Axial centering force versus displacement.

proportional to $\sqrt{U}$. Since the leakages from the high-pressure port to the low-pressure port and to the case are proportional to h^3, it follows that the expression for the leakage will include a term containing the $\frac{3}{2}$ power of the speed. This $\frac{3}{2}$-power variation has been observed experimentally with pumps of the construction shown.

A second example of the application of the theory of the thermal wedge is the calculation of the force that tends to center the rotor and vanes of a vane pump, or the gears of a gear pump, axially between the pressure and wear plates. Figure 4.25 gives the nondimensional centering force in terms of the displacement ratio e/h_0, where e is the axial displacement from the center position and h_0 is the clearance at each side when the rotor is centered.

It has been found experimentally that vane and gear pumps follow fairly accurately the theory which assumes constant clearance; this fact indicates strongly that the centering process just described is fairly effective.

5

F. D. Ezekiel
H. M. Paynter

Fluid-Power Transmission

In transmitting fluid power from one point to another in space, a number of challenging problems confront the user of fluid-power-control equipment. The conduit is subjected to many types of stress and strain owing to the pressure of the fluid it carries or to end conditions, intermediate mounting conditions, or simply to mechanical vibration. Many types of fittings and joints have been devised for assembling the components of fluid-power-transmission lines, yet no completely satisfactory scheme has been devised to transmit fluid power from power source to load. The mechanical problems associated with providing a dependable conduit for fluid-power transmission are not considered in detail because, if the fluid pressures and mounting conditions are known, it is usually possible to obtain a reasonable conduit design based on available strength theories and knowledge of applied mechanics. Instead, an attempt is made to understand and characterize the dynamic behavior of the fluid in the line in order to study interactions between the line and associated components, to predict pressure transients in the transmission line, to determine the effects of the dynamic characteristics of the line on the over-all performance of complete systems, and to learn which factors may tend to limit the power transmissibility of fluid lines.

5.1. BASIC CONCEPTS

From the standpoint of power and energy transfer, the various elements in any control system may be classified as follows:

1. Energy transducers, such as pumps, motors, and actuators.
2. Energy modulators, such as valves and throttles.
3. Energy transmitters, such as fluid pipes and tanks.

Figure 5.1a shows schematically the interrelationship of these elements in a simple type of hydraulic control system.

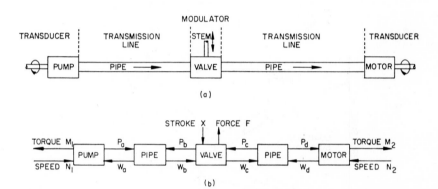

Fig. 5.1. Schematic and block-diagram representations of a hydraulic control system.

The available mechanical energy of a fluid stream at a given point can be measured in terms of the total pressure, P, as indicated by Eq. 3.35 of Chapter 3,

$$P \overset{\Delta}{=} P_s + \rho g z + \frac{\rho V^2}{2} \tag{5.1}$$

where P_s is the static pressure at the given point in the flow, ρ is the fluid density, and g is the acceleration of gravity. Because changes in elevation, z, and stream velocity, V, constitute negligible fractions of the energy content of the stream in most high-pressure control systems, usually measurement of only the static pressure in a line is needed to determine its total pressure. However, changes in elevation and stream velocity may be important in some instances, and their effects should not be overlooked.

The rate of mechanical-energy transfer or fluid-power transfer across any normal section of a fluid stream is given by

$$H = \frac{PW}{\rho g} \tag{5.2}$$

where H is the power and W is the weight rate of flow in pounds per second.

Equation 5.2 deals only with available fluid power and neglects internal-energy and heat-transfer effects. In other words, systems in

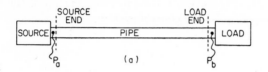

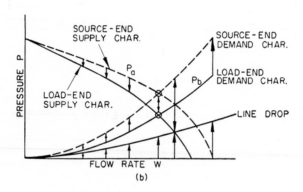

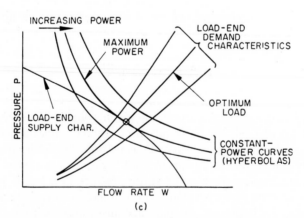

O INDICATES STEADY-STATE OPERATING POINT

Fig. 5.2. Effect of steady-state pipe losses on system characteristics.

which heat inputs or losses produce significant changes in available energy are not being considered. Fluid friction and dissipation losses are assumed to take the form of a decrease in available energy. Thus, the discussion is limited to liquid systems and to gaseous systems for which fluid-density changes are small.

Figure 5.1b shows one of several possible block diagrams for the physical system of Fig. 5.1a.

Under steady-flow conditions, the effect of a pipe can be taken into account in the way shown in Fig. 5.2. The characteristics of the general source and load system shown in Fig. 5.2a can be modified by line drop to yield a new set of characteristics that are applicable either to the load end or to the source end of the system, as shown in Fig. 5.2b. Figure 5.2c shows graphically how an ideal load characteristic can be selected to derive maximum power from a given source and a given line. However, this procedure yields design criteria covering static performance only and does not necessarily ensure satisfactory dynamic characteristics for transient operation. Satisfactory transient behavior for small changes can be obtained by using the slopes of characteristic curves and certain pipe characteristics.

A transmission system can be represented by a four-terminal element having two inputs and two outputs. Figure 5.3 shows four different configurations that can be used, depending on whether each of the four

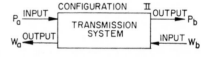

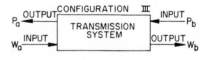

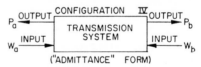

Fig. 5.3. Block diagram of fluid transmission systems.

quantities involved is to be considered an input or an output. In all cases, however, either the total pressure, P, or the flow, W, (but not both) can be considered as an input at each end of the system. These representations can be used to describe an actual transmission system for both steady-state and transient operation. The selection of a particular configuration and its contents depends on many factors, some of which are discussed and analyzed in the following sections.

5.2. LUMPED-PARAMETER TRANSMISSION LINES

As a first approximation in analyzing a transmission line from the point of view of steady-state as well as transient operation, its various parameters frequently can be lumped. For example, the fluid capacitance, inertance, and resistance in a transmission line can be assumed to be concentrated at single locations and their interactions can be assumed to be negligible; these quantities then can be considered separately.

5.21. Fluid Inertance

If the fluid is assumed incompressible and frictionless, that is, if it is assumed to have only inertia, application of Newton's second law of motion to the fluid column yields

$$\Delta P \overset{\Delta}{=} P_a - P_b = \frac{l}{Ag}\frac{dW}{dt} = I\frac{dW}{dt} \tag{5.3}$$

where P_a and P_b = the upstream and downstream pressures, respectively, $lb/in.^2$

  l = line length, in.

  A = line cross-sectional area, $in.^2$

  I = line inertance defined as l/Ag, $sec^2/in.^2$

The inertance, I, can be determined experimentally by measuring the fluid acceleration, that is, the rate of change of flow rate, in the line when a given pressure difference is slowly applied across its ends. If the change is applied too rapidly, compressibility effects cannot be neglected, and the flow rate is not uniform throughout the line.

The characterization of such an ideal line by inertance only is shown in Fig. 5.4a both in circuit-diagram and operational-block-diagram form. An example of the importance of fluid inertance is given in a paper by S.-Y. Lee [1] where "damping length" is discussed in connection with transient forces in control valves.

5.22. Fluid Capacitance

For a line in which only the fluid capacitance is considered, that is, in which the fluid density is negligible, Eq. 5.4 relates the incoming and

---

[1] S.-Y. Lee and J. F. Blackburn, "Contributions to Hydraulic Control—2, Transient Flow Forces and Valve Instability," *Trans. ASME*, Vol. 74 (1952), pp. 1013–1016.

outgoing flow rates with the time rate of change of the pressure:

$$W_a - W_b = \frac{\gamma A l}{\beta} \frac{dP}{dt} = C \frac{dP}{dt} \qquad (5.4)$$

where $\gamma = \rho g$ = specific weight of the fluid, lb/in.^{3}
$\quad\;\; \beta$ = fluid bulk modulus in the line, lb/in.^{2}
$\quad\;\; C$ = line capacitance, defined as $\gamma A l/\rho$, in.^{2}

The capacitance, C, can be determined experimentally by measuring the weight of fluid that must be introduced slowly into the line to produce

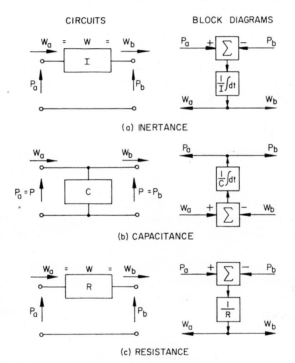

Fig. 5.4. Block diagram of fluid elements.

a unit pressure rise. If the fluid is introduced too rapidly, fluid-density effects cannot be neglected, and the pressure is not uniform throughout the line.

A characterization of capacitance, similar to that for inertance, is given in Fig. 5.4b. An analysis by Shearer [2] gives an example in which fluid capacitance only is considered in a hydraulic conduit.

[2] J. L. Shearer, "Dynamic Characteristics of Valve-Controlled Hydraulic Servo-motors," *Trans. ASME*, Vol. 76, No. 6 (August 1954), pp. 895–903.

5.23. Fluid Resistance

The effects of resistance can be lumped by establishing a hypothetical "resistance joint" where all the losses are assumed to occur. These losses manifest themselves in the form of a pressure drop in the direction of flow. Figure 5.4c shows how such a resistance element can be characterized. The exact function, R, in Fig. 5.4c that relates the flow to the pressure drop is difficult to establish for either steady-state or transient operation. However, the pressure drop usually is assumed to be proportional to the square of the flow rate. Thus

$$\Delta P = C_l \frac{\rho}{2} |V| V \qquad (5.5)$$

where the loss coefficient, C_l, is given in Table 5.1. Because the pressure drop is in the direction of fluid flow, the velocity term is written $|V| V$ instead of V^2 to allow for reverse flows.

Table 5.1.　Loss Coefficients for Fittings

Fitting	Loss Coefficient, C_l
45° elbows	0.3 to 0.6
90° elbows (standard radius)	0.7 to 1.2
Sudden expansion to infinitely large pipe	1
Sudden contraction from infinitely large pipe	0.5
Straight pipe of diameter d and length l	0.01 l/d to 0.04 l/d

The block diagrams shown in Figs. 5.4a, b, and c are by no means the only possible ones. Depending on which quantities are considered inputs and outputs, several others can be constructed. Differentiation should be avoided if lumped elements are to be used for computation.

Because the various parameters have been analyzed separately, a complete transmission-line model having lumped inertance, capacitance, and resistance now can be set up by assembling the various elements in series. Figure 5.5a shows such a hypothetical model, and Fig. 5.5b shows one corresponding four-terminal representation in the form of an operational block diagram.

The most important effect of lumping the various elements in a line is to replace a system having an infinite number of degrees of freedom with a system having only one degree of freedom. Although in many cases the error involved is not excessive, in others the model bears little resemblance to the actual line. In Figs. 5.5a and b, the sequence of the three elements was chosen arbitrarily; the resulting system may be a suitable model when a pressure input is applied at the left end and a flow

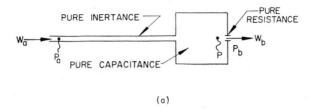

(a)

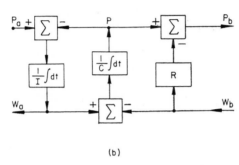

(b)

Fig. 5.5. Lumped model of a fluid conduit.

input is applied at the right end. The choice of the sequence of the elements also should be governed by the type of terminations to which the line is subjected.

5.3. DISTRIBUTED-FLUID SYSTEMS

All fluid-transmission systems extend through space and time and may be considered concentrated at one or more points in space only when the spatial variations in flow variables are negligible with respect to temporal changes. Thus, a system may be characterized by means of lumped models whenever the significant wave lengths of all variables are large compared with the physical dimensions of the system. Otherwise, the actual distributed nature of the system may produce appreciable effects not present in the lumped model. Nevertheless, the lumping approximations frequently prove to be very useful and convenient in the analysis of relatively short lines.

5.31. Lossless Distributed-Parameter Lines

In Chapter 3, which deals with the fundamentals of fluid flow, four time-difference equations (Eqs. 3.97 through 3.100) relating the flow

and the pressure in a distributed-parameter line are derived. By adding
Eqs. 3.97 and 3.98 and subtracting Eq. 3.100 from Eq. 3.99, we obtain
a new pair of equations:

$$P_b + Z_sQ_b = P_a(t - T_e) + Z_sQ_a(t - T_e) \qquad (5.6)$$

$$P_a - Z_sQ_a = P_b(t - T_e) - Z_sQ_b(t - T_e) \qquad (5.7)$$

where Z_s is the line surge impedance, Q_a and Q_b are volumetric rates of
flow, t is time, and T_e is the wave travel time along the pipe. These
two equations now can be represented by suitable block diagrams in
terms of four different configurations, as shown in Fig. 5.6. In these

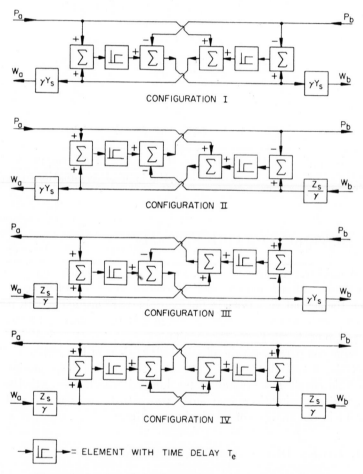

Fig. 5.6. Models of distributed-parameter lossless conduits

block diagrams, the weight rate of flow, W, is used instead of the volumetric rate, Q, with appropriate modifications of the impedance, Z_s, and the admittance, Y_s. The general representation of a line with distributed parameters is useful, especially when computers are used to solve a complicated problem in which the line plays only a small part. Of course, the analytical solution is available in terms of hyperbolic functions of the pressure-flow relationship in a line with distributed parameters (see Eqs. 3.93 through 3.96 of Chapter 3). For very simple configurations, the analytical solution may be preferred.

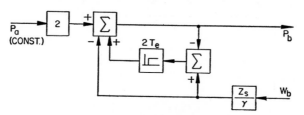

Fig. 5.7. Special case of a lossless conduit with constant upstream pressure.

The configurations shown in Fig. 5.6 can be simplified if either the pressure or the flow at one end is held constant.[3] For example, when P_a is constant and the magnitude of Q_a is not required, configuration II simplifies to that shown in Fig. 5.7, where only one time-delay element of duration $2T_e$ is used instead of two separate ones having T_e each. For a composite line, that is, a line made up of two sections having different diameters and different lengths, two similar representations of either configurations II or III can be connected together. The final representation takes into account all reflected waves, even though this feature cannot be seen readily from the block diagram.

5.32. Distributed-Parameter Lines with Resistance

As yet, no exact, simple block-diagram representation of a line having distributed inertance, capacitance, and resistance is available. However, when frictional effects must be included, they may be lumped at either end, or both ends, of any of the configurations shown in Fig. 5.6. For many engineering applications, such a procedure is adequate.

[3] F. D. Ezekiel, *Effect of a Hydraulic Conduit with Distributed Parameters on Control-Valve Stability*, Sc. D. Thesis, Department of Mechanical Engineering, Massachusetts Institute of Technology, 1955.

5.4. FREQUENCY-RESPONSE CHARACTERISTICS

The formulas in the preceding sections have immediate application to the determination of the transfer characteristics of a fluid pipeline subjected to sinusoidal inputs. For example, for a lossless line, the

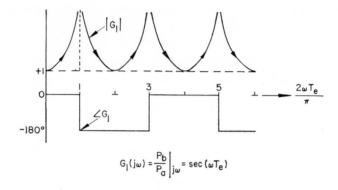

$$G_1(j\omega) = \frac{P_b}{P_a}\bigg|_{j\omega} = \sec(\omega T_e)$$

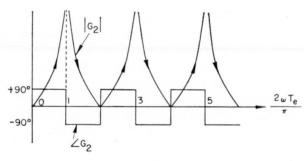

$$G_2(j\omega) = \frac{P_b}{Z_s W_b}\bigg|_{j\omega} = -j\tan(\omega T_e)$$

Fig. 5.8. Frequency response of a lossless conduit.

situation depicted as configuration II in Fig. 5.6 is governed by the equation

$$P_b(t) = [\operatorname{sech} T_e D]P_a(t) - [Z_s \tanh T_e D]W_b(t) \tag{5.8}$$

where D is the differential operator d/dt. In terms of sinusoidal inputs of angular frequency, ω, represented by $P_a(j\omega)$ and $W_b(j\omega)$, the response $P_b(j\omega)$ can be determined simply by substituting $D = j\omega$ to obtain

$$P_b(j\omega) = \sec(\omega T_e)P_a(j\omega) - j\tan(\omega T_e)Z_s W_b(j\omega) \tag{5.9}$$

This equation may be written

$$P_b(j\omega) = G_1(j\omega)P_a(j\omega) + G_2(j\omega)Z_sW_b(j\omega) \qquad (5.10)$$

where the functions G_1 and G_2 are, respectively,

$$G_1(j\omega) \equiv \left.\frac{P_b(j\omega)}{P_a(j\omega)}\right|_{W_b(j\omega)=0}$$

and

$$G_2 \equiv \left.\frac{P_b(j\omega)}{Z_sW_b(j\omega)}\right|_{P_a(j\omega)=0}$$

These transfer ratios are plotted in Fig. 5.8, where the resonances at all odd harmonics are identical with the acoustic resonances in organ pipes.

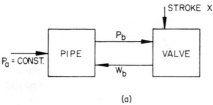

(a)

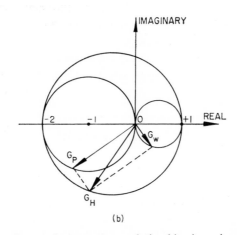

(b)

Fig. 5.9. Pressure, flow, and power phase relationships in a simple conduit-valve system.

The significance of these relations is illustrated by an example where the pipe described in Fig. 5.9, which is supplied from a constant-pressure source (P_a = constant), is terminated by a valve assumed to be governed by the relation

$$W_b = KX\sqrt{P_b} \qquad (5.11)$$

where K is a constant and X is the valve stroke. Because the downstream power transfer is

$$H = \frac{W_b P_b}{\rho g} \tag{5.12}$$

small incremental changes result in

$$\frac{\Delta W_b}{W_i} = \frac{\Delta X}{X_i} + \frac{1}{2}\frac{\Delta P_b}{P_i} \tag{5.13}$$

and

$$\frac{\Delta H_b}{H_i} = \frac{\Delta W_b}{W_i} + \frac{\Delta P_b}{P_i} \tag{5.14}$$

where the subscript i refers to the initial steady-state condition. By combining Eqs. 5.10, 5.13, and 5.14, frequency-dependent transfer ratios can be determined for variations in power, H_b; flow, W_b; and pressure, P_b, due to sinusoidal changes in the valve stroke, X. These may be defined in the form

$$G_H \stackrel{\Delta}{=} \frac{\Delta H_b/H_i}{\Delta X/X_i} \qquad G_W \stackrel{\Delta}{=} \frac{\Delta W_b/W_i}{\Delta X/X_i} \qquad G_P \stackrel{\Delta}{=} \frac{\Delta P_b/P_i}{\Delta X/X_i}$$

and are plotted as polar frequency loci in Fig. 5.9b. This graph shows that, even in the simplest situation, the power transfer, H_b, varies widely as the frequency changes. Moreover, there are frequencies at which the power transfer is in direct phase opposition to the valve motion, X. These frequencies are precisely the resonant frequencies indicated in Fig. 5.8.

NOMENCLATURE

Symbol *	Definition	Unit
A	line cross-sectional area	in.^{2}
C	line capacitance	in.^{2}
C_l	loss coefficient	
D	differential operator (d/dt)	sec^{-1}
d	pipe diameter	in.
G	function	
g	acceleration of gravity	in./sec^2
H	power	in. lb/sec
I	line inertance	sec^2/in.^{2}
K	constant	
l	conduit length	in.

Symbol *	Definition	Unit
P	total pressure	psia
P_s	static pressure	psia
Q	volumetric rate of flow	in.³/sec
T_e	wave travel time along a pipe	sec
t	time	sec
V	fluid velocity	in./sec
W	weight rate of flow	lb/sec
X	valve stroke	in.
Y_s	line surge admittance	in.²/sec
Z_s	line surge impedance	sec/in.²
z	height above a datum level	in.
β	fluid bulk modulus	lb/in.²
γ	fluid specific weight	lb/in.³
ρ	fluid density	lb sec²/in.⁴
ω	frequency	radians/sec

* Subscripts a, b, and i denote upstream, downstream, and initial steady state, respectively.

ADDITIONAL REFERENCES

Evangelisti, G., *La Regolazione delle turbine idrauliche*, N. Zanichelli, ed., Societá Tipografica Editrice Bolognese, **Bologna**, Italy, 1947, Chapter 4.

Ezekiel, F. D., and H. M. Paynter, "Computer Representations of Engineering Systems Involving Fluid Transients," *Trans. ASME*, Vol. 79, No. 8 (Nov. 1957), pp. 1840–1850.

Fredriksen, E., *Pulsating Air Flow in Pipe Systems*, I. Kommission Hos Teknisk Forlag, Copenhagen, 1954.

Parmakian, J., *Waterhammer Analysis*, Prentice-Hall, New York, 1955.

Paynter, H. M., "Electrical Analogies and Electronic Computers: Surge and Water-Hammer Problems," *Trans. Am. Soc. Civil Engrs.*, Vol. 118 (1953), pp. 962–1009.

Soroka, W. W., *Analog Methods in Computation and Simulation*, McGraw-Hill Book Co., New York, 1954.

6

J. F. Blackburn

On Valve Control

6.1. INTRODUCTION

So far in this book we have discussed the fundamentals of flow, the characteristics of fluids, and the methods of transforming mechanical to fluid power and vice versa and of transmitting fluid power from one place to another. We have also mentioned very briefly (and shall discuss further in Chapter 15) the control of fluid power by controlling the rate at which it is generated, but have neglected the method of control which involves placing a restriction—a control valve—in the fluid transmission line. In practice, particularly in recent years, the latter method has been used much more widely than the former, and the following eight chapters will be devoted to it. Before considering the details of valves, however, we must define what they are called upon to do. This chapter is an attempt at such a definition. It is very easy to give a general statement of the task of a fluid drive; it is to cause a given load to move in a prescribed fashion. The difficulty arises in the giving of the load and the prescribing of the fashion.

6.2. DESCRIPTION OF THE LOAD

Actual mechanical loads, as encountered in practice, may have widely varying characteristics, and it is not always obvious how to describe them most usefully. Probably the best description in most cases is to specify the required force in terms of the position, velocity, and acceleration of the load itself. This description may be in analytical form, but it is often very useful, especially for purposes of visualization, to use graphical representation also.

Omitting for the moment those components of the force which depend upon position or acceleration, and therefore upon the storage of energy, the dissipative portion of the load may be written

$$F_d = F_s + F_c + F_v + F_w \qquad (6.1)$$

where F_d = total "frictional" load

F_s = "stiction" = force required to break the load loose and get it moving

F_c = Coulomb friction, which is independent of velocity

F_v = viscous friction, proportional to velocity

F_w = windage, proportional to (velocity)^2

There may be other terms, such as higher-order velocity effects, but the four types given are the commonest ones.

Fortunately, stiction, by its very nature, is effective only at zero velocity, or at least only at "very small" velocities, and therefore may be omitted from consideration as soon as the load starts moving. With this omission, the friction becomes

$$F_d = F_c + b_1 v + b_2 v^2 \qquad (6.2)$$

where the values of F_c and the b's depend upon the particular load.

In one actual case—a fairly high-speed rotating antenna—the numbers were

$$F_s = 175 \text{ in. lb}$$

$$F_c = 105 \text{ in. lb}$$

$$b_1 = 12.25 \text{ in. lb/rps}$$

$$b_2 = 9 \text{ in. lb/(rps)}^2 \text{ in still air}$$

and, for completeness, the moment of inertia, $J = 450$ in. lb sec^2. In this example, the load rotated instead of moving linearly; the translation from linear to rotary motion should be obvious.

The "constants" are often far from constant, and the designer must often include a margin of safety to take care of this variation. Thus, stiction is very erratic and unpredictable. Coulomb friction is usually less erratic than stiction but is not accurately predictable. Viscous friction is usually fairly consistent but may vary greatly with temperature if it is caused primarily by shear of a liquid in narrow clearance spaces. Quasi-viscous friction caused by grease in bearings may be relatively little affected by temperature changes; for example, in a particular optical tracker, the train bearing used several hundred small-diameter rollers on a wide training ring and was lubricated with a light grease. In spite of the large mass of the mount (about 8 tons) the inertia force was negligible and the load looked like almost pure viscous friction up to at least 10 cycles per second. The coefficient of viscous friction changed little with temperature; measured torques were nearly the same

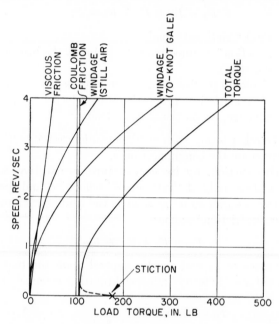

Fig. 6.1. Load curves for Example 1—rotating antenna.

in winter as in summer, with a temperature change at the bearing of about 150°F. Again, windage is affected by the velocity of the wind past the rotating object; in the case of the antenna previously mentioned the windage coefficient was about 9 in still air but doubled in a 70-knot gale. In this case, the speed of rotation was such that the windage was by far the largest term.

The behavior of the frictional part of a load can be most easily visualized by plotting the torque (or force) against the rotational (or translational) velocity. In Figs. 6.1 and 6.2, this has been done for the two cases just mentioned. It can be seen that for the antenna the windage is the important term even in still air, that Coulomb friction and stiction are appreciable, and that viscous friction is rather small. For the tracker before modification the stiction was very large, Coulomb friction moder-

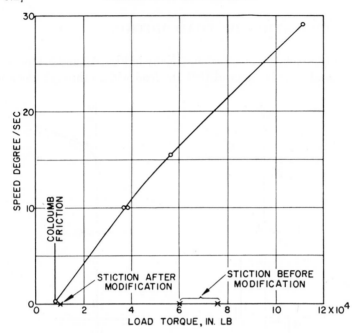

Fig. 6.2. Load curves for Example 2—optical tracking instrument.

ate, windage negligible, and viscous friction predominant. For the antenna the stiction was unimportant since there was no requirement for operation at very low velocities. For the tracker this was the all-important feature and the very high stiction caused by the original weather seal would have seriously degraded the performance. Replacement of this seal by a light felt dust ring reduced the stiction to a tolerable amount.

The use of diagrams such as those of Figs. 6.1 and 6.2 is of considerable value as an aid to visualization of the load, and as will be seen later, the same representation can help greatly in designing the drive.

The two examples just given represent probably the simplest types of loads from the standpoint of the drive designer, since the required accelerations were low and offered no difficulty. In many cases, however, this is not true. As soon as energy-storing elements—masses or springs—are introduced, the system becomes much more complicated and the design of the drive more difficult. This difficulty arises in part from the fact that the force required at a given instant depends no longer merely upon what the system is now doing, but also upon what it has been doing and what it must do in the future. It therefore becomes necessary to prescribe the desired motion in considerable detail.

6.3. SPECIFICATION OF LOAD MOTION

Load-locus plots are given schematically in Figs. 6.3 and 6.4 for several types of loads. It is assumed that the load will go through a complete

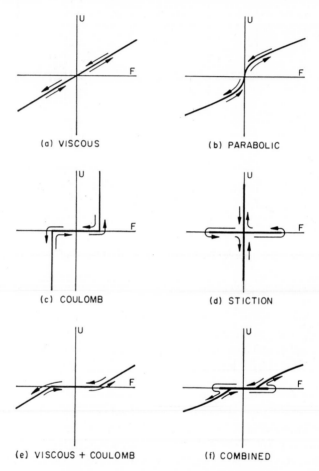

Fig. 6.3. Load loci—no energy storage.

cycle of accelerating from zero to a maximum positive velocity, then back through zero to maximum negative, then back to standstill, as indicated by the small arrows. As is the case throughout these chapters, velocity (or its equivalent) is made the ordinate and force the abscissa.

In Fig. 6.3, as for the previous examples, there is no energy storage. Figure 6.3a is the designer's ideal, pure viscous friction. Figure 6.3b is the parabolic plot of pure windage or square-law friction. Figure 6.3*c* represents pure Coulomb friction; the force is independent of the velocity and reverses direction when the velocity goes through zero. Figure 6.3*d* represents the hypothetical case of pure stiction; the force is indeterminate between two limits so long as the velocity is zero but drops to zero as soon as there is a finite velocity. The last two sketches rep-

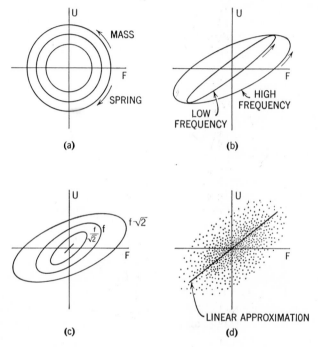

Fig. 6.4. Load loci—energy storage present.

resent somewhat more realistic cases where two or more types of friction are simultaneously active.

Pure energy storage results in a closed loop with equal areas in the four quadrants, as shown in Fig. 6.4*a*. If the motion is sinusoidal and the elements are linear, the locus is a circle for some particular frequency which depends upon the ratio of the scales of ordinate and abscissa. (If the amplitude is kept constant, the velocity will vary as ω and the force as ω^2, where ω is the angular frequency in radians per second, so the circle will become elliptical as the frequency is changed. This effect can be eliminated by using U/ω and F/ω^2 as ordinate and abscissa.) For a mass load the load point traverses the circle counterclockwise; for a spring it goes clockwise.

If we combine mass (or spring) with viscous friction, we have a case analogous to the combination of pure inductance and pure resistance in the electrical example, and the figure is again elliptical. As the frequency increases, the effect of the mass becomes relatively greater and the ellipse becomes fatter; Fig. 6.4b shows the effect of varying the frequency at constant peak velocity, and Fig. 6.4c the effect with the amplitude kept constant.

In actual practice the motion of the load is often incompletely specified, sometimes very much so, and the designer must guess what the eventual requirements will be. As an aid to determining whether the requirements have been met, it is necessary to predict what the response of the system will be to hypothetical input signals. The techniques of making this prediction are discussed in Chapter 14, and only some general remarks need be given here.

6.31. Types of Inputs

The servomechanism art and science originally received its principal impetus from communication engineers, for whom it was natural to use the familiar method of frequency analysis. This method makes the implicit assumption that the load will be forced to move sinusoidally, at a varying frequency but with an amplitude sufficiently small so that the system can be assumed to be linear. This is a poor representation of the actual state of affairs in many practical cases, but it does permit the use of the numerous and powerful tools that have been developed by the communications engineer. In the past, most servos have been designed by these methods, and on the whole they have worked pretty well.

A somewhat more recent technique assumes that the input remains constant at one value until at a particular instant it jumps suddenly to a second value and remains constant thereafter. The output attempts to follow the input, but since in a physical system it cannot do so exactly, the designer tries to make it go from the initial to the final state as rapidly and as smoothly as possible. For linear systems, it can be shown that the second method is exactly equivalent to the first, and nowadays most servo engineers are trained to think in both systems and to translate easily from the language of one to that of the other. For nonlinear systems, the equivalence is not demonstrable, and most of the tools of both methods become inapplicable. For this case about all that could be done in the past was to pretend that the system was linear or nearly so, and to hope that the errors introduced by this assumption would not be too great. In many cases they were not; even such violently nonlinear systems as contactor-type servos can be successfully designed by linear methods.[1,2] In many important cases, however, the linear approximation simply is not good enough.

One method which makes no linearizing assumptions and which offers considerable promise is the use of random signals or "white noise" as a test input, rather than the sinusoids or step functions of the older methods.[3,4] It is still too new to permit complete evaluation, but it has worked very well so far.

With a random input, or equally with a poorly defined one, the load locus becomes indeterminate and degenerates into a patch of fog without sharp edges, somewhat like the picture of an elementary particle so beloved by certain current "popularizers" of theoretical physics. This fog patch is symbolized by Fig. 6.4d, and at first sight would seem to have no features useful to the designer.

This is not necessarily true, however, simply because if a mechanism can be made which is capable of operation at all points on the boundary of some area in the U-F plane, it will also be capable of operation at all points within that boundary unless excluded from parts of the interior by additional constraints. Since this is so, we need worry only about the boundary, and it is usually possible to establish performance limits and to design to them. The three most useful maxima are the peak horsepower, the runaway speed, and the stalled torque (or force). Usually the power requirement is the most difficult to meet and therefore the most important.

[1] R. J. Kochenburger, "A Frequency Response Method for Analyzing and Synthesizing Contactor Servomechanisms," *Trans. AIEE*, Vol. 69 (1950), Part I, pp. 270–284.

[2] For other applications of linear methods to nonlinear systems, see:

E. C. Johnson, "Sinusoidal Analysis of Feedback-Control Systems Containing Nonlinear Elements," *Trans. AIEE*, Vol. 71 (1952), Part II, Applications and Industry, pp. 169–181.

E. I. Reeves, "Analysis of the Effects of Nonlinearity in a Valve-Controlled Hydraulic Drive," *Trans. ASME*, Vol. 79, No. 2 (February 1957), pp. 427–432.

J. G. Truxal, *Automatic Feedback Control System Synthesis*, McGraw-Hill Book Co., New York, 1955, Chapter 10, "Nonlinear Systems and Describing-function Analysis" and Chapter 11, "Phase-plane Analysis."

[3] W. W. Seifert, "Experimental Evaluation of Control Systems by Random-Signal Measurements," *Convention Record of the IRE 1953 National Convention*, Part 1—Radar and Telemetry, pp. 94–98.

[4] R. C. Booton, Jr., "An Optimization Theory for Time-Varying Linear Systems with Nonstationary Statistical Inputs," *Proc. IRE*, Vol. 40 (August 1952), pp. 977–981.

6.4. TYPES OF DRIVES

It is essential to point out at this point that throughout this book the term "drive" implies the controllability and variability at will of the position, velocity, or acceleration of the load; constant-speed drives are not considered. With this limitation of the field of interest, nearly all drives will fall into one or another of the following classes:

TYPES OF DRIVES

Mechanical
Clutch-controlled: dog clutches, friction clutches, fluid or powder clutches
Variable-speed friction drives: wheel-and-disc, and so on
Electrical
Direct-current
Rheostatic control: ohmic rheostats, high-vacuum series tubes, thyratrons, magnetic amplifiers plus rectifiers
Generative control: Ward-Leonard drives, Amplidynes, and so on
Alternating-current
Rheostatic control, as with direct current
2-phase motor with controlled excitation of one phase
Variable-frequency excitation
Hydrokinetic
Fluid drive, torque converter, pump plus turbine
Positive-displacement (either hydraulic or pneumatic)
Generative control using variable-displacement pump
Rheostatic control using control valve

The available space does not permit an adequate discussion of the relative merits of the various types of drives, but a few comments may be made.

In general, except for essentially on-off control, as in the automobile, mechanical drives have been little used. Positively acting clutches such as the dog clutch can usually be engaged only at very low speeds because of their high accelerations. Ordinary friction clutches of various types, particularly air-loaded clutches, are occasionally useful, especially at high power levels, when the number of engagements is small; and the clutch servo probably has the highest potential torque-to-inertia ratio of all types of drives, but the relatively short life and the highly variable and generally undependable nature of the friction process make the clutch servo unattractive. The magnetic-fluid or magnetic-powder clutch is smoother and more controllable than the ordinary type, but it has a short life and very high quiescent losses since the ratio of maximum to minimum drag is comparatively small.

Variable-speed friction drives are very useful in certain instrument applications and can be made to have accurately repeatable performance, but their poor load-carrying characteristics and their susceptibility to damage by overload eliminates them from the power-drive field. Variable-ratio belt drives might be considered a special variety of this class; they are very useful as speed-setting devices but their controllability is rather poor, especially with respect to the ratio of maximum to minimum speed.

The general availability and the flexibility of application of electrical power have produced a host of variable-speed electrical drives. Among the direct-current drives we may distinguish two main classes, rheostatic and generative control. Rheostatic control is accomplished by placing in series with the line and the armature or field (or sometimes both) of the motor a variable resistance element. This element may be a mechanically operated ohmic rheostat, or it may be a triode or other vacuum tube, the motor current passing through the plate circuit and the control being accomplished by varying the grid voltage. This gives very good control but is expensive and comparatively inefficient because of its use of large high-vacuum tubes. These two disadvantages may be mitigated at some cost of controllability by using thyratrons or magnetic amplifiers plus rectifiers.

Generative control of direct-current motors may be accomplished by the use of the Ward-Leonard system or any of its various relatives such as the Amplidyne. These systems give excellent control but are expensive and comparatively slow to respond to input signals. A system that might be considered either rheostatic or generative is the use of a continuously-variable-ratio transformer plus a rectifier as a source of control current. This is limited by the transformer to a few tens of kilowatts output, but within its limits it is very useful. A line of such devices in various ratings is sold by the General Radio Company under the name of Variac Motor Speed Controls.

A-c motors may be controlled by most of the methods used for d-c motors, plus a few others. Rheostatic control may be accomplished by a series resistor, or the control may be placed in series with the rotor windings of a wound-rotor motor, in which case it forms a fairly effective limited-range torque control. The speed of an a-c motor may be varied by varying the frequency of the current which supplies it; this scheme allows accurate speed control but is very expensive because of the high cost of the variable-frequency generator. One of the commonest and most useful a-c motor controls is the system whereby one phase of a two-phase motor is supplied with exciting current from the line, and the current for the other phase comes from a control device, such as a servoamplifier.

This scheme is very expensive in the higher power ratings—though this situation may be changing with the more general availability of magnetic amplifiers—but is extremely useful for instrument servos.

One of the principal limitations of the electrical drive is the limited torque of any electromagnetic device. This limits the torque-to-inertia ratio and therefore the available acceleration and the speed of response. A similar limitation applies to the hydrokinetic drive, wherein a fluid, usually a liquid, is driven at high velocity but with comparatively small change in pressure against the blades of a turbine whose shaft is the output member of the drive. This device has found very wide application as the fluid drive or torque converter of the modern American automobile but has been comparatively little used in other fields. As an automobile transmission it is not usually a control device, since there is nothing that corresponds to the input shaft of a rheostat, for example, but in some industrial applications the effective torque can be varied by varying the amount of the circulating liquid. The hydrokinetic drive appears to be susceptible of considerable development, but in its present form it is suitable only for special applications.

The final class of drive, with which the rest of the book will be concerned, is the positive-pressure mechanism. It consists of a positive-displacement motor, either rotary or linear, driven from a source of fluid and either with or without a rheostatic control in the supply line. If the rheostat is absent, the control must be generative, by varying the displacement of the fluid-supply pump. This type of control is discussed in Chapter 15, and will be ignored in most of the rest of the book, though it has many excellent characteristics and has been much used. Its principal disadvantages are somewhat higher cost and relatively slow response. Rheostatically-controlled fluid-power systems depend upon some kind of control valve to vary the fluid supply to the motor, and most of the book will be devoted to the valve and its applications.

6.5. VALVE-CONTROLLED DRIVES

In the past, valve-controlled drives have not been used as much as they might have been, both because most performance specifications could be met with other types of drives and because the essential system element, the valve, was very generally not understood or misunderstood. Demands for higher performance have increased rapidly over the last few years, and valves are now better understood than they were. This chapter and the seven which follow are intended to make this understanding more general and to furnish data which will be of value to the designer of valve-controlled systems.

There are two main classes of valve-controlled systems, depending on the nature of the fluid supply. Probably the commonest is operation at constant pressure, though in practice the supply pressure usually fluctuates somewhat. In such a system the input pressure to the control valve is kept constant; this type of operation is analogous to the customary constant-voltage operation of most electrical power-consuming devices. The other class operates with a constant rate of flow from the source; therefore the pressure at any instant depends upon the conditions of operation at that time. This second class is much less popular than the first, both because it is much less amenable to the operation of several independent drives from the same source (analogous to a constant-current-series street-lighting system) and because with conventional valves operation is much less linear than constant-pressure operation. Nevertheless, there are certain applications for which constant-flow operation has advantages, as will be seen in Sec. 7.7.

The principal fields of application of valve control are where high speeds of response or high power amplifications are required, and where the combination of a number of subsystems into an over-all system makes operation from a common oil supply attractive. Where extremely high speeds of response are needed, the valve-controlled system has no real competitor at present, though perhaps some day the clutch servo may take over. It is fairly easy to get an unloaded frequency of 50 cps from a hydraulic servo rated at a few horsepower, and 100 cps or higher is not uncommon in practice.

The power efficiency of the valve-controlled system is considerably less than that of the hydraulic transmission (variable-displacement pump plus motor). Numerical comparisons are impossible unless the conditions of operation, including the duty cycle, are completely specified, but the theoretical maximum efficiency at maximum power output is 66.7 per cent for any valve that acts effectively like a series orifice. If there is also a shunt path, the efficiency is far less. As with any device that is not running at full load, the light-load efficiency falls off badly. It is possible to increase the average efficiency by operating with comparatively small pressure drops across the valve, but controllability suffers badly and operation becomes very nonlinear.

The over-all efficiency of the system depends not only on the load and the duty cycle, but also on the nature of the power supply.[5] It is worst when the supply is a constant-displacement pump running at a constant speed high enough to take care of the peak flow, and dumping the excess fluid through a relief valve. Here the power input is constant and nearly all of it goes into heating up the oil as it goes through the relief valve.

[5] R. Hadekel, *Hydraulic Systems and Equipment*, Cambridge University Press, 1954, Chapter 4, "The Idling Problem."

If the highest peaks are of short duration and if some imperfection in pressure regulation can be tolerated, a hydraulic accumulator can be added to the system to store up enough energy to take care of these short peaks. The pump can then run considerably slower, with a much smaller waste of power. An accumulator is useful to smooth out pressure pulsations anyway, and it often makes the system operate much more smoothly and quietly than it would otherwise, though under certain conditions it may make the system unstable, as shown in Chapter 17.

If the system only has to do work occasionally and is idle or nearly so between the demand peaks, it is usually possible to store enough energy

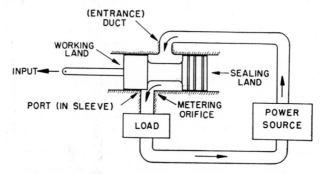

Fig. 6.5.   Two-way valve with power source and load.

in a large accumulator to permit the pump to be turned off between demands, or at least to be unloaded so that it runs idle and does not have to pump against the system pressure. In some cases, where the total amount of energy required will be small, the pump can be dispensed with and the system run from an accumulator alone, or from one pressurized by an air bottle or by a pyrotechnic cartridge. This one-shot type of operation is usually desirable only for a military device such as a guided missile. Finally, especially for large systems, fairly economical operation can be obtained from a variable-displacement pump which is stroked by a spring-loaded piston or equivalent device so that the pump output is made dependent upon the demand of the system for oil, an accumulator being used to supply the fast peak loads until the pump can catch up with them.

It is even more difficult to estimate the efficiency of a constant-flow system than that of a constant-pressure one. The former system can, however, be operated in such a way that the standby loss is small because the standby pressure is low; the over-all pressure is then allowed to rise during the peak demand periods. This variable-pressure operation is inherently very nonlinear but can be made linear within limits by modi-

fication of the valve design. This point will be discussed in Sec. 7.7.
One attractive feature of this type of system is that it permits the design
of an extremely simple and inexpensive servo of fair efficiency and
medium performance. Such a "poor man's hydraulic servo" has been
needed for a long time.

The motion of a load can be controlled by many different arrangements of valves and motors or rams.[6-8] A typical schematic diagram of

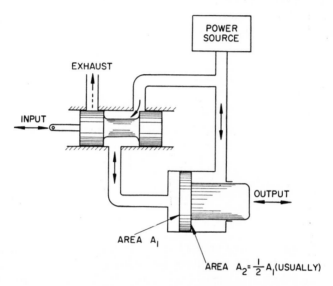

Fig. 6.6. Three-way valve and double-area ram.

a two-way valve with a power source and a load is shown in Fig. 6.5.
The valve introduces into the line a restriction which uses up part of
the pressure available from the power source, and by varying the degree
of this restriction the amount of the power flowing to the load can be
controlled.

The flow of power from source to load can be controlled from a maximum down to zero, but the direction of motion of the load cannot be
reversed so long as only the single flow-restricting orifice of a two-way
valve is used. The simplest valve that permits reversing the load is the
three-way arrangement, shown in Fig. 6.6. Here a double-acting ram

[6] R. Hadekel, *Hydraulic Systems and Equipment,* Cambridge University Press,
1954, Chapters 3, 5, and 11.

[7] Ian McNeil, *Hydraulic Operation and Control of Machines,* Ronald Press, New
York, 1955, Chapter 4.

[8] Hugh Conway, *Fluid Pressure Mechanisms,* Pitman, London, 1949, Chapters
11–17.

is used to drive the mechanical load; the right face of the piston has an effective area of only half that of the left face and is subjected at all times to the full supply pressure. Moving the valve spool to the right connects the left end of the cylinder to the supply pressure also, and the piston moves to the right against the load because of the larger area of its left face. If the valve is moved to the left of center, fluid is allowed to escape to the return line and the piston is driven to the left by the supply pressure acting on its right face. Naturally any mechanically equivalent method of getting a constant force can be used in place of the right half of the cylinder; one very common one is the use of the

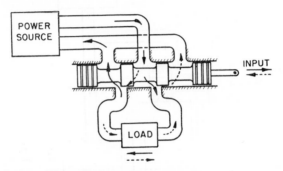

Fig. 6.7.   Four-way valve.

gravity return of a heavy press platen or jack, where the load is driven up by the pump but down by its own weight.

The principal disadvantage of the three-way arrangement just shown is the fact that it can be used with a rotary motor only by some clumsy and uneconomical expedient such as the use of a second reverse-connected motor of half the effective torque. There are also two minor disadvantages compared with the conventional four-way valve: the differential ram is bulkier and heavier than the equivalent single-area ram, and under some conditions the stiffness of the differential ram as seen by the load is less. These last two objections are relatively unimportant; there are many high-performance servos in operation which are based on the three-way valve. The principal advantage of the three-way over the four-way valve is that it has only one critical axial dimension instead of three, but the hole-and-plug construction described in Sec. 9.54 offers a means of getting accurate dimensions with such ease that this argument has lost much of its force. In high-performance systems it is usually important that each of the system elements be as linear as possible. To obtain linearity in a valve it is important that the advantages of symmetry be exploited to the utmost, and this requires the use of the four-way valve.

The commonest solution to the problem of controlling a reversible load is the use of a four-way valve with four metering orifices, as shown in Fig. 6.7. In most respects this is the best method; its one major disadvantage is the fact that in its usual form it has three critical axial dimensions on the spool and three on the sleeve. With modern machining and gaging methods these dimensions can be held to ±0.001 in. without great difficulty, but for short-stroke valves used in critical applications, the maximum permissible tolerances are perhaps ten times

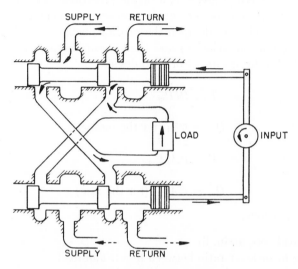

Fig. 6.8. Split four-way valve.

smaller, and the fabrication of satisfactory valves in production becomes a difficult and expensive task.

One way to decrease the number of critical dimensions is to replace a four-way valve with two three-way valves, mechanically ganged but with the lengths of the connecting links adjustable. Although no linkage arrangement can ever be as stable dimensionally as a single-piece valve spool, in practice this "split-piston" construction is very satisfactory if properly designed and adjusted. It would be possible to eliminate all critical axial dimensions by ganging four two-way valves in a similar fashion, but it is doubtful if the benefits would be worth the added complication.

One very satisfactory construction from the point of view of stability and the reduction of the effects of both steady-state and transient flow forces is shown in Fig. 6.8. The two half-pistons are hung from opposite ends of the driver armature, and the ports are arranged so that flow forces need not be transferred around the linkage from one piston to the other.[9]

6.6. TYPICAL SIMPLE DRIVES

The remainder of this chapter will be devoted to the calculation of the size of the motor (or ram) and the gear (or lever) ratio which will be required to move a given load in the prescribed fashion, and of the size of the valve and power supply necessary to control that motion. The basic method depends upon the relations between torque (or force) and speed (or linear velocity) and the pressure across or the flow rate through the motor or ram. For ideal devices,

$$T = R\delta P_m \tag{6.3}$$

and

$$Q_m = R\delta n \text{ (for the motor)} \tag{6.4}$$

or

$$F = RA_p P_m \tag{6.5}$$

and

$$Q_m = RA_p U \text{ (for the ram)} \tag{6.6}$$

In these equations,

T = load torque, in. lb
R = gear or lever ratio between motor and load
δ = motor displacement *per radian*, in.^{3}
P_m = pressure drop across motor or ram, psi
Q_m = rate of flow through motor or ram, in.^{3}/sec
n = motor shaft speed, radians/sec
F = load force, lb
A_p = effective piston area, in.^{2}
U = load velocity, in./sec

Of course motors are not ideal and rams are still less so. Their deviations from ideality can be allowed for by adding their friction to the load force or torque, and their leakage flow and probably also their compressibility flow (charging flow) to the load flow. If the compressibility flow is large, as it is for gases, its effect is large and the whole

---

[9] An experimental valve of this type was briefly described in S.-Y. Lee and J. F. Blackburn, "Transient-flow Forces and Valve Stability," *Trans. ASME*, Vol. 74 (August 1952), pp. 1013–1016. Similar valves, both with and without flow-force compensation, are available commercially. See also Sec. 12.42.

problem becomes considerably more complicated. Pneumatic-drive design is discussed in Chapters 8, 16, 17, 19, and 20.

The four equations just quoted permit the transformation from T-n (or F-U) space to P-Q space, and vice versa. Since power is invariant under these transformations, peak power is a very convenient parameter to use when it is applicable. The method of use may be purely graphical, as it must be when the valve characteristics are empirical, or it may be supplemented by analytical operations when characteristic equations are available, as they often are. The characteristic equations of several types of valves are given in Chapter 7. Even in the absence of the equations, the charts are useful as an aid to visualization.

6.61. Use of the Load-Locus Charts

In the following discussion we shall make the somewhat unrealistic but very general assumption that the complete boundary of the region of the F-U plane which may be swept out by the load is defined and shall call this boundary the load locus. Usually, of course, only a few points of this boundary are defined, and there is seldom any reason for defining any more points. This discussion will therefore include these more realistic and less generalized cases.

In general, any characteristic curve of a drive, whether it be an equivalent valve curve or any other, which completely encloses the load locus will represent a drive that will do the job. The region which is enclosed by the drive curve but not by the load locus represents uneconomical overdesign without any corresponding benefit, and the designer's task is to minimize the area of this region in the most practical fashion. This fashion usually means that a characteristic drive curve is taken which represents a desired type of drive, and that this curve is expanded or shrunk in the two coordinates until it is approximately tangent to the load locus at one or more points and yet does not bulge excessively at other points. In most cases the point of tangency should be at or near the point of maximum instantaneous power into the load. If this point is also the point of maximum absolute output for the drive, the drive and the load may be considered to be matched, and this matched condition is usually the best design compromise.

The very general statement just made can best be illustrated by a specific example. Suppose that the load locus is represented by the heavy curve of Fig. 6.9. The peak instantaneous power is taken by the load at point C, through which the dashed curve of constant power has been drawn. Now assume that the load is to be driven by a ram controlled by a zero-lap valve, which has the parabolic characteristic of

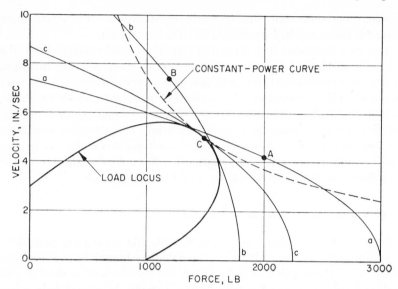

Fig. 6.9. Linear F-U chart.

Case P1, Sec. 7.31. This is not only the simplest case analytically, but much the most important in practice. Any of the infinitely many parabolas which can be circumscribed about the load locus will be possible solutions, but usually that parabola which takes in the least outside territory is the best.

The dimensions of the parabolas depend upon the supply pressure and upon the parameters of the valve and of the ram or motor, as follows:

1. Increasing the supply pressure moves the whole parabola to the right without changing its shape.

2. Increasing the maximum area of the valve metering orifices broadens the parabola without moving the vertex.

3. Increasing the ram area or motor effective displacement (actual displacement times motor-to-load gear ratio) moves the vertex to the right and makes the parabola narrower without moving the peak-output-power point.

Thus in designing the drive we have three parameters to juggle and essentially two conditions to satisfy. Which parameter will be fixed arbitrarily or by other constraints will naturally depend upon the particular circumstances. For a rotary motor, usually an available motor will be chosen and this will fix δ. This does not necessarily reduce the number of free parameters, since the gear ratio, R, must still be chosen, but this ratio is often fixed by considerations such as the maximum rated speed of the motor versus the required top speed of the load. Availability will often dictate the choice of the valve also, which fixes the shape of the parabola. The tangency condition then determines the supply pressure.

Rams are not usually used with multiplying levers, though they may be, and thus R is usually unity. This leaves A_p (which may be fixed

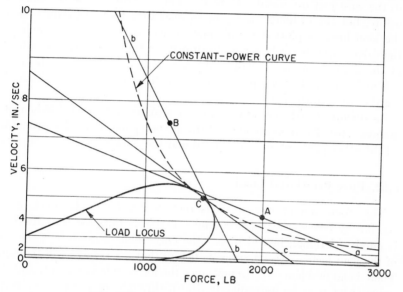

Fig. 6.10. Special F-U chart for P1 valves.

by the use of a commercial ram), valve area, and supply pressure, P_s. In any case probably the simplest procedure is to choose a tentative supply pressure, often either 1000 or 3000 psi, and then to choose a motor or ram and a valve with adequate but not excessive peak horsepower ratings at that supply pressure. When these are chosen, the supply pressure can be recalculated if an accurate "match" is desired. Usually it is sufficient to ensure that the drive will be strong enough for the job without being absurdly oversized.

The effects of mismatch are shown by the three drive curves of Fig. 6.9. In curve a the valve is too small or the ram too big, and an excessive supply pressure is required. The speed regulation and controllability are very good but the efficiency is poor. In curve b the ram is too small or the valve is too large, the supply pressure is low, the efficiency still lower, and speed regulation, controllability, and linearity are all bad. Curve c represents the optimum condition, for which the load peak-power point coincides with the valve peak-output point. This is usually the preferred solution to the problem.

It is rather laborious to plot parabolas which must be tangent to a given curve; however, the procedure can be greatly simplified by the use of a warped coordinate system so chosen that the drive curves become straight lines. This has been done in Fig. 6.10, which applies to the case just discussed. Here the scale of abscissas is the same as before, but the ordinates are proportional to velocity squared. After the load locus is plotted in this new coordinate system, the choice of the drive curve is easily made by drawing a tangent to the load locus at the desired point. Usually this tangent should intersect the force axis at $\frac{3}{2}$ times the peak-power force, which is the condition for matching the drive to the load.

As already stated, the supply pressure remains to be fixed. When this is chosen, usually upon the basis of the use of a particular valve or motor, all the other parameters can be fixed and the calculations are complete.

6.62. Pure Frictional Load

The discussion of the previous section can be made much more specific by considering some actual examples. Suppose that the load is purely frictional, in which case the only point of interest is the end of the load locus. Also assume as before that a P1 valve will be used. There is no problem of tangency since the locus ends in a point, and the valve curve may be immediately drawn through the peak-power point and the point at $\frac{3}{2}$ peak-power force and zero velocity. If we use the same numbers as in Figs. 6.9 and 6.10,

$$\text{Force at peak power} = 1500 \text{ lb}$$

$$\text{Velocity at peak power} = 5 \text{ in./sec}$$

$$\text{Stalled force} = \tfrac{3}{2} \times \text{peak-power force} = 2250 \text{ lb}$$

$$\text{Runaway velocity} = \sqrt{3} \times \text{peak-power velocity} = 8.67 \text{ in./sec}$$

We assume that for practical reasons we must use a supply pressure of 1000 psi. Then to obtain the predicted stalled force the ram area must be 2.25 in.^{2}, and at a runaway speed of 8.67 in./sec the maximum flow will be $8.67 \times 2.25 = 19.5$ in.^{3}/sec or about 5 gal/min. The pump horsepower is the product of this flow and the supply pressure, namely, 19,500 in. lb/sec or about 3 hp. The maximum system efficiency is attained at maximum mechanical output, since the input power is constant, and is $7500/19,500 = 38$ per cent. This low efficiency is the price which must be paid for the performance that can only be obtained from a valve-controlled system.

From the peak flow and the supply pressure the valve size can be determined. It will be shown in Sec. 9.46 that $a_{max} = q_{max}/(70\sqrt{P_s})$. In the present case $q_{max} = 19.5$ and $P_s = 1000$, whence $a_{max} = 0.0088$ in.[2] This assumes the use of a four-way valve with fluid of 0.85 specific gravity, which is about right for petroleum-base oils.

6.63. Pure Mass Load

As a second design example, consider the case of a load which is pure mass, with negligibly small friction at the maximum frequency at which the load is to be moved. This is not quite so unrealistic as might be supposed since it is sometimes necessary to move massive loads very quickly and the acceleration force becomes much greater than the dissipative force. Assume also that the peak-power point is the same as

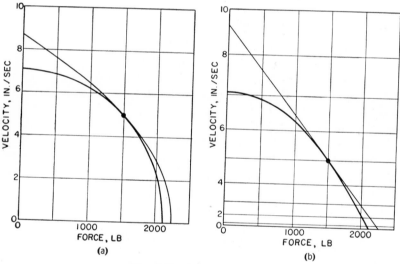

Fig. 6.11. Pure mass load.

before, namely, 1500 lb at 5 in./sec. If the scales of ordinates and abscissas are correctly chosen, the load locus will be a circle, as in Fig. 6.11a. Replotting this circle on the distorted coordinate system of Fig. 6.10 gives Fig. 6.11b. The tangent at the peak-power point is now the valve curve, and on the linear plane of Fig. 6.11a is the parabola. Naturally, since the peak-power point is the same as for the previous case, the drive is the same. Here, however, we have a tangency condition to satisfy, and the figure shows that it is satisfied since the load locus lies wholly within the valve curve. There is little waste space between the two curves, so in this sense the drive is very "efficient" and well matched to the load.

6.7. EFFECTS OF NONIDEALITIES OF THE DRIVE

So far nothing has been said about the effects of departures from ideal performance of the drive or of the fluid. These effects are very important in certain cases, and two such effects will be considered here. One is the effect of the compressibility of the fluid upon the dynamics of the system, which in a sense inserts a spring between the ram or motor and the load. The other is the somewhat more complicated case of the partly open valve, which decreases the stiffness of this spring and also introduces damping.

6.71. The Hydraulic Spring

Consider the case of an ideal frictionless, leakless hydraulic motor connected to a valve by two lines, the entire fluid space within the motor, lines, and valve being filled with a liquid at a uniform fairly high pressure and the valve being tightly closed. If the liquid is truly incompressible, the motor shaft will be immovably fixed. If it is somewhat compressible, as all real liquids are, the shaft can be turned somewhat and its motion will make the motor act as a pump, raising the pressure in one line and lowering it in the other. The torque required to produce this effect will be proportional to the pressure difference, and the angle through which the shaft turns will be proportional to the volume increase on one side and the decrease on the other. To the extent that the bulk modulus of the liquid is constant the shaft of the motor will act as though it is connected to a stiff linear spring, and the effective spring constant can be calculated.

It can easily be shown that for a ram

$$C = \frac{\kappa V_t}{4 A_p{}^2} \tag{6.7}$$

and for the rotary motor

$$C_h = \frac{\kappa V_t}{4\delta^2} = \frac{\pi^2 \kappa V_t}{D^2} \tag{6.8}$$

where C = linear compliance, in./lb

κ = compressibility, in.^{2}/lb

V_t = *total* trapped-fluid volume on both sides of valve, actuator, and lines

A_p = effective piston area, in.^{2}

C_h = torsional compliance, radians/in. lb

δ = motor displacement, in.^{3}/radian, or

D = motor displacement, in.^{3}/revolution

In deriving the above formulas it is assumed that the pressure is not lowered below zero on either side of the ram or motor, so that cavitation does not occur, and that the trapped volume is the same on the two

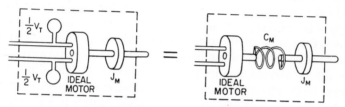

Fig. 6.12. The hydraulic spring.

sides. In practice the first assumption is nearly always true, but the second will be false for a decentered ram. In this last case, the stiffnesses (reciprocal compliances) must be calculated separately for the two sides and added. The formulas were given in terms of compliance since compliances add linearly for springs in series and the compliance of the trapped oil is in series with the load and the load usually has some compliance of its own. This series connection is illustrated by the equivalence diagram of Fig. 6.12.

In a system for which the speed of response is small and for which stiction is not a problem, the effect of the liquid compliance is usually unimportant. (It is always a major problem in pneumatic systems.) For a sticky, massive system, such as a large machine-tool table, the fluid compliance tends to make the load move in jerks since the valve must pass an appreciable volume of fluid before the pressure difference builds up enough to break the stiction, and then the load moves ahead under the lower running friction for some distance before the pressure difference becomes equalized and the motion stops, permitting the load to stick anew. This problem is greatly aggravated by the use of a fluid of low bulk modulus and a large trapped volume. In heavy-machine-tool practice this sometimes eliminates the ram as a table drive in spite of its apparent simplicity.

In a system for which the requirements for speed of response are severe, it frequently happens that the ultimate limit on obtainable response is given by the mechanical natural frequency of the drive and the structure. If the fluid compliance is appreciable with respect to the mechanical compliance, it may seriously lower the natural frequency and thus degrade the performance.

The calculation of the trapped volume of a system is usually laborious and not very accurate, and its measurement is surprisingly troublesome, so at least for initial-design purposes a simple approximate rule is useful. For piston-type rotary motors of small and medium sizes, it is convenient to use an empirical factor, k_v, which is simply the ratio of the total trapped volume to the displacement per revolution, or V_t/D. The theoretical minimum value of k_v is 0.5 since for any shaft position the average of the piston positions is half stroke. Such a low factor is never approached because of the considerable volume of the ports and connections in the motor, of the lines from motor to control valve, and of motor ducts, and so on, in the control-valve body. With Vickers motors at least, a factor of $k_v = 4$ is very good, even when care is taken to make the lines as short and small as possible, while 6 to 8 is much more likely even with fairly short lines. Motor housings could be redesigned, of course, or the control valve could even be built into the motor head, but the demand for such special motors has never been great enough to justify their production.

One reason for this is that the rotary motor is almost always geared to its load, and as seen by the load the compliance of the motor is divided by the square of the gear ratio. This can be shown as follows.

Assume a spring load connected to a motor by an ideal gear box of ratio G. When the motor exerts a torque T_M on the input shaft, the output shaft will exert a torque $GT_M = T_L$ on the load. This will cause the load to deflect through an angle $\theta_L = T_L C_L = GT_M C_L$, where C_L is the compliance of the load. But this motion will permit the input shaft and therefore the motor to move through an angle $\theta_M = G\theta_L$, so the motor will see a spring of apparent compliance $C_M = G^2 C_L$. By a similar argument it can be shown that the load inertia as seen by the motor is diminished by a factor of G^2. This matter will be discussed again in Sec. 6.81.

As a practical example of the effects of oil compliance, consider the DACL flight table, which will be further discussed in Sec. 6.82. In this table in its original form, the four gimbals were geared to identical drives by gear trains of ratios 37.5, 136.5, 190.8, and 195.3 for the X, Y, Z, and A axes, respectively. For these four axes the oil compliances amounted to 68, 12, 7, and 7 per cent of the total compliances. This would have been serious for the X axis except for the fact that its relatively low inertia permitted a reasonably high natural frequency in spite of the excessive compliance. The inertias of the other three axes were much greater, and an expensive program of modifications had to be carried out before the table could be used. If the load on the X servo had been more massive, that drive also would have had to be altered, probably by using a higher gear ratio. As it was, the substitution of a new valve and shorter lines considerably reduced the oil compliance on the X axis also.

It should be mentioned that the use of an inner pressure-feedback loop around the valve can effectively eliminate the effect of the hydraulic spring in certain applications. This artifice must be used with caution, however, since it often leads to instability of the servo as a whole.

6.72. The Partly Open Valve

In the previous section it was assumed that the control valve completely closed off the motor lines. This is only approximately true for most valves when they are centered, and the motor looks to the load like a somewhat dissipative spring. If the valve is slightly opened, more oil leaks past as the shaft is twisted, and the dissipation rises rapidly. At the same time, the motor begins to see the supply and exhaust lines, and the concept of trapped volume becomes less and less meaningful.

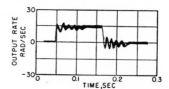

Fig. 6.13. Effect of partly open valve.

Thus the mechanical impedance of the motor shaft changes from a spring to a square-law resistor as the valve opens wider and wider.

This effect has been observed experimentally [10] and is shown by Fig. 6.13. The differences between the opening and closing transients are visible, if not impressive. Careful measurements on this and on other similar oscillograms yielded approximate values of the effective natural frequency and the damping of the system for the two conditions:

	f_n (cps)	ζ_h
Valve opening	89	0.73
Valve closing	114	0.06

The effect on the frequency is considerable, but that on the damping is

[10] D. G. O'Brien, "Synthesis of a High-Speed Flight Simulator," S.M. Thesis, Electrical Engineering Department, Massachusetts Institute of Technology, 1951, pp. 106–108.

very much greater. It cannot be overemphasized that calculations of system performance for one operating condition in a system as non-linear as a fluid-power system may be grossly in error for other and not very different conditions.

6.8. MORE COMPLICATED SYSTEMS

The systems that have been discussed so far have been fairly simple in the sense that the characteristic equations of the loads have been simple. Probably this is true of most loads encountered in practice, but not infrequently it is necessary to design a drive for a multiply resonant load. The best method to use in such a case depends upon the particular circumstances, but it may be worth while to give two examples of such loads.

6.81. Two-Mesh Systems

Consider the system shown schematically in Fig. 6.14a. A rotary motor is connected to its control valve by lines of appreciable volume and to a load of inertia J_L in series with a compliance of C_L by a gear box of ratio G. The moment of inertia of the motor is J_M and its effective mechanical compliance is C_M. The gear box can be eliminated by replacing it and the actual motor inertia and compliance with an effective inertia of $J_M G^2$ and an effective compliance of C_M/G^2. In order to simplify the mathematics, dissipative effects will be neglected; of course, in an actual case they should at least be checked to make sure that this neglect does not invalidate the analysis. With these assumptions, the actual system can be replaced by the equivalent mechanical system of Fig. 6.14b.

The factor that eventually limits the speed of response of a closed-loop system such as a servo is normally the lowest resonant frequency of the system, since as resonance is approached, the phase shift increases rapidly, and the loop gain must be decreased safely below unity before the phase shift becomes too great, in order to have a stable system. In the present case, the pertinent phase shift is that between the input shaft and the load inertia. Without taking dissipation into account it is impossible to calculate the actual phase shift, but at least a knowledge of the value of the undamped resonant frequency will set an upper limit to the attainable bandwidth of the system. The resonant frequencies may be calculated as follows.

Many, perhaps most, engineers compute the dynamic characteristics of relatively complicated systems with the help of analogies, and one of the most useful of these is the Firestone mobility analogy. In this method, which is the inverse of the so-called impedance analogue, voltage is taken as the analogue of velocity, and current as the analogue of

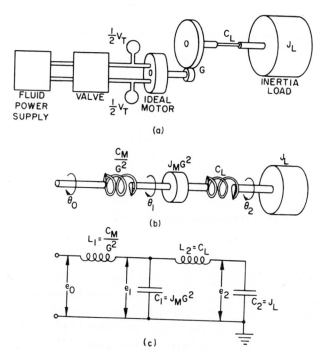

Fig. 6.14. Two-mesh system.

force or torque. Thus a spring becomes an inductance and a mass becomes a capacitance. Since masses refer to inertial space, their analogous capacitors all have one terminal grounded. The equivalent diagram then becomes that of Fig. 6.14c. The inductances may be given in radians/in. lb instead of henries, and the capacitances in in. lb sec^2 instead of farads, since any self-consistent system of units will work.

The quantity of interest here is the input-to-output transfer function, ϕ, the ratio of output voltage (velocity) to input voltage (velocity). For a linear system this ratio can be written down by inspection since the over-all ratio is the product of the voltage ratios for each mesh:

$$\phi \overset{\Delta}{=} \frac{e_2}{e_0} = \frac{e_2}{e_1} \cdot \frac{e_1}{e_0} = \left(\frac{Z_4}{Z_3 + Z_4}\right)\left(\frac{\dfrac{Z_2(Z_3 + Z_4)}{Z_2 + Z_3 + Z_4}}{Z_1 + \dfrac{Z_2(Z_3 + Z_4)}{Z_2 + Z_3 + Z_4}}\right) \tag{6.9}$$

where the Z's are the impedances of the four circuit elements, that is,

$$Z_1 = j\omega L_1 \tag{6.10}$$

$$Z_2 = \frac{1}{j\omega C_1} \tag{6.11}$$

$$Z_3 = j\omega L_2 \tag{6.12}$$

$$Z_4 = \frac{1}{j\omega C_2} \tag{6.13}$$

After some algebra these equations give

$$\frac{1}{\phi} = \omega^4 L_1 L_2 C_1 C_2 - \omega^2 (L_1 C_1 + L_2 C_2 + L_1 C_2) + 1 \tag{6.14}$$

At resonance ϕ will be a maximum. In a truly lossless system it would be infinite, and $1/\phi = 0$. The undamped resonant frequencies will therefore be the roots of the above quadratic in ω^2. Because it is a quadratic, it will in general have two roots, and there will be two resonant frequencies.

The physical interpretation of Eq. 6.14 can be aided by some substitutions. Let

$$L_1 C_1 = \frac{1}{\omega_M{}^2} = \lambda^2 L_2 C_2 \tag{6.15}$$

$$L_2 C_2 = \frac{1}{\omega_L{}^2} \tag{6.16}$$

$$L_1 C_2 = \mu^2 L_2 C_2, \text{ or } L_1 = \mu^2 L_2 \tag{6.17}$$

and

$$\eta = \frac{\omega}{\omega_L} \tag{6.18}$$

In these equations

ω_M = angular natural frequency of motor = $2\pi f_M$

ω_L = angular natural frequency of load, radians/sec

η = ratio of system frequency to load frequency

λ = ratio of motor frequency to load frequency

μ^2 = ratio of motor equivalent compliance to load compliance as seen at output of gear box

Substituting Eqs. 6.15 through 6.18 into Eq. 6.14 and putting $1/\phi = 0$ gives

$$\eta^4\lambda^2 - \eta^2(1 + \lambda^2 + \mu^2) + 1 = 0 \qquad (6.19)$$

This equation leads to certain physical conclusions:

1. If the motor inertia is negligibly small, $C_1 = 0$, $\lambda = 0$, and the system frequency is the load frequency times $1/\sqrt{1 + \mu^2}$. The motor compliance is added to the effective load compliance and the frequency is reduced.

2. If the load compliance is infinite or the load inertia zero, the load ceases to exist as far as the motor is concerned and the system frequency is the motor frequency.

3. Usually the load frequency is much lower than the motor frequency and the load compliance is much smaller than the motor compliance. If this is true, λ is small, μ is large, and the lower of the two system frequencies may be much lower than the load natural frequency.

This is best illustrated by a numerical example. Suppose that an inertia load is to be driven through a gear box whose ratio is to be chosen later, by a Vickers 3909 motor using MIL-O-5606B oil, and that the motor lines are fairly short so that $k_v = 5$. The system constants will be

$$L_1 = C_M = 7 \times 10^{-4} \text{ radian/in. lb}$$
$$C_1 = J_M = 1.02 \times 10^{-3} \text{ in. lb sec}^2$$
$$f_M = 188 \text{ cps}$$

Assume that the load characteristics will be

$$C_L = 7 \times 10^{-6} \text{ radian/in. lb}$$
$$J_L = 10.2 \text{ in. lb sec}^2$$
$$f_L = 18.8 \text{ cps}$$

The load inertia is that of a steel flywheel about 18 in. in diameter and 1.35 in. thick, and the compliance that of a 3-in. steel shaft 26 in. long.

The motor frequency is ten times the load frequency, so $\lambda = 0.1$. The compliance ratio depends upon the gear ratio, so $\mu^2 G^2 = 100$. With these constants the upper system frequency will be approximately $\sqrt{2}$ times the load frequency and can be neglected. The lower system frequency, however, will depend upon the choice of G, as follows:

G	μ	η	f, cps
∞	0	1.0	18.8
100	0.01	0.9949	18.7
31.6	0.1	0.9530	17.9
10	1	0.6998	13.2
3.16	10	0.3015	5.7
1	100	0.0995	1.88

It can be seen that it is desirable to have as high a gear ratio as possible between the motor and the load. With a less massive and less stiff load the effect of lowering the gear ratio would not have been so great, but it is important in any case.

6.82. Effects of Compliant Structure

In a relatively complex structure with several lumped loads coupled by compliances, the resonance pattern may be very complicated. One good example is the DACL flight table shown in Fig. 6.15. This device consists of four concentric gimbals, each driven by a hydraulic servo mounted on the next gimbal out. The four are denoted X, Y, Z, and A, counting outward. The moments of inertia seen by the four drives were originally as follows:

$$J_{xx} = 23 \quad J_{yx} = 81 \quad J_{zx} = 577 \quad J_{ax} = 1890$$
$$J_{xy} = 45 \quad J_{yy} = 145 \quad J_{zy} = 382 \quad J_{ay} = 1181$$
$$J_{xz} = 61 \quad J_{yz} = 214 \quad J_{zz} = 249 \quad J_{az} = 845$$

In this table the unit is the in. lb sec^2. The subscripts denote first the gimbal casting including all the equipment mounted on it, and second the axis about which the inertia is defined; thus J_{yx} is the inertia of the Y gimbal about the X axis. All gimbals are assumed to be in their normal positions.

The structural compliances were originally the following, using the same subscripts and 10^{-6} radian/in. lb as the unit:

$$C_{xx} = 0.360 \quad C_{yx} = 0.100 \quad C_{zx} = 0 \quad C_{ax} = 0$$
$$C_{xy} = 0 \quad C_{yy} = 0.307 \quad C_{zy} = 0.037 \quad C_{ay} = 0.127$$
$$C_{xz} = 0 \quad C_{yz} = 0 \quad C_{zz} = 0.315 \quad C_{az} = 0.0275$$
$$C_{xh} = 0.531 \quad C_{yh} = 0.044 \quad C_{zh} = 0.023 \quad C_{ah} = 0.0244$$

Fig. 6.15. The DACL flight table.

The compliances in the last row are the equivalent oil compliances referred to the one-speed shafts.

For the purposes of analysis, it is assumed that the system is lossless and that the equivalent circuits are as given in Fig. 6.16. Here the generator symbol denotes the drive, and the load inertia is starred. Since certain of the compliances are effectively zero because the gimbals are very stiff about certain axes, the four equivalent circuits of Fig. 6.16a can be simplified to those of Fig. 6.16b, giving finally two two-mesh and two three-mesh circuits. The poles and zeros of the resulting transfer functions are plotted in Fig. 6.17, with the upper row in each case representing the frequencies before modification of the table, and the lower row the later values. The measured frequencies of the four drives agree surprisingly well for the X, Y, and A drives, and less well for the Z drive. The reason for this last discrepancy is not known, but the neglect of dissipation in the system and the other simplifying assumptions are enough to make the calculations only rough approximations at best.

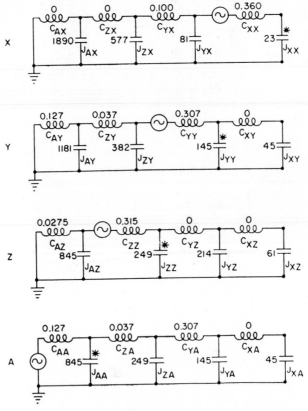

Fig. 6.16(a). Gimbal equivalent circuits. Prototypes.

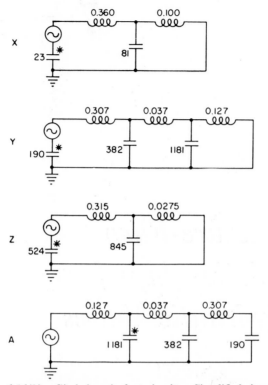

Fig. 6.16(b). Gimbal equivalent circuits. Simplified circuits.

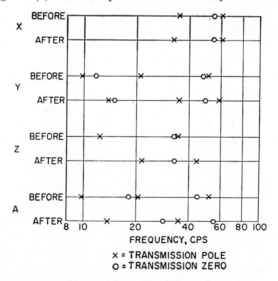

Fig. 6.17. Gimbal drive transmission poles and zeros.

7

J. F. Blackburn

Pressure-Flow Relationships for Hydraulic Valves

7.1. INTRODUCTION

So far we have assumed that we know just what a valve is and what it does. Actually the description of a valve and its characteristics in quantitative terms is not simple. For example, a control valve is ordinarily said to be used to "control flow." This somewhat meaningless statement is equivalent to the (also common) statement that a rheostat is used to "control current." Occasionally someone comes up with the idea that pressure (usually meaning pressure across the load) should be controlled instead of flow. Actually the valve itself does not know what happens at the load; the only way in which a "pressure-control" or a "flow-control" valve can be made to function is to measure the pressure or the flow in question with some sort of primary measuring element and to feed its output back into the mechanical input of the valve, thus setting up a closed-loop controller. This may be done in many different ways, some not immediately obvious, such as the use of a twin-jet flapper valve, which has built-in pressure feedback that can be fairly accurate if the valve is properly designed. What any valve does is to vary restriction which it introduces into the line in which it is installed.

178

This "restriction" must be defined in terms of a group of relationships among the various flows and pressures of interest, the most important of which are the supply pressure (or flow) and the pressure across and flow rate through the hydraulic motor, or load. There are many ways of presenting such groups, but one that has been found to be particularly useful is a nondimensionalized plot of motor flow, Q_m, as ordinate against motor pressure, $\mathcal{P}_m$, as abscissa, with the valve-stem position or some analogous quantity as a parameter. The actual flow, pressure, and valve stroke are nondimensionalized by dividing by appropriate reference quantities, which are chosen anew for each case to be discussed.

These $\mathcal{P}_m$-Q_m plots, which are analogous to families of plate characteristics of thermionic vacuum tubes, contain a lot of information and can be used in various ways by the system designer, but it would be much more convenient if a single quantity could be defined for each valve which would characterize it uniquely and completely. Such a quantity does not exist, but certain differential coefficients can be defined which are first approximations to it. The accuracy of the approximation and the usefulness of the coefficient depend upon the linearity of the valve characteristics. As will be seen, some types of valves are very linear while others are not.

In discussing the linearity of a valve, or of anything else that is a function of more than two variables, it is necessary to specify what is linear with respect to what. An ideal orifice (which many valves approximate) has a linear relationship between flow rate and orifice area if the pressure drop across the orifice is held constant, but the relationship between flow and pressure for constant area is parabolic. Orifice area varies linearly with spool displacement for a spool valve with rectangular ports, but parabolically for one with triangular ports. (Incidentally, this latter variation will *not* compensate for the nonlinear flow-pressure relationship.)

The flow-pressure curves and the differential coefficients can be obtained from measurements on actual valves, but for many configurations they can also be derived analytically. The analysis involves four assumptions:

1. The fluid is ideal.
2. The fluid source is ideal.
3. The geometry of the valve is ideal.
4. Steady-state conditions prevail.

The assumption of an ideal nonviscous and incompressible fluid is close to the truth under most conditions. At the highest flow rates and especially with relatively viscous fluids, the effective supply pressure at the valve intake may fall off because of pressure drop in the supply lines, but this fall should be no more than 5 to 10 per cent in a well-designed system. The assumption of incompressibility is also justified *as far as phenomena inside the valve are concerned.* At normal working pressures, the finite compressibility of real liquids has only a negligible effect upon the flow through the metering orifices, although it is often very important in other parts of the system.

An "ideal" constant-pressure (CP) fluid source supplies fluid to the intake of the valve at a constant pressure, independent of flow rate. Similarly, an ideal constant-flow (CQ) source puts out fluid at a constant flow rate, independent of pressure fluctuations. It is possible to build sources for which the assumption of ideality is fairly accurate even for rapidly fluctuating loads, but even where the regulation of the source is not extremely good, the effect on the system is usually not excessive.

"Ideal geometry" implies that the edges of the metering orifices are sharp and that the working clearance is zero (for a slide valve) so that the geometry of the orifice is independent of valve-stem position. This assumption is usually valid except for displacements below 1 or 2 thousandths of an inch. The results given in this chapter are therefore somewhat incorrect for very-short-stroke slide valves.

As in the case of compressibility, the assumption of steady-state conditions is valid for the valve alone, but transient effects in the rest of the system may be large. These effects are particularly important with regard to the instability of valves, where the system may react on the valve in such a way as to make it squeal or oscillate, sometimes with a large amplitude. Some of the numerous causes of valve instability will be discussed in Chapter 12. For a stable valve and system, however, the steady-state assumption is fairly accurate.

In most engineering and scientific work, the use of analogues is a very valuable tool. In the hydraulic field, the analogue is usually the corresponding electrical circuit. This analogy is reasonably good with respect to switching operations, and to the correspondence of flow rate to current, pressure to voltage, inertia to inductance, and compressibility to capacitance, but the resistance analogue must be used with care. Most electrical resistances obey Ohm's law: the current is proportional to the applied voltage. In the hydraulic field, the analogous law is obeyed for laminar flow, but the very large temperature coefficient of viscosity of all practical liquid hydraulic media must always be kept in mind; the case is as if electrical resistors were made not of metal but of high-coefficient thermistor material.

Most of the hydraulic resistances we will be considering are nonohmic and follow at least approximately the orifice equation:

$$q = C_d a \sqrt{2p/\rho} \tag{7.1}$$

where q = rate of flow
C_d = discharge coefficient
a = area of valve orifice
p = pressure drop
ρ = density of liquid

The discharge coefficient, C_d, is nearly constant for pressure drops above a certain value, which depends both on the characteristics of the fluid and on the orifice geometry. Let us define the Reynolds number for flow through a narrow slit by the equation

$$N_R = \frac{2vx}{\nu} \tag{7.2}$$

where N_R = Reynolds number (dimensionless)
v = average velocity through slit, in./sec
x = width of slit, in.
ν = kinematic viscosity, in.^{2}/sec

(The factor 2 is included to make the definition consistent with that customarily used for a circular opening.)

For geometries such as are found in most sliding-type control valves at small openings, C_d is fairly constant above the critical value $N_R = 260$. If the orifice edges are sharp, $C_d = 0.60$ to 0.65; if they are rounded or truncated by even a small flat, C_d will be 0.8 to 0.9 or even more and will vary somewhat with x. Orifice shape seems to have very little effect; C_d will be nearly the same for a long, narrow slit as for a circular hole.

For subcritical values of N_R, the behavior of C_d depends upon the geometry of the conduit upstream from the orifice. In general, as N_R decreases, C_d may decrease somewhat, then will rise to a maximum, and finally will drop rapidly for low values of N_R. The variation will be roughly like that shown in Fig. 7.1.[1]

As in other cases of flows intermediate between the laminar and the turbulent types, the portions of the curves for intermediate values of

[1] V. A. Khokhlov, "Hydraulic Loss and Fluid Discharge Coefficients Through the Orifices of a Cylindrical Spool-Valve Hydraulic Performance Mechanism," *Avtomat. i Telemekh.*, Vol. 16, No. 1 (1955), pp. 64–70. (In Russian: English translation available from M. D. Friedman, 67 Reservoir Street, Needham Heights, Mass.) Figure 7.1 is recalculated from Fig. 4 of the article, but too much reliance should not be placed on the numerical values since the figures in the mimeographed translation are not accurately drawn.

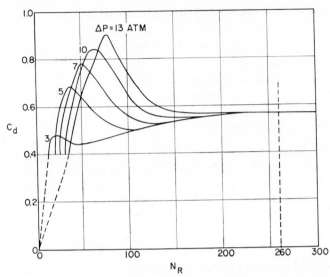

Fig. 7.1. Discharge coefficient versus Reynolds number for various pressure drops.

N_R are not well established or understood. For low values of N_R, however, the flow is almost purely laminar and is therefore much more consistent. The results of one experiment are plotted in Fig. 7.2. As can be seen, *for the conditions of this particular experiment,* the plot of C_d versus $\log_{10} N_R$ is remarkably linear down to about $N_R = 5$.

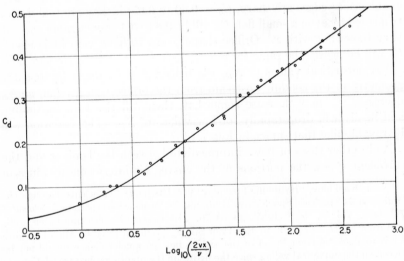

Fig. 7.2. Discharge coefficient at low values of Reynolds number.

The reasons for the discrepancies, if they actually exist, between the results of Fig. 7.1 and those of Fig. 7.2 are not known. In actual fluid-power practice, the pressure drops across the valve orifices are usually high enough and the fluid viscosities low enough so that N_R is considerably above the critical value, and C_d can be assumed constant at about 0.625. (Khokhlov gives a value of 0.57, which is somewhat lower than we have found.) Other effects in the system are usually large enough to mask small variations in C_d, but sometimes, particularly in airborne operation at very low temperatures, the viscosity may increase sufficiently to cause a significant decrease in C_d. This will be particularly true if the supply pressure is low.

Even if both temperature and pressure are high, however, there are other factors that can cause the actual performance of a valve to deviate from that predicted by the theory of this chapter. Probably the commonest effect is that of regulation of the pump and drop in the lines; at large flows, the pressure available to force oil through the valve and motor may decrease considerably, and the actual flow may be less than the calculated value. This effect is sometimes referred to as "saturation."

At large openings, where the orifice area is no longer small compared with the area of the upstream conduit, the variation of C_d with N_R becomes appreciable. At very small openings, on the other hand, x becomes comparable with the radial clearance of the valve spool, and the variation of effective area with stem position is no longer that assumed by the theory, and also the total flow is no longer large compared with the leakage flow. Again there will be a discrepancy between the measured and the calculated flows. This effect is small with closely fitted valves and increases rapidly with clearance; in this connection, it is necessary to warn against neglecting the effects of "silting," which are discussed in Sec. 10.24.

All these and other effects may cause measured valve characteristics to differ from calculated values, but in most cases, the discrepancies will not be serious. In the worst case, the measured maximum flow with the valve wide open might be down 50 per cent from the calculated value; this is primarily an indication that the valve should not have been opened so wide, or that the pump and lines are too small. In a properly designed system using high-quality parts, the discrepancies should be only a few per cent of full flow. As an actual example, the curves of Fig. 15.3b show a set of measured flow-pressure characteristics for a typical high-performance rectangular-port spool valve. The individual curves approximate closely to the parabolic form predicted by the theory and given later in Fig. 7.5, but some saturation effects are apparent, particularly for $x = 0.005$ in.   The open-center measured curves of Fig. 15.3a show much less saturation and conform fairly closely to the theoretical curves of Fig. 7.8.

For convenience, we may define a quantity that we will call the hydraulic conductance of an orifice, which will be proportional to the orifice area.   Equation 7.1 can be written $q = g\sqrt{p}$, where the conductance, g, is given by

$$g = aC_d\sqrt{2/\rho} \qquad (7.3)$$

Conductance is chosen rather than resistance since it is proportional to area, and area of the metering orifices is what is varied directly by valve-stem motion.   This convention makes the algebra much simpler and will be used throughout this chapter.

Conductances in parallel add directly, just as capacitances do.   Conductances in series add as the reciprocal squares, so the effective conductance of a series set is

$$g_{\text{eff}} = \sqrt{\frac{1}{(1/g_1{}^2) + (1/g_2{}^2) + \cdots}} \qquad (7.4)$$

where g_1, g_2, $\cdots$ are the conductances of the several orifices.   Thus the conductance of two unit orifices in series is 0.707, instead of 0.5 as it would be in the linear case.

As the restricting orifice departs farther and farther from the ideal sharp-edged orifice and approaches the form of a long slot of very small height normal to the direction of flow, the pressure-flow relationships become more and more linear for small Reynolds numbers.   The limiting case is that of a truly laminar flow-restricting element, which is usually impractical even if its linearity makes it attractive on paper.   Linear systems and elements will be given comparatively little attention in this book because the techniques of handling them are easily available elsewhere.   Intermediate cases are seldom amenable to mathematical analysis and must be handled on a purely empirical basis at present, though they comprise most of the cases met in practice.   Sharp-edged valves, which follow the orifice equation fairly accurately, are of great importance, and the rest of this chapter will be devoted to them.

7.2. GENERAL VALVE EQUATIONS

Control valves of the types with which we are concerned consist of arrangements of orifices, variable and fixed, which are so connected that they can vary the restriction which they offer to the flow of liquid from source to load, in the desired fashion, as the moving member of the valve is positioned by some external agent. For the time being, we will assume that we know nothing whatever about the load and that p_m and q_m, the pressure across the load and the flow through it, may independently assume any values whatever up to certain maxima. Our task is to draw up an equivalent circuit for each valve type and condition of operation and to derive from this circuit a characteristic equation in terms of p_m, q_m, the valve-stem position, x, (or other input quantity), and the known quantities. This characteristic equation, which represents a surface, can be plotted as a set of p_m-q_m curves for the particular valve. In some cases, it can also be differentiated and the differential coefficients can be derived. In other cases, however, the order of the characteristic equation is so high that it is impracticable to obtain general expressions for these coefficients, though they may still be evaluated for certain unique points such as the origin.

The curves and the differential coefficients can be made much more widely useful by nondimensionalizing. This is done by dividing p_m and q_m, the actual pressure and flow, by reference pressures and flows which are appropriately chosen for each case. The mechanical input is also normalized by using the fractional opening; if it is assumed that the valve ports are rectangular, this quantity will be equal to the fractional stroke, x/X, where X is the maximum stroke of the valve.

The special graphical symbols are given in Fig. 7.3, and the notation used is summarized and the quantities are defined in Table 7.1. Hy-

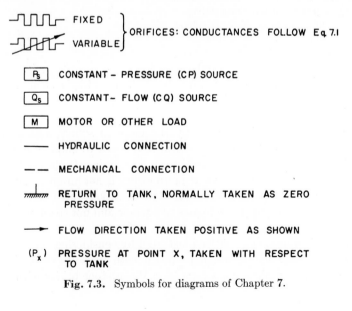

Fig. 7.3. Symbols for diagrams of Chapter 7.

Table 7.1. Notation for Chapter 7

p = pressure drop across an orifice
q = flow rate
g = hydraulic conductance of an orifice
a = effective area of an orifice
x = mechanical input—usually valve-stem displacement

For these quantities, lower-case letters denote instantaneous values of variables and upper-case letters denote constant values, usually maxima. Subscripts 1, 2, 3, and 4 refer to the four arms of the general hydraulic bridge, s to the supply, and m to the motor or hydraulic load. Subscripts are omitted when not required to avoid ambiguity.

Script $\mathcal{P}$'s and $\mathcal{Q}$'s are nondimensionalized by dividing actual values by reference pressures and flows. Thus $\mathcal{P}_m = p_m/P_r$, and so on.

For CP operation, $P_r = P_s$ and $Q_r = G\sqrt{P_s}$.

For CQ operation, $P_r = Q_s^2/G^2$ and $Q_r = Q_s$.

y = fractional conductance; $y = g/G = a/A$. For rectangular ports $y = x/X$ also.

Other symbols are:
CP = shorthand for "constant-pressure" source
CQ = shorthand for "constant-flow" source
$\mathcal{S}_m$ = auxiliary variable defined in Eq. 7.43
k_{q0} = flow sensitivity or "gain" of a valve
k_{p0} = pressure sensitivity of a valve
k = slope of load line on $\mathcal{P}_m$-$\mathcal{Q}_m$ chart
f_m = load force
v_m = load velocity
k_f = viscous-friction constant of load
A_m = effective area of ram, assumed to be ideal. Ram friction, and so on, may be lumped with load characteristics

draulic conductance is denoted by the German square zigzag symbol rather than by the common electrical saw tooth to emphasize the fact that it is parabolic ("square-law") and not ohmic. It should be stated again that the electrical analogue of an orifice is not an ordinary resistor but a nonlinear varistor and that a control valve, which can be varied by a mechanical input, is analogous to a triode and not to an ohmic rheostat.

The method used can best be illustrated by considering the case of the general four-way valve, which is drawn with its equivalent circuit in Fig. 7.4. The circuit is that of a loaded Wheatstone bridge with square-

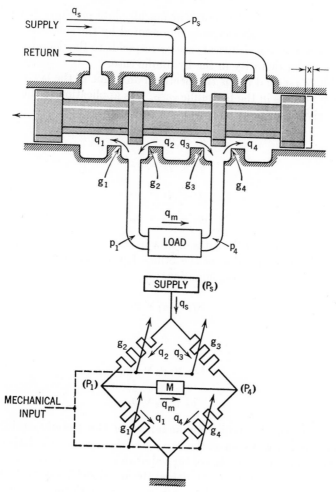

Fig. 7.4. The general four-way valve and its equivalent circuit.

law arms. In the general case, all four arms are simultaneously variable, as indicated by the arrows and their dashed-line connections. The general valve equations are set up in Table 7.1.

The application of Kirchhoff's first law to the system gives

$$p_1 + p_2 = p_s \tag{7.5}$$

$$p_3 + p_4 = p_s \tag{7.6}$$

$$p_1 - p_4 = p_m \tag{7.7}$$

$$p_3 - p_2 = p_m \tag{7.8}$$

In these equations the p's represent not absolute pressures but pressure drops across the circuit elements represented by the subscripts. Kirchhoff's first law in the electrical case corresponds to the definition of pressure as a scalar quantity in the hydraulic case.

Kirchhoff's second law is the electrical analogue of the law of the conservation of mass in the hydraulic case. Its application to the circuit of Fig. 7.4 gives four more equations for the flows:

$$q_1 + q_4 = q_s \tag{7.9}$$

$$q_2 + q_3 = q_s \tag{7.10}$$

$$q_2 - q_1 = q_m \tag{7.11}$$

$$q_4 - q_3 = q_m \tag{7.12}$$

Here the q's represent the rates of flow in the various branches.

Finally, the application of the orifice law to the metering orifices of the valve yields four more equations:

$$q_1 = g_1 \sqrt{p_1} \tag{7.13}$$

$$q_2 = g_2 \sqrt{p_2} \tag{7.14}$$

$$q_3 = g_3 \sqrt{p_3} \tag{7.15}$$

$$q_4 = g_4 \sqrt{p_4} \tag{7.16}$$

The g's are the hydraulic conductances of the metering orifices. In general, all four will vary as the valve stem is moved. The laws of variation depend upon the design of the valve and must be given for each case.

Equations 7.5 through 7.16 define the relationships between pressure and flow for a particular valve when combined with the equations for its g's. It should be noted that two of the groups of equations are redundant; actually there are ten rather than twelve independent equations. Of the various quantities involved, the g's and either p_s or q_s are assumed to be known. The p's and q's with the number subscripts are to be eliminated, leaving a single equation in p_m, q_m, and the knowns.

Even in the linear case, the equation for the loaded bridge is rather complicated; in the present case, the squaring operations required to eliminate the four radicals of Eqs. 7.13 through 7.16 make the general equation almost completely unmanageable. It is therefore expedient to introduce the auxiliary conditions for the special cases at this point rather than to try to operate on the general equation.

As will be seen, the characteristics of a valve are radically altered by changing from a CP to a CQ supply. For this reason, we will consider the CP cases first, as being much more widely used, and then take up the CQ cases. When we confine ourselves to CP operation, we will ignore Eqs. 7.9 and 7.10 in the derivation of the characteristic equations. CQ operation similarly omits the use of Eqs. 7.5 and 7.6.

7.3. CONSTANT-PRESSURE OPERATION

The most general configuration of major interest is the four-arm bridge discussed in the previous section. Undoubtedly the most used form of this bridge is the four-way spool valve in which two opposite arms increase in conductance and the other two decrease as the valve stem is moved. Often four-way valves are called "open-center" when all four arms are partially open with the spool centered, and "closed-center" when they are all closed. Closed-center valves may be given a large amount of overlap when it is desired to hold leakage to a minimum, but this overlap introduces a dead zone that makes the valve completely unsuitable for high-performance closed-loop operation. When overlap must be used in a servo valve, it should always be held to the smallest possible value; the highest order of performance can be gotten only from a valve with zero effective lap.

7.31. Case P1—The Series Circuit

In a valve with zero or positive lap, only two of the arms are open at any one time, and the configuration becomes a simple series circuit. This greatly simplifies the derivation of the characteristic equation, which can be written down by inspection. In Fig. 7.4, suppose that the valve stem has been moved to the left so that g_1 and g_3 are both zero. (A hydraulic conductance is never negative.) Then the flow through all of the elements of the circuit is the same, since we have assumed that the fluid is incompressible.

Let us assume also that all four valve ports are rectangular and identical so that the valve is symmetrical. The effects of asymmetry will be discussed in Sec. 7.5. Now $g_2 = g_4$, and we can drop the number subscripts as unnecessary. Also, for rectangular ports the fractional area equals the fractional stroke so that

$$g = \frac{Gx}{X} \tag{7.17}$$

Then by inspection we can write

$$p_m = P_s - \frac{2q^2}{g^2} \qquad (7.18)$$

as the characteristic equation of the "ideal" valve.

This equation can be made more useful by nondimensionalizing. Let

$$\mathcal{P}_m = \frac{p_m}{P_s} \qquad (7.19)$$

$$\mathcal{Q}_m = \frac{q}{G\sqrt{P_s}} \qquad (7.20)$$

and

$$y = \frac{g}{G} \qquad (7.21)$$

where G is the conductance of *each* open orifice when $y = 1$. If these equations are combined with Eq. 7.18, we get

$$\mathcal{P}_m = 1 - \frac{2\mathcal{Q}_m{}^2}{y^2} \qquad (7.22)$$

which is the equation of a family of parabolas. It is plotted in Fig. 7.5.

Actually, of course, Eq. 7.22 represents only the upper half of Fig. 7.5, but since movement of the valve stem to the right replaces g_2 and g_4 by g_1 and g_3 and reverses the flow through the load, with no change in the

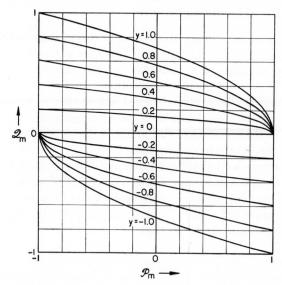

Fig. 7.5. Characteristics for Case P1—the series circuit.

form of the equation, the complete characteristic will be symmetrical about the origin and will include negative values for y and Q_m.

The figure shows that the "ideal" valve is decidedly nonlinear. In spite of this nonlinearity, it is the most widely used of the several types which will be discussed because it has in the highest degree most of the other properties that are usually desired in a control valve.

7.32. The Half Bridge

As soon as we abandon the series circuit and introduce orifices shunting the load, the valve algebra becomes much more complicated. It is convenient to derive the equations connecting q_m and p_1 for several cases since they are of interest in themselves, and it is easy to derive the expression for a complete bridge from those for the two halves. The half bridges and the other configurations of interest for CP operation are shown in Fig. 7.6.

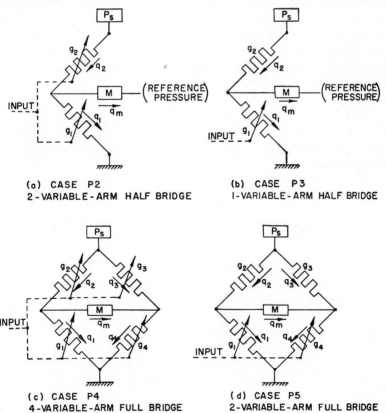

Fig. 7.6. Equivalent circuits for CP operation.

7.321. Case P2—Two Arms Variable, the Underlapped Three-Way Valve.

Suppose we consider the left half of the valve of Fig. 7.6a with its metering orifices, g_1 and g_2, and that we grind off the faces of the piston land so that there is appreciable underlap when the spool is centered. Suppose also that we will operate entirely within the region of underlap so that the maximum spool displacement, X, is equal to the underlap. Then, except for the extreme positions, there will always be a shunt path, and the equivalent circuit will be that of Fig. 7.6a. If we nondimensionalize, using Eqs. 7.19, 7.20, and 7.21, it is easy to derive one form of the characteristic equation as

$$Q_m = (1 + y)\sqrt{1 - \mathcal{P}_1} - (1 - y)\sqrt{\mathcal{P}_1} \qquad (7.23)$$

This is a good place to sound a warning about the use of conventional mathematical notation in this kind of derivation. The square-root symbol implicitly puts a "$\pm$" sign in front of any term in which it appears, and only one or the other sign has a physical meaning. When the pressure pattern changes in such a way as to reverse the flow through any orifice, the sign of the radical over the pressure drop across that orifice also changes. In the present case, the equation as written is valid over the range $0 \leq \mathcal{P}_1 \leq 1$, which is the region of primary interest. If the load is sucking oil out of the valve so rapidly that $\mathcal{P}_1 < 0$, we should write

$$Q_m = +(1 + y)\sqrt{|1 - \mathcal{P}_1|} + (1 - y)\sqrt{|\mathcal{P}_1|} \qquad (7.24)$$

while if it is forcing oil rapidly back into the valve so that $\mathcal{P}_1 > 1$, we should write

$$Q_m = -(1 + y)\sqrt{|1 - \mathcal{P}_1|} - (1 - y)\sqrt{|\mathcal{P}_1|} \qquad (7.25)$$

In the former case, the flow has reversed through g_1, and in the latter, it has reversed through g_2, and the sign of the term in each case must be chosen according to the direction of flow. The absolute value of pressure drop is always used under the radical.

Equation 7.23 is valid as it stands, and a family of flow-pressure curves can be plotted from it, but it is often more useful to have an explicit expression for $\mathcal{P}_1$. This can be derived by successive squaring and rearrangement operations, which give

$$\mathcal{P}_1 = \frac{(y + 1)^2(y^2 + 1) - 2yQ_m{}^2 - (1 - y^2)Q_m\sqrt{2(1 - y)^2 - Q_m{}^2}}{2(y^2 + 1)^2}$$

$$(7.26)$$

Obviously Eq. 7.23 is better adapted to computing the characteristic curves, which are plotted in Fig. 7.7.

As it stands, Eq. 7.26 gives $\mathcal{P}_1$, which is always positive in normal operation. The motor pressure, $\mathcal{P}_m$, is equal to $\mathcal{P}_1$ minus some counter-pressure, which may be real, as in the differential ram of Fig. 6.6, or may be only virtual, as in the case of a motor working against a gravity load. With a full bridge, the counterpressure is $\mathcal{P}_4$.

7.322. Case P3—One Arm Variable, the Single-Jet Flapper Valve. Consider the arrangement of Fig. 7.8, in which there is a fixed upstream orifice and a variable orifice downstream of the load take-off point. This represents an arrangement which is very widely used in control devices, though usually on low-pressure compressed air rather than on liquid. It has been used as a pilot stage for two-stage

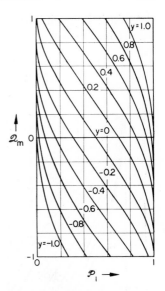

Fig. 7.7. Characteristics for Case P2 — two-variable-arm half bridge.

power-control valves, however, and is worth analyzing. The equivalent circuit is given in Fig. 7.6b; the characteristic equation is the same

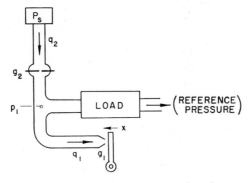

Fig. 7.8. Single flapper-and-nozzle valve.

as Eq. 7.23 except that the coefficient of the first term is 1 instead of $(1 + y)$, so

$$Q_m = \sqrt{1 - \mathcal{P}_1} - (1 - y)\sqrt{\mathcal{P}_1} \qquad (7.27)$$

It is plotted in Fig. 7.9.

Again it is possible to get an explicit but complicated expression for $\mathcal{P}_1$. It is

$$\mathcal{P}_1 = \frac{2y - y^2(1 - \mathcal{Q}_m{}^2) - 2(1 - y)\mathcal{Q}_m\sqrt{2 - 2y + y^2 - \mathcal{Q}_m{}^2}}{(2 - 2y + y^2)^2} \quad (7.28)$$

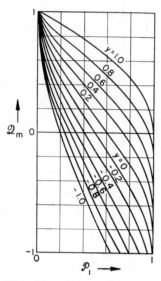

Fig. 7.9. Characteristics for Case P3 — single flapper-and-nozzle valve.

7.33. The Full Bridge

The two configurations just discussed are of interest in themselves, but they can also be combined in various ways to give several types of complete four-arm bridges. These are of considerably greater interest than the partial bridges since in general the four-way valves are much more used and have more desirable characteristics than the somewhat simpler configurations.

7.331. Case P4—Four Arms Variable, the Symmetrical Underlapped Valve. One important bridge is formed by combining two two-variable-arm half bridges to get a bridge with all four arms variable. If the bridge is symmetrical, so that $g_1 = g_3$ and $g_2 = g_4$, the same is true for diagonally opposite p's and q's, and we have the symmetrical underlapped four-way valve. Its characteristic equation may be obtained by subtracting $\frac{1}{2}$ from Eq. 7.26 and doubling the result. It is

$$\mathcal{P}_m = \frac{-2y(1 + 2y) + 2y\mathcal{Q}_m{}^2 - (1 - y^2)\mathcal{Q}_m\sqrt{2(1 - y)^2 - \mathcal{Q}_m{}^2}}{(1 + y^2)^2} \quad (7.29)$$

It is also possible to start with the basic equations, inserting the symmetry conditions, and to derive an expression for $\mathcal{Q}_m$ in terms of $\mathcal{P}_m$. It is

$$\mathcal{Q}_m{}^2 = 1 + y^2 - 2y\mathcal{P}_m - (1 - y^2)\sqrt{1 - \mathcal{P}_m{}^2} \quad (7.30)$$

Neither of these equations is particularly simple, but they represent a family of characteristic curves, which is shown in Fig. 7.10. As can be seen, the individual curves are remarkably straight, parallel, and evenly spaced over most of the area of the figure, so the underlapped

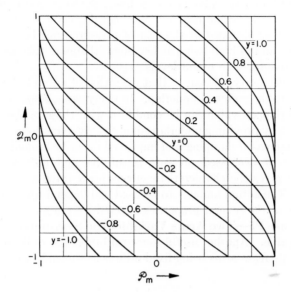

Fig. 7.10. Underlapped four-way valve.

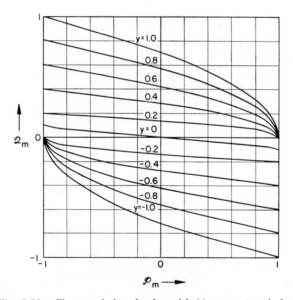

Fig. 7.11. Characteristics of valve with 20 per cent underlap.

symmetrical four-way valve is a very linear device as hydraulic devices
go. This linearity is the result of the high degree of symmetry of the
bridge, but it is obtained at the cost of pressure sensitivity (to be discussed later) and of high losses at low flows; therefore this form of valve
is unattractive for high power levels.

It is not necessary, of course, that operation of an underlapped valve
be confined to the region of underlap. In many cases, some of the virtues
of the underlapped valve can be combined with those of the ideal zero-
lapped valve by providing underlap equal to only a fraction of full stroke.
This buys increased damping and improved linearity around zero stroke
at the cost of decreased pressure sensitivity and increased leakage.
Figure 7.11 gives the characteristic curves for a valve with 20 per cent
underlap. The actual amount to be used will depend upon the relative
importance of the several properties affected by the underlap.

**7.332. Case P5—Two Arms Variable, the Symmetrical Two-Jet
Flapper Valve.** Another useful four-arm bridge has only two variable
arms, the upstream arms being fixed orifices as shown in Fig. 7.12. The

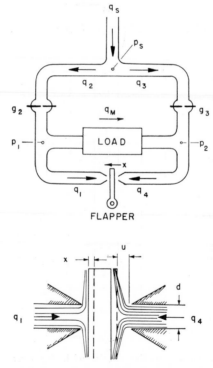

ENLARGED SECTION OF FLAPPER AND JETS

Fig. 7.12. Twin flapper-and-nozzle valve.

equation for this case can be obtained by writing Eq. 7.29 for $\mathcal{P}_1$, and the same equation with the signs of y and $\mathcal{Q}_m$ reversed to get $\mathcal{P}_4$, and then subtracting the second from the first. The resulting equation in standard form is so complicated as to be hardly worth writing. For computational

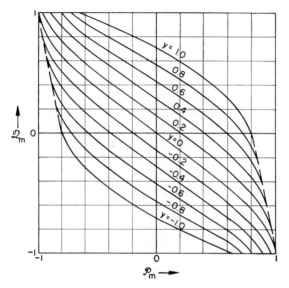

Fig. 7.13. Characteristics for Case P5—two-variable-arm full bridge.

purposes, a reasonable set of equations is

$$\mathcal{P}_1 = \frac{1}{\alpha^2}[\alpha + (\alpha - 2)\mathcal{Q}_m{}^2 - 2\mathcal{Q}_m(1 - y)\sqrt{\alpha - \mathcal{Q}_m{}^2}] \qquad (7.31)$$

$$\mathcal{P}_4 = \frac{1}{\beta^2}[\beta + (\beta - 2)\mathcal{Q}_m{}^2 + 2\mathcal{Q}_m(1 + y)\sqrt{\beta - \mathcal{Q}_m{}^2}] \qquad (7.32)$$

$$\alpha \overset{\Delta}{=} y^2 + 2 - 2y \qquad (7.33)$$

$$\beta \overset{\Delta}{=} y^2 + 2 + 2y \qquad (7.34)$$

$$\mathcal{P}_m = \mathcal{P}_1 - \mathcal{P}_4 \qquad (7.35)$$

$$\mathcal{Q}_m = \frac{q_m}{G\sqrt{P_s}} \qquad (7.36)$$

The resulting characteristic curves are plotted in Fig. 7.13. Again the linearity is good, but the gain is only about half that of Fig. 7.10, and the efficiency is much worse. The simplicity of construction of the flapper-and-jet valve does make it attractive for pilot-valve service.

7.333. Other Cases. There are several other possible configurations, but they are of little or no interest. Single- and three-variable-arm bridges apparently have no advantages over the other types to offset their poorer linearity. There is a second two-variable-arm bridge in which the variable arms are both in the left half-bridge while the right half is fixed. Its equations can be derived from Eq. 7.26 by obtaining the equation for $\mathcal{P}_4$ by reversing the sign of Q_m and setting $y = 0$. The characteristics of the resulting bridge are very similar to those of the other two-variable-arm bridge, and it seems to have no advantage over the latter, which is easier to make since the discharge from the flapper chamber can be at the return pressure. The only other possible symmetrical configuration would be one with all four arms fixed, in which case there would be no control.

7.4. CONSTANT-FLOW OPERATION

When the hydraulic fluid is derived from a CQ instead of from a CP source, an entirely different set of characteristics is obtained. CQ operation is occasionally suggested as more economical than CP operation, apparently on the basis that a gear or vane pump is cheaper than a variable-displacement pump, but even this alleged advantage is somewhat doubtful if the comparison is made on the basis of the complete system, including all necessary auxiliary equipment. CQ operation is also poorly adapted to operating several independent systems from the same supply, for the same reason that makes constant current operation rare in electrical systems; the systems must be operated in series, and the whole system has to be made strong enough to stand a peak pressure equal to the sum of the peak pressures of the several parts. In a CP system, the peak demand is for a high flow, and especially if the peaks are of short duration, this is relatively easy to obtain.

The simple series circuit is useless for CQ operation since the motor will turn at a constant speed and the series-restricting elements will have no control over it; they will serve only to waste power. The same comment applies to the upper arms of Fig. 7.6a and b, or to any other restriction through which the whole supply flow passes. The configurations of interest for CQ operation will be those of Fig. 7.14.

CQ operation is definitely of interest for one class of service, however. There is a need, which has gone largely unfilled, for very cheap servos of only fair performance characteristics which would nevertheless be con-

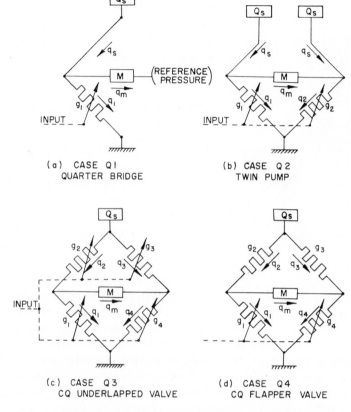

Fig. 7.14.   Equivalent circuits for CQ operation.

siderably better than the best that can be economically obtained from electrical devices.   Certain CQ systems seem to be very attractive for such applications.

7.41. Case Q1—The Quarter Bridge

The simplest possible control will be given by the combination of a CQ supply and a simple two-way valve, as in Fig. 7.14a.   The equation of this combination can be written down by inspection as

$$q_m = q_s - G(1 - y)\sqrt{P_1} \qquad (7.37)$$

For CQ devices, it is convenient to nondimensionalize by using Q_s and

Q_s^2/G^2, respectively, as the reference flow and pressure. When this is done, Eq. 7.37 becomes:

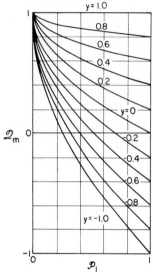

Fig. 7.15. Characteristics for Case Q1—quarter bridge.

$$Q_m = 1 - (1 - y)\sqrt{\mathcal{P}_1} \qquad (7.38)$$

This is the equation of a family of parabolas, which are plotted in Fig. 7.15.

7.42. Case Q2—Twin-Pump Configurations

If two separate identical CQ sources are provided, plus a three-way valve, we have the arrangement of Fig. 7.14b. Its characteristic equation is

$$\frac{\mathcal{P}_m}{4}(1 - y^2)^2 = (1 + Q_m^2)y - Q_m(1 + y^2) \qquad (7.39)$$

which is plotted in Fig. 7.16.

As can be seen, this arrangement is highly nonlinear. The economy of equipment involved makes it very attractive, however, particularly since the required tolerances on the valve dimensions can be made very wide. It will be shown in Sec. 7.7 that for certain types of loads the valve ports can be so shaped that a fairly linear response can be obtained, and this offers the possibility of making inexpensive servos with fairly good response.

A second twin-pump configuration, of rather less interest, is obtained by replacing the three-way valve with two independent two-way valves so arranged that only one is decreased in area at a time, while the other remains constant. Since we no longer have the skew symmetry of the previous case, the characteristics are less linear, as would be expected. This arrangement seems to have no advantages over the previous one, and the valve is rather more complicated.

7.43. Full Bridges

In CQ operation from a single source, it is no longer permissible to split the bridge into right and left halves and to consider each half independently because the flow from the source will not, in general, divide in any constant ratio between the two halves. It becomes necessary to work from the complete set of bridge equations, and the algebra becomes

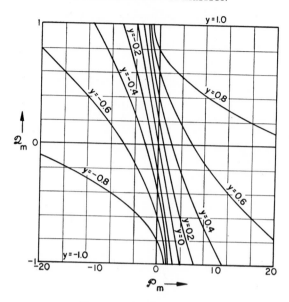

Note change of scale of $\mathcal{P}_m$

Fig. 7.16. Characteristics for Case Q2—twin pump.

rather complicated. Since the three cases to be discussed are of rather restricted interest, only the results will be given.

7.431. Case Q3—Four Arms Variable, the CQ Underlapped Valve. When the symmetrical underlapped valve of Fig. 7.14c is fed from a CQ instead of from a CP source, its characteristics are completely altered. The equation is almost the same as Eq. 7.39, or

$$\mathcal{P}_m(1 - y^2)^2 = (1 + \mathcal{Q}_m{}^2)y - \mathcal{Q}_m(1 + y^2) \tag{7.40}$$

Comparison of Fig. 7.16 (disregarding the factor 4 in the scale of $\mathcal{P}_m$, which applies to the previous case but not to this one) with Fig. 7.10 for the CP underlapped valve shows that the most linear valve of all has become violently nonlinear simply because of changing the type of supply. The underlapped CQ valve is not particularly important in itself, but it is worth mentioning as a possible alternative to the twin-pump arrangement of Sec. 7.42.

7.432. Case Q4—Two Arms Variable, the CQ Flapper Valve. When the twin-jet flapper valve of Fig. 7.6d is operated from a CQ supply, its characteristics again are altered for the worse, though not so radically as in the previous case. The characteristic equation is rather complicated and is most economically given as

$$\mathcal{P}_m(1 - y^2)^2 = (1 + S_m^2)y - S_m(1 + y^2) \tag{7.41}$$

where

$$S_m \overset{\Delta}{=} \mathcal{P}_m + 2\mathcal{Q}_m \tag{7.42}$$

It is plotted in Fig. 7.17. It is not quite so nonlinear as Fig. 7.16 but would seem to have little practical value.

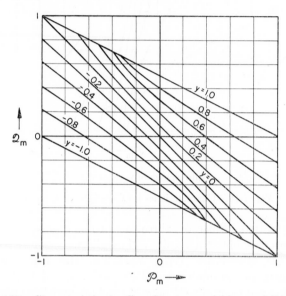

Fig. 7.17.   Characteristics for Case Q4—two-variable-arm full bridge.

A variant of this case would be to put the two variable arms in the left half-bridge and the two fixed orifices in the right. This requires that the flapper chamber (if the flapper construction is used rather than a conventional three-way valve) be at the pressure p_1 rather than at the return pressure. It seems to have no compensating advantages.

7.5. EFFECTS OF ASYMMETRY

So far, we have assumed as high a degree of symmetry as possible in each case. It is of interest to determine the effects on the valve characteristics of departures from this symmetry. In general, such departures always make a valve more nonlinear, and may affect other properties also, usually in an adverse fashion. It would consume an excessive amount of space to discuss these effects in detail for each case, but a few general remarks may be useful.

The most linear of all the cases we have discussed is the CP under-lapped valve of Sec. 7.331, and it is linear because it is highly symmetrical. It departs from linearity only when the ratio of the areas of the orifices differs radically from unity, and even then, the skew symmetry of the valve limits the amount of the departure.

Symmetry of the right kind produces linearity because it permits the positive curvature of the characteristic of one orifice to be balanced by

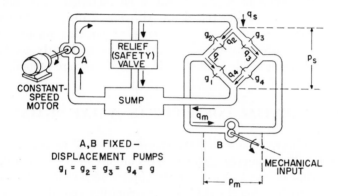

Fig. 7.18.   Linear hydraulic resistance.

the negative curvature of that of another, acting differentially. A close analogue is the cancellation of even-order harmonics in the output current of a push-pull vacuum-tube amplifier.

In the hydraulic case, cancellation can be made complete and the device can be made perfectly linear (within limits) by making all four bridge arms identical and feeding from a CQ supply, as in Fig. 7.18.[2] The equation can be derived from any of several of the CQ cases already discussed by letting $y = 0$; it is

$$p_m G^2 = -q_s q_m \qquad (7.43)$$

Since G and q_s are constant, p_m is directly proportional to q_m, and the device is linear. Actually, this linearity persists only so long as $|q_m| > |q_s|$; when this condition is violated, the flow through two of the orifices reverses, and the characteristic becomes parabolic. This is an example of the need for the warning given in Sec. 7.321 about using the equations blindly.

[2] J. F. Blackburn, "Contributions to Hydraulic Control, 4: Notes on the Hydraulic Wheatstone Bridge," *Trans. ASME*, Vol. 75 (1953), pp. 1171–1173.

The minus sign in Eq. 7.43 means that the device is a power absorber; the load must be "negative." It is not a control device since there is nothing that corresponds to the input or stem position of an ordinary valve, but it might be useful as a high-powered "viscous" damper which does not depend upon the viscosity of the fluid. One interesting property is that the losses in the system are constant; as it begins to take power from the "load," it takes less from the pump, and the sum of the two powers remains the same.

The opposite extreme from the standpoint of linearity is the single-orifice arrangement of Sec. 7.41, whose characteristics are plotted in Fig. 7.15. Comparison of this figure with Fig. 7.10 is an adequate demonstration of the effects of asymmetry. Even this amount of nonlinearity is not excessive for some applications, and fairly good servos can be made from highly nonlinear elements if they are properly used.

7.6. DIFFERENTIAL COEFFICIENTS—VALVE "GAIN"

So far we have represented the characteristics of valves only in the form of families of curves. These curves are useful and permit drawing semiquantitative conclusions as to the applicability of a particular valve to a given problem, but it would be convenient if we could establish some kind of a figure of merit in the form of a single number that would allow direct and quantitative comparisons to be made. It is unreasonable to expect that anything so complicated as a valve can be completely characterized by a single number, but certain quantities can be defined that are very useful even if they do not tell everything about a valve.

Probably the most useful of these is a number that tells us the "gain" of a valve, which is that quantity that will be used in computing the forward gain of a hydraulic servo loop. The input can be defined unambiguously; it is merely the displacement of the valve stem or other operating member from its normal position, which is usually the center. The output, however, is not quite so obvious, since it is necessary to decide what a valve does.

Bearing in mind that by hypothesis we know nothing as yet about the nature of the load, the one quantity which is most intimately and simply related to the input is the rate of flow through the load, or q_m. The valve "wants to" establish a particular rate of flow, and it does so insofar as the load allows it to. In terms of the position of the motor output member, the valve is an integrator, since the *position* of the output is the time integral of the *position* of the input.

If we regard the valve as a device that controls flow rate in accordance with stem position, the characteristic we want is the relation between the two, or in the language of the calculus, the derivative of flow with respect to stem position, pressure drop being kept constant. This is what is ordinarily meant by the gain of the valve. In general, it varies from one point to another on the $\mathcal{P}$-$\mathcal{Q}$ chart, and since operation is usually around the origin, it is most conveniently evaluated there. We will define the flow sensitivity or gain of the valve as

$$k_{q0} \overset{\Delta}{=} \frac{dq_m}{dx}\bigg|_{p_m=\text{const}} \tag{7.44}$$

A second important characteristic is the pressure sensitivity, which may be defined in an analogous fashion as

$$k_{p0} = \frac{dp_m}{dx}\bigg|_{q_m=\text{const}} \tag{7.45}$$

Its magnitude is a measure of the ability of the valve-motor combination to start a high-inertia or sticky load, and it is precisely because the attainable values of this quantity are so high that hydraulic servos are so useful. Both k_{p0} and k_{q0} are given in Table 7.2 for the various cases discussed in this chapter.

Table 7.2. DIFFERENTIAL VALVE COEFFICIENTS

Case	Section	k_{q0}	k_{p0}
P1	7.31	$\dfrac{G}{X}\sqrt{P_s/2}$	(see Note)
P2	7.321	$2\dfrac{G}{X}\sqrt{P_s/2}$	$\dfrac{P_s}{X}$
P3	7.322	$\dfrac{G}{X}\sqrt{P_s/2}$	$\dfrac{P_s}{X}$
P4	7.331	$2\dfrac{G}{X}\sqrt{P_s/2}$	$2\dfrac{P_s}{X}$
P5	7.332	$\dfrac{G}{X}\sqrt{P_s/2}$	$\dfrac{P_s}{X}$
Q1	7.41	$\sqrt{2}\dfrac{Q_s}{X}$	$2\dfrac{Q_s^2}{G^2 X}$
Q2	7.42	$2\dfrac{Q_s}{X}$	$4\dfrac{Q_s^2}{G^2 X}$
Q3	7.431	$\dfrac{Q_s}{X}$	$\dfrac{Q_s^2}{G^2 X}$
Q4	7.432	$\tfrac{1}{2}Q_s/X$	$\dfrac{1}{2}\dfrac{Q_s^2}{G^2 X}$

Note: For the ideal valve k_{p0} is theoretically infinite. In practice it is easy to get 10^6 psi per in. for supply pressures in the 1000-psi range.

These coefficients apply only for rectangular ports.

Various other "constants" could be derived and are of use in special cases, but the two just defined are the most important. They are analogous to two of the dynamic characteristics of a triode; k_{q0} to the transconductance and k_{p0} to the amplification factor. The third of the trio, which is analogous to plate resistance, could be called the "stiffness" of the valve; it is the negative reciprocal of the slope of the curve of constant y on any of the $\mathcal{P}$-$\mathcal{Q}$ charts already given. It is equal to the ratio k_{p0}/k_{q0}.

7.7. PORT SHAPING

As stated above, the gain of a valve varies with the position of the operating point on the $\mathcal{P}$-$\mathcal{Q}$ diagram. For rectangular ports and for most valve types, it is constant on the $\mathcal{P}_m = 0$ line, so if the load required no force, the valve would be an ideal linear integrator. This is usually a trivial solution, however, and if we have an actual load, the gain may vary radically with operating point.

Even if it were possible, which it usually is not, to define a load and its duty cycle accurately, it would still be hopeless in most cases to marry the load equation and the valve-characteristic equation and to come up with anything which could be handled analytically. If, however, we make the assumption that the load is massless and viscous, as in Sec. 6.62, we can derive some useful results. The assumption is unjustified in that few loads even approximate linearity except over very restricted ranges of operating conditions, but it is justified to the extent that a linear load is a better approximation to the truth than a zero load.

Suppose that we have a pure viscous load for which

$$f_m = k_f v_m \tag{7.46}$$

where f_m = instantaneous force applied to load

and

v_m = resulting instantaneous velocity

Suppose that this load is caused to move, in some fashion, by an ideal double-acting ram of effective area A_m. (The parallel derivation for the rotary case follows obviously.) Then in terms of p_m and q_m, $f_m = A_m p_m$ and $q_m = A_m v_m$ so that

$$p_m = \frac{k_f}{A_m{}^2} q_m \tag{7.47}$$

This equation may be nondimensionalized exactly as were the previous equations of this chapter, by dividing by appropriate reference flows and pressures. If these quantities are called P_r and Q_r, the equation becomes

$$\mathcal{P}_m = k\mathcal{Q}_m \tag{7.48}$$

where

$$\mathcal{P}_m = \frac{p_m}{P_r} \tag{7.49}$$

$$\mathcal{Q}_m = \frac{q_m}{Q_r} \tag{7.50}$$

$$k = \frac{k_f Q_r}{A_m{}^2 P_r} \tag{7.51}$$

Thus, a load which is linear in f-v space is also linear in $\mathcal{P}_m$-$\mathcal{Q}_m$ space, and by the proper choice of A_m, P_r, and Q_r we can make its slope anything we please. This choice of slope will be useful in improving the linearity of the valve-load combination.

The simplest example of the technique to be described is its application to the ideal valve, for which the characteristic equation is

$$\mathcal{P}_m = 1 - \frac{2\mathcal{Q}_m{}^2}{y^2} \tag{7.22}$$

If this is combined with Eq. 7.51, we get a quadratic in $\mathcal{Q}_m$ with k as a parameter:

$$2\mathcal{Q}_m{}^2 + ky^2\mathcal{Q}_m - y^2 = 0 \tag{7.52}$$

whose solution is

$$\mathcal{Q}_m = -\frac{y}{4}\left(ky \pm \sqrt{k^2 y^2 + 8}\right) \tag{7.53}$$

This equation is plotted in Fig. 7.19 for three values of k. When $k = 0$, the graph is a straight line, but for the other two values, it is convex upward. The value $k = \sqrt{8/3}$ was chosen because it is the condition for getting the absolute maximum power from the valve, which occurs when $y = 1$, $\mathcal{P}_m = 2/3$, and $\mathcal{Q}_m = \sqrt{1/6}$. The value $k = 1$ was chosen merely because it makes the arithmetic simpler; it reduces the obtainable power only about 10 per cent and makes the port shape simpler, as we shall see.

Even the most convex of the three curves of Fig. 7.19 is fairly straight and at first glance would seem to represent a fairly linear device. The gain of the valve, however, is proportional to the *slope* of the curve, and the slope is a direct factor of the forward gain when the valve is used in a closed-loop system. A generally accepted rule of thumb in servo design is that loop gain should not be allowed to vary more than 2:1 over the operating range. Since other elements of the loop may also be somewhat nonlinear, it is desirable to minimize the change in gain of the valve

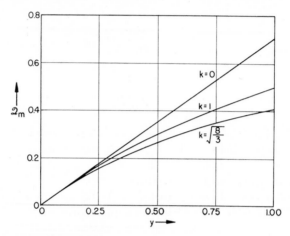

Fig. 7.19. Q_m-versus-y curves for Case P1.

and therefore to straighten out the curve of Fig. 7.19 which corresponds to the desired value of k.

The gain of a valve, as defined by Eq. 7.45, is proportional to the peripheral width of the several ports at the particular stem position, and we can modify the gain-versus-position relationship by modifying the shapes of the ports. In the present case, the gain is highest for small values of y and decreases (the curve becomes more nearly parallel to the y-axis) as y increases. This effect could be canceled for any one value of k by making the port narrow for small y's and widening it for large ones so that the width is made inversely proportional to the slope of the Q_m-versus-y curve at each point. Profiles of the port shapes required for constant-gain operation for the three k's of Fig. 7.19 are given in Fig. 7.20.

It will be seen that $k = 0$ implies a rectangular port (as it does for many types of valves), $k = \sqrt{8/3}$ requires a port with wide flare and strongly curved sides, and $k = 1$ needs a port with much less flare and straighter sides, which could be fairly well approximated by a trapezoid.

Conversely, both Figs. 7.19 and 7.20 show that for a given port shape the linearity is strongly dependent on the value of k; for a rectangular port, k should be small.[3]

The ideal valve has been chosen as an example because of the sim-

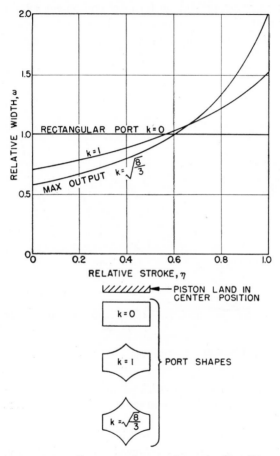

Fig. 7.20. Constant-gain port shapes for Case P1.

plicity of its characteristic equation. As a practical matter, the change of gain is not excessive for most purposes except for large values of k, which should ordinarily be avoided. Also, the gain is greatest at the

[3] For another approach to the same conclusion, see F. C. Paddison and W. A. Good, "A Method for the Selection of Valves and Power Pistons in Hydraulic Servos," ASME Paper No. 55-S-10, presented at the ASME Baltimore Meeting, April, 1955. This approach was briefly outlined in Paddison's discussion of the Lee and Blackburn paper, *Trans. ASME*, Vol. 74 (1952), pp. 1009–1011.

origin and falls off seriously only for large values of y, which is not a bad state of affairs. There are other valves, however, for which the converse is true and which can be greatly improved by judicious port shaping.

One such case is the twin-pump servo of Sec. 7.42. This is about the simplest arrangement possible in terms of equipment required, but with rectangular ports it is so nonlinear as to be rather useless for most purposes. For the higher values of k, not only is the gain change large

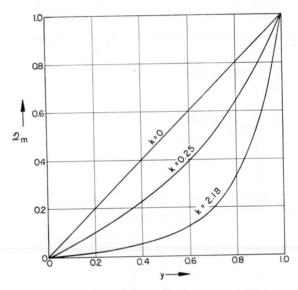

Fig. 7.21. Q_m-versus-y curves for Case P2.

but the gain is least near the origin and increases with y. It is necessary to keep other factors of loop gain small enough to prevent oscillation when y is large, which means that the gain will be very low near the center and the servo performance will be poor.

If we use the procedure just followed for the ideal valve, the twin-pump servo will give the curves of Q_m versus y that are shown in Fig. 7.21. Two of the values of k were chosen as before; the other, 2.18, gives the maximum power output for a maximum y of 0.8 and also means that with a peak pressure of 1000 psi, for example, the idling pressure would be only $P_s/8k$, or about 57 psi. This is a very desirable property since it means that the stand-by losses are only about 6 per cent of the peak power and the over-all efficiency is fairly high for a low-duty-cycle system. There is nothing significant about the actual number chosen, but it is roughly the value preferred by the makers of automotive power-steering gears and similar devices.*

Again it is possible to improve the performance by shaping the ports. Port profiles for the three values of k are given in Fig. 7.22. They are

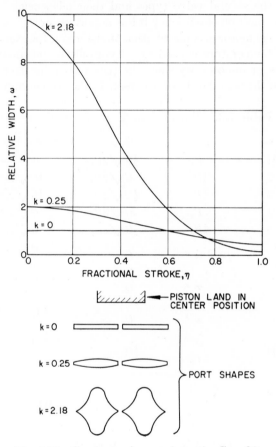

Fig. 7.22. Constant-gain port shapes for Case Q2.

strongly curved near the ends, but if operation is confined to values of y no larger than 0.8, and if it is remembered that the approximation of the linear load is a poor one at best, a trapezoidal or even a triangular port would be good enough. It might even be possible to get fairly good operation using a segment of a circle, which would enable the use of drilled ports and make the valve much cheaper to manufacture.

* See Chapter 18 for a discussion of hydraulic automotive steering gears.

7.8. POWER OUTPUT AND EFFICIENCY

The final characteristics to be discussed in this chapter are the power outputs of the several valve types and their efficiencies. For several reasons, the question of efficiency is less important in this case than with a squirrel-cage motor or a distribution transformer, for example. In the first place, temperature control of a hydraulic or pneumatic device is usually easy or unnecessary since the power medium itself serves as a coolant, and the heat developed by losses in the system may be removed

Table 7.3. POWER AND EFFICIENCY FORMULAS

Case	$\mathcal{P}_{m\ max}$	$\mathcal{Q}_{m\ max}$	Quiescent Power Loss
P1	P_s	$\frac{1}{2}G\sqrt{2P_s}$	0
P2	$\frac{1}{2}P_s$	$G\sqrt{2P_s}$	$\frac{1}{2}G\sqrt{2P_s^3}$
P3	$\frac{1}{2}P_s$	$\frac{1}{2}G\sqrt{2P_s}$	$\frac{1}{2}G\sqrt{2P_s^3}$
P4	P_s	$G\sqrt{2P_s}$	$G\sqrt{2P_s^3}$
P5	$\frac{4}{5}P_s$	$\frac{1}{2}G\sqrt{2P_s}$	$G\sqrt{2P_s^3}$
Q1	∞ *	Q_s	$\dfrac{Q_s^3}{G^2}$
Q2	∞ *	Q_s	$2\dfrac{Q_s^3}{G^2}$
Q3	∞ *	Q_s	$\dfrac{1}{2}\dfrac{Q_s^3}{G^2}$
Q4	$\dfrac{Q_s^2}{G^2}$	$\frac{1}{2}Q_s$	$\dfrac{1}{2}\dfrac{Q_s^3}{G^2}$

Case	Condition for $W_o = max$	$W_{o\ max}$	Efficiency When $W_o = max$
P1	$p_m = \frac{2}{3}P_s$	$\sqrt{\frac{2}{27}}\,GP_s^{3/2}$	$\frac{2}{3}$
P2	$p_m = \frac{2}{3}P_s$	$\sqrt{\frac{2}{27}}\,GP_s^{3/2}$	$\frac{2}{3}$
P3	$p_m = \frac{2}{3}P_s$	$\frac{1}{2}\sqrt{\frac{2}{27}}\,GP_s^{3/2}$	$\frac{2}{3}$
P4	$p_m = \frac{2}{3}P_s$	$2\sqrt{\frac{2}{27}}\,GP_s^{3/2}$	$\frac{2}{3}$
P5	$p_m = 0.1214P_s$	$0.1922GP_s^{3/2}$	0.1671
Q4	$q_m = \frac{1}{4}Q_s$	$\dfrac{1}{8}\dfrac{Q_s^2}{G^2}$	0.1671

* In cases Q1, Q2, and Q3 (and in any other series circuit with a CQ supply) the peak pressure is theoretically infinite and the maximum power occurs at maximum pressure, with infinite output and unity efficiency. Obviously, this is not a useful mode of operation.

at any convenient point by a suitable heat exchanger.   In an electrical device, on the other hand, the medium has no such effect, and the removal of heat becomes a major problem.   Then, too, several types of hydraulic power sources are essentially constant-power devices; they put power into the oil at a constant rate, and any power not used in the load is wasted as heat, either in the valve or elsewhere.   With such a power supply, there is relatively little point in worrying very much about efficiency in the power-using or -controlling device.   Finally, average efficiency is usually high only in a system with a relatively constant load; in systems with poor load factors and especially with high-performance servos, the average efficiency is determined more by the nature of the load and especially by what the load is made to do than by the design of the servo itself.

The foregoing remarks are probably somewhat too strong.   Obviously, if a device is only a few per cent efficient, it is uneconomic to use it at high power levels.   The real point is that a difference between say 50 and 75 per cent efficiency is usually unimportant for the devices under discussion.

The input power is obviously the product of the input flow and the supply pressure.   One of these quantities is assumed to be known in each of the cases already discussed, but in general the expression for the other is as complicated as that for the output quantity, and it hardly seems worth while to compute families of input-flow-versus-pressure curves analogous to the $\mathcal{P}_m$-$\mathcal{Q}_m$ curves already given.

It is important, of course, to know the maximum values of flow and pressure that the pump will have to furnish, and these and various other quantities of interest are given in Table 7.3.

8

F. D. Ezekiel
J. L. Shearer

Pressure-Flow
Characteristics of
Pneumatic Valves

8.1. INTRODUCTION

When a compressed gas rather than a pressurized liquid is used as the working medium in a fluid-power-control system, the nature of the flow through orifice-type control valves is somewhat different from that which holds for liquid flows. The greatest difference is that the fluid-density changes which occur in the case of a compressed-gas flow are many times greater than in the case of a liquid flow. Also, compressed gases usually have less density and much less viscosity than most liquids. In addition to the greater density changes which occur with gases for given pressure changes, the velocity of sound in gases is much lower than that in liquids, and it is possible to attain sonic velocity in gaseous flow through an orifice when the pressure drop is about 50 per cent of the upstream pressure. Once sonic flow has been attained, any further drop in downstream pressure does not change the weight rate of flow through the orifice (see Chapter 3, Fundamentals of Fluid Flow). Furthermore, it is more convenient in many cases to use weight rate of flow instead of volume rate of flow when dealing with pneumatic systems because the continuity and energy equations are more familiar to most engineers when expressed in terms of weight (sometimes called mass) rates of flow. It should be noted that temperature is an important variable in pneumatic systems since the density of the fluid is a strong function of both temperature and pressure.

Although it is in some ways more difficult to analyze the character-istics of pneumatic valves than it is to analyze the characteristics of hydraulic valves, a variable orifice is desirable for pneumatic-power control because of about the same reasons that it is widely used in hydraulic-power control (see Chapter 9, Valve Configurations and Constructions). In addition, the presently available means of varying the rate of pneumatic-power generation are slow and cumbersome com-pared with variable-displacement hydraulic pumps.

In the discussion which follows, the pressure-flow characteristics are derived from equations that hold for frictionless flow of perfect and semiperfect [1,2,3] gases. Because the viscosity of most commonly used gases is at least two orders of magnitude less than that of commonly used hydraulic fluids, the effects of friction in control-valve orifices are almost always negligible. Commonly used gases such as air, nitrogen, products of combustion, and so on, are usually used in a range of states that is sufficiently far enough away from their critical points and liquid states for them to be considered as perfect or semiperfect gases. This is true, for instance, of air at pressures below 3000 psi and temperatures above 350°R. Also, many of the significant conclusions that come out of work with perfect and semiperfect gases may be applied qualitatively to cases where the behavior of the gas departs considerably from the perfect-gas equations. This is true, for instance, when steam is used without much superheat, as long as none of it condenses to liquid water.

8.2. FLOW THROUGH A SINGLE ORIFICE

When it is necessary to control the flow of gas to a load in such a way that the direction of load flow does not change, a single variable orifice is usually the most satisfactory type of pneumatic valve to use.

The basic equation for the flow of a perfect or semiperfect gas through

[1] A. H. Shapiro, *The Dynamics and Thermodynamics of Compressible Fluid Flow—Vol. I*, Ronald Press, New York, 1953.

[2] J. H. Keenan, *Thermodynamics*, John Wiley and Sons, New York, 1941.

[3] H. W. Liepmann and A. E. Puckett, *Introduction to Aerodynamics of a Compressible Fluid*, John Wiley and Sons, New York, 1947,

a single orifice is of the following form: [1,3]

$$W = C_d A_o f\left(P_u,\ T_u,\ \frac{P_d}{P_u}\right) \tag{8.1}$$

where W = weight rate of flow, lb/sec
C_d = discharge coefficient of the orifice, dimensionless
A_o = orifice area, in.[2]
P_u = upstream stagnation pressure, psia
T_u = upstream stagnation temperature, °R
P_d = downstream pressure, psia
f = a function of P_u, T_u, and $\dfrac{P_d}{P_u}$

When P_d/P_u is greater than $(P_d/P_u)_{\text{crit}}$,

$$f\left(P_u,\ T_u,\ \frac{P_d}{P_u}\right) = \frac{W}{C_d A_o} = C_1 \frac{P_u}{\sqrt{T_u}}\left(\frac{P_d}{P_u}\right)^{1/k}\sqrt{1 - \left(\frac{P_d}{P_u}\right)^{(k-1)/k}} \tag{8.2}$$

where k is the ratio of the specific heats, C_p/C_v. But when P_d/P_u is equal to or less than $(P_d/P_u)_{\text{crit}}$,

$$f_{\max}\left(P_u,\ T_u,\ \frac{P_d}{P_u}\right) = \left(\frac{W}{C_d A_o}\right)_{\max} = C_2 \frac{P_u}{\sqrt{T_u}} \tag{8.3}$$

For air at pressures below 3000 psia and temperatures above 350°R, $k = 1.4$ and $R = 2.48 \times 10^5$ in.^{2}/sec^2 °R, so the following values may be used:

$$\left(\frac{P_d}{P_u}\right)_{\text{crit}} = \left(\frac{2}{k+1}\right)^{k/(k-1)} = 0.528 \tag{8.4}$$

$$C_1 = g\sqrt{\frac{2k}{R(k-1)}} = 2.06\,\frac{\sqrt{°R}}{\text{sec}} \tag{8.5}$$

$$C_2 = g\sqrt{\frac{k}{R\left(\dfrac{k+1}{2}\right)^{(k+1)/(k-1)}}} = 0.532\,\frac{\sqrt{°R}}{\text{sec}} \tag{8.6}$$

The equation for weight rate of flow may also be written as follows:

$$W = C_d A_o\left[\frac{\dfrac{W}{C_d A_o}}{\left(\dfrac{W}{C_d A_o}\right)_{\max}}\right]C_2\frac{P_u}{\sqrt{T_u}} \tag{8.7}$$

Dividing Eq. 8.2 by Eq. 8.3 gives for $(P_d/P_u) > 0.528$

$$\frac{\left(\dfrac{W}{C_d A_o}\right)}{\left(\dfrac{W}{C_d A_o}\right)_{max}} = \frac{A^*}{C_d A_o} = f_1\left(\frac{P_d}{P_u}\right) = \frac{C_1}{C_2}\left(\frac{P_d}{P_u}\right)^{1/k}\sqrt{1 - \left(\frac{P_d}{P_u}\right)^{(k-1)/k}} \qquad (8.8)$$

and when $(P_d/P_u) < 0.528$, $f_1(P_d/P_u) = 1.0$.

The function $f_1(P_d/P_u)$ is shown graphically in Fig. 8.1. This function is identical with the function A^*/A given in tabular form in the

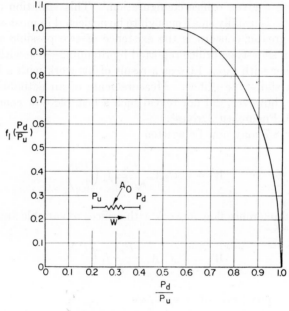

Fig. 8.1. Plot of $f_1(P_d/P_u)$.

Gas Tables.[4] It is significant to note that the pressure at the *vena contracta* of an orifice never falls below $0.528P_u$ and that when the downstream pressure, P_d, falls below $0.528P_u$, it no longer affects the rate of flow in any way. Although the static temperature of the air passing through the orifice falls as its velocity increases, its stagnation temperature, stagnation enthalpy, and stagnation internal energy remain constant throughout the flow so that it is usually unnecessary to be

[4] J. H. Keenan and J. Kaye, *Gas Tables*, John Wiley and Sons, New York, 1948.

concerned with this temperature unless problems of heat transfer and temperature gradients in the area of the high-velocity flows become important. In a control valve the length of high-velocity flow stream is necessarily quite short, and experimental data that have been obtained to date do not reveal any significant effects of heat transfer to the locally cold regions of orifice flow. In fact, the downstream stagnation temperature of the fluid is usually slightly lower than anticipated from a perfect-gas flow because of the Joule-Thompson effect.[5]

The discharge coefficient, C_d, does seem to vary considerably with the geometry of the orifice. Observations to date indicate that C_d may vary from 0.6 to 1.0 for various orifice configurations and seems to increase slightly with decreasing pressure ratio (P_d/P_u) when $P_d/P_u < 0.5$ for flow through a given orifice configuration.[6] This variation of C_d with pressure ratio is usually small enough to be neglected in most engineering work. Of greater concern is the existence of two possible stable flow regimes in a slide-type orifice reported by Stenning—one with $C_d \approx 0.8$ and the other with $C_d \approx 1.0$—as a result of his work with a large scale model of a slide-type orifice. Measurements of an actual-size sliding plate valve, however, did not reveal such a bistable flow condition (see Chapter 16, Pneumatic Drives).

Equation 8.7 may now be written

$$W = C_d C_2 A_o \frac{P_u}{\sqrt{T_u}} f_1 \left(\frac{P_d}{P_u} \right) \tag{8.9}$$

and a useful reference flow in work with orifices of variable areas is given by

$$W_i = C_d C_2 A_i \frac{P_s}{\sqrt{T_s}} f_1 \left(\frac{P_e}{P_s} \right) \tag{8.10}$$

where W_i = initial rate of flow, lb/sec
A_i = initial orifice area, in.^{2}
P_e = constant exhaust pressure, psia
P_s = constant supply pressure, psia
T_s = constant stagnation temperature of supply, °R

By neglecting variations in C_d and assuming that T_u is equal at all times to T_s (see Chapter 16), Eq. 8.9 may be divided by Eq. 8.10 to give a nondimensional flow equation:

[5] J. H. Keenan, *Thermodynamics*, John Wiley and Sons, New York, 1941.
[6] A. H. Stenning, "An Experimental Study of Two-Dimensional Gas Flow Through Valve Type Orifices," ASME Paper No. 54-A-45.

$$\frac{W}{W_i} = \gamma \frac{P_u}{P_s} \frac{f_1(P_d/P_u)}{f_1(P_e/P_s)} \qquad (8.11)$$

where $\gamma = \dfrac{A_o}{A_i}$

In some cases it is convenient to use another form of this equation, namely

$$\frac{W}{W_i} = \gamma \frac{P_d}{P_e} \frac{(P_u/P_d)f_1(P_d/P_u)}{(P_s/P_e)f_1(P_e/P_s)} = \gamma \frac{P_d}{P_e} \frac{f_2(P_d/P_u)}{f_2(P_e/P_s)} \qquad (8.12)$$

where $f_2\left(\dfrac{P_d}{P_u}\right) = \dfrac{P_u}{P_d} f_1\left(\dfrac{P_d}{P_u}\right)$

$\qquad f_2\left(\dfrac{P_e}{P_s}\right) = \dfrac{P_s}{P_e} f_1\left(\dfrac{P_e}{P_s}\right)$

The function $f_2(P_d/P_u)$ is given by the graph in Fig. 8.2.

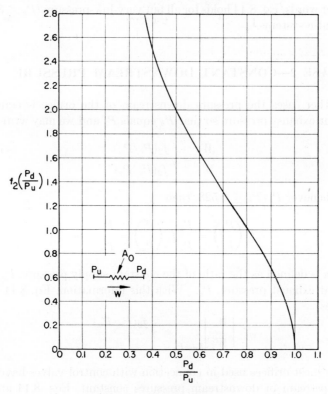

Fig. 8.2. Plot of $f_2(P_d/P_u)$.

8.3. CASE 1—CONSTANT UPSTREAM PRESSURE

In many cases the pressure upstream of an orifice is equal to a constant supply pressure so that we may write

$$\frac{W}{W_i} = \gamma \frac{f_1(P_d/P_s)}{f_1(P_e/P_s)} \tag{8.13}$$

Note that $f_1(P_e/P_s)$ is constant when P_e is constant and that it is equal to 1.0 when $P_e/P_s < 0.528$.

In a pneumatic system which exhausts to atmospheric pressure of approximately 15 psia, the following equation holds whenever $P_s > 30$ psia.

$$\left[\frac{W}{W_i}\right]_{P_u=P_s} = \gamma f_1\left(\frac{P_d}{P_s}\right) \tag{8.14}$$

In other words, Eq. 8.14 holds for all but very-low-pressure ($P_s < 30$ psia) pneumatic systems.

8.4. CASE 2—CONSTANT DOWNSTREAM PRESSURE

In other cases the pressure downstream of the orifice is equal to a constant exhaust pressure so that P_d equals P_e and we may write

$$\frac{W}{W_i} = \gamma \frac{f_2(P_e/P_u)}{f_2(P_e/P_s)} \tag{8.15}$$

And whenever $P_e/P_s \leq 0.528$, then

$$f_2\left(\frac{P_e}{P_s}\right) = \frac{P_s}{P_e}f_1\left(\frac{P_e}{P_s}\right) = \frac{P_s}{P_e} \equiv n \tag{8.16}$$

where n is defined as the ratio of the constant supply pressure, P_s, to the constant exhaust pressure, P_e. With this information, Eq. 8.11 simply becomes

$$\left[\frac{W}{W_i}\right]_{P_d=P_e} = \gamma \frac{f_2(P_e/P_u)}{n} \tag{8.17}$$

Since most orifices used in connection with control valves have either their upstream or downstream pressures constant, Eqs. 8.14 and 8.17 form the basic building blocks for the following work.

8.5. STEADY FLOW THROUGH TWO ORIFICES IN SERIES

Figure 8.3 is a sketch of two variable orifices in series having a pressure, P_a, between them. The flow through the upstream orifice, W_1/W_i, is determined from Eq. 8.14 since P_s is constant. The flow through the downstream orifice, W_2/W_i, is similarly determined from Eq. 8.17. In order to be able to see graphically the relationship between the two flows and the variables involved, it is found useful to obtain a plot that combines both flows as a function of a single variable. This operation can be achieved as follows:

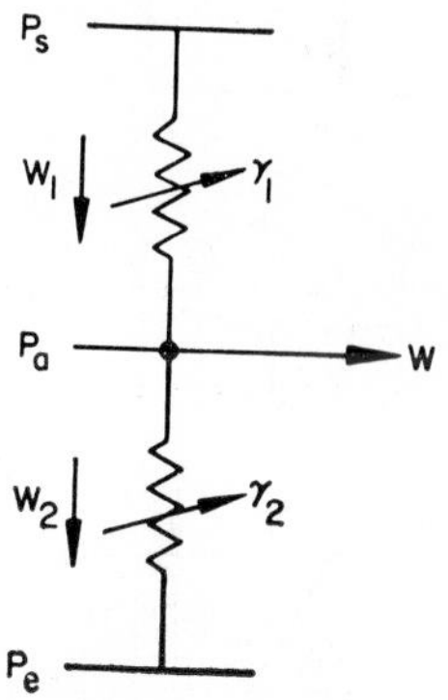

$$\frac{W_1}{W_i} = \gamma_1 f_1 \left(\frac{P_a}{P_s}\right) \tag{8.18}$$

$$\frac{W_2}{W_i} = \gamma_2 \frac{f_2(P_e/P_a)}{n} = \gamma_2 \frac{f_2(P_s/nP_a)}{n} = \gamma_2 f_2' \left(\frac{P_s}{P_a}\right) \tag{8.19}$$

Fig. 8.3. Two variable orifices in series.

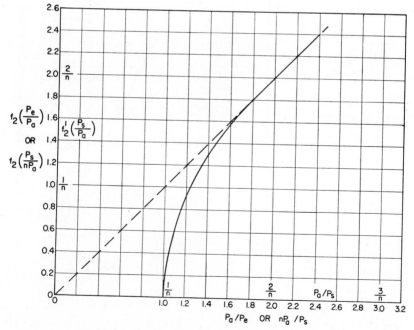

Fig. 8.4. Transformation of $f_2(P_e/P_a)$ into $f_2'(P_s/P_a)$.

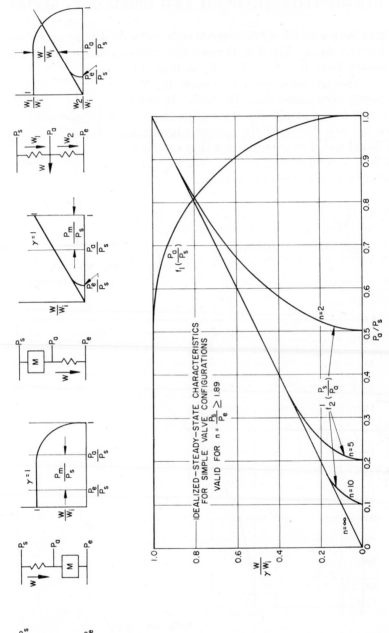

Fig. 8.5. Flow determinations in various simple orifice configurations.

Figure 8.4 plots $f_2(P_e/P_a) = f_2(P_s/nP_a)$ as a function of $P_a/P_e = nP_a/P_s$. By dividing both coordinates of Fig. 8.4 by n one can visualize the same plot as being $f_2(P_s/nP_a)/n = f_2'(P_s/P_a)$ as a function of P_a/P_s. With this transformation, unity in Fig. 8.4 will be $1/n$ units in the new plot as shown in the same figure. The advantage of this transformation is quite significant, as can be seen when the two flows (both functions of P_a/P_s) are plotted in Fig. 8.5 for various values of n.

8.6. STEADY-FLOW CHARACTERISTICS OF SINGLE-ORIFICE CONTROL

When a single orifice is used to control the flow of gas through a motor, it may be placed either upstream or downstream of the motor. Figure 8.6a shows a sketch of the first arrangement. Figure 8.6b shows how Fig. 8.5 can be used to obtain typical characteristics similar to the ones shown in Fig. 8.6c.

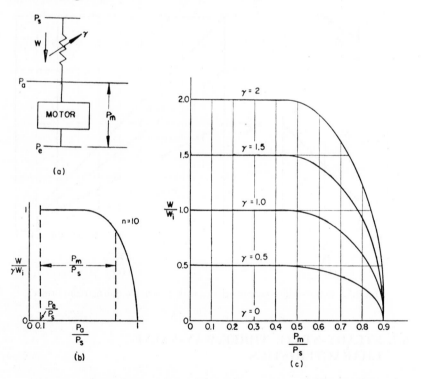

Fig. 8.6. Single-orifice characteristics with a variable upstream orifice.

Figure 8.7a represents a sketch of the second arrangement. Figure 8.7b shows how Fig. 8.5 can be used to obtain typical characteristics similar to the ones shown in Fig. 8.7c. It should be noted here that the temperature, T_a, leaving the motor is not likely to be equal to T_s, and if it departs appreciably from T_s, the nondimensionalized equations cannot be used. However, since the flow varies inversely as the square root of T_a, where T_a is in degrees Rankine, the effect of a variation of T_a by as much as 100°F from T_s results in less than 10 per cent change in flow. For this reason, T_a will be taken equal to T_s. Figures 8.6 and 8.7 use 10 for the value of n. It is important to note the significant difference between upstream and downstream valve operation. This difference does not exist in a hydraulic system.

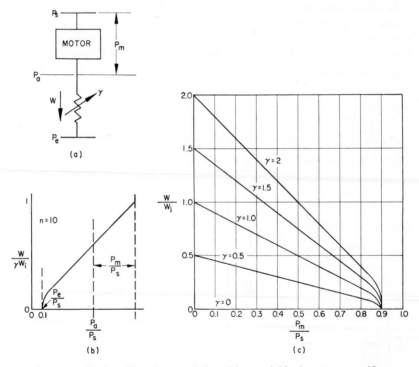

Fig. 8.7. Single-orifice characteristics with a variable downstream orifice.

8.7. STEADY-STATE THREE-WAY-VALVE CHARACTERISTICS

In order to be able to control the direction as well as the rate of flow to a motor, three-way valves are used similar to the ones shown in Figs. 8.8,

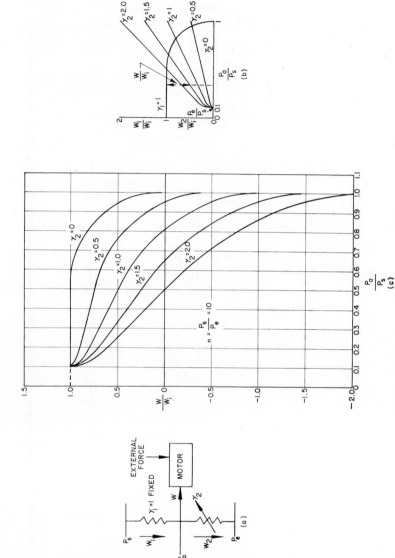

Fig. 8.8.   Three-way-valve characteristics with a fixed upstream orifice and a variable downstream orifice.

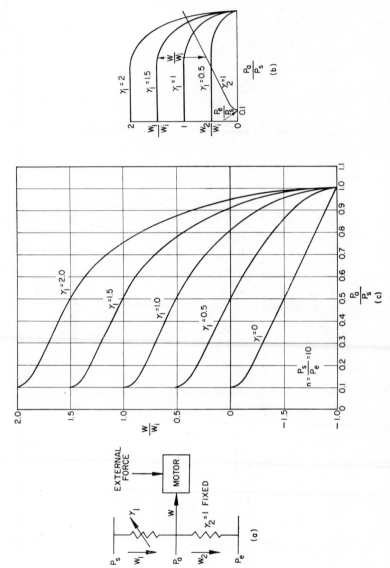

Fig. 8.9. Three-way-valve characteristics with a variable upstream orifice and a fixed downstream orifice.

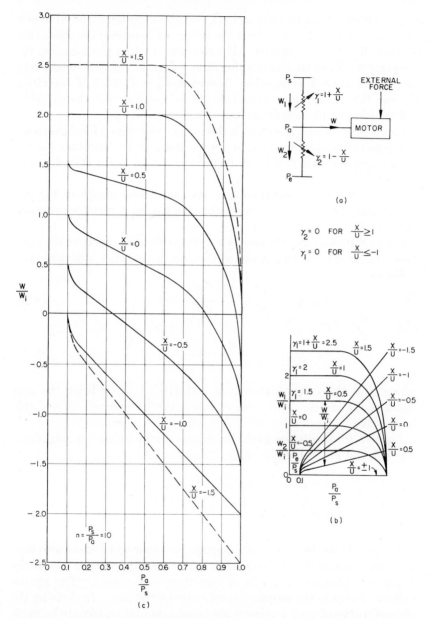

Fig. 8.10. Three-way-valve characteristics with variable upstream and downstream orifices.

8.9, and 8.10.   A spring force or back pressure must be employed to provide the energy required to drive the motor in the reverse direction, that is, when W becomes negative.   For the same reasons given earlier, the temperature, T_a, of the fluid flowing from the motor will be assumed to be equal to T_s when W is negative.   However, $T_a = T_s$ when W is positive because the stagnation temperature of the gas does not change when it flows through the supply orifice.

Figure 8.8a shows a sketch for a three-way valve with a fixed upstream orifice and a variable downstream orifice.   Figure 8.8b shows how Fig. 8.5 can be used to obtain typical characteristics similar to the ones shown in Fig. 8.8c.

Figure 8.9 shows correspondingly the same items shown in Fig. 8.8 except that they are for a three-way valve having a variable upstream orifice and a fixed downstream one.

Figure 8.10 describes the case of varying both orifice areas at the same time—increasing the area of one as the area of the other is decreased.

For this case, let

$$\gamma_1 = \frac{A_1}{A_i} = \frac{A_i + wX}{A_i} = 1 + \frac{X}{U} \tag{8.20}$$

and

$$\gamma_2 = \frac{A_2}{A_i} = \frac{A_i - wX}{A_i} = 1 - \frac{X}{U} \tag{8.21}$$

where X is the valve displacement and $U = A_i/w$, w being the port width of the orifice.   Such area variation can be achieved by a simple spool valve.

Figure 8.10b shows how to obtain characteristics for such a valve from Fig. 8.5, and Fig. 8.10c gives a typical example.

8.8. STEADY-STATE FOUR-WAY-VALVE CHARACTERISTICS

Figure 8.11 shows a schematic diagram for a four-way valve controlling a piston in a cylinder.   It can be seen that the configuration of a four-way valve is essentially made up of two three-way valves.   The piston-cylinder motor is the coupling agent between the two.   In deriving the characteristics of such a valve, a few more assumptions have to be made in order to reduce the complexity of the problem.   Some of these assumptions are:

1. There is no friction between the piston and cylinder walls.
2. There is no leakage of gas past the piston.

3. There will be enough heat transfer to maintain constant temperature on either side of the piston.

4. The gas in each chamber is homogeneous.

5. The pressures on each side of the piston do not vary with time.

The most unrealistic assumption of the above is No. 3. However, since many factors are involved when the ram is in motion, it is felt that it is rather important to derive the characteristics for the hypothetical constant-temperature case before analyzing more complicated ones.

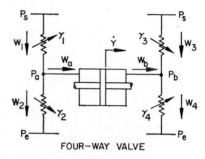

Fig. 8.11. Schematic diagram for a four-way valve controlling a ram in a cylinder.

Having made these assumptions, it can now be deduced that the volumetric rate of flow in and out of the ram must be the same. This deduction can be expressed mathematically as follows:

$$\frac{W_a}{\rho_a}\frac{\rho_s}{W_i} = \frac{W_b}{\rho_b}\frac{\rho_s}{W_i} \quad (8.22)$$

Since all the temperatures involved are assumed to be equal to T_s, by using the perfect-gas law, Eq. 8.22 can be written in terms of the pressures rather than the densities.

$$\frac{W_a}{P_a}\frac{P_s}{W_i} = \frac{W_b}{P_b}\frac{P_s}{W_i} = \frac{\dot{Y}}{\dot{Y}_i} \quad (8.23)$$

where $\dot{Y}$ is the velocity of the ram and $\dot{Y}_i = W_i/\rho_s g A_r$.

The flow equations can now be written as follows:

$$\frac{W_a}{W_i} = \gamma_1 f_1\left(\frac{P_a}{P_s}\right) - \gamma_2 f_2'\left(\frac{P_a}{P_s}\right) \quad (8.24)$$

$$\frac{W_b}{W_i} = \gamma_4 f_2'\left(\frac{P_b}{P_s}\right) - \gamma_3 f_1\left(\frac{P_b}{P_s}\right) \quad (8.25)$$

Multiplying Eq. 8.24 by P_s/P_a and Eq. 8.25 by P_s/P_b and using Eq. 8.23 result in

$$\frac{W_a}{W_i}\frac{P_s}{P_a} = \frac{P_s}{P_a}\left[\gamma_1 f_1\left(\frac{P_a}{P_s}\right) - \gamma_2 f_2'\left(\frac{P_a}{P_s}\right)\right] = \frac{\dot{Y}}{\dot{Y}_i} \quad (8.26)$$

$$\frac{W_b}{W_i}\frac{P_s}{P_b} = \frac{P_s}{P_b}\left[\gamma_4 f_2'\left(\frac{P_b}{P_s}\right) - \gamma_3 f_1\left(\frac{P_b}{P_s}\right)\right] = \frac{\dot{Y}}{\dot{Y}_i} \quad (8.27)$$

The above two equations represent the nondimensional ram velocity as calculated to satisfy continuity on both sides of the ram.

8.81. Closed-Center Valve

For a closed-center valve only two diagonal orifices (Fig. 8.11) are open at any one time. Assuming $\gamma_1 = \gamma_4 = \gamma$ and $\gamma_3 = \gamma_2 = 0$ for the case when only orifices 1 and 4 are open, Eqs. 8.26 and 8.27 simplify to

$$\frac{P_s}{P_a}\gamma f_1\left(\frac{P_a}{P_s}\right) = \frac{P_s}{P_b}\gamma f_2'\left(\frac{P_b}{P_s}\right) = \frac{\dot{Y}}{\dot{Y}_i} \tag{8.28}$$

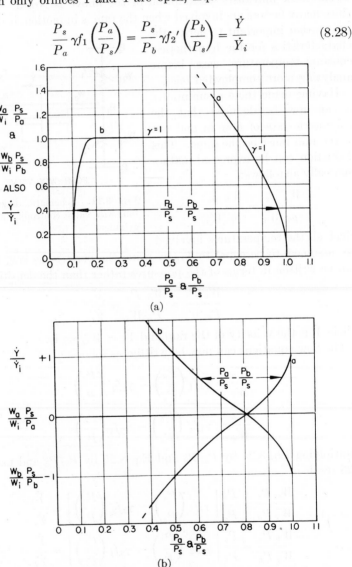

(a)

(b)

Fig. 8.12. Graphical procedure for determining the characteristics for a four-way valve.

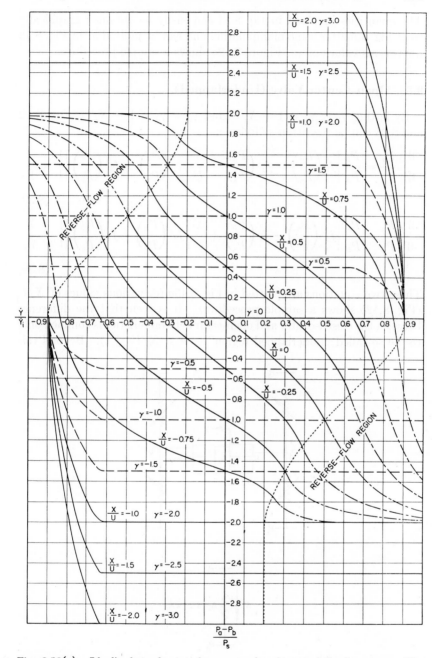

Fig. 8.13(a).   Idealized steady-state four-way-valve characteristics for compressible fluids at constant temperature.

The quantities involved in Eq. 8.24 can easily be obtained graphically from Fig. 8.5 after some manipulation and can be plotted as in Fig. 8.13 for various values of γ. Figure 8.12a shows a sample procedure.

8.82. Open-Center Valve (Underlapped)

By letting $\gamma_1 = \gamma_4 = 1 + X/U$ and $\gamma_2 = \gamma_3 = 1 - X/U$, Eqs. 8.22 and 8.23 can be rewritten as follows:

$$\frac{W_a}{W_i}\frac{P_s}{P_a} = \frac{P_s}{P_a}\left\{f_1\left(\frac{P_a}{P_s}\right) - f_2'\left(\frac{P_a}{P_s}\right) + \frac{X}{U}\left[f_1\left(\frac{P_a}{P_s}\right) + f_2'\left(\frac{P_a}{P_s}\right)\right]\right\} = \frac{\dot{Y}}{\dot{Y}_i}$$

(8.29)

$$\frac{W_b}{W_i}\frac{P_s}{P_b} = \frac{P_s}{P_b}\left\{f_2'\left(\frac{P_a}{P_s}\right) - f_1\left(\frac{P_b}{P_s}\right) + \frac{X}{U}\left[f_2'\left(\frac{P_b}{P_s}\right) + f_1\left(\frac{P_b}{P_s}\right)\right]\right\} = \frac{\dot{Y}}{\dot{Y}_i}$$

(8.30)

These two equations are plotted in Fig. 8.13a for various values of X/U (Fig. 8.12b shows a sample procedure). It is seen that the difference between the closed-center and open-center characteristics is confined only to the so-called "underlapped region" $(-1 \leq X/U \leq 1)$. Beyond this

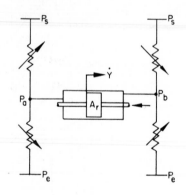

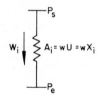

Fig. 8.13(b). Configurations for Fig. 8.13a.

region both characteristics coincide. The areas in Fig. 8.13 that are marked "reverse-flow region" represent flow to the supply source, a condition that will be discussed in the next paragraph.

8.9. A METHOD FOR HANDLING REVERSE FLOWS

In certain cases one may expect the gas to flow to the supply source or even some gas to flow from the exhaust side into the system. For these cases the methods outlined earlier can be extended. Figure 8.14 represents the flow-pressure characteristics of two orifices in series. In

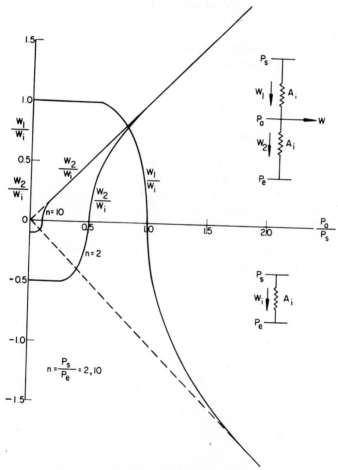

Fig. 8.14. Reverse flows through two orifices in series.

this plot a negative flow signifies a flow in the reverse direction from the defined one. Figure 8.14 can be looked upon as a more complete picture of Fig. 8.5 and can be obtained by extending the idea that the upstream orifice, 1, has a constant downstream pressure, P_s, when P_a is larger than P_s. Similarly a negative flow through orifice 2 signifies that P_a is less than P_e; hence this orifice can be considered as having a constant-pressure supply, P_e, for this special situation.

With the help of Fig. 8.14 all the previously obtained characteristics for various valve configurations can be extended to take into account reverse flows.

NOMENCLATURE

Symbol	Definition
A	orifice area, in.^{2}
A^*	critical orifice area, in.^{2}
A_r	ram area, in.^{2}
C_d	coefficient of discharge
g	acceleration of gravity, 386 in./sec^2
k	ratio of specific heats
n	ratio of supply pressure to exhaust pressure
P	absolute pressure, psia
T	absolute temperature, °R
U	initial valve area per unit of port width, in.
w	valve port width, in.
W	weight rate of flow through an orifice, lb/sec
X	valve displacement, in.
$\dot{Y}$	ram velocity, in./sec
γ	ratio of valve area to an initial fixed area
ρ	fluid density, lb sec/in.^{4}

Subscripts	
1, 2, 3, 4	various orifices
a, b	points intermediate between exhaust and supply
d	downstream
e	exhaust
s	supply
u	upstream

9

S.-Y. Lee
J. F. Blackburn

Valve Configurations and Constructions

9.1. INTRODUCTION

So far we have considered the control valve in terms of what it does and how it is used in a system, but have largely ignored the questions of how it is constructed and how it is made to do what we want it to do. These questions will be discussed in the following five chapters, starting with a summary of various useful valve constructions and with considerations of design and manufacture.

9.2. BASIC VALVE CONFIGURATIONS

Valves are assemblies of one or more basic flow-restricting elements, and nearly all of these basic elements belong to one of three main classes: sliding, seating, and flow-dividing elements. The three types are shown schematically in Fig. 9.1. They will be described very briefly here; additional information is given in Chapter 10.

9.21. Sliding Elements

The sliding flow-restricting element of Fig. 9.1a may be defined as one in which the geometry of the flow-controlling members (the metering orifice) is essentially that of a pair of shear blades; the fluid flows be-

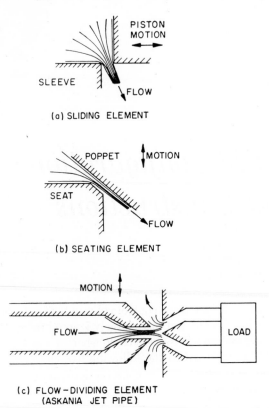

(a) SLIDING ELEMENT

(b) SEATING ELEMENT

(c) FLOW – DIVIDING ELEMENT
(ASKANIA JET PIPE)

Fig. 9.1. Basic flow-controlling elements.

tween a pair of sharp edges, and the width of the opening can be varied from some maximum down to zero. Ideally, the geometry of the opening is independent of the position of the moving member, but in practice the necessity of providing a finite working clearance prevents the attainment of this ideal at very small openings.

Insofar as it is attained, the discharge coefficient of the orifice remains constant and the flow for a given pressure drop is accurately proportional to the area of the orifice. For a constant area the flow is proportional to the square root of the pressure drop, being given by the orifice equation which has already been discussed in Sec. 7.1. This nonlinearity is important since it renders hydraulic circuits essentially different from electrical circuits, in which the resistances are largely or entirely linear. If the electrical analogy is to be employed, it must be kept in mind that a valve is analogous to a triode rather than to an ohmic rheostat.

If the edges of the opening are not sharp, there are two effects: the orifice geometry is dependent upon the valve setting, and the discharge coefficient increases appreciably. In this case it is impossible to make quantitative statements without a detailed analysis of the particular case, but fortunately most sliding valves have fairly sharp edges and follow the orifice law well except at very small openings.

9.22. Seating Elements

The second principal class of control elements includes those in which the stream is not cut off by a shear but is pinched off by a pair of opposing blunt edges. The sliding type of valve is represented by the gate valve or the common plug cock; the seating type by the poppet valve or the globe valve.

The seating element of Fig. 9.1b has two major advantages: it is fairly immune to dirt, and it can provide an almost perfectly tight shutoff, while even with the best attainable fit there is always some leakage through a sliding element. On the other hand, the seating element has several major disadvantages which have prevented its general use in servo applications: it is very difficult to balance completely against static pressure differences, it usually requires large and somewhat unpredictable operating forces, its flow-versus-displacement curve tends to be embarrassingly nonlinear at small openings, and it has a small but sometimes significant time lag that may handicap it for high-speed applications.

9.23. Other Configurations

Other basic configurations are possible beside the two just mentioned, but it will be sufficient to mention only one other, the flow-dividing element shown in Fig. 9.1c. This type makes comparatively little change in the rate of flow in the supply pipe, but changes the ratio in which that flow is divided between two or more paths. The twin needle valve sometimes used with Pelton water wheels operating at very high heads might be considered to belong to this class; probably the best known member is the Askania jet pipe, which (like all open-center valves) is comparatively inefficient in its use of power but has proved very useful in low-power applications. One of its greatest advantages is its comparative immunity to small particles of dirt in the fluid.

9.3. VALVE CONSTRUCTIONS

A hydraulic control valve usually consists of an assembly of several individual flow-controlling elements operated more or less in unison. There are many possible constructions for these elements and it would be impracticable to spend much space describing them, but the following summary will give an idea of the commoner types. The four-way valve will be used as an example in most cases.

9.31. The D-Valve

Probably the oldest four-way valve in common use is the D-valve as applied to the steam engine. It is shown in Fig. 9.2. Suppose that the D is moved to the right. This motion moves lip 1 of the D off of the cylinder port M_1 and admits high-pressure steam to the left side of the piston. Simultaneously lip 2 is moved off of port M_2, allowing the steam from the right side of the piston to flow out to the exhaust port, R. Motion of the D to the left will produce a similar action except that the functions of the two cylinder ports will be interchanged and the piston will move to the left.

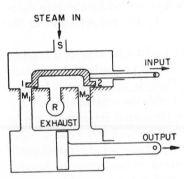

Fig. 9.2. D-valve as used in steam engine.

The moving member of the D-valve is pressed downward against its seat by a large force, equal approximately to the projected area of the D on the seating plane times the difference of pressure between supply and exhaust. This ensures good seating and negligible leakage, but it makes the friction between valve and seat so great that this construction is practical only for applications such as the engine where large forces are available to move the valve. In most applications it is desirable to have static pressure balance on the moving member of the valve.

9.32. The Spool Valve

One common construction—probably the commonest in hydraulic power practice—in which static pressure balance is easy to obtain is the spool valve shown in Fig. 9.3. Here the action is the same as with the D valve, except that it is usual to feed the high pressure in at the center and take the return from the outer ends, for ease in sealing against external leakage.

For any except the crudest applications, however, the spool valve must be very accurately made since it depends upon closeness of fit between spool and sleeve to hold the leakage down to an acceptable value. This matter of dimensional tolerances will be discussed in Sec. 9.52, but it must be emphasized here that it is one of the most important factors in determining the characteristics of a spool valve, and certainly the one

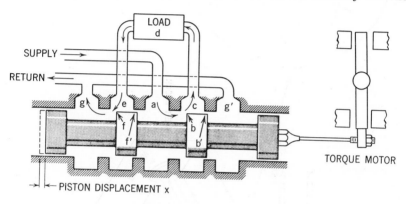

Fig. 9.3. Spool valve.

which contributes in greatest measure to the high cost of a good valve. In spite of this major disadvantage, however, the spool valve at the present time is all but universally used for the continuous control of fluid power, as contrasted with on-off applications, and in most of this book it is the type which is discussed, even though the discussion may also apply to other types as well.

9.33. The Plate Valve

The feature that makes a spool valve hard to manufacture is the fact that the inner surface of the sleeve is comparatively inaccessible so that it is difficult to finish to high accuracy and very difficult to measure. If the mating surface of spool and sleeve could be unwrapped and flattened into a plane, this feature would disappear, and in effect this is done by using a plate valve rather than a spool valve. One type of plate valve is shown in Fig. 9.4. Beside ease of manufacture, the plate valve has several other important advantages, such as the possibility of repair and renewal after wear or damage in service and a great freedom for

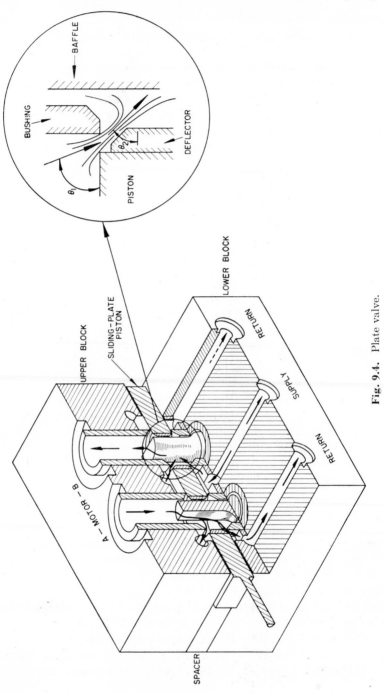

Fig. 9.4. Plate valve.

design modifications. Plate valves may be of many kinds; the ones described in this book are only a few of the possible varieties. A given basic design can be modified in various ways to give widely differing types and ranges of characteristics, by changing only such parameters as the widths and shapes of the milled ducts, the sizes of the holes, and the mode of motion of the plate. In addition, as will be seen in Sec. 10.321, the plate valve can be rather more easily and completely force-compensated than the spool valve, and it is more easily adapted to operation with gases or other nonlubricating fluids.

9.34. The Suspension Valve

One modification of the plate valve is the suspension valve of Fig. 9.5, which is discussed later in this chapter and also in Chapter 13. Here the

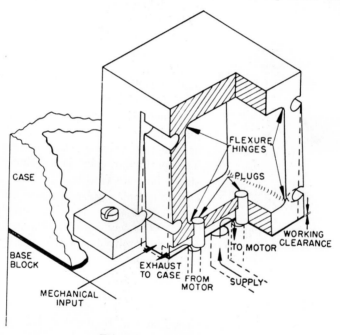

Fig. 9.5. Suspension valve.

proper clearance is maintained by a mechanical suspension, and there is never any metal-to-metal contact. This feature assures that friction will be negligibly small (in the absence of large dirt particles, which will cause trouble in any closely fitted mechanism) and allows the valve to be used interchangeably with lubricating or nonlubricating fluids, including gases. The use of the hole-and-plug technique described in Sec. 9.54 provides close dimensional tolerances at minimum expense, and the valve can be economically made in very small sizes where conventional constructions are prohibitively expensive and usually work very poorly. The suspension-valve construction is probably best adapted to these small sizes, though fairly large ones have been made and work very well.

9.35. Rotary-Plug Valves

Probably the oldest type of valve is the ordinary plug cock, which is an example of another basic construction, the rotary-plug valve. This

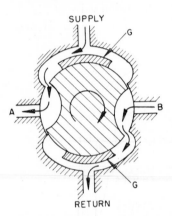

Fig. 9.6. Section of unbalanced rotary-plug valve.

construction is nearly always used in crude forms, such as the glass stopcock of the laboratory or the service entrance cock for domestic water and gas supplies, but it can also be used in critical applications. When properly constructed and lubricated, and particularly when the mating surfaces are conical, it gives a very tight shutoff but requires relatively large operating forces. It can easily be made in multiway form; Fig. 9.6 shows a four-way valve of this construction. This particular valve is unbalanced, but static balance can be obtained at the cost of complexity by doubling the number of ports and connecting diametrically opposite pairs of ports together. Four-way zero-lap valves of this construction have been made experimentally but have usually been abandoned in favor of other constructions.

9.36. Poppet Valves

The valves we have just discussed have all been of the sliding type, in which the motion of the operating member is parallel to the plane of the metering orifice. There is also a class of seating valves, in which the motion is normal to the plane of the orifice and therefore parallel to the force exerted by the static pressure drop across the valve. The commonest seating valves are probably the globe valve and the domestic water faucet, but another very common type, which is better adapted to power control, is the poppet valve, of which one type is shown in Fig. 9.7.

It is used in some steam and practically all internal-combustion engines, and in many types of shutoff applications in hydraulic systems. It is fairly easy and inexpensive to make and its great virtues are its relative immunity to dirt and the fact that it gives a very tight shutoff unless it is grossly abused. Its disadvantages for precise control are also serious, so much so that it is seldom used for really critical continuous-control applications. If it is not balanced against static pressure differences, it requires relatively enormous operating forces. In its "balanced" form

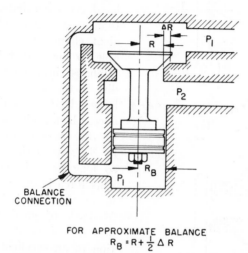

FOR APPROXIMATE BALANCE
$$R_B = R + \tfrac{1}{2} \Delta R$$

Fig. 9.7. Piston-balanced conical poppet valve.

(the balance is usually far from complete), although the valve may not leak, the balance piston usually does. Its flow-versus-displacement and its force-versus-displacement curves are violently nonlinear at small openings. It is subject to a peculiar kind of unsymmetrical "stiction" which in effect introduces a highly variable lag in operation. This last effect is negligibly small in most applications but may be serious if very fast operation is required and if large operating forces are not available. Attempts have been made to use ganged poppet valves in high-performance servo applications, but these have seldom if ever been successful. For shutoff applications, however, it is the most generally useful type.

9.37. Flapper-and-Nozzle Valves

A second class of seating valve is the flapper valve, which is much used as a pilot valve in industrial control instruments. Two types of flapper valve must be distinguished, as their properties are very different. The sharp-edged flapper valve of Fig. 9.8a is essentially a pure orifice even at small openings, and so long as the maximum opening, x, is not greater than about one-fifth the diameter, d. Its chief weakness is its liability to damage if the flapper is allowed to hit the nozzle with appreciable force. On the other hand, it is somewhat easier to make than most sliding valves of equivalent characteristics, it is adaptable to low power levels, and it is almost immune to damage by dirt if the particles are not too large. It does have static unbalance, which is almost exactly equal to the pressure drop multiplied by the area of the nozzle opening, so long as x is not too large. This force is often useful because it provides a built-in pressure feedback of good accuracy. It also makes possible the construction of an excellent variable-area flowmeter.

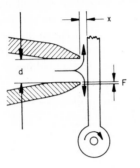

PREFERABLY, $F < x_{min}$ AND $x_{max} < \frac{1}{5} d$

(a)

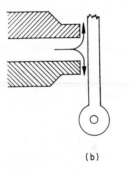

(b)

Fig. 9.8. Flapper-and-nozzle valves.

The flat-faced nozzle of Fig. 9.8b radically alters the characteristics of the flapper valve, especially at small openings. As soon as the resistance to flow of the radial sheet of fluid between the flapper and the nozzle end becomes appreciable compared to that of the nozzle orifice, the law of flow departs from the true orifice law and begins to approach the linear laminar-flow law. As soon as the laminar portion of the resistance is the controlling factor, the flow varies as x^3 instead of x and also becomes inversely proportional to the viscosity of the fluid. The force on the flapper increases considerably also, and becomes unpredictable because it depends upon very small details of the flapper and nozzle surfaces. In general, if it is protected from damage, the sharp-edged nozzle-and-flapper combination is a very useful and predictable hydraulic element; the flat-faced nozzle is not.

9.38. Jet-Pipe Valves

In spite of a history going back at least to 1889,[1] the jet-pipe valve remains an almost completely empirical device, and at the present time it seems likely to remain so because the complexity of the flow phenomena which determine its characteristics is so great that a useful and meaningful analysis is almost impossible. As an empirical device it has been used by several firms, particularly by Askania-Werke in Germany and by the Askania Regulator Company of Chicago. In nearly all cases it has been used at low pressures, usually not above about 100 psi, and it is sometimes stated that it cannot be used at high pressures. This is definitely untrue since there is at least one jet-pipe pilot valve in production that works very well at 2000 psi. It is true, however, that a given design seems to have a definite maximum pressure above which it refuses to work, though the reason for this is unknown. It is also stated, especially in sales literature, that the jet-pipe valve has no tangential component of flow force; this also is false. The flow force may be considerable, and of such a nature as to cause serious instability. Again the reasons are unknown and the behavior is unpredictable.

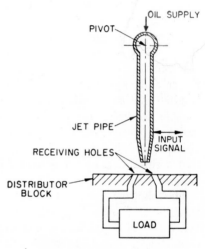

Fig. 9.9. Jet-pipe valve.

In spite of its empirical nature the jet pipe is a useful engineering device for certain applications. It can be used with any type of fluid, including contaminated ones, and its relatively large clearances permit fairly wide tolerances and therefore moderate manufacturing costs. The required force level (when the valve is properly used) is low so that it can be controlled by actuators of low power. As a null device, especially as a substitute for the single flapper and nozzle in an industrial controller, it has proved to be very flexible in application and dependable in operation. One type of jet-pipe arrangement is shown schematically in Fig. 9.9.

[1] Beauchamp Tower, "An Apparatus for Providing a Steady Platform for Guns, etc., at Sea," *Trans. Inst. Naval Architects,* Vol. XXX, pp. 348–361 and Plates XXXII–XXXIV, 1889.

9.4. VALVE DESIGN

As with any other type of design, this is a very large subject and it will be impossible to do it justice in the space available. Also, as is true of all design, many of the designer's decisions are based on intangibles—on instinct and hunch derived from his personal experience—and another designer might arrive at a different but equally good solution to a given problem. The design criteria given below are those which have been found important at DACL for the projects on which it has been engaged, and they are not necessarily those which should be given the greatest weight in a different situation.

Chapter 6 has already discussed a number of factors which determine the choice of other elements of the drive than the valve, and it will be assumed that the drive design has established the values of peak horsepower and supply pressure and the required speed of response, that suitable motors or other components have been selected, and that it remains to design the control valve and to choose its actuator. The design of electromagnetic actuators is covered in Chapter 11 and of electrohydraulic actuators in Chapter 13; in this chapter we shall assume that a suitable actuator is available but must be chosen by the valve designer.

The following discussion will assume that a symmetrical zero-lap four-way valve is to be designed, for which the characteristics are given in Sec. 7.31. This type has been chosen both because of the simplicity of its characteristic equations and because it is the commonest and most important type in current hydraulic practice.

9.41. Ratings

There seems to be general agreement upon the definition of the horsepower rating of a valve; it is the maximum power which the valve will deliver to the optimum load with the maximum rated supply pressure. The conditions for maximum power delivery can be derived as follows.

From Sec. 7.31, the characteristic equation of the valve for the P1 case is

$$p_m = P_s - \frac{2q^2}{g^2} \tag{7.18}$$

where p_m = pressure drop across motor (hydraulic load), psi
   P_s = supply pressure (constant), psi
   q = flow rate, in.^{3}/sec
   g = hydraulic conductance of *each* of the two metering orifices

The quantity g is the important parameter of the valve and is defined as

$$g = aC_d\sqrt{2/\rho} \tag{7.3}$$

where a = area of each metering orifice, in.^{2}

C_d = discharge coefficient, taken as 0.625 hereafter. It varies somewhat but is usually within about ± 3 per cent for most valves under normal conditions of operation.

ρ = fluid density, lb sec^2/in.^{4}

If the appropriate values are inserted, $g = 91.3a\sqrt{\sigma}$ for a single orifice, where σ is the specific gravity of the fluid. Most petroleum-base hydraulic fluids have $\sigma = 0.85$ approximately, and in this case $g = 100a$ fairly closely.

The power into the load will be the product of load pressure and flow, or

$$H_m = qp_m = qP_s - \frac{2q^3}{g^2} \tag{9.1}$$

This power will be a maximum when $dH_m/dq = 0$, or

$$\frac{dH_m}{dq} = P_s - \frac{6q^2}{g^2} = 3p_m - 2P_s = 0 \tag{9.2}$$

Therefore the power will be a maximum when $p_m = \frac{2}{3}P_s$, and this maximum power will be $26a_{max}P_s^{3/2}/\sqrt{\sigma}$ in. lb/sec or $3.94 \times 10^{-3}a_{max}P_s^{3/2}/\sqrt{\sigma}$ horsepower. For oil with $\sigma = 0.85$, $H_{max} = 27.3a_{max}P_s^{3/2}$ in. lb/sec or $4.14 \times 10^{-3}a_{max}P_s^{3/2}$ horsepower when a_{max} and P_s are given in inch and pound units.

Two other quantities of interest are immediately obtainable from the known values of P_s and a_{max}. The stalled pressure is simply P_s for all valve openings since the valve pressure drop is zero when there is no flow. The maximum no-load flow occurs with the valve wide open and the full supply pressure across the two orifices in series, and is therefore given by

$$q_{max} = g_{max}\sqrt{P_s/2} = 64.7a_{max}\sqrt{P_s/\sigma} \tag{9.3}$$

or

$$q_{max} \doteq 70a_{max}P_s \text{ (for oil)} \tag{9.4}$$

We have assumed that P_s is given by the drive designer. Actually the choice of a supply pressure is made on the basis of many factors and is not usually critical. In valve design it is customary to provide sufficient strength in the valve body to keep the pressure-induced distortions down to a reasonably small value for the highest supply pressure that is likely to be encountered. In present practice this is often 3000 psi, though 1000 and 5000 psi are also frequently used design figures. The only reason for mentioning it here is to emphasize the importance of providing a sufficiently rigid body. This is not always easy, especially when attempting to minimize over-all weight for airborne operation, or when a large-diameter built-up sleeve is inserted into the body and subjected to full supply pressure. Another frequent source of distortion of the valve body is the use of tapered-thread tubing or pipe fittings, since the tendency of every assembler is to wind them into the body with all his strength. Fittings will be mentioned in Sec. 9.45.

9.42. Choice of Maximum Stroke

For closed-loop operation it is usually necessary that a valve have constant gain, as has been stated in Sec. 7.6. We shall assume, therefore, that the valve to be designed will have the rectangular ports necessary for constant gain, and we shall use w to denote the total peripheral width and x to denote the axial length of each metering orifice. (A metering orifice, or port, may be divided into several individual orifices distributed around the periphery of the valve sleeve and all opening simultaneously; w is then the total width of the several orifices in the same transverse plane.) The effective area of the orifice is then $a_{max} = wx_{max}$. Choosing x_{max} will now fix w.

The choice of x depends upon a balance between several opposing factors. One weighty one is the fact that a large x_{max} permits correspondingly large tolerances in the axial dimensions of the valve parts. Another one, which is often the most weighty, is that an existing tractive-magnet type of actuator must be used for reasons of availability, and in this case x_{max} is usually made equal to the maximum stroke of the actuator. Other considerations are the following:

1. The changes in the discharge coefficient at low values of Reynolds number which were discussed in Sec. 7.1 cause an undesirable variation in the gain of the valve for small values of x and high-viscosity liquids. If $x \ll w$, the Reynolds number can be defined as $R_e = (2ux\rho)/\mu$, where u is the average velocity of the fluid through the orifice and μ the absolute viscosity. Combining this expression with the orifice equation yields

$$R_e = C_d \sqrt{8\rho}\, \frac{x}{\mu}\, \sqrt{\Delta p} \tag{9.5}$$

where Δp is the pressure drop across the orifice. This expression is actually meaningless in a way, because of the dependence of C_d upon

R_e, but it does point out the fact that R_e and therefore C_d and the valve gain can be small at small values of x.

This change in gain will be obscured by other nonlinearities for values of x less than 1 or 2 mils even in high-grade valves. If we assume that the allowable gain change is a factor of 2 at $x = 0.001$ in., $C_d = 0.312$, which corresponds to $R_e = 45$ according to the curve of Fig. 7.2. If we assume a drop across each orifice of 1000 psi and use Univis J-43, which is a standard aircraft hydraulic fluid, the maximum permissible kinetic viscosity turns out to be about 0.07 in.^{2}/sec, or 110 cs, which is reached at $-5°$F. For the same conditions, if the temperature is reduced to $-40°$F, as may easily happen in flight, the viscosity increases to 466 cs and the effective C_d goes down to well below 0.1, or off the limits of the chart, and the gain has decreased by a factor of at least 10.

These calculations are very rough and serve only to illustrate the possibility of trouble when viscous fluids are used with small valve openings. For most conditions of operation the effect can be ignored.

2. The foregoing effect indicates a preference for a longer stroke. The same preference is indicated by a consideration of the rate of leakage flow past the spool. Doubling the stroke and all other axial dimensions and halving the diameter will divide the leakage by 4, which is desirable. This argument, however, is weak because of the great effect of the clearance, which must be roughly independent of diameter. Doubling the clearance will multiply the viscous flow rate by 8.

3. One of the strongest arguments for a short stroke is the matter of the power required to operate the valve. Actually, since the valve is an integrator, its actuator is a positioning device only, and unless the duty cycle is completely defined, it is impossible to state either the input power or the power gain. It is apparent, however, that a longer stroke requires more power, since the inertial force increases as x^2, and the viscous friction force as x. (The Bernoulli force is independent of x for constant output power.) The output of an electromagnetic actuator is never embarrassingly high, and even a moderate increase in stroke requires a disproportionate increase in size and driving power and a sometimes serious decrease in speed of response.

4. Obviously the ports cannot extend over more than 360° of the periphery of the sleeve, which is another way of saying that the choice of w sets a minimum diameter for the piston. It is preferable to use 360° ports for several reasons. Perhaps the most important is the matter of dirt. To a considerable extent a 360° port is self clearing; a piece of dirt which lodges against the upstream side of a piston land will probably be swept through the port and out of the valve as soon as the port opens. With a narrow port, however, this will happen only to the dirt that lodges

opposite the port, and the other dirt particles will be dragged back and forth between the piston land and the closed part of the sleeve, greatly increasing the friction and eventually damaging the valve. This argument is less than conclusive since there are other areas of the piston to which it does not apply, but it does seem to be true that there is less dirt trouble in valves with 360° ports. This is probably because the much lower pressure gradient across sealing lands has less tendency to cause the dirt to wedge into the clearance space.

If force compensation is to be used, the 360° port is very desirable since it eliminates lateral spreading of the jet and the accompanying weakening and nonlinearity of the compensation.

5. It frequently happens that other considerations limit the port width to a value that requires an unreasonably small diameter for the spool if 360° ports are used with square lands. One way of allowing the full ports to be maintained with a larger piston is the use of tapered lands; instead of a plane, the land face is a slowly tapering cone, and the stroke required to get a given opening is increased as the secant of the cone semiapical angle. Conical land faces have two important disadvantages, however: they considerably increase the Bernoulli force, and they introduce an additional and very difficult concentricity tolerance. They also invite trouble from viscosity of the fluid since the metering orifice looks more like a passage than an orifice. For these reasons they have not been popular.

If conical land faces are not used, and if spool diameter is to be greater than w/π, the ports must occupy only part of the periphery. The construction of accurately dimensioned and accurately located rectangular ports in a valve sleeve or body (if the sleeve is integral with the body) is not an easy task, and a number of manufacturing methods may be used to accomplish this task. Some of them are discussed in the last sections of this chapter.

One criterion for the use of partial- or full-periphery ports is derived from a consideration of the necessary dimensions of the access passages through which the fluid enters and leaves the metering orifices. If these passages are too small, at high flow rates an excessive pressure drop will occur across them and the flow to the load will be unduly restricted. This phenomenon is often referred to as "saturation" of the valve. A good rule of thumb is that the passages should have everywhere at least four times the maximum area of the metering orifices. This specification is hardest to meet in the spaces between the piston lands. Assuming that the diameter of the stem of the piston is half the bore diameter (and if it is not, the piston is probably insufficiently stiff), it is easy to show that the critical value is $w/x_{max} = 67$ approximately. If x_{max} is fixed by other considerations and the required value of a_{max} is such that $a_{max}/x_{max}^2 = w/x_{max} > 67$, 360° ports can be used. If it is less, they cannot since the passage area will be too small.

6. A long stroke obviously means a slower-acting valve since it takes longer to push a piston a greater distance. This consideration is of importance for very-high-speed systems, but probably not for medium- or slow-speed ones. Incidentally, practically all commercially available electromagnetic actuators are designed primarily for very-high-speed operation and are not necessarily the best that could be made for the usually slower-speed requirements of industry. Presumably heavier and slower units could be made which would be more powerful and (hopefully) much cheaper. They are badly needed.

9.43. Port Design

As stated already, closed-loop operation of a valve usually demands that the valve gain, which is proportional to the peripheral width of the ports, be independent of spool position, or in other words that the ports be rectangular. From the manufacturing standpoint this requirement is unfortunate since it is much more expensive to make a square hole than a round one. Nevertheless, constancy of gain is so important that practically all servovalves do have rectangular ports. In some cases, such as the XA and MX valves described in Chapter 13, these ports were broached in the valve body itself, which served both as body and sleeve. In most cases, however, the ports are machined into a separate sleeve, which is then inserted in the valve body. The ports themselves can be formed by milling through the wall from the outside in such a way that the intersection of the cut and the bore forms a rectangular port. This is perhaps the commonest method, but the accuracy of the milling process is usually insufficient to stay within the tolerances required for valves of the highest quality. In this case the sleeve is usually built up of a series of cylindrical sections (usually five) with parallel end faces, the ports being formed by cuts on these faces. The high accuracy attainable with a good surface grinder makes it comparatively easy to hold the required tolerances, and the use of abrasive techniques for the final machining allows the use of very hard materials for the sleeve, which is also highly desirable.

When the requirement of constancy of gain is somewhat less stringent, it may be permissible to replace a rectangular port with one consisting of a series of round holes, suitably overlapped as to axial position. Naturally the gain is zero when the piston land just starts to uncover the first hole, but it rapidly increases as the center of the hole is approached.

The variation in gain as one hole starts to become completely uncovered and another is still opening depends upon the percentage of the diameter by which the holes are overlapped, as well as upon the relative diameters and the number of holes which the land edge simultaneously intersects. After the initial build-up, the variation in gain can be held to a minimum by using many holes in parallel, but this scheme is not particularly useful because as their number increases, the holes become too small to drill accurately and economically, and the viscous effects increase excessively.

It is sometimes necessary to use special port shapes, such as those described in Sec. 7.7. These are usually made most economically either by broaching (if the valve is large enough) or by pantograph milling from a large master. Pantograph milling can also be used for making rectangular ports if the tolerances are not too tight and if a small change in gain caused by the necessary rounding of the corners of the port can be tolerated. A really good pantograph mill can hold tolerances well under 0.001 in. if it is properly maintained and operated and if the cuts are not too heavy.

9.44. Friction and Sticking

One of the commonest troubles encountered by users of valves is excessive friction and sticking. This subject will be discussed in Chapter 10 and it is only necessary to refer to it in passing. If trouble of this sort is encountered with a particular design, it may usually be cured by grooving the piston lands. In more stubborn cases, the lands should be tapered or stepped, and as a last resort a hydrodynamic bearing may be used. This type of bearing is easy to provide in a plate valve—indeed this fact is one of the plate valve's principal advantages—and somewhat more difficult but still practical in a spool valve. If properly designed, the increased leakage of the bearing may be largely or completely compensated by the decrease of leakage (ideally $2\frac{1}{2}$ times) caused by keeping the spool approximately centered. The hydrodynamic bearing is, of course, almost essential in a valve for use with gases or other nonlubricating fluids.

In laying out the dimensions of the piston and sleeve it is also necessary to keep in mind the requirements for stability of the valve, which are discussed in Chapter 12. Never use a two-land valve unless the actuator is very rigid, and always make the damping length positive, or at least zero, if trouble is to be avoided when the actuator is as soft as the average electromagnetic "torque motor." Even though the force available from the actuator may be sufficient for normal operation, its stiffness as seen by the valve spool may still be sufficiently small to permit certain kinds of instability, particularly if the natural frequency of the piston against the actuator and Bernoulli-force springs bears the wrong ratio to the natural frequency of the load or of a connecting pipe. This type of resonant instability is discussed in Sec. 12.5, where it is also shown that one of the best preventives of instability is a stiff actuator.

9.45. Ducting and Connections

It has already been pointed out that the ducts in the valve body which lead to the metering orifices must be large enough to avoid the phenomenon of "saturation." The 4-to-1 ratio which was suggested is none too generous from the standpoint of steady-state pressure drop, and it should be larger wherever possible. In a power valve (as opposed to a pilot valve), especially when designing for low pressures and large flows, it may be difficult to exceed the 4-to-1 ratio in the piston passages, and sometimes in the valve sleeve itself, but it should at least be made larger in the valve-body passages. On the other hand, undue increase in the *volume* of the motor lines should also be avoided, so it is best to lay them out first, making them as short and direct as possible. The supply and return lines may be as large as desired.

So far as may be, the ducts should be straight and smooth, not complicated and branching, though this requirement is usually very difficult to satisfy and is probably of only minor importance. The added pressure drop due to extra corners will probably be negligible, but if a passage is tortuous and is formed by cross-drilling and plugging, there will be one or more blind pockets left where dirt and chips are almost certain to lurk, waiting to drift downstream into the mechanism at the most inconvenient moment. It is very difficult to get these pockets thoroughly clean, or to clean them out again if dirty oil has ever been passed through the valve.

Some care should be given to the method of sealing off what unused drill holes are left after forming the passages. Sometimes short lengths of taper pin are used, of such a diameter as to require heavy pressure and actual deformation of the metal for insertion. We have found this practice to be dangerous, even when the taper plug is driven well below the surface and the corners of the hole are peened over behind it. The seriousness of the injuries that can result from a blown plug, or even from a high-pressure oil jet, is sufficiently great so that we have tabooed this kind of seal in DACL. Parallel-sided dowel pins may be better, and some people have suggested the use of balls, heavy press fits being used in either case. The only really dependable method of sealing, however, is to use a plug backed up by a real mechanical retainer such as a screw and binding wire, or its equivalent.

Many commercial control valves are supplied with taper-tapped holes in which any desired type of compression-tubing fitting may be inserted. This is not good practice because of the almost inevitable distortion of the valve body when the fittings are screwed in. When a sleeve and piston have a working clearance of 0.0001 in., the enclosing valve body does not have to be sprung or wedged very far to distort the sleeve enough to make the piston rub or stick. There are two remedies: use of straight-thread fittings or use of straight butt joints. In both cases, the seal is made leak-tight by using O-rings or their equivalent. The straight-thread "A-N" fittings have been much used in military equipment, and are excellent if properly installed, but like other plumbing of the same general class have a tendency to make the whole installation bulky and somewhat clumsy looking. The technique which seems to be preferred by many machine-tool builders is to make a single "christmas-tree" manifold by machining one or more flat plates, and then to bolt the valves and other elements of the system to this plate, depending again upon O-rings for sealing the butt joints. This technique is impractical for the usual experimental setup, but makes a very neat, trouble-free, and comparatively inexpensive assembly for an industrial installation.

9.46. Example of Valve-Design Calculations

The following is an example of the quantitative design of an actual valve, and will furnish the information necessary for the detail design.

The pertinent system specifications are as follows.

The load locus is as shown in Fig. 9.10. At the point of maximum instantaneous power,

$$F_w = 1500 \text{ lb}$$

$$V_w = 3 \text{ in./sec}$$

MIL-O-5606 oil of 0.85 specific gravity is to be supplied at

$$P_s = 1000 \text{ psi}$$

Maximum permissible internal leakage rate $Q_l = 0.1$ gpm $= 0.385$ in.^{3}/sec.

The minimum natural frequency of the actuator loaded by the valve spool is to be

$$f_n = 180 \text{ cps}$$

Hysteresis plus dead zone is to be held to 3 per cent maximum.

The calculation is carried out in the following steps:

1. Assuming that the maximum-power condition occurs at maximum output for the valve, and that the valve is a normal four-way valve so that the equations of Sec. 7.31 apply, the stalled force will be $1.5F_w = 2250$ lb, and the runaway speed will be $\sqrt{3}\,V_w = 5.2$ in./sec. On the

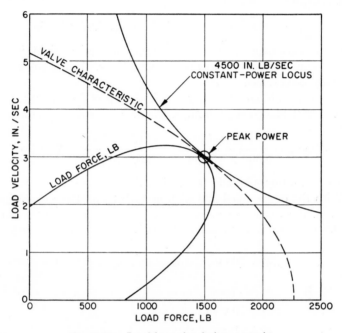

Fig. 9.10.　Load locus for design example.

plot of the load locus draw a parabola with vertex at (2250 lb, 0 in./sec) and passing through the maximum-power point (1500 lb, 3 in./sec) and the runaway point (0 lb, 5.2 in./sec), as shown by the dashed line of Fig. 9.10. If no part of the load locus lies outside this parabola, the valve will satisfy the requirements. (This use of the load locus is discussed in Chapter 6.)

2. We now select the effective piston diameter for the ram. Since the force is to be 2250 lb with the full 1000-psi supply pressure applied to the ram, the effective area must be at least 2.25 in.^{2} To this must be added the area of the piston rod. If we use a ½-in. rod, with an area of 0.196 in.^{2}, the total piston area is 2.446 in.^{2}, which requires a piston diameter of 1.74 in. In the interests of using standard tooling we should

specify a 1.750-in. piston with a 0.500-in. rod, for which the effective area will be $A_p = 2.27$ in.^{2} This increases the stalled force to 2270 lb, which is less than a 1 per cent increase and certainly unobjectionable.

3. With this revised piston area we calculate the flow at maximum output,

$$Q_w = V_w A_p = 3 \times 2.27 = 6.81 \text{ in.}^3/\text{sec}$$

and the runaway flow,

$$Q_{max} = \sqrt{3}\, Q_w = 11.8 \text{ in.}^3/\text{sec, or } 3.07 \text{ gal/min}$$

The latter figure suggests the use of a 3-gpm pump if a slight decrease in pressure is permissible under runaway conditions. If it is not, a 3.5-gpm pump may be specified.

Assuming a full-load efficiency for the pump of 80 per cent, which is pessimistic for high-quality pumps, the drive motor horsepower will be

$$H_{drive} = \frac{Q_{max}P_s}{6600 \times \text{efficiency}} = \frac{11.8 \times 1000}{6600 \times 0.80} = 2.24 \text{ hp}$$

A 2½- or 3-hp motor should be specified.

4. With the oil specified and with no low-temperature operation contemplated, it will not be necessary to consider operation at very low Reynolds numbers, as discussed in Sec. 9.42. For oil of the given density,

$$q = 70a\sqrt{p_v}, \text{ approximately}$$

whence

$$a_{max} = \frac{q_{max}}{70\sqrt{p_v}}$$

or for the runaway condition,

$$a_{max} = \frac{q_{max}}{70\sqrt{P_s}} = \frac{11.8}{70\sqrt{1000}} = 0.0054 \text{ in.}^2$$

5. As already stated, the choice of the valve stroke is usually determined by the characteristics of the available actuator. One available unit has the following characteristics:

Mid-position force	F_{mid}	$= 6.6$ lb
Spring stiffness	K	$= 200$ in. lb/radian
Drive radius	r	$= 0.750$ in.
Stroke	X	$= 0.011$ in.
Rotor inertia	J_r	$= 64 \times 10^{-6}$ in. lb sec^2
Hysteresis		$= 1$ per cent max

Presumably we should use all the available stroke so that $x_{max} = 0.011$ in. Then

$$w = \frac{a_{max}}{x_{max}} = \frac{0.0054}{0.011} = 0.49 \text{ in.}$$

This completes the calculations for the port dimensions, but additional calculations must be made as checks before detail design can begin.

6. Check for duct area, per Sec. 9.42:

$$\frac{w}{x_{max}} = \frac{0.49}{0.011} = 45 < 67$$

Therefore full 360° ports cannot be used. If they were used, the piston diameter would be about $\frac{5}{32}$ in. Try a quarter-inch piston with each port split into two 0.245-in. ports. These will cover about 0.4 of the periphery, which is a reasonable fraction even if a one-piece sleeve is used. (We shall see that the axial tolerances are too tight to permit a one-piece milled sleeve, however.)

7. Check for natural frequency. Pending detail design, assume that the piston has the same mass as a solid rod of a length equal to about 6 diameters, or 1.5 in. Such a rod of steel would weigh 0.021 lb and have a mass of 54×10^{-6} lb sec^2/in. When attached at a $\frac{3}{4}$-in. radius, its effective moment of inertia would be $54 \times 10^{-6} \times (\frac{3}{4})^2 = 30 \times 10^{-6}$ in. lb sec^2. Adding this to the rotor inertia gives a total inertia of 94×10^{-6} in. lb sec^2.

The natural frequency of the actuator when loaded by the piston will be

$$f_n = \frac{1}{2\pi} \sqrt{\frac{K}{J_{total}}} = \frac{1}{2\pi} \sqrt{\frac{2 \times 10^2}{94 \times 10^{-6}}} = 232 \text{ cps}$$

This is satisfactory since it is well above the 180-cps minimum specification.

8. Check Bernoulli force on piston. The Bernoulli force will be a maximum under runaway conditions, when the valve is wide open and the load pressure is zero. At this time

$$F_t = 0.5 P_s a_{max} = 0.5 \times 1000 \times 0.0054 = 2.7 \text{ lb}$$

(The computed constant is 0.436 instead of 0.5, but the latter errs on the safe side.) This is to be compared with the force available from the actuator at full stroke, which is usually about half the mid-position force, or 3.3 lb in the present case. The excess force is 0.6 lb, which is probably adequate but not comfortably so. If a greater margin of safety seems advisable, a larger and more powerful actuator can be used, or else the valve may be flow-force compensated as suggested in Sec. 10.321.

The equivalent spring stiffness of the Bernoulli force is added to the torque-motor spring and merely helps to increase the natural frequency of the valve *provided the flow is in phase with the valve motion.* If it is not (and it will not usually be accurately in phase), there is a possibility that the valve may be unstable, as discussed in Sec. 12.5. The stiffness of the Bernoulli spring is $K_B = 0.5 P_v w$, and is a maximum under runaway conditions, for which $K_B = 245$ lb/in. Since it is applied to the rotor at a radius of 0.750 in., the equivalent torsional stiffness will be $K_B/r^2 = 410$ in. lb/radian, or about twice the mechanical-spring rate.

9. Determine maximum underlap. The axial tolerances of the valve ports and piston lands are determined by the permissible leakage, which limits the allowable underlap, and the dead zone, which limits the overlap. The leakage will be a maximum with zero load pressure and the valve centered, when full supply pressure is applied to a series-parallel circuit of four supposedly identical orifices. The specifications give 0.385 in.^{3}/sec as the maximum leakage rate, whence the underlap, u_{max}, will be

$$u_{max} = \frac{Q_l}{2 \times 70 w \sqrt{P_s/2}} = \frac{0.385}{140 \times 0.48 \sqrt{500}} = 2.5 \times 10^{-4} \text{ in.}$$

This assumes zero clearance and perfectly sharp land corners, and any deviation from these conditions will tend to increase the leakage. Since the clearance will be roughly 0.0001 in., a maximum underlap figure of the same magnitude would not be unreasonable.

An overlap maximum is hard to establish quantitatively because of the effects of the finite valve clearance but can be "guesstimated" by the following argument. The specifications gave 3 per cent as the maximum dead zone and 1 per cent as the hysteresis of the torque motor. This leaves 2 per cent for overlap, or about 0.0002 in. of a full stroke of 0.011 in. Since we already had a maximum underlap of 0.0001 in., we may set the overlap at the same figure for safety, which leaves a total tolerance of only ±0.0001 in. to be shared between spool and sleeve, with three axial dimensions of each adding up. The resulting tolerances are exceedingly tight, but they are realistic if the given performance specifications are realistic. This situation can be improved by using a longer stroke, which demands a much larger and slower actuator, by relaxing some of the performance specifications, particularly the leakage and dead-zone figures, which requires a battle with the customer (or the boss), or by adopting some other construction which permits the required accuracy to be obtained more easily and cheaply. One such construction will be described in Sec. 9.54.

To recapitulate this section, the detail designer may now be given the following dimensions and tolerances, based on the calculations of the foregoing section:

Actuator as specified, mounting dimensions given on drawing furnished by manufacturer.

Piston diameter—0.2499, −0.0001, +0.

Sleeve-bore diameter—0.2500, −0, +0.0001.

Axial positions of all port and land edges to correspond within ±0.0001, as shown in Fig. 9.11, Sec. 9.52.

Each port to consist of two or more identical rectangular ports equally spaced around the axis, of total peripheral width 0.49 ± 0.010 in.

Sealing lands must be provided. All lands must be grooved or relieved except within 0.020 in. of ends of active lands in order to minimize friction and hydraulic lock. (See Sec. 10.21 for a discussion of grooving and tapering piston lands.)

9.5. EXPERIMENTAL FABRICATION OF VALVES

The adjective "experimental" has been added to the title of this section in order to emphasize the fact that our experience in making valves has been entirely that of the research laboratory and not of the large-scale producer. For this reason our designs, our machining facilities, and our testing methods will need to be modified in the light of production experience, and of the requirements, facilities, and traditions of the particular producer. Nevertheless, we hope that the suggestions given in the following sections—for they are only suggestions—will be useful, not only to other experimenters in the field, but also to full-scale manufacturers. At least, we have found a number of ways to get into trouble, and these ways are even more dangerous to the production department than to the model shop.

9.51. Choice of Materials

Until now nearly all servovalves have been intended for use with petroleum-base hydraulic fluids which are relatively noncorrosive and have fairly good lubricating qualities. In this case steel is the logical material, and has been almost universally used, although aluminum alloys have been much used for valve bodies in airborne applications.

Whatever material is chosen for the pistons and sleeves—or the analogous parts of valves of other constructions—should be relatively brittle and not ductile. This does not mean that it should be weak and easily shattered, like a poor grade of glass, but it should never be plastic or ductile like copper or annealed mild steel. The two principal reasons for this statement are that the material will probably be finish-machined by abrasive methods and that a ductile material will give a great deal of trouble in service because of the action of abrasive dirt in the fluid or of the tendency of such materials to gall.

It is certainly possible to machine valve parts from materials that are soft enough to be turned, drilled, milled, and broached on conventional machines, and to avoid all grinding, honing, and lapping. This procedure may permit the maintenance of tolerances that are tight enough for low-performance valves, but for the grade of performance that is now demanded in most applications, the necessary tolerances can only be held by using abrasive processes, and these processes work well only on comparatively brittle—that is, nonductile—materials. It is possible to grind soft copper to accurate dimensions, but it is not easy. It is easy, with proper equipment, to hold very tight tolerances when grinding hardened tool steel. The same statement applies more strongly to honing, and still more strongly to lapping, which is almost universally used to take off the last ten-thousandth of an inch of material, if not more. The same advantage of easier machinability of brittle materials applies also to some of the newer methods such as ultrasonic machining, spark machining, and precision liquid- or vapor-blasting.

Even if satisfactory machining methods are available, a ductile material should not be used for the mating parts of valves because of the effects of grit in the fluid. The gritty portion of most dirt is largely angular fragments of quartz, which is harder than most materials of construction. When a quartz particle gets into the clearance space between the cylinder and piston, it will plow a groove in both of them. If they are made of comparatively brittle material, the chips will break away and be carried off downstream, and eventually the grit particle will also break up and disappear, leaving a narrow canyon between flat-topped mesas on both the mating surfaces. This topography may increase the leakage somewhat, but has little other effect. On the other hand, if the materials are ductile, the particle will probably embed itself in one part and be dragged back and forth against the other, plowing a groove and throwing up ridges on each side, since the chips of the ductile material will not break off harmlessly. As before, the groove itself is of little consequence, but the ridges will scrape away the boundary layer of lubricant, causing metal-to-metal contact, high friction, welding, galling, and comparatively rapid failure of the mating surfaces.   This process is outlined in Sec. 10.22.

A valve material must not only be sufficiently hard and not too ductile; it must also hold its dimensions without warping or "growing" in service or during storage.   Among the steels, a good nondistorting die or gage steel, properly heat-treated, is very good.   We have used Stentor and Ketos among the die steels.   We have also used SAE 52100 ball-bearing steel, which is good.   When corrosion is a problem, as with air valves, we have used hardenable stainless steels.   For high-temperature operation we have used sintered carbides such as Kennametal, though in that case the problem is somewhat different in that our design (described in Chapter 20) did not involve sliding friction.   We have made inconclusive experiments with beryllium copper, with deep-anodized aluminum alloys, and with sapphire.   Sintered boron carbide apparently will not hold a sharp edge.   Some of the newer dense, strong ceramics would probably be suitable, especially for corrosive conditions and high temperatures, though they are difficult to machine.

Besides being sufficiently hard, nonductile, and corrosion-resistant for the particular application, a valve material must also be dimensionally stable, and must be so treated as to attain the maximum stability.   When a piston is as closely fitted into its mating bore as the average servovalve spool must be, even a very small warping or bending will lead to trouble from increased friction, and any large distortion will cause complete locking, even without the added effects of dirt.   Many otherwise suitable steels must be eliminated for this reason, and even the best ones often give trouble if they are not properly heat-treated. One procedure which we have found useful with the "nondistorting" die steels such as Ketos is to harden, quench, and draw to Rockwell C-55/ C-60 and then to stress-relieve, all before final grinding.   After grinding, the part is stabilized by a heat-cycle treatment; it is dipped alternately into a dry-ice bath (using acetone or other suitable liquid as a heat-conducting medium) and then into hot oil.   The part remains in each bath long enough to attain the bath temperature, and three cycles are sufficient for most purposes.   The temperatures are not critical; the hot oil should be as hot as is convenient but not hot enough to soften the part.   This treatment is useful for spools and sleeves, and almost mandatory for plate-valve parts.   Finish lapping is done after the stabilization.

9.52. Required Dimensional Tolerances

The example given in Sec. 9.46 has shown that the axial and radial tolerances of high-performance valves are uncomfortably tight, and it is worth while to repeat and emphasize this fact. The following discussion will refer specifically to spool valves but can be easily translated to other types. Two types of dimensions are involved: diametral dimensions, which determine the working clearances, and axial dimensions, which determine the flow-versus-displacement characteristics at small openings.

Experience tends to indicate that when a light oil such as the usual commercial hydraulic fluid is used as the pressure medium, the minimum practical radial clearance is about 0.00005 in., or say 1 micron, and that it is usually preferable to have somewhat more than this even with carefully filtered fluids. If the fluid is more viscous or if the working pressures are low, the clearances can be opened up somewhat without seriously increasing the leakage, but the figures just given represent good practice in the aircraft and machine-tool field. It is not difficult for a good machinist to fit a plug into a hole within a diametral tolerance of ±0.0001 in., but it is considerably more difficult to finish the bore of a valve sleeve to this accuracy at all points, particularly at the very edges of the ports, where the fits must be closest of all if the valve is to work properly. The job can be done and it is being done every day, but it seems that a long period of education with its accompaniment of numerous rejects is necessary in every shop that tries to enter the business.

One reason for this difficulty lies in the fact that the mechanical gaging techniques of the present day are not good enough to give the necessary control of the finishing process, particularly for the bores. Plug gages are useless since they only tell the diameter of the tightest spot, not the diameter at the port edges. Air gages will not work close to a corner. Some types of expanding plug gages can be used to give a rough survey of the profile of the bore and to detect gross errors, but they are not accurate enough for really good work. Presumably it would be possible to build a special measuring device to give an accurate and highly magnified profile of the inside of a bore, but at present there seems to be no such device on the market. The best available technique seems to be hydraulic flow measurements, which are very sensitive to dimensional imperfections but are at best difficult to interpret.

For really critical applications the axial dimensions must have tolerances about as tight as those of the radial fit, or say ±0.0001 in. In standard four-way valves there are three critical axial dimensions, which are shown in Fig. 9.11. Here the four port edges a, b, c, and d must correspond simultaneously within the given tolerance to the positions of the piston land edges A, B, C, and D. If the tolerance is ±0.001 in. or more, the problem is not particularly difficult; if it is ±0.0001 in., it is practically impossible to use conventional gaging practice or to achieve interchangeability of parts. The usual practice is to leave excess metal on the piston faces, which is ground off according to indications obtained from experimental curves of flow versus displacement for each of the four metering edges separately.

With a three-way valve, two of the edges are no longer functional, so

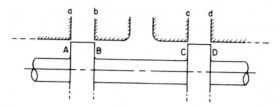

Fig. 9.11. Critical axial dimensions of four-way valve.

$$\text{Lengths } \left.\begin{array}{l} a - b = A - B \\ b - c = B - C \\ c - d = C - D \end{array}\right\} \text{within } \frac{\pm x}{3}$$

where x = maximum over- or underlap for any one edge.

Tolerance on one length may be relaxed if other tolerances are correspondingly tightened. Dimensioning may be done by any of several equivalent methods.

the only dimensions which must correspond are one length on the sleeve and one on the piston. Naturally this makes manufacture much easier.

9.53. Spool Valves

Many of the factors involved in manufacturing spool valves have already been discussed, and this section will be in part a recapitulation for emphasis. Generally speaking, the piston of a spool valve is no harder to make than is any other part with similarly tight tolerances; the difficulty comes in making the sleeve, and particularly in holding tolerance for the ports and for the portions of the bore immediately adjacent to the ports.

Practically all high-performance systems use negative feedback and obtain their performance by using large values of loop gain. These large gains can be obtained without making the system unstable only if each element of the system is nearly linear; or if nonlinear, its nonlinearity must be compensated in some fashion. Insofar as the valve is concerned, the linearity requirement applies to the relationship between flow rate and valve displacement; that is, for a given pressure drop across the valve, equal increments of displacement should produce equal increments of flow. This requirement in turn demands that the valve ports be of constant effective width; with the spool valve, it is required that they be rectangular.

Even with accurately rectangular ports, the gain of the valve falls off considerably near the origin if the cutoff is not sharp. This unsharpness, shown in Fig. 9.12, may arise from several causes, particularly from excessive clearance or underlap, or from rounding or bellmouthing of the port or land edges. These effects are roughly independent of the size of the valve; for example, as has already been stated, the minimum practi-

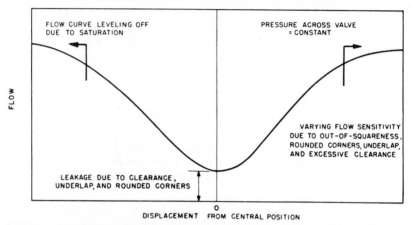

Fig. 9.12. Flow-travel relationships of typical closed-center control valve with no overlap.

cable clearance seems to be approximately 1 micron (40 μin.), and at least for servovalves up to some tens of horsepower it should not exceed about 5 microns. Similarly, in practice it seems to be very difficult to maintain a land corner perfectly sharp; there is always some erosion and rounding, but this decreases to a negligible amount (if the oil is well filtered) after the radius has increased to a few ten-thousandths of an inch.

Assuming, then, that the ports must be rectangular, there are several ways in which they can be made. One way is to mill across the sleeve, allowing the cut to break through into the bore. This is satisfactory if the percentage of peripheral width of the ports is not so great as to make the sleeve fragile, nor so small that the depth of the cut becomes excessively critical, and also if the dimensional tolerances are not so tight as to rule out milling altogether.

A second method is to drill round ports and then to broach them to the desired shape. This is sometimes useful, and was chosen for the two-stage valve shown in Fig. 13.5. With care, broached holes can be made accurate both as to dimensions and position. Both milling and broaching, however, must be performed upon materials of comparatively low hardness, and this usually means that the parts must be hardened after machining and before finishing. If the heating is done in air, the resultant scaling will ruin the accuracy of the broached ports, but if a good controlled-atmosphere furnace is used, the method is quite practical. In the valve of Fig. 13.5, incidentally, the pistons were hardened but the valve sleeves, which were integral with the bodies, were either cold-rolled steel or Meehanite cast iron. This use of soft materials violates what has been said in Sec. 9.51, but we got away with it, largely because of very careful filtering of the oil and because the eventual life of the valve was not intended to be long.

Sleeves can also be made from materials that are hard initially if some of the newer machining techniques are used, such as spark machining and ultrasonic drilling. Perhaps electrolytic machining can also be used. These methods are too new to have been completely evaluated, but they seem applicable where the tolerances are not too tight. They are not applicable when dimensions must be held to a few microns, at least in their present state of development. They are very well adapted to the making of holes of irregular shapes, which would be very difficult to make by other techniques.

To date, most valve sleeves with really tight tolerances have been built up from separate short sections, the ports being milled into the end faces of the appropriate sections. If the depth of the cut can be maintained with sufficient accuracy, a four-way valve sleeve can be made from three parts, but it is usually better to use five parts, two for the two ports and three for the center and end sections. The mating faces of the sections are finished on a precision surface grinder, and it is not particularly difficult to hold the thicknesses within the required tolerances.

The assembly of a multipart sleeve is a critical operation. In a valve intended for high-pressure operation the mere shrinking-in of the sleeve parts into the body, depending on friction to hold them, is useless, and pressing them in is worse. They must be adequately retained against endwise movement, and the forces available to move them may amount to tons. These forces will separate the sleeve parts, thus deforming and scraping away the bore of a valve body, especially if the outer sleeve corners are not broken. Aluminum-alloy bodies, which are particularly beloved of the airplane designer, are especially bad in this respect, and magnesium bodies are probably worse. We once made a large valve using an outer sleeve of mild steel, and it was a sad mistake. It was almost impossible to get the inner sleeve sections into place, and they would not stay put after they were there. We ruined several sleeves before we got one satisfactory one. The only proper way to do the job seems to be to use a fine-thread screwed retaining ring to put plenty of pressure on the stack of sleeve parts, and to make these parts a *very light* press fit in the outer sleeve or the body. Also it is inadvisable to use the standard snap rings which are stamped from sheet metal; they simply will not take the stresses without excessive deformation. Round-section snap rings are better, but only if they are very carefully fitted, and sufficiently heavy. Rectangular-section split rings or threaded retainers are best, but any retaining means must be accurately made so that it will bear all around the end of the sleeve. If it bears at only one point, there will be a wedging and bending action on the sleeve that will make it useless. The tolerances of the retainers could be eased at the cost of further complexity by the use of floating spherical-surfaced washers such as are used with self-aligning thrust bearings. All this seems like a lot of complication, but when one considers that the sleeve parts must be held in position over wide temperature ranges and against forces measured in tons, within perhaps a ten-thousandth of an inch, the reasons for the precautions become apparent.

The complexity of the mechanically retained multipart sleeve makes the single-piece sleeve look very good, and fortunately there is one good way to make it to the necessary tolerances. After the sleeve parts are finish-ground to the desired thickness, they are *very* lightly copper-plated, the plating being considerably less than 0.0001 in. thick. They are then assembled under light pressure in a suitable jig and copper-brazed in an atmosphere of dry hydrogen, being quenched directly from the brazing furnace. With suitable techniques it is possible to get a sleeve that is dimensionally accurate and as strong and rigid as though it had been made in one piece.

If the ports in a finished sleeve are slightly undersized, it is sometimes possible to correct them by removing small amounts of material. If the material is not "glass-hard," it can be filed by using special files—or perhaps they should be called hand-broaches—which are provided with flat bearing surfaces and used in suitable jigs, though naturally the process is expensive. Round holes, incidentally, can be "moved" by lapping with an abrasive-loaded bent piano wire in a sensitive drill press. One valve manufacturer uses this process extensively and has had very good results, though naturally the round holes are not applicable to valves which must have constant gain down to zero opening. Perhaps an analogous process could be developed for rectangular ports, but all such salvage processes are makeshifts at best.

After the ports are finished, the bore must be worked out to the final diameter. Here the difficulty lies not in the shape or in the tight tolerance, but in preventing the removal of excess metal close to the edges of the ports. Even a ten-thousandth of an inch of bellmouthing at this point will cause an appreciable degradation of the performance of the valve. Honing of the bore, usually followed by lapping, is probably the most popular technique because a honing machine is not expensive and can give very good results on bores which do not have ports. The honing stones are surprisingly flexible, however, particularly the smaller ones. The stone for a $3/16$-in. hole is hardly larger than a toothpick, and if the operator tries to force the job, as he will if he is trying to get out production, the unsupported portion of the stone will tend to bulge into the port, and the bore will become bellmouthed. This may seem like an exaggeration, but the results are all too easily visible in the flow curve. The same effect is obtained in the lapping process if an excess of lapping compound is used, particularly on a somewhat flexible lap. Here, however, the result may be corner-rounding rather than just bellmouthing.

Undoubtedly the best method of finishing a bore is internal grinding— *if* the grinder is good enough. This means not only a rigidly built machine with accurate ways and smooth feeds, but also a quill of really extraordinary accuracy and capable of very high speed. Such machines are now available from both U.S. and foreign manufacturers, but they are not cheap and they must be used properly. It is now possible to grind a ported bore $1/8$ in. in diameter and a couple of inches deep to an over-all tolerance of ± 0.0001 in., and to do it in production, but there are not many shops yet that can do this at a reasonable price. Presumably the increasing demand for valves and for other devices requiring small close-tolerance holes will create a source of supply for such holes.

After the sleeve is finished it must be mounted in the body, unless the finishing is done after mounting. In any case it is desirable, if it can be done, to make the sleeve fit fairly loosely in the outer bore and to depend upon O-rings or the like for preventing leakage between the several ports and their connecting ducts. Since both ends of the sleeve will be at the same pressure, at least in most designs, retention of the sleeve in the body is not a serious problem. One problem which is serious, especially when low-temperature operation is required, is differential expansion. This problem is greatly aggravated by the use of high-expansion materials, such as aluminum, and by shrinking or pressing the sleeve into the body. It should be obvious that the admission of hot oil to a valve which has been squeezed down by a cold outer body is the signal for trouble. Even if the piston is still movable in the sleeve before the oil hits it—and often it will be so tightly gripped as to be immovable—it will follow the temperature of the oil fairly rapidly and will very probably expand enough to bind before the body can warm up and relieve the situation. This should be obvious, but maybe it is not. At any rate, expansion troubles with aluminum-housed valves are all too common. The remedy *is* obvious.

9.54. Plate Valves [2]

Much of the difficulty of making a high-precision valve arises from the difficulty of machining the inside of a small hole and would disappear if

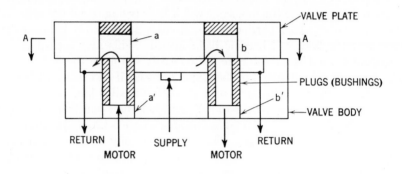

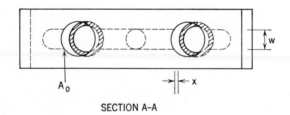

SECTION A-A

Fig. 9.13.   Single-sided plate valve.

the hole were somehow unwrapped so that its inside surface became accessible. In the plate valve this has been done, in effect, and the resulting advantages are shared by several different constructions. Some of these will be described in the following section.

The principle is perhaps most clearly shown in Fig. 9.13. The two parts, the movable valve plate and the fixed valve body, are initially clamped together and the holes, a, a' and b, b' are drilled, bored, and finish-lapped through both parts at the same time. This ensures that the two pairs of holes will be accurately aligned. A groove of width w

[2] Condensed from S.-Y. Lee, "Contributions to Hydraulic Control—6 New Valve Configurations for High-Performance Hydraulic and Pneumatic Systems," *Trans. ASME*, Vol. 76 (August 1954), pp. 905–911.

is milled in the body and then bushings are pressed into the lower holes, a' and b', and finished off flush with the upper surface of the body. The upper ends of the holes a and b are plugged. The edges of the bushings which are exposed by the longitudinal groove, together with the corresponding portions of the upper holes, form the four metering orifices of a four-way valve. Because the holes and bushings are matched accurately as to diameter and alignment, the valve has essentially zero overlap and underlap. Actually there is normally a very slight overlap since the bushings are sufficiently larger than the holes to give a light press fit, but this geometric overlap is in part compensated by the effect of the necessary working clearance between plate and body.

When the plate is displaced longitudinally, as shown in Fig. 9.13, two orifices are opened. Each of these orifices is bounded by two equal circular arcs and by two parallel straight lines, as shown in Fig. 9.14, so that they form curved parallelograms. The areas of the two are equal and independent of the diameter of the holes a and b so long as this diameter is greater than w. Also, if w is constant, the areas of these orifices are proportional to the displacement x, just as with rectangular ports. As already stated, this is very desirable for high-performance systems.

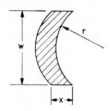

Fig. 9.14. Area of opening.

The drawing in Fig. 9.13 is merely schematic; obviously, means must be provided to permit the valve to move only in the desired direction and to maintain the necessary small clearance between plate and body. These restraints have been applied successfully to two general types of construction; in one, reed-type suspensions support the plate against the unbalanced pressure and permit longitudinal motion only, and in the other, a symmetrical construction balances out the pressure force and other constraints assure occurrence of only the desired motion.

The unbalanced pressure force which acts on the valve plate in Fig. 9.13 can be opposed by various types of mechanical supports. One successful type which is fairly easy to make was shown in Fig. 9.5. It consists of a pair of parallel reeds to which the valve plate is attached at the bottom and which are attached to a fixed support at the top. The reeds are thinned in four places to form flexure hinges, permitting the valve plate to move laterally much like an old-fashioned lawn swing. This motion is parallel to the slot which is milled in the lower face of the valve plate, and the intersections of this slot and the plugs form four metering edges; the other four are the adjacent edges of the holes in the base plate.

The reeds are made sufficiently rigid in compression to keep within reasonable limits the upward movement of the valve plate under working pressure. The vertical separation of the hinges determines the stiffness of the system to lateral deflections and also the amount of lifting at the end of the stroke. The latter is usually too small to be of any importance; for example, for a hinge separation of 1 in. and a lateral deflection of 0.010 in. the lift is only 50 μin.

The metering holes in the valve plate are drilled and lapped after it is assembled and doweled to the valve body. After the holes are finished, the valve-plate assembly is removed, the plugs are pressed in, and the whole lower face of the valve plate, its support, and the plugs, is ground flat. Any desired clearance may be ground at the same time.

One of the important properties of this construction is that there is never a metal-to-metal contact between plate and body, with the result that much of the rather unpredictable friction associated with most valves is eliminated. This low and fairly constant friction, together with the ease of manufacturing the valve in very small sizes, makes this construction ideal for pilot-valve service, where the input power is small, and also for pneumatic systems where lubrication is a serious problem. The cost of manufacture is much less than that of a spool valve of comparable characteristics. Chapter 13 describes in considerable detail a complete hydraulic servo which uses the suspension valve.

The construction just described is somewhat difficult to apply to large high-pressure valves, where the pressure force on a large-area valve plate would be excessive. For such applications it is preferable to use a symmetrical construction in which the pressure force on one side of the plate is balanced by an identical force on the other. A valve of this type was shown in Fig. 9.4. The valve body is made up of two blocks and two spacers which are doweled and screwed together, and a valve plate which moves in the space between the blocks. The metering holes (and the pivot-pin hole, if one is used) are drilled and finished as before, in a single setup before the final grinding of the interior surfaces. The pressure and return cavities are milled after disassembly, care being taken to have them of the same area and located directly oppositely in the upper and lower blocks. If this is not done, the pressure forces will be unbalanced and the friction will be excessive.

Hollow sleeves or bushings are pressed into the metering holes, usually four of them into the two blocks, though the inverse construction with the bushings and cavities in the valve plate can also be used. The bushings are bored out to permit the oil to pass through into the motor lines, and it is usually advisable to chamfer the inner ends of the bores in order to cut down flow resistance at this point. The dimensions of the chamfer are unimportant so long as some flat is left on the end of the bushing.

Since there are now two metering orifices in parallel, one above and one below the valve plate, the effective port width for this valve is twice the width of the milled grooves. The valve is statically pressure-balanced and has practically zero lap.

The plate valve can be made in many different configurations, both sliding and rotary. Fig. 9.15 shows the flow path for a rotary-plate valve, Fig. 9.16 a photograph of a disassembled valve of this type, and Fig. 9.17 a typical flow-versus-displacement curve for such a valve.

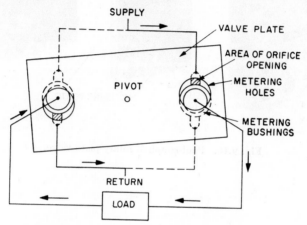

Fig. 9.15. Schematic drawing of rotary-plate valve.

The basic problems of sliding- and rotary-plate valve friction are essentially the same as those of spool valve friction and will be discussed in Chapter 10, but the more accessible construction of the plate valve makes practical several antifriction devices that are difficult to apply to other configurations. Most friction problems are caused either by the lodgment of dirt in the close working clearances of the valve or by unbalanced pressure distribution, which in turn may be caused either by dirt or by departures from ideal geometry of the mating surfaces.

The dirt problems of the plate valve are essentially the same as those of any other close-clearance device that operates under the same conditions, and may be solved in the same way. The problems of unbalanced pressure distribution, however, may be somewhat more serious for the plate valve than for the spool valve since it is more difficult to ensure perfect symmetry. This maldistribution of pressure can be improved considerably by providing small holes through the valve plate, which function in the same manner as the grooves usually provided on hydraulic pistons and valve spools. The provision of a few such holes in the proper locations permits a reasonable amount of dissymmetry in the milled cavities, or other manufacturing imperfections that otherwise

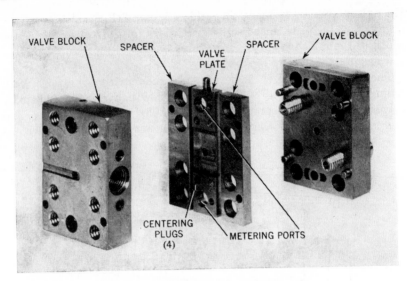

Fig. 9.16. Photograph of rotary-plate valve.

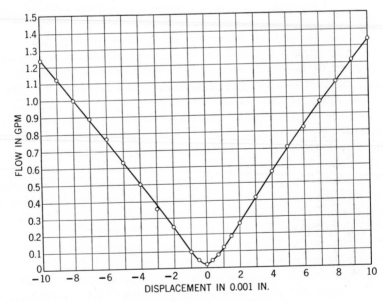

Fig. 9.17. Flow-travel curve.

would greatly increase the friction. It also helps to relieve all mating surfaces that are not actually required for bearing or sealing purposes. Experience has shown that these schemes, plus reasonable but not excessive care in manufacturing, result in plate valves that have friction levels at least as low as those of comparable spool valves when used in oil.

When the valves must have exceedingly low friction or when they are used on air or some other nonlubricating medium, it becomes necessary to use a more effective means of centering the valve plate. One simple and practical scheme is shown in Fig. 9.18. Small plugs are pressed into

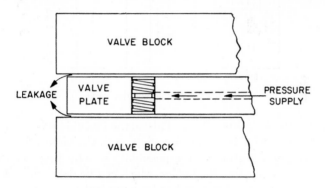

Fig. 9.18. Self-centering device.

holes in the valve plate, and the centers of these holes are connected with the pressure supply. The plugs are provided with small grooves which have flow resistances roughly equal to those of the leakage paths outward through the clearance spaces. If the valve plate moves upward, for example, the average pressure above the plate increases and that below it decreases, thus producing an unbalanced force which recenters the plate effectively. The depth and width of the grooves in the pins will be approximately 0.003 to 0.005 in. Spiral grooves of fine pitch are perhaps preferable because they are longer and therefore may be somewhat larger in diameter for the same resistance, and thus are less likely to be blocked by dirt. In most cases, as additional insurance against blocking, it is advisable to provide small filters in the supply duct. This scheme has proved to be very effective in a number of valves used on both air and oil.

The hydrodynamic force caused by the flow of fluid through the unmodified metering orifices is essentially the same for a plate valve as for a spool valve and is discussed in Chapter 10, which also discusses methods of compensation. The first scheme of Sec. 10.321, which uses a specially profiled downstream chamber, can also be applied to a plate valve but is less attractive than in the case of the spool valve because of the difficulty of making the profiled stream deflectors. As in the case of the spool valve, this form of compensation depends critically upon the clearance and the sharpness of the metering edge.

A second method, also described in Sec. 10.321, uses very simple and noncritical stream deflectors, plus baffles which are fixed to the valve

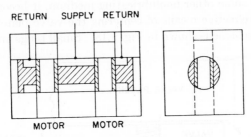

Fig. 9.19. Sliding-cylinder valve.

body, as illustrated in Fig. 10.17. This method would be very difficult to apply to a spool valve but is ideally adapted to the plate valve. One of its most attractive features is the possibility of adjusting the degree of compensation by rotating the baffles, which can be done while the valve is in service. This construction has been applied to a number of plate valves and works very well.

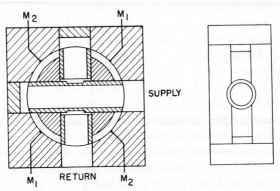

Fig. 9.20. Rotating-cylinder valve.

Obviously, the hole-and-plug technique of valve manufacture may be applied to other constructions than the flat-plate and the suspension valves. Two other applications are suggested in Figs. 9.19 and 9.20. The first is a sliding-cylinder design analogous to the sliding-plate types shown in Figs. 9.4 and 9.13, and the second is a rotating-cylinder type that might be considered a precision rotary-plug valve. The cylinders must be constrained, from rotating in the first case and from axial movement in the second. Self-centering devices and flow-force compensation can be applied to these designs also, as well as to the many other variants of the basic construction.

9.6. SPECIAL FEATURES OF PNEUMATIC VALVES

The design of valves for gaseous media involves the same considerations as design for liquids, but the relative weights of the various factors are considerably different. Most of what has already been said on valve design applies to pneumatic valves, and the following section will list the principal differences.

Probably the most important difference between a liquid and a gas, at least from the valve designer's point of view, is that the gas is much less viscous than a liquid, by a factor of roughly a thousand. This means that leakage flows in pneumatic valves are large and very important, and usually considerable effort must be taken to minimize them. Clearances must be held to the practicable minimum, and clearance paths must be as long and as narrow as possible. Underlap must be eliminated in sliding valves so far as possible, and overlap should be used when it can be permitted, as it usually cannot. This requirement for close tolerances, both axial and radial, is very difficult to satisfy in many cases and is one factor that has kept high-pressure pneumatics from being much used in the past.

Probably the feature of second importance is the fact that gases are extremely poor lubricants. This usually rules out constructions in which mating pieces are merely allowed to slide on each other, since the friction and wear will be excessive. In some cases a combination of materials can be used which has sufficiently good dry-friction properties, but usually some type of mechanical or hydrodynamic suspension must be provided. The plate-valve and suspension-valve constructions of the previous section are very well adapted to pneumatic applications.

The effects of dirt are much the same whether the fluid be gaseous or liquid. With a gas, once the dirt finds its way into the valve clearance it may cause more trouble than it would with a liquid, but on the other hand the low viscosity and density of the gas make the filtration problems much easier.

Somewhat allied to the dirt problem is the question of condensation. A small amount of water in oil usually does not have much effect on the operation of a hydraulic system, except possibly for an increase in corrosion, and as a rule the water is easy to eliminate. In a pneumatic system, on the other hand, the total amount of water vapor in the air may be considerable, and it is likely to condense out in many places. A valve handling high-pressure air usually becomes very cold during operation, and the water is likely to condense in the form of ice crystals, which may plug small openings or even freeze the valve shut and render it inoperable. Even if this does not happen, the valve almost certainly will become externally covered with condensed moisture or frost, and unless it is suitably protected it may corrode. Internal corrosion is equally likely, and naturally much more serious. One alleged advantage of pneumatic systems, especially for airborne applications, is the possible elimination of the return line. This is very poor practice, for the dehydration equipment necessary for satisfactory operation of the system in most climates will weigh and cost far more than any return line. Once the air is dried and purified it is too valuable to waste.

The valve of a closed-loop pneumatic system should be many times faster acting than would be expected from the system response requirements alone. This is because of the relatively great compressibility of the gaseous medium, even at high working pressures. If there is any stiction, Coulomb friction, or mass in the load, a considerable volume of air will have to be passed through the valve and into the motor or ram before the pressure changes enough to get the load moving. With a low-compressibility liquid the volume to be passed is much smaller and the valve speed requirements are less severe. The valve and its actuator must open rapidly to allow the large mass of gas to pass through into the ram, and then must close quickly, or even reverse, to prevent overshooting and oscillation. This factor of compressibility makes the design of a pneumatic servo different from and appreciably more difficult than that of a hydraulic one. Pneumatic-servo design will be discussed in Chapter 16.

The flow of a gas through an orifice differs essentially from that of a liquid, particularly in the occurrence of the phenomenon of critical flow. This phenomenon, together with the high compressibility of gases, makes it desirable to design a pneumatic valve so that the downstream orifices are larger than those upstream. Figure 9.16 showed a rotary-plate air valve, and in the face of the rear block, the upstream passages to the right of the bushings can be seen to be only about half as wide as the downstream passages to the left. The characteristics of pneumatic valves have been discussed in Chapter 8.

One advantage of the use of a gaseous medium is that the mass flow rate is much less than that of a liquid for the same controlled horsepower. Since the flow forces are proportional to the mass flow rate, not to the volumetric flow rate, this means that for the same power a smaller actuator can be used on an air valve, or conversely, that the same actuator can handle a considerably larger power. It also means that flow-force compensation is not usually needed in a pneumatic valve, which of course means a simpler and cheaper valve.

The choice between a pneumatic and a hydraulic servo must of course be made on the basis of many factors beside the valve and its requirements. These factors are discussed at greater length in Chapter 19, and also in a paper by Lee and Shearer.[3]

[3] S.-Y. Lee and J. L. Shearer, "Development of Valves for the Control of Pneumatic Power," presented at the Tenth Annual National Conference on Industrial Hydraulics, Chicago, Illinois, October 14–15, 1954

10

J. F. Blackburn

Steady-State
Operating Forces

10.1. INTRODUCTION

The operating member of a valve—the spool for example—is subjected to a number of different forces while it is in operation. There is, of course, the force supplied by the valve-operating device, but in addition other forces may either aid or oppose the operating force. These forces have various origins and various characteristics, and it is necessary to know their magnitudes and natures in order to design for successful operation of the valve. In many cases, these forces are large enough to cause erratic operation, or even to prevent operation completely, if suitable precautions are not taken.

It is not always possible to calculate valve forces accurately, and with some types of valves even the existence of certain types of forces cannot be predicted with certainty. In many cases, however, the analysis can be made fairly completely and accurately. The examples of the following chapter are useful in themselves, and the methods described will be found applicable to many other cases. In most of what follows, the derivations and the text will apply primarily to spool valves, but the formulas will apply to other constructions as well, although they may have to be modified appropriately.

The forces which act upon a valve spool may be classified according to their direction of action as peripheral, lateral, and axial. The first class is usually negligibly small, though in a few cases high peripheral forces have given trouble in V-port process-control valves. Lateral forces are not directly important since they act normal to the direction of motion of the spool, but in practice they may be very large so that the friction of the spool against the sleeve becomes excessively great. This friction may be so large in serious cases that the spool becomes immovable, hence the name "hydraulic lock," often applied to the phenomenon. Axial forces are naturally important since they directly affect the operation of the valve. They are often considerably greater than the inertial and frictional forces, and hence, they directly determine the design of the valve-operating device. In addition, they often cause the valve to oscillate or "sing," and this tendency alone may make an otherwise good valve useless for a particular application.

In the past, valve forces have been poorly understood or often not understood at all, and the remedies for the various troubles just alluded to have been arrived at purely empirically. Even now, the forces for many configurations have not been studied, and in no case is the complex of phenomena completely understood. Enough is known, however, to permit the design of spool valves and certain other types with reasonable confidence of success in most applications. The following chapter will summarize much of what is known about valve forces.

10.2. LATERAL FORCES—FRICTION AND HYDRAULIC LOCK [1]

Valve sticking or excessive friction, although usually greatly aggravated by the presence of dirt in the oil, frequently occurs in spite of all precautions to exclude dirt. It has long been known that the condition is worst with plain pistons and that it can be greatly alleviated by adding peripheral grooves around the piston lands. The standard explanation for the effect of the grooves is that they act to equalize the pressure distribution around the land. This is obviously correct, but it is of interest to develop a quantitative theory of the pressure distribution and the resulting land force for various land configurations.

[1] Sections 10.21 and 10.22 are quoted with slight modification from: J. F. Blackburn, *"Contributions to Hydraulic Control, 5—Lateral Forces on Hydraulic Pistons," Trans. ASME*, Vol. 75 (August 1953), pp. 1175–1180, by permission of the American Society of Mechanical Engineers.

In this section, such a theory [2] will be developed, and in the next, it will be supplemented by a qualitative discussion of the physics of the rubbing of the piston against the cylinder wall.

10.21. Origin of the Lateral Force

Throughout the following analysis, let it be assumed that the cylinder bore is perfectly straight and true but that the piston land is not; clearly the opposite case, or any equivalent combination of the two cases, would lead to the same results. The effects of any ports opposite the lands will be neglected, but the extension of the theory to include these effects is straightforward even if somewhat involved. Inertial and gravitational effects will also be neglected; these effects will be unimportant for any hydraulic piston with reasonably close fits. Let it also be assumed that all the flow is parallel to the axis and that the peripheral component of flow is negligibly small. This assumption is fairly well justified for short pistons with close clearances, but leads in other cases to excessive and, therefore, safe estimates of the lateral force. With these assumptions, the flow becomes laminar and two-dimensional. In addition, the assumption will be made that fully developed laminar flow is present throughout the length of the flow path.

First consider the case of a truly cylindrical piston in a true bore, with the axes of the piston and bore parallel but noncoincident. Consider the flows dq_1 and dq_2 through two elements of the peripheral clearance space at the top and bottom of the piston, the piston being displaced upward as in Fig. 10.1. Each element has a peripheral width, $dz = a \, d\theta$, normal to the plane of the figure.

Since for each elementary conduit the cross-sectional area, $y_1 \, dz$ or $y_2 \, dz$, is constant over the length, l, of the piston land, the pressure gradient $dp/dx = (P_1 - P_2)/l = $ constant for both conduits. Thus

[2] Similar suggestions undoubtedly have been made many times in the past. As examples, see:

Discussion by F. H. Towler, *Proc. Inst. Mech. Engrs. (London)*, Vol. 156 (1947), pp. 295–296.

Discussion by J. F. Alcock, *Proc. Inst. Mech. Engrs. (London)*, Vol. 158 (1948), pp. 203–205.

An unpublished memorandum by C. E. Grosser of Hughes Aircraft Co., November 16, 1952. (We are indebted to the Hughes Aircraft Co. for a copy of this memorandum.)

D. C. Sweeney, "Preliminary Investigation of Hydraulic Lock," *Engineering*, Vol. 172 (1951), pp. 513–516 and 580–582.

Alcock and Grosser both show diagrams similar to Fig. 10.6. Sweeney gives a brief outline of a derivation similar to that leading to Eq. 10.5.

the curve of pressure versus x is a straight line between the points P_1 and P_2 at the ends of the piston for each of the two conduits. The downward force on the piston is

$$df_1 = \int_0^l p\, dz\, dx = \frac{P_1 + P_2}{2} l\, dz \tag{10.1}$$

The upward force, df_2, is exactly the same; the df's balance by pairs all around the piston and the net lateral force is zero for any case in which the surface of the piston is parallel with the adjacent wall of the cylinder.

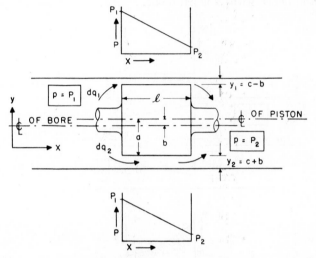

Fig. 10.1. Cylindrical decentered piston.

The net force is also zero for any configuration possessing axial symmetry, but this is not necessarily true for decentered pistons, where the axial symmetry is destroyed. Probably the simplest case of this type is that of a decentered conical piston, shown in Fig. 10.2. Here the cross-sectional area and, therefore, the pressure gradient are no longer constant over the length of the piston, and the force per unit peripheral width, dz, must be found by a double integration. The details of the calculation will not be given, but the resulting formulas are as follows.

For an elementary length, dx, of the conduit of width dz and height y,

$$\frac{dp}{dx} = -\frac{12\mu}{y^3}\frac{dq}{dz} \tag{10.2}$$

where μ is the absolute viscosity and dq/dz is the flow through a conduit of width dz. With integration and insertion of the boundary conditions,

when
$$x = 0, \quad y = C_1, \quad \text{and} \quad p = P_1$$
and when
$$x = l, \quad y = C_1 + t, \quad \text{and} \quad p = P_2 = P_1 - \Delta p$$
we get

$$p = P_1 - \frac{\Delta p(C_1 + t)^2}{t(2C_1 + t)}\left(1 - \frac{C_1{}^2}{y^2}\right) \tag{10.3}$$

where t is the radial taper of the piston. Thus, the variation of pressure with (y and hence with) x is parabolic.

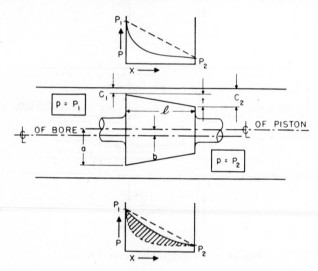

Fig. 10.2. Decentered conical piston.

The actual deviation of this parabolic pressure curve from the linear curve of a uniform conduit depends essentially upon the taper *and upon the closeness of approach* of the large end to the cylinder wall. For the upper duct of Fig. 10.2, the large end approaches closely, the pressure gradient near the large end is high, and the p–x curve drops rapidly from the P_1 point. For the lower duct, the ratio of the y's at the upstream and downstream ends is much nearer unity, and the departure from linearity is much less.

The radial force on an elementary area, $dx\,dz$, of the piston will be $p\,dx\,dz$. Integration gives the total radial force on the elementary strip of width dz as

$$df = P_1 - \left(\frac{(C_1 + t)\,\Delta P}{(2C_1 + t)}\right)l\,dz \tag{10.4}$$

The total lateral force F on the piston is obtained by integrating df around the periphery, when account is taken of the fact that only the component of df parallel to the displacement b is of interest since the normal components will cancel because of the symmetry of the configuration. Thus, if dF is the component of df parallel to b,

$$dF = -la\left(P_1 - \Delta p\,\frac{C + t + b\cos\theta}{2C + t + 2b\cos\theta}\right)\cos\theta \qquad (10.5)$$

where C is the radial clearance at the large end with the piston centered. Integration with respect to θ from 0 to 2π gives the total side-thrust on the piston:

$$F = \frac{\pi a t\,\Delta p}{2b}\left(1 - \frac{2C + t}{\sqrt{(2C + t)^2 - 4b^2}}\right) \qquad (10.6)$$

Since the second term in the parentheses is greater than 1, the force will be negative, that is, *away from* the larger opening, and is therefore a *decentering* force. Thus the equilibrium of a centered tapered piston with the higher pressure applied to the larger end is unstable, and the piston will be forced into contact with the wall. Conversely, if the higher pressure is applied to the smaller end, the piston will center itself in the bore.

Calculations have been made for one land of a very small pilot piston which stuck persistently. The resulting curves for F versus b are plotted

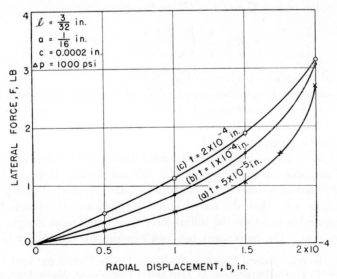

Fig. 10.3. Lateral force versus displacement for a small piston land.

in Fig. 10.3. The values of the various quantities involved are:

Axial length, l = 0.094 in.

Radius, a = 0.063 in.

Radial clearance at large end, C = 0.0002 in.

Pressure drop, Δp = 1000 psi

Radial taper, t = 0.00005 in. for curve (a)

= 0.0001 in. for curve (b)

= 0.0002 in. for curve (c)

From Fig. 10.3, it can be seen that the principal effects of varying the amount of taper are: (1) the rate of increase of decentering force for small displacements from center decreases rapidly with decreasing taper, and (2) the force holding the piston against the wall (when $b = 2 \times 10^{-4}$ in., in the present case) changes comparatively little.

A curve of fairly universal applicability can be made by imagining a fictitious uniform pressure, P_c, which when applied to the laterally projected area, $2la$, of the piston would produce the force F_c. (The subscript c indicates that F_c is the force when the large end of the piston just contacts the cylinder wall.) Since P_c is directly proportional to Δp, it is most useful to eliminate the latter by forming the ratio $R = P_c/\Delta p$. Substitution of $b = C$ and $\tau = t/C$ in Eq. 10.6 gives

$$R = \frac{\pi}{4}\tau\left(\frac{2+\tau}{\sqrt{4\tau+\tau^2}} - 1\right) \qquad (10.7)$$

A plot of R versus τ is given in Fig. 10.4.

It is seen that R increases very rapidly as τ increases from zero to 0.2, goes through a flat maximum at about $\tau = 0.9$, and thereafter decreases slowly. With reasonably good valve design and workmanship, τ would ordinarily be much less than 1.

A second case which can be handled analytically, although at the cost of considerable labor, is that of a stepped piston. Both cases have been checked experimentally [3] and the results are plotted in Fig. 10.5.

It can be seen that the forces in the stepped and the tapered cases are comparable but that the initial force rises much more steeply in the stepped case. The agreement between experiment and theory is good,

[3] H. E. Weber, "Lateral Forces on Hydraulic Pistons Caused by Axial Leakage Flow," S.M. Thesis, Department of Mechanical Engineering, Massachusetts Institute of Technology, Cambridge, Mass., 1951, pp. 30–34.

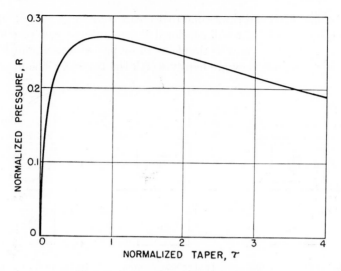

Fig. 10.4. Pressure ratio versus taper.

especially for the tapered piston. The deviations from the theoretical curves are due principally to two causes—the fact that an appreciable distance from the entrance is required to establish true laminar flow, and the fact that the theory neglects the small peripheral flow. The first cause tends to increase the force for low values of β (the normalized

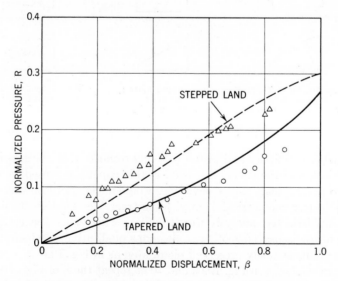

Fig. 10.5. Dimensionless lateral force versus displacement.

displacement, b/C) and the second to decrease it for large values. Rigorous solutions of the three-dimensional flow equations can possibly be obtained, at least in cases as simple as those just discussed, but from a practical standpoint they are hardly worth the trouble. The machinist

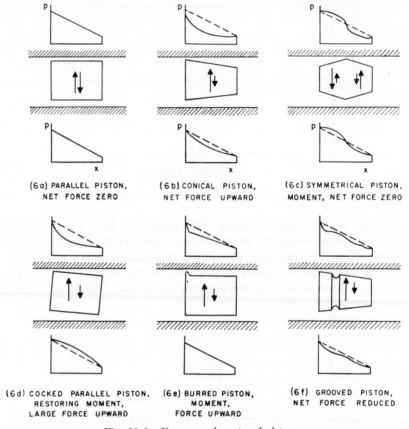

(6a) PARALLEL PISTON,
NET FORCE ZERO

(6b) CONICAL PISTON,
NET FORCE UPWARD

(6c) SYMMETRICAL PISTON,
MOMENT, NET FORCE ZERO

(6d) COCKED PARALLEL PISTON,
RESTORING MOMENT,
LARGE FORCE UPWARD

(6e) BURRED PISTON,
MOMENT,
FORCE UPWARD

(6f) GROOVED PISTON,
NET FORCE REDUCED

Fig. 10.6. Forces on decentered pistons.

who makes a cylinder and piston is trying to make them both perfectly cylindrical, and the effects discussed in this paper are caused by deviations from this ideal. These deviations would never approach sufficiently closely to any geometrical form yielding tractable equations to make a more sophisticated analysis worth while. This being the case, it is sufficient to discuss a few other cases from a qualitative standpoint. The results are most easily presented pictorially, as in Fig. 10.6.

The cases shown in Fig. 10.6a and 10.6b repeat those of Figs. 10.1 and 10.2. Figure 10.6c represents a double-coned piston, which might be considered a rough approximation of the barrel shape that sometimes results from improper lapping technique. So long as the profile is symmetrical about the midplane, there will be no net force, though there will be a moment normal to the piston axis that might be important in some cases. The results in the rather unlikely case of an hourglass-shaped piston would be similar except that the moment would be oppositely directed and that F would be greatly affected by the ratio of the clearances at the two ends.

If the moment shown in Fig. 10.6c, or any other normal moment, can produce a cocking of the piston in the bore, a case similar to Fig. 10.6d will result. Here a large decentering force is produced, plus a moment which tends to restore parallelism of the piston and the bore. This case, analyzed and also verified experimentally by Weber,[4] is important because unbalanced external moments are very common—they may be caused, for example, by a bent push-wire or a poorly designed or constructed linkage system—and if this moment is greater than the hydraulic restoring moment, a very large lateral force may result. An even more frequent source of trouble, especially with slender pistons, is a bent or sprung piston, which produces the same effect at the end lands, though the contact points are likely to be on the convex side of the center lands.

Another important case is that of a piston with a local protuberance such as the burr shown on the upstream corner of Fig. 10.6e. Such a protuberance, wherever located, casts a low-pressure "shadow" downstream, and if this shadow is not balanced by an equivalent shadow on the opposite side of the piston, the protuberance is forced against the wall. The magnitude of the force depends on the area and "darkness" of the shadow, and is greater the farther upstream the bump is located and the greater its peripheral width.

Even if both piston and cylinder are geometrically true, a case similar to the last can be caused by dirt particles in the fluid. Particles too large to pass through the clearances lodge at the upstream end of the clearance, casting low-pressure shadows downstream and pulling the piston radially toward the side where they have lodged. The effect of dirt may be mitigated, however, by two fortunate circumstances. In the first place, the rate of lodgment or "silting" depends on the rate of leakage flow, with the result that large particles are much more likely to lodge on the side of the piston with large clearance, where they tend to recenter it. Second, since the dirt is not tightly attached to the piston, a comparatively small movement of the piston often causes it to be

[4] Weber, *op. cit.*, pp. 25–29.

washed away and the piston is free again. In some British experiments [5] it was found that, for the particular piston used, sludge or fine dirt, even up to several times the radial clearance in particle size, caused no permanent effect although the oil was black with it; large chips, however, and particularly burrs sheared from screw threads on careless making up of pipe joints, disabled the valve completely.

The obvious methods of minimizing side forces are (1) to make both piston surface and bore as truly cylindrical as possible and (2) to provide some means of lightening the shadows. The latter is most easily accomplished by the old but useful scheme of providing peripheral grooves, as shown in Fig. 10.6f. The size and form of the grooves are unimportant as long as the flow resistance of a groove is small compared to that of the clearance space between the piston and the bore; thus for ordinarily close fits, grooves only a few thousandths of an inch wide and deep would suffice. It is probably desirable that the walls of the groove meet the outer piston surface at right angles in order to decrease the probability that dirt particles will become wedged between piston and cylinder.

Some experimental work has been done on the effect of the number and location of the grooves.[6] In general, it is preferable to have a single wide groove, leaving only a narrow fin at the upstream and downstream edges of the land, but this is usually impracticable since it results in excessive leakage and insufficient wearing area. If a single narrow groove is used, it should be placed as near as practicable to the upstream edge; additional grooves will help, but their effect is somewhat smaller than that of the first one.

Similar results were obtained by Dr. Sweeney of Birmingham University.[7-9] He found that adding a single groove at the center of the land decreased the locking force (the axial force necessary to free the piston after it had stood for a while with pressure applied) to 40 per cent

[5] Discussion by J. F. Alcock, *Proc. Inst. Mech. Engrs. (London)*, Vol. 158 (1947), pp. 203–205.

[6] A. Goldberg, "Elimination of Causes and Effects of Lateral Forces on Hydraulic Piston," S.M. Thesis, Department of Mechanical Engineering, Massachusetts Institute of Technology, Cambridge, Mass., 1951, pp. 43–49.

[7] D. C. Sweeney, "Preliminary Investigation of Hydraulic Lock," *Engineering*, Vol. 172 (1951), pp. 513–516 and 580–582. (I am indebted to Mr. H. G. Conway of Short Brothers and Harland, Ltd., for sending photostats of this article and for discussions of the problem of hydraulic lock.)

[8] D. C. Sweeney, "Out-of-Balance Reactions in Hydraulic Piston Type Control Valves and a Preliminary Investigation of Hydraulic Lock," Ph.D. Thesis, University of Birmingham, 1949.

[9] J. Manhajm and D. C. Sweeney, "An Investigation of Hydraulic Lock," *Proc. Inst. Mech. Engrs.*, Vol. 169 (1955), pp. 865–879.

of that for an ungrooved piston; three equally spaced grooves reduced it to 6.3 per cent, and seven, to 2.7 per cent.

There are other methods that may not be so widely applicable as the grooves but that are probably more effective when they can be used. One such method is that employed in the pressurized bearing.[10] As applied to a cylindrical piston, this procedure might involve providing thin transverse slots in the valve sleeve opposite the piston lands, each slot having a suitable flow resistance and fed with high-pressure fluid from an outer manifold. This scheme is very effective, but it is rather complicated and may be subject to trouble from clogging. It has been applied to plate valves with excellent results, as described in Sec. 9.54.

A much simpler scheme takes advantage of the fact that a decentering force becomes a centering force if the small end of the tapered or stepped piston land is placed toward the higher pressure. As shown already, the height of the step or the degree of the taper involved is small, and in many cases the increase in leakage entailed by this scheme is permissible. The method cannot be applied to lands across which the pressure drop may reverse, and it is not easily applicable to valving lands when a sharp cutoff of flow is required. If the piston is reasonably stiff, however, it is often satisfactory to use a taper or step on the outer lands only, depending on their centering action to keep the inner valving lands centered.

10.22. Friction between Piston and Cylinder

The previous discussion has shown that deviations of either cylinder or piston from the ideal cylindrical form may result in forces of considerable magnitude which tend to press the piston laterally into the wall of the cylinder. This section outlines briefly what happens after the piston begins to make contact with the wall.

If the piston is somehow constrained to remain centered, it does not touch the cylinder and can be moved axially with a very small force. The friction is purely hydrodynamic, and the frictional force is proportional to the viscosity of the fluid and to the piston velocity. This is the ideal case, and it sometimes can be closely approximated in practice.

If the piston is not constrained to stay in the center but approaches the wall, a succession of complicated phenomena takes place. If there is no boundary layer on the metal surfaces (which will probably never be the case), the piston moves sidewise until one of its high points makes contact with a high point of the cylinder. If true metal-to-metal contact is established, the axial force required to move the piston suddenly in-

[10] Weber, *op. cit.*, pp. 40–47.

creases enormously and becomes independent of velocity but more or less proportional to the lateral force which holds the piston against the wall. This represents the case of dry or Coulomb friction.

Almost all practical cases lie somewhere between the two extremes just outlined. Some kind of boundary layer is usually present, and the nature and magnitude of the friction depend upon the properties of this layer and also upon the nature of the surface finish of the metal parts and upon the magnitude of the lateral force.

One grossly simplified picture of the boundary layer is that it consists of a single layer of molecules held fairly tightly to the metal surface by molecular forces. These molecules may be oxygen, in which case the layer may be very thin, or they may be lubricity additives such as oleic acid, in which case they will line up with their long axes normal to the metal surface, giving a boundary layer which has been compared to a field of wheat. In any case, if the boundary layer is only one molecule thick, it is very thin compared to the roughness of any attainable surface finish.

It is highly unlikely, however, at least when liquids are used as working fluids, that the boundary layer is ever only one molecule thick. Even when a liquid surface is in contact only with its own saturated vapor, the surface molecules are highly oriented, forming a two-dimensional crystal. If the liquid surface is in contact with a solid, so that this "crystalline" layer is not required to bend, it will probably be about as strong as any other crystal, and in effect will make the solid one molecular layer thicker. The oriented monolayer will also have a field which orients the second layer of liquid molecules almost as perfectly, the second layer orients the third, and so on for a considerable distance into the liquid. The alignment of the molecules becomes more and more random with the increasing distance from the surface until the medium no longer supports a finite shearing stress; it is a true liquid. The rate of change from a quasi-solid to a true liquid depends greatly on the composition of the liquid and in many cases upon that of the bounding medium also. There is good evidence from several fields of investigation that in some cases,[11] such as oil in contact with steel, the effective thickness of the boundary layer is many microinches, large compared to the roughness of a reasonably smooth surface and comparable to practicable dimensional tolerances and clearances.

If the foregoing concept of the structure of the boundary layer is accepted, the sequence of events as the piston approaches the cylinder

---

[11] J. C. Henniker, "The Depth of the Surface Zone of a Liquid," *Revs. Modern Phys.*, Vol. 21 (1949), pp. 322–341. This excellent article presents evidence from a great variety of sources and has an extensive bibliography.

will be somewhat as follows. Initially, the separation is large, there is a layer of true liquid between the surfaces, friction is viscous, and the rate of approach is fairly high. As the liquid layer becomes thin, the rate of approach decreases rapidly because it is harder to squeeze a liquid out of a thin crack than out of a wide one. The problem of the rate of approach will be discussed briefly in Sec. 10.23; for perfectly smooth surfaces and no boundary layer, the velocity of approach decreases to zero asymptotically so that ideally the surfaces never come into contact. Actually, however, the rate of approach is greatly affected in its final stages by the character and magnitude of the surface roughness and the boundary layer.

In the presence of a boundary layer of appreciable thickness, the friction begins changing from its large-separation character as the higher spots on the boundary layers of piston and sleeve begin to touch. The magnitude and character of the change depend upon the surface profile and especially upon the characteristics of the layers. Since the change in properties is presumably gradual from true liquid through a stiffer and stiffer jelly to a crystalline solid, the friction changes in a continuous manner from viscous to Coulomb friction, probably with a rapid increase in the apparent coefficient of friction as the surfaces near contact. If the "effective" thickness (however it may be defined) of the boundary layer is large compared to the height of the surface roughness, the layer will be able to support a considerable load and the friction will be low, even though it may be much higher than for hydrodynamic lubrication.

If the surface has small-area peaks that stick out through the boundary layer, however, even a small lateral force from any cause will be enough to rupture the layer over the peak and produce metal-to-metal contact. The immediate effect of this contact is a sudden large increase in frictional force, but its ultimate consequences depend upon the lateral force and upon the natures of the metal surfaces and the fluid. The most unfavorable case is probably that of similar surfaces of a soft tough metal that pressure-welds easily [12] and a low-viscosity chemically inactive fluid; the surfaces weld and tear apart at the peaks, with a rapid galling and roughening until the piston becomes immovable. If the material is brittle, the effect is plucking and spalling, but the results are much the same. On the other hand, if the fluid contains suitable reactive compounds, such as the conventional extreme-pressure lubricating-oil additives, the peaks are converted into low-melting halides, sulfides, and so

[12] A preliminary investigation of the effects of piston and sleeve materials is reported in N. M. Edelson, "Friction Reduction in a Hydraulic Servo Valve," S.B. Thesis, Department of Mechanical Engineering, Massachusetts Institute of Technology, Cambridge, Mass., 1951.

on, which flow off into the adjacent valleys, leaving a fairly smooth topography over which the boundary layer can form an effective cushion.

10.23. The Time Factor

So far, nothing has been said about the rate at which the piston approaches or recedes from the cylinder wall under the influence of the lateral forces. It is obvious that it will take time to squeeze an oil film out of the thin crack between the approaching surfaces, and the equations of motion for simplified cases can be solved. One excellent analysis of the problem has been given by Sweeney.[13] He shows that for an ideal Newtonian fluid and for geometrically perfect surfaces the rate of approach becomes zero as the separation becomes zero, so the surfaces never would make actual contact. Physical surfaces are never geometrically perfect, however, and he takes account of the effects of surface roughness by assuming that the final "contact" position, which is determined by the elastic deformation of the solid surfaces, is some large fraction of the sum of the heights of the surface roughness of piston and cylinder. If this quantity is called s, the final expression for the time to make contact is

$$ t = \frac{12\pi\mu r^3}{Rc^{3/2}\sqrt{2s}} \tag{10.8} $$

where μ = viscosity of fluid
R = lateral force per unit axial length of piston
r = radius of piston
c = radial clearance

Sweeney carried out a thorough experimental investigation of the time factor in hydraulic lock, and his results verify his theory very well if reasonable values of s are chosen. His values are consistent with his measured values of surface roughness. His values of t range from 8 to 600 sec.

Sweeney's apparatus was designed to study the phenomenon of hydraulic lock and, therefore, eliminated some of the variables that make it a complicated problem in actual hydraulic practice. His oil apparently contained no additives, his pistons and cylinders were smooth and accurately machined, and his operating pressures were not over 200 psi. With higher pressures, somewhat dirty oil containing additives, and rougher, less closely fitted surfaces, the times to contact may vary over a very much wider range. In practice, it is nearly always found that the

---

[13] J. Manhajm and D. C. Sweeney, *op. cit.*

frictional or "locking" force builds up rapidly at first and then more slowly, approaching a maximum value which is usually proportional to the pressure drop along the piston. If the piston is moved, even slightly, it breaks free and the build-up process must start over again. Thus, one means of preventing lock is to keep the piston in motion, either by "dithering" it axially or by rotating it. Axial dither is frequently employed, for this and other reasons. Rotation is less common as it requires a separate drive and seems to cause excessive wear in some cases. Either dither or rotation will cause wear anyway, since if a decentering lateral force exists, there will be at least a small amount of metal-to-metal friction unless the boundary layer is very thick and very tough.

Once established, hydraulic lock will persist until it is forcibly broken by motion of the piston or until the oil pressure is removed. Even then, the piston may remain tightly attached to the cylinder in extreme cases, but usually the force necessary to break it loose will decrease gradually. The unlocking curve is roughly similar to the inverse of the locking curve, though usually with a longer time constant. If there is no mechanical force available to pry the contacting surfaces apart, the oil must creep in, aided by molecular forces only, and since these are comparatively weak, the unlocking process is slow.

This lag on unlocking has one interesting consequence for seating-type valves. Consider a poppet valve whose seat has appreciable width. On closing, the poppet surface approaches the seat closely and begins to see a large resistance to its motion because it has to squeeze out the last drops of oil from the thin remaining clearance. At this time, the valve is essentially closed, since the flow is exceedingly small compared with that when the valve is open even a mil or two. As the closing force continues, the valve gradually approaches and finally reaches metal-to-metal contact with the seat, with a practically perfect shutoff. Now suppose the valve actuator tries to open the valve. Unless the applied force is very large and is applied through a very stiff spring, there will be an appreciable lag before the valve unlocks and rises the rest of the way. Thus, there will be very little lag on closing the valve, but there may be an appreciable one on opening. The phenomenon is complicated by the finite but highly variable "tensile" strength of the liquid, which may be large enough to require a considerable force to cause mechanical rupture of the film. Rough estimates of the lag in actual valves with smooth seats and seating-surface widths of a few hundredths of an inch show lags of the order of a millisecond or more. In most applications, these are negligibly small, but in very-high-speed operation, they may be serious. This is one more argument against using poppet valves in servo loops.

10.24. The Effects of Dirt

This section cannot tell what is implied by the title since essentially nothing quantitative is known about the effects produced by various kinds of "dirt" in the hydraulic fluid. This lack of quantitative information is not surprising in view of the extreme complexity of the phenomena and the fact that there is not even a good definition of what dirt is. About all that can be done is to make some unsupported assertions. These assertions seem to be in agreement with the experience of most operators of hydraulic systems, however, and it is hoped that they will have practical utility if not theoretical foundation.

"Dirt" in a hydraulic system may have various characteristics. Much of it is grit, usually fine quartz sand, which may enter the system through the breather pipe on the tank, or otherwise. Much more is very finely divided metal or metal oxides, the products of wear of moving parts of the system, particularly the pump. There is often much organic solid or semisolid matter, which usually forms a more or less heavy black sludge. And finally, there may be, and nearly always are, large metal chips derived from machining or assembly operations on various parts of the system.

Probably the greatest trouble in an adequately maintained system occurs because of sticky operation of a closely fitted valve with a comparatively weak actuating device. Here, even textile fibers from cleaning rags can cause serious trouble, and a sticky or gummy sludge, even in rather small concentrations, may make the valve inoperative. If the actuator is reasonably powerful, however, or if maintenance is adequate to eliminate fibers and sludge, much of the trouble will be caused by fine grit. This can be eliminated by adequate filtration, but the filtration must be adequate and adequately maintained. A given filter is only useful for a comparatively short time. When it is new, it may let too much grit through. After some time, the holes begin to plug up with dirt and the filtering action improves. As the plugging continues, the pressure drop through the filter increases until particles or agglomerates of particles are forced through the filtering medium and emerge downstream. The filter obviously should be replaced somewhat before this begins to happen, and one of the best criteria for replacement is the pressure drop.

Filters for use in hydraulic systems are usually rated by the manufacturers as "10-micron" or "5-micron" or according to some other system that is usually interpreted by the buyer to mean that they will stop all particles larger in diameter than the rated size. This assumption is fallacious, and at the present time so little is known about either the exact mechanism of the filtering action or about the particle size and shape distributions of hydraulic-system dirt that the ratings are almost meaningless. One conclusion can be drawn, perhaps; a "5-micron" filter from any reputable manufacturer will stop more dirt and will pass smaller dirt than a similar "10-micron" filter, and so on for the other ratings. Until a few years ago, the 10-micron size was the finest hydraulic filter available, but now other ratings can be had down to $\frac{1}{2}$ micron. Probably the 10- and 5-micron types are the most popular, and the latter is adequate for most applications.

Given reasonably adequate maintenance and filtering, plus sufficiently powerful valve actuators, the principal effect of fine grit in the fluid is its abrasive action. This shows up at rubbing surfaces, such as the pistons and cylinders of pumps and motors, and also as a sand-blast effect at points of high fluid acceleration, particularly at the corners of valve metering orifices. Piston abrasion slowly increases leakage, which is not usually a serious effect. The rounding of the corners of orifices is more serious, however. It greatly increases the quiescent leakage of the valve and reduces both the flow sensitivity and the pressure sensitivity. Also, if the valve has force compensation which depends upon the maintenance of a sharp corner, as in the Lee type of Sec. 10.321, even a small amount of wear will greatly decrease the effectiveness of the compensation and will probably limit the useful life of the valve.

Another common but less serious effect of dirt in the fluid is usually called "silting." This takes place at small valve openings, usually about 1 to 2 mils or less. At first, after the opening is closed down to a narrow slit from some larger width, the flow is what would be expected for the pressure drop and area of opening. After a short time, the flow begins to decrease, and after some minutes may even stop almost completely. A slight movement of the spool is sufficient to restore the flow to its normal value, but the roughly exponential decrease of flow with time occurs again and again. It is caused by the valve itself acting as a filter; a dam of coarse grit particles builds up on the upstream side of the narrow orifice, and the chinks between the larger particles become filled by smaller ones. If the oil is not extremely well filtered, this phenomenon will always be observed at very small valve openings. In a system with only "10-micron" filtering and with a valve slit width of about 1 mil, the flow will decrease by half in from a few seconds to a minute or so. The substitution of a 5-micron filter will practically eliminate the effect for 2-mil openings but will have little effect for openings of a few tenths of a mil. Silting is a very annoying phenomenon for the person who is trying to take flow curves on a valve, or in systems requiring accurate regulation of very small flows. It is considerably less important in most servos and similar applications because even very brief openings of the valve allow the silt dam to be swept away downstream and restore normal operation. Silting is usually more of an annoyance than a serious handicap, at least in reasonably-well-filtered systems.

Sludge, which is almost always the result of poor operating practices, and textile fibers from cleaning rags may cause serious deterioration of system performance but are easily eliminated by any reasonably adequate filter. Even an exceedingly coarse strainer would eliminate large chips, screws, ball bearings, pieces of rag or paper, and chunks of paint, caulking compound, and wood. It is just this type of "dirt" that causes the most trouble in practice, however. It is impossible to over-emphasize the necessity of extreme cleanliness and care in assembling a hydraulic system, and it is almost equally impossible to put this point across sufficiently well with most people. Even after the rags, wood, and other obviously foreign objects are eliminated, a large chip from some machining operation will often remain in hiding, only to detach itself and float downstream to the point where it will do the most harm. If it is upstream from a filter, no harm will be done, but the perversity of Nature usually arranges to have it hidden just upstream from the valve, with the inevitable consequences. For this reason some valve manufacturers include a small strainer just inside the supply connection in the valve block. The strainer should have a considerably coarser mesh than the system filter, so that it need be cleaned but rarely, but it should stop the chips. If it is made difficult for the maintenance man to remove, so much the better; he will be less likely to leave it out.

While on the subject of chips, it should be pointed out that some manu-facturers design their valves to be "self-clearing" in that, if the valve is disabled by a chip, a mechanical means is provided whereby a large force is applied to the piston, and the chip is sheared. This is a good selling point, and undoubtedly this property might be both useful and permissible in special cases, but there are two compelling arguments against it. In many cases, the resulting transient disturbance to the system while the valve is shearing the chip and then restoring things to normal would be very serious. If the valve is controlling the feed of a machine tool, for example, about the least that could be expected would be the spoiling of the work piece and the breaking of the tool. Secondly, if the chip is of steel, as it probably would be in most systems, the chances are that the valve would be seriously damaged anyway. Good practice demands that land and port edges be made of very hard materials, and if the valve is used as a shear, it is very probable that the metering edges will be chipped or nicked, and the valve will be ruined for critical applications. It is far better to get the chips out of the system in the first place.

10.3. AXIAL FORCES

The axial forces that operate on a valve spool include the inertial force, which is readily calculated and is usually small, the frictional forces, which are not accurately calculable but which should also be small in a properly designed and properly applied valve, and various forces which owe their existence to the fact that fluid is flowing through the valve and exerting force on the spool. Knowledge of these flow-produced forces is still incomplete, and every new basic valve design presents new problems, but for certain common types of valves, the principal forces are known, and their magnitudes can be calculated with good accuracy.

10.31. Static Pressure Unbalance

In any valve, under normal conditions there will be a greater or less pressure difference between the upstream and downstream connections, and with many constructions, this pressure difference exerts a force on the operating member of the valve. This is particularly true of seating-type valves, which are difficult to balance completely; most sliding-type valves are inherently balanced. In hydraulic-power practice, the pressure differences are so great that the resulting unbalance forces are very large for poppet and similar valves. The same is true of the unbalanced valves, which are often intermediate between sliding and seating types, that are used for the control of process streams. Manually controlled valves, especially in the larger sizes, and servovalves which are controlled by electromagnetic actuators are nearly always of the sliding type, and, especially in the latter, considerable care must be taken to eliminate static unbalance and to reduce friction.

10.32. Steady-State Flow Forces

The following analysis of the steady-state flow force on a spool valve is believed to be both accurate and adequate for practical application, since it has been verified by a large amount of experimental investigation and operating experience. Insofar as the configurations of the metering orifices are similar, the formulas can be applied to valves of other constructions, such as the sliding-plate and the suspension valves of Chapter 9. Even if the formulas do not apply, the same basic methods of analysis can often be applied to other cases.

The origin of the flow force can most easily be understood from a brief qualitative explanation. In Fig. 10.7, which represents one metering orifice of a spool valve, with the upstream and downstream chambers and the connecting ducts, the cross-sectional area of the stream near the orifice is much smaller than it is everywhere else, and therefore, the velocity of flow is small except near the orifice edges. According to

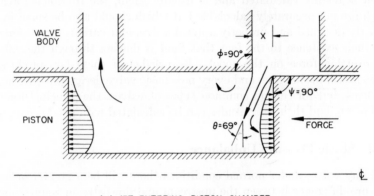

(a) JET ENTERING PISTON CHAMBER

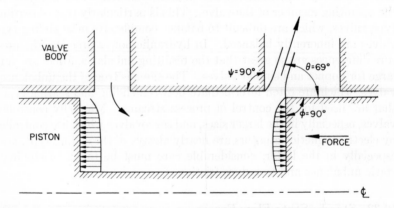

(b) JET LEAVING PISTON CHAMBER

Fig. 10.7. Origin of the flow force.

Bernoulli's principle, this region of high velocity is also a region of low pressure, and this lowering of pressure is indicated symbolically by the pressure curve. The force on the piston face is proportional to the area of the curve, and this force tends to open the valve.

The pressure against the downstream face is independent of position because the adjacent fluid is moving slowly, and its pressure curve is therefore a rectangle with a height equal to the pressure in the downstream chamber. The area it encloses is greater than that enclosed by the first curve, the closing force is therefore greater than the opening force, and the flow tends to close the valve.

A quantitative solution of this case can be found as follows.[14] Consider Fig. 10.8, which represents the case just discussed, and assume:

1. The fluid is nonviscous and incompressible; this is approximately true in most practical cases.

2. The peripheral width of the orifice is large compared with its axial length so that the flow can be considered to be two-dimensional; this also is true for small piston displacements in all valves and is approximately true for other cases.

3. The flow is assumed to be quasi-irrotational in a region immediately upstream from the orifice.

This last assumption requires some justification. Throughout most of the volume of most practical hydraulic power systems, particularly in

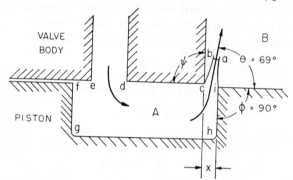

Fig. 10.8. Square-land chamber configuration.

something as complicated geometrically as a control valve, the flow will usually be highly turbulent. As the stream approaches a small orifice, however, the flow lines converge strongly and the average velocity increases greatly so that the turbulent eddies are damped out, and the kinetic energy of the fluid becomes energy of bulk transport rather than kinetic energy of rotation of eddies. This condition is essentially that of irrotational flow, and this fact together with the previous two assumptions permits the solution of the problem by means of the methods of classical hydrodynamics. In any case, the assumptions are justified by the excellent agreement between the theory and the experimental results.

With the assumptions that the flow is two-dimensional, irrotational,

[14] S.-Y. Lee and J. F. Blackburn, "Contributions to Hydraulic Control, 1—Axial Forces on Control-Valve Pistons," *Trans. ASME*, Vol. 74 (August 1952), pp. 1005–1011.

nonviscous, and incompressible, the solution of the flow pattern in the region upstream from the orifice becomes a solution of Laplace's equation for the configuration shown in Fig. 10.8. This solution has been found by von Mises,[15] and it is found that, for a square-land valve when the valve opening, x, is small compared with the other dimensions of the upstream chamber, the angle, θ, which the axis of the stream makes with the piston axis is $\cos^{-1} 0.36$, or $69°$. This value has been experimentally verified both on actual valves and on a two-dimensional valve model with glass sides. For lands with angles other than $90°$, the value of θ can also be found either experimentally or theoretically. When θ is known, the axial force on the piston can be derived in the following manner.

The axial force on the piston equals the axial component of the net rate of efflux of momentum through the boundary a-b-c-d-e-f-g-h-i-a in Fig. 10.8, where a-b is the *vena contracta* of the jet. In an actual valve, the area of a-b is much smaller than that of d-e where the fluid enters the upstream chamber, A. Since the velocities are inversely proportional to the areas, the influx of momentum through d-e is negligibly small compared with the efflux at a-b, which is equal to $QU\rho$, where

$$Q = \text{total rate of flow}$$

$$U = \text{velocity of jet at } \textit{vena contracta}$$

$$\rho = \text{density of fluid}$$

By Bernoulli's equation,

$$u = \sqrt{(2\,\Delta p)/\rho}$$

and the net axial force is

$$F_{AB} = F_{fg} - F_{hi} = QU\rho \cos \theta \qquad (10.9)$$

This equation can be transformed into the more useful form

$$F_{AB} = 2C_q wx\, \Delta p \cos \theta \qquad (10.10)$$

where

$C_q = \text{coefficient of discharge}$

$$= \frac{Q}{wx\sqrt{(2\,\Delta p)/\rho}}$$

$w = \text{peripheral width of orifice}$

$x = \text{axial length of orifice}$

$\Delta p = \text{pressure difference between chamber } A \text{ and chamber } B$

[15] R. von Mises, "Berechnung von Ausfluss- und Ueberfallzahlen," *Z. Ver. deut. Ingr.*, Vol. 61 (1917), pp. 447–452, 469–474, 493–498 (especially pp. 494 and 495).

Since θ is always less than 90° for valves of the type under discussion, F_{AB} of Eq. 10.10 is always positive and tends to close the valve. If the direction of flow is reversed so that the jet flows into chamber A instead of out of it, the equation is still valid, and if the jet angle θ is the same for both cases, the force will be the same. It should be noted that for Fig. 10.8 the jet angle depends upon the angle ϕ but is independent of ψ; for the reversed flow, the opposite is true. In either case, it is the configuration of the *upstream* chamber that determines the force.

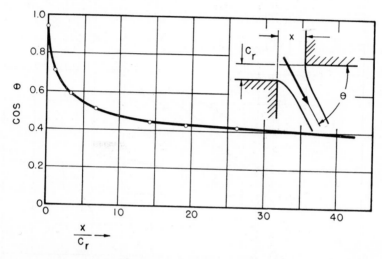

Fig. 10.9. Effect of radial clearance on $\cos \theta$.

Equations 10.9 and 10.10 apply only to an ideal valve with perfectly sharp corners and zero radial clearance. Since such an ideal valve cannot be built, it is important to investigate the effects of these imperfections. For an orifice with radial clearance, C_r, as shown in Fig. 10.9, Eq. 10.10 becomes

$$F_{AB} = 2C_q w\, \Delta p \sqrt{x^2 + C_r{}^2}\, \cos \theta \qquad (10.11)$$

Here, $\cos \theta$ is no longer a constant but varies with the quantity x/C_r, as shown by the curve.

If the net force on the piston is plotted against the displacement for a constant value of pressure drop, a curve similar to those of Fig. 10.10 is obtained. Here, the curve for an ideal valve is shown dashed. It can be seen that the effect of radial clearance is to make the curve rise abnormally rapidly for small openings, after which it sags downward and approaches the ideal curve asymptotically.

The effect of finite radius of the land edges is difficult to calculate theoretically but is very similar to that of radial clearance. Qualitatively, the variation of θ can be visualized with reference to Fig. 10.11. In Fig. 10.11a, the ideal case of zero radial clearance and sharp land edges is represented. The figure is skew-symmetrical about the center of the opening; it is the skewness of the symmetry that causes the axis of the

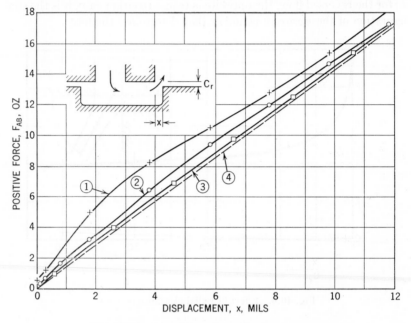

Fig. 10.10.   Experimental positive-force curves.

Curve	①	②	③	④ (theoretical)
C_r, in.	0.0003	0.00005	0.0001	0
r, in.	0	0.0003	0	0

jet to deviate from the vertical, and it is this deviation and the resulting x-component of momentum that give rise to the axial force on the piston. Figure 10.11b represents a valve with the axial displacement, x, set equal to the radial clearance, C_r. The figure is symmetrical about the two 45° lines, and the jet angle, θ, is 45° because of this symmetry. This is true also for any value of radius r of the land edges so long as the r's are the same for both edges. If the valve opening, x, is varied, however, the symmetry will be destroyed, and θ will no longer equal 45°. For large values of x, the behavior will approach that of an ideal valve, and θ will approach 69°; for small values of x, the jet will approach parallelism with the axis, and the force will be greater than that for an ideal valve.

In a valve with zero radial clearance but finite curvature of the land edges, as in Fig. 10.11c, the symmetry is skewed as for the ideal valve but not to such a great extent; the angle θ is less than 69°; the piston force is greater than for an ideal valve; and θ varies with x. For large values of x, θ approaches 69°; for negative values of x, θ approaches 0 as x approaches $-r$. In an actual valve in which neither C_r nor r is zero, θ approaches zero at small or negative values of x, and the slope of the

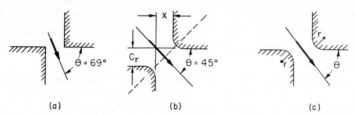

Fig. 10.11. Land configurations.

curve of force versus x increases at most by a factor of $1/0.36 = 2.78$. This is not particularly serious, of course, since in most cases the curve drops off and approaches that for an ideal valve before the actual value of the force is very large; that is, the phenomenon occurs essentially only for small values of x.

In an actual four-way valve, there are two identical orifices in series and, therefore, there is twice the force on the piston. Equation 10.10 then becomes

$$F_T = 2F_{AB} = 2Q\sqrt{\rho P_v}\cos\theta \qquad (10.12)$$

or for a particular fluid

$$F_T = K_1 Q\sqrt{P_v} \qquad (10.13)$$

where F_T is the total flow force on the piston, and the total pressure drop across the valve for the two orifices in series, $P_v = 2\,\Delta p$. For petroleum-base fluids with specific gravities of about 0.85, $K_1 = 0.0064$ when Q is given in cubic inches per second and P_v in pounds per square inch.

Two useful variants of Eq. 10.13 may be obtained by combining it with the flow equation. For the same units and fluid,

$$Q = 67A\sqrt{P_v} \qquad (10.14)$$

where A is the area of each orifice in square inches. Eliminating P_v from Eq. 10.13 gives

$$F_T = 9.6 \times 10^{-5}\frac{Q^2}{A} \qquad (10.15)$$

and eliminating Q gives

$$F_T = 0.43AP_v \qquad (10.16)$$

To illustrate the order of magnitude of the forces ordinarily encountered, computation of the magnitude for a typical case may be useful. The maximum horsepower which a valve with a given orifice area can deliver to a load is $3.9 \times 10^{-3}AP_s^{3/2}$, and this power is delivered when two-thirds of the supply pressure, P_s, appears as drop across the load and one-third as drop across the valve. For this condition of maximum output and the common supply pressure of 3000 psi, the force on the piston is 0.67 lb/hp.

So far, nothing has been said about the radial component of the force on the piston. It could be calculated by substitution of sin θ for cos θ in Eq. 10.10, but it is not usually of interest, since it can be easily canceled by the simple scheme of dividing the port peripherally into two or more equal parts equally spaced around the sleeve.

10.321. Flow-Force Compensation. Steady-state flow forces of the type just discussed have been recognized for years, and various schemes have been suggested for reducing or eliminating them. Some of these methods have been completely unworkable, and others have been partially successful, but there seems to be no means which will give complete compensation for all conditions of flow and pressure drop. One which works excellently for valves with strokes of about 0.025 in. or more, and which gives a useful if imperfect compensation down to 1 or 2 mils, is the following.[16]

In the valve construction to be described, the axial component of efflux of momentum from the piston chamber is made greater than the influx so that a negative (opening) force is developed which is proportional to x for a given pressure drop, and this negative force is balanced by the positive force generated at a conventional square-land orifice. It has been found experimentally that except for very small openings, the two forces can be balanced perfectly for all pressures and flows. Here, the effects of radial clearance and of rounding of the land edges operate to give a positive force which at its maximum is only a very small fraction of the maximum for a conventional valve.

It should perhaps be noted that this method can be used to give a zero-force port, but since the special port construction is somewhat more difficult to manufacture than a conventional square port and since it may lead to the transient instability discussed later, the balancing scheme is preferable.

[16] Lee and Blackburn, *op. cit.*, pp. 1007–1009.

The construction of the negative-force port is shown schematically in Fig. 10.12. It differs from a conventional port in two ways: the chamber in the piston is shaped somewhat like a turbine bucket, and the sleeve is cut out to form an extension of the downstream chamber. Here, the jet through the orifice enters the piston chamber at an angle of θ_1, is reflected from the chamber wall near d, and leaves at an angle of θ_2. If

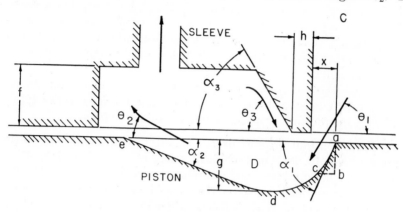

Fig. 10.12. Negative-force port.

the cross section of the flow were constant and the frictional force negligible, the resulting force on the piston would be

$$F_{CD} = QU\rho(\cos \theta_1 - \cos \theta_2) \tag{10.17}$$

Thus, the positive force of Eq. 10.9 may easily become a negative force, since $\cos \theta_2$ is ordinarily considerably larger than $\cos \theta_1$.

Furthermore, an eddy is produced which moves clockwise (in the case shown), and an additional flow enters the piston chamber at an angle of θ_3, which increases the negative force still more.

The total resultant force on the piston is thus rather complicated. It depends upon numerous variables such as α_1, α_2, α_3, ρ, μ, w, and d, where ρ is the density and μ the viscosity of the fluid, w the peripheral width of the port, and d the piston diameter. A theoretical derivation of the force would be practically impossible.

The port width, w, occupies only a portion of the circumference of the piston in many valves. Although it is usually large compared with x so that the flow can be treated as two-dimensional at the orifice, it is small compared with the dimensions of the chambers. When the incoming jet of fluid hits the piston near the point d and is reflected upward, it tends to spread out normal to the plane of the figure, and the condition of constancy of cross section assumed in Eq. 10.17 no longer holds. The actual exit angle, θ_2, thus tends to be greater than α_2, and the jet tends to separate from the piston surface. Experiment shows that the spreading tendency is greatly dependent on the angle of incidence of the jet on the piston surface. If $\alpha_1 = \theta_1$, the angle of incidence is zero, and the negative force is a maximum; α_1 should normally be made equal to 69°, the influx angle for a square land. It is found, however, that a square step such as a-b-c in Fig. 10.12 has little effect on the force if it is not too deep, and it is sometimes desirable for manufacturing reasons that such a step be provided. In the case for which $w = 360°$, there would be no lateral spreading.

Although the actual force for a given case is too complicated to calculate, it is found that for a given valve profile the force varies with x and with the pressure drop, Δp, across the orifice as follows:

1. For a constant pressure drop, F_{CD} varies with x according to the curves of Fig. 10.13. For very small values of x, the force is positive, going through a maximum and returning to zero when $x = a$. The distance a is directly proportional to the unavoidable errors of manufacture, that is, to C_r and r. It is also somewhat affected by the shape of the piston profile. The explanation of the positive force is analogous to that already given; when x is small, θ_1 is small, and the x-component of momentum is a maximum for a given flow. Furthermore, as the angle of incidence on the piston and the resultant spreading of the jet increase, θ_2 increases, and the negative force is reduced. The additional spreading decreases the velocity of the return flow, θ_3, and its contribution to the negative force. If C_r and r could be made zero, θ_1 would be independent of x, and the positive hump would vanish; the curve of F_{CD} plotted against x would be as shown by the dashed curve of Fig. 10.13. The effects of changing C_r and r are shown in the solid curves.

2. When x is large compared to C_r and r, F_{CD} is directly proportional to Δp for a given value of x. This indicates that for an ideal valve it would be proportional for all values of x. These two experimental results can thus be combined into the equation

$$F_{CD} = -Kx\,\Delta p \qquad (10.18)$$

where K is a constant for an ideal port of a given geometrical design but is a function of x/C_r and of x/r for an actual valve. As previously stated, the modified piston profile a-b-c-d-e of Fig. 10.12 gives results almost identical with those of the profile a-c-d-e as long as the angle e-a-c is equal to or a little greater than 69°.

The coefficient K depends upon both α_2 and α_3 of Fig. 10.12. For manufacturing reasons, α_3 is somewhat less easy to vary than α_2, but

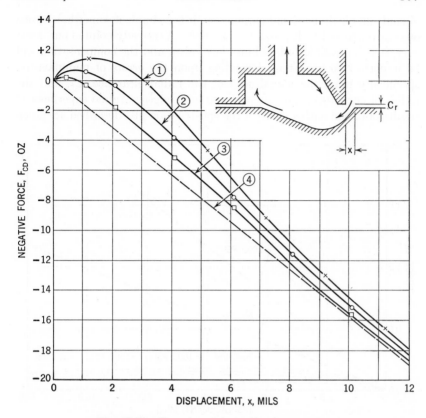

Fig. 10.13. Experimental negative-force curves.

Curve	①	②	③	④ (theoretical)
C_r, in.	0.0006	0.0006	0.0004	0
r, in.	0.0003	0	0	0

K can be controlled easily by varying the latter. The value of K also depends to some extent upon the dimensions f, g, and h. For maximum compensation, f should be as large and h as small as possible.

Since in an ideal four-way valve there are always two identical orifices in series and since C_q, w, x, Q, and Δp are the same for both, the positive force of the square-land port will be equal to the negative force of the modified port if F_{AB} of Eq. 10.9 is equal to F_{CD} of Eq. 10.18. This will be the case if

$$K = 2C_q w \cos \theta \qquad (10.19)$$

or if

$$K = 0.72 C_q w \qquad (10.20)$$

since $\cos \theta = 0.36$ for an ideal valve. Since K depends upon a number of controllable dimensional parameters, it is a simple if somewhat laborious matter to satisfy Eq. 10.20 in practice. As already pointed out, however, it is impossible to build an ideal valve, and the compensation will be imperfect to an extent depending upon the errors of manufacture, specifically upon the magnitudes of the radial clearance, C_r, and the radius of curvature, r, of the port and land edges. Actual experimental results obtained with a valve of the construction just described are given

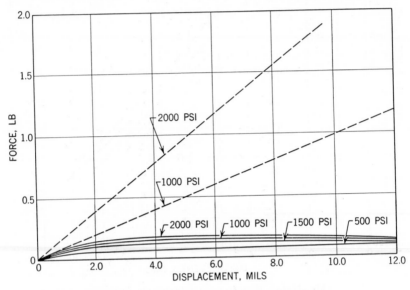

Fig. 10.14. Performance of force-compensated valve.

in Fig. 10.14, where curves are also sketched for an ideal uncompensated valve of the same dimensions.

It can be seen from the figure that the force for a given pressure drop is everywhere less than for an uncompensated valve and that instead of rising linearly with valve opening, the force reaches a low maximum (at least for the higher pressures) and then decreases slowly.

One possible measure of the effect of the modified construction would be the ratio of the maximum forces on the pistons of a conventional and a modified valve. For the example given, this ratio reaches a value of about 18 at 2000 psi and 15-mil displacement. Since it is almost directly proportional to displacement, however, the ratio has little meaning except as a measure of maximum displacement. If the valve had been tested at 50-mil displacement, for example, which is about average for the DACL power pistons used, the improvement ratio would have been about 50; for a displacement of 5 mils, such as is used in the pilot valves, the ratio would have been only 6. The quantity of greatest importance is the maximum force required. The curves show that this maximum force is greatly decreased by the use of the modified construction. Further substantial reductions can be made by minor adjustments of the dimensional parameters and by improved workmanship. A sketch of a four-way valve of this construction is given in Fig. 10.15. The results illustrated are by no means the best that can be attained; optimum

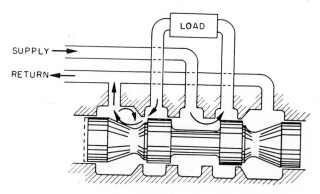

Fig. 10.15. Four-way valve with force compensation.

design and improved workmanship will reduce the remaining stroking force considerably.

The method of flow-force compensation that has just been outlined is nearly perfect in theory, and should be excellent in practice for long-stroke valves, as stated previously. Up to now, unfortunately, it has been applied largely to short-stroke valves, with results that have left something to be desired. Even if the compensation is initially excellent, the inevitable erosion of the land corners by dirt in the oil acts to make it less effective after the valve has been in service for a while, and this tends to restrict its application to short-life or expendable devices, such as missiles. Furthermore, even if wear and life are not important factors, the radical nonlinearity of the relationship between the valve force and displacement at small openings makes it difficult to use the valve in many high-performance systems. This nonlinearity is evident from Fig. 10.14 but can be seen better in Fig. 10.16,[17] which represents the axial forces for a particular compensated 8-hp valve. (It should be no-

[17] Reproduced by permission from an internal report of Midwestern Instruments of Tulsa, Oklahoma. We are indebted to Messrs. D. G. O'Brien and S. J. Jatras for this and other information, and for stimulating discussions of flow-valve theory and practice.

ticed that the scale of ordinates is purposely chosen to accentuate the nonlinearities; even in the worst case, the actual force is no greater at any point than the force on an uncompensated valve.) Finally, several manufacturers who have tried to put this type of valve into production feel that it is excessively expensive. This last point is certainly debatable, since other shops have been able to make the same valve with

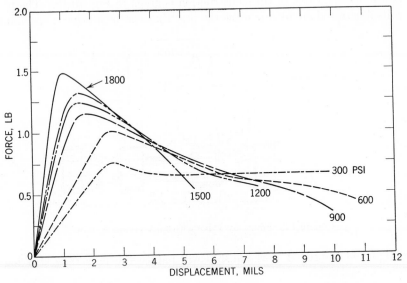

Fig. 10.16. Axial force versus displacement for compensated 8-hp valve.

comparatively little trouble. On the whole, the basic compensation method seems to be very useful, but it is by no means a panacea and it has not yet been given a fair trial in the applications to which it is best suited.

Most of the schemes that have been proposed for the compensation of flow forces fall into one or the other of two categories: either the incoming jet is deflected in such a way that it carries out of the piston chamber the momentum that it brought in (or more), as in the previous case, or else this momentum is caught and transferred to the sleeve by some sort of vane or other member which projects into the piston chamber. In a spool valve, the provision of the vane leads to annoying if not serious problems in fabrication and assembly, but the plate valve of Sec. 5.33 is very well adapted to this type of compensation. One arrangement (due to Mr. J. L. Coakley of DACL) is shown schematically in Fig. 10.17. Here, the holes in the valve plate or "piston" are provided with deflectors, which force the entering jet to make an angle of θ_2 with

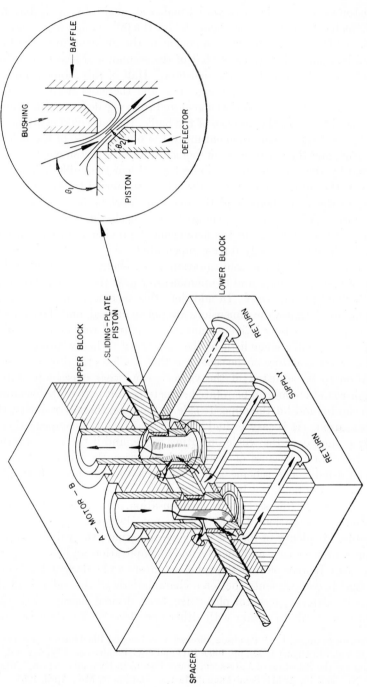

Fig. 10.17. Force-compensated plate valve.

the piston axis. The downstream jet makes an angle of θ_1 with the axis, which, as has been shown, is about 69° for the 90° configuration shown. If θ_2 is made 45°, the angular component of the momentum of the first jet will be approximately twice that of the second, and also twice what it would have been without the deflector. Half of this momentum has been acquired from the supply chamber, and therefore from the valve block, but the other half has been acquired from the deflector, and hence from the piston. In the absence of the baffle, the jet would continue on to strike and transfer its momentum to the opposite side of the hole in the piston, and the net transfer of momentum to the piston would be unchanged by the presence of the deflector. When the baffle is inserted, however, the total momentum of the jet is transferred to the baffle and thence to the valve body, and the upstream port becomes a negative-force port. This negative (opening) force cancels the positive force of the downstream port, as in the arrangement previously discussed, and the valve is approximately force-compensated.

In practice, the force compensation is excellent over most of the range of valve openings, and the nonlinearity near the origin is no worse than in the previous case. In addition, this scheme is much easier to manufacture because the dimensions are not so critical, and the compensation appears to be relatively immune to wear of the metering edges and therefore will change little as the valve ages. Finally, probably the most attractive feature of the scheme is that it allows the degree of compensation to be adjusted; if the valve is somewhat overcompensated, the half-round baffles can be rotated so as to permit part of the incoming jet to slip by and hit the opposite side of the hole, thereby reducing the compensation. It is easy to find an angle at which the compensation is nearly perfect over the whole operating range, and if the compensation ever deteriorates because of wear or other causes, the baffles can easily be readjusted. The scheme is still too new to have been thoroughly tried out in the field, but it looks very good.

In concluding this section on compensation, it is pertinent to point out that compensation of flow forces is not always necessary, or even desirable. If plenty of force is available to actuate the valve, compensation is unnecessary and merely adds complication and expense. In other cases, it may be that the valve actuator can be designed in such a way that it complements the force characteristic of the valve, and the composite element, actuator plus valve, has a desired characteristic that would be difficult to obtain with either one alone.[18] The principal dis-

---

[18] See, for example, F. C. Paddison, discussion to Lee and Blackburn paper, *Trans. ASME*, Vol. 74 (August 1952), pp. 1009–1011 and F. C. Paddison and W. A. Good, "A Method for the Selection of Valves and Power Pistons in Hydraulic Servos," ASME Paper No. 55-S-10, ASME Semi-Annual Meeting, Baltimore, Md., April, 1955.

advantages of this scheme are that it is usually applicable only to a very restricted range of operating conditions, that it is very difficult to design and mechanize, and particularly that it is very desirable to have as strong an actuator as possible in order to have some margin of force to take care of excessive friction or other parasitic effects. Also, even if the force available from the actuator cannot be increased, it is still very desirable to have the actuator as stiff as possible in the interests of valve stability. Neither of these last two aims is normally compatible with the scheme of tailoring the actuator characteristics to fit the valve-force curve.

10.4. FORCES ON FLAPPER VALVES—PRESSURE FEEDBACK

The flapper-plus-nozzle valve has been widely used as a pilot for two-stage valves and as an element of industrial process-control devices. In the latter field, it is nearly always used with low-pressure air rather than with liquid, but the device is of sufficient importance in the hydraulic field to deserve discussion here. As has been stated, the sharp-edged and the flat-faced flapper valves are two very different devices. The former will be discussed first because it is simpler and because it is much more important as an element for the continuous control of liquid flow.

Consider the configuration of Fig. 10.18a, in which a sharp-edged nozzle is located close to a flat flapper, and assume for convenience that the hole in the end of the nozzle is circular, with a radius of r. Because the edge of the orifice is sharp, it will define the point of separation of the jet, and the device will follow the orifice law for small values of x/r, where x is the distance between jet and flapper. Under these conditions, the effective orifice is the area of the cylinder of diameter $2r$ and length x, or $2\pi rx$, and the flow will be given by the orifice equation,

$$q = C_d 2\pi rx \sqrt{(2\,\Delta p)/\rho} \qquad (10.21)$$

where Δp is the pressure drop between the mouth of the nozzle and the external chamber and ρ is the density of the fluid. The discharge coefficient, C_d, is approximately 0.6 for the usual geometry.[19]

Now consider three successive sections across the stream, one well upstream of the nozzle where the average velocity is very small, the second at the nozzle, and the third at the *vena contracta*, where the pressure at all points on the section is equal to that in the external chamber.

[19] von Mises, *op. cit.*, p. 496, Table 5.

Let us call the pressures at these three sections P_s, P_1, and P_2, and let us take the last as zero by definition. Let the velocities at the first section be zero and at the other two be V_1 and V_2, respectively. Then, by

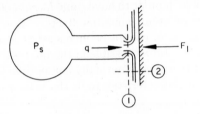

(a) FLAPPER CLOSE TO NOZZLE

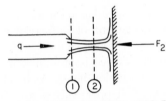

(b) FLAPPER FAR FROM NOZZLE

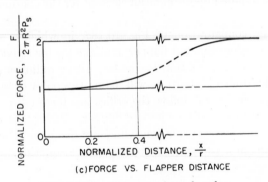

(c) FORCE VS. FLAPPER DISTANCE

Fig. 10.18. Flapper-and-nozzle valve.

Bernoulli's equation, at section 1

$$P_1 = P_s - \frac{V_1^2}{2} \tag{10.22}$$

and, by hypothesis, at section 2

$$P_2 = P_s - \frac{V_2^2}{2} = 0 \tag{10.23}$$

By the continuity equation,

$$V_1 \pi r^2 = V_2 C_d \cdot 2\pi r x \tag{10.24}$$

The total force on the flapper will be equal to the static pressure, P_1, times the area of the nozzle, plus the momentum transferred from the jet to the flapper. This will be equal to the total flux of momentum at the mouth of the nozzle, since the jet leaves radially and carries away no axial momentum so that

$$F = P_1 \pi r^2 + q\rho V_1 \tag{10.25}$$

But $q = \pi r^2 V_1$, and by combining the equations to eliminate the unknowns we finally obtain

$$F = \pi r^2 P_s \left[1 + \left(\frac{2C_d x}{r} \right)^2 \right] \tag{10.26}$$

Thus, we see that the force will be initially the static pressure force and will rise parabolically as the flapper moves away from the nozzle.

The foregoing derivation will obviously be valid only for small values of x/r, where the radial orifice is the controlling one. When this is no longer the case, the exact law of force is unknown, but a limiting value for large separations can be obtained as follows.

In Fig. 10.18b, the pressure at the *vena contracta*, section 2, will be zero by hypothesis, as before, so that Eq. 10.23 holds as in the previous case. The static pressure term of Eq. 10.25 will now be zero because $P_2 = 0$ so that the force on the flapper is entirely that due to the momentum of the jet. This momentum will be

$$F = q\rho V_2 \tag{10.27}$$

But again $q = \pi r^2 V_1$, and $V_1 = C_c V_2$, where C_c is the contraction coefficient for this configuration; it will not be the same as the C_d of the previous case. Combining these equations, we get

$$F = 2\pi r^2 C_c P_s \tag{10.28}$$

For most nozzles, the hole diameter is constant for some distance upstream from the mouth so that C_c will be nearly unity and

$$F = 2\pi r^2 P_s \tag{10.29}$$

Thus, the force on the flapper will approach twice the static pressure force when the flapper is far from the nozzle. The curve of force versus position will be somewhat like that given in Fig. 10.18c, though further investigation will be necessary to establish it accurately.

The fact that the force on the flapper is accurately proportional to the supply pressure for small values of x/r is frequently useful because it provides a built-in pressure feedback. An obvious application would be the control of pressure across a load; the valve actuator supplies a force proportional to the desired pressure, and the valve opens or closes as required to maintain that pressure. If the load pressure is to be maintained with respect to ground, a single nozzle is all that is required; if a differential pressure controller is wanted, a second nozzle is provided on the opposite side of the flapper, and the two pressures in question are connected to the two nozzles. If the load is a mass (actuated by a ram), a force on the flapper calls for an acceleration of the mass, since the combination of valve and ram becomes a force multiplier. If the mechanical load is a spring, a force on the flapper calls for a deflection of the load. With suitable design, the response of the system valve actuator—flapper valve—ram (or rotary motor)—load (linear or rotary) can be made fairly accurate without resorting to feedback from the load. In many applications, a simple open-loop system such as this is very satisfactory.

The flapper-and-nozzle configuration can also be used to make an excellent flowmeter. If the flow, q, is held constant and the force is also kept constant, as by a weight, it can be shown that

$$q = 2C_d x \sqrt{\frac{2\pi W}{\rho(1 + 2y^2)}} \tag{10.30}$$

where the symbols are as before except that

$$y = 2C_d \frac{x}{r}$$

and

$$W = \text{the constant force on the flapper}$$

If we use inch-pound-second units and take $C_d = 0.6$, and restrict the flapper motion so that $2C_d^2 y^2 \ll 1$, then, approximately,

$$q = 290x\sqrt{W/\sigma} \tag{10.31}$$

where $\sigma = $ specific gravity of the fluid. For $\sigma = 0.85$, as for the usual petroleum-base hydraulic fluid, $q = 314x\sqrt{W}$.

The use of a weight to supply the constant force, W, is undesirable, both because the added mass slows down the response and because the meter becomes sensitive to acceleration and leveling. The requirement of constancy of force implies a zero (or at least a very low) resonant frequency and a consequent sluggish response of the meter were it not for the equivalent spring provided by the gradient of the flow force. It can be shown that

$$\left(\frac{\partial F}{\partial x}\right)_{q=\text{const}} = -\frac{2W}{x(1 + 2y^2)} \tag{10.32}$$

so that the spring becomes infinitely stiff at zero flow. In an actual flow-meter, the stiffness of the hydraulic spring is considerable; in one experimental model, the natural frequency of the flapper assembly without oil was 2 cps, while with oil flowing it was of the order of 100 cps. As expected, it was much higher at low than at high flow rates.

The accuracy which can be attained with the flapper flowmeter is excellent; the output is proportional to the flow rate within a few parts per thousand, and further development work would probably improve this somewhat. This high accuracy, plus the fast response and the fact that the linearity makes averaging and totalizing easy, makes it a most valuable device for measuring pulsating flows, which are very difficult to meter with conventional instruments.

All the foregoing remarks apply to the sharp-edged nozzle as used with liquids. Unfortunately, an accurate sharp-edged flapper-and-nozzle unit is not easy to make, especially in the smaller sizes that are required for most applications with high supply pressures. Many nozzles are therefore made with flat ends of various diameters surrounding the central hole. These are easier to make and much less liable to damage if the flapper strikes the nozzle, but they depart radically from the characteristics of the sharp-edged nozzle as soon as the resistance to flow of the radial slit between nozzle face and flapper becomes appreciable. Again, the intermediate case is not susceptible of theoretical analysis, but for very small values of x, if we assume perfect flatness and parallelism between nozzle face and flapper, the flow will be given by

$$q = \frac{\pi x^3 \, \Delta p}{6\mu \ln \, (r_2/r_1)} \tag{10.33}$$

where Δp is the pressure drop across the slit, r_1 is the hole radius, and r_2 is the outer radius of the nozzle face. The total force on the flapper will be

$$F = \frac{\pi(r_2{}^2 - r_1{}^2)}{2 \ln \, (r_2/r_1)} \Delta p \tag{10.34}$$

Thus, for very small values of x the flow increases as the cube of x rather than linearly, as with the sharp-edged nozzle, and also the force is greater—sometimes much greater—than before. For a nozzle with an outside diameter equal to twice the hole diameter, the force is up by a factor of 2.16 over that for the sharp edge.

The foregoing equations assume that the whole area of the nozzle flat sees fully developed laminar flow. This is approximately true only for exceedingly small values of $x/(r_2 - r_1)$; for larger values, the laminar-flow assumption is not satisfied; and at still larger values, separation takes place at the corner of the hole, and Eq. 10.26 applies, providing x/r is still small. The onset of separation is quite unpredictable and erratic, and it is found experimentally that in the intermediate range of values of x the force and the flow both vary rapidly and unpredictably with time, the variation being so large as to make the flat-faced nozzle useless for many applications. It is still useful with gases, or as a null device where for equilibrium it always returns to the same x, or for essentially on-off service,[20] but it simply is not a good quantitative device of predictable performance in the continuous control of liquid flow.

10.5. FORCES ON OTHER TYPES OF VALVES

It would be convenient if analyses similar to those already given could be presented for other commonly used and important types of valves, such as the poppet and the Askania jet-pipe, but unfortunately this appears to be impossible at the present time since these configurations have not been adequately investigated. This short section may, however, serve a useful purpose if it suggests further research, and also if it warns against fallacious statements that have occasionally appeared in the literature.

Figure 10.19 shows schematically a so-called "balanced" poppet valve. If the balance piston is omitted, the total downward force will be $P_1\pi(R + \Delta R)^2$, and the upward force will be $P_2\pi R^2 + F_A$, where P_1 and P_2 are the pressures above and below the valve and F_A is the axial component of the total force exerted on the seating area by the fluid confined between the mating faces. The addition of the piston will furnish an additional upward force of $P_1\pi R_B{}^2$ which will give an approximate balance at the cost of additional complexity and leakage.

Unfortunately, it is almost impossible to get a very accurate balance since the distribution of pressure over the seating area is highly variable and indeterminate. Just as with the pressure distribution on the side of a piston, which was discussed in Sec. 10.21, the actual pressure distribution will be greatly affected by exceedingly small deviations from ideal geometry and by dirt in the liquid and other accidental causes. This variation in pressure will cause a corresponding variation in F_A and make true balance impossible.

It is sometimes stated that the pressure all over the annular seating surface will be the higher of the two pressures P_1 and P_2. It is difficult

[20] J. G. Chubbuck, "Acceleration Switching Hydraulic Servo," Johns Hopkins University, Applied Physics Laboratory Report CM-843, Silver Spring, Md., 1955.

to see how this idea originated since there seems to be no logical basis
for it.

The absolute magnitude of the residual unbalance force can be minimized by making ΔR as small as possible.   In the limit, the seat would
become a true knife edge, but this is impractical for valves that must
operate at all rapidly; a minimum seating area must be provided to withstand impact on closing.   Perhaps the most practical solution is to make
the poppet angle slightly more acute than the seat angle, which will en-

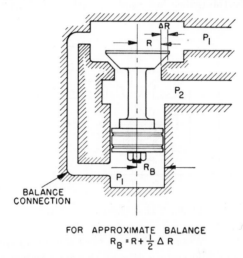

FOR APPROXIMATE BALANCE
$$R_B = R + \tfrac{1}{2} \Delta R$$

Fig. 10.19.   Piston-balanced conical poppet valve.

sure closure from the inside outward.   The balance piston could then be
made integral with the poppet, with a consequent reduction in problems
of alignment and concentricity, since the piston radius could then be
equal to that of the inner edge of the seat ($R = R_B$).   If the difference in
angles is sufficiently small, the seat should be able to stand up in service.
Whether such a design would work in a particular application would depend largely on the precision of manufacture and on the severity of the
impacts between valve and seat, and would have to be determined by
trial.

The use of a balance piston entails both friction and leakage, and therefore partially nullifies two of the principal advantages of the poppet
valve.   A construction that would give perfect static balance, at least
theoretically, is shown in Fig. 10.20.   Here, two identical poppets with
45° seats are mounted on a common stem so that they can be adjusted to
seat simultaneously.   It is not difficult to arrive at a construction that

would permit this adjustment to be made easily and accurately, but it is questionable if the adjustment would be effective in service since it would be deranged both by differential thermal expansion and by extension or compression of the stem under the very large forces transmitted along it from one poppet to the other. Again, only extensive service trials would permit evaluation of this construction.

The laws of flow of the poppet valve are little known and are much more complicated than those of the sharp-edged orifice. At large openings, presumably the orifice law applies, with a discharge coefficient of

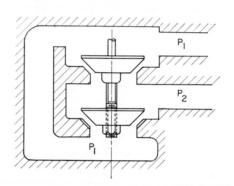

Fig. 10.20. Twin-balanced-poppet valve.

perhaps 0.8 or more. When the lift was very small compared to the seat width, the flow would be viscous, though this condition would be of little importance in practice. The theory of the flow at intermediate lifts is excessively complicated by the transition from one flow regime to another and by the comparatively large number of significant parameters.

Essentially nothing is known about flow forces on the poppet valve, either steady-state or transient. Obviously, the Bernoulli effect will produce a closing force for the usual geometries, but its magnitude and the effects of impact of the jet on downstream extensions of the poppet such as the balance piston of Fig. 10.19 cannot be estimated until the flow pattern is solved. Some work has been done on the whole problem [21]

[21] M. Tidor, "A Study of Pressure and Flow in a Poppet-Valve Model." S.M. Thesis, Department of Mechanical Engineering, Massachusetts Institute of Technology, Cambridge, Mass., 1954.

J. A. Stone, "Design and Development of an Apparatus to Study the Flow-Induced Forces in a Poppet-Type Flow Valve," S.B. Thesis, Department of Mechanical Engineering, Massachusetts Institute of Technology, Cambridge, Mass., 1955; "An Investigation of Discharge Coefficients and Steady State Flow Forces for Poppet Type Valves," S.M. Thesis, Department of Mechanical Engineering, Massachusetts Institute of Technology, Cambridge, Mass., 1957.

and more is planned, and we hope eventually to obtain a reasonably complete set of design data.

Even less is known about the jet-pipe [22] than about the poppet valve. It is known that with suitable proportions of the jet nozzle and of the orifices in the distribution block the pressure recovery with the load blocked can approach unity, but few if any curves of load pressure versus load flow seem to have been made public. Still less is known about the forces on the jet-pipe itself. Sales literature and some journal articles assert that the tangential component of force is zero and that even a very weak actuator is adequate. This statement is definitely false; tangential forces are appreciable and may lead to instability at high pressures. To date, the standard remedy for a misbehaving jet-pipe valve has been a somewhat random attack on the distribution block orifices with a small scraper; the knowledge of just how to scrape seems to be both empirical and esoteric, a predicament which does not inspire much confidence in the procedure.

Some progress has been made in attacking the jet-pipe problem, theoretically, by model experiments, and by measurements of the characteristics of commercial units, but the variables are even more numerous than for the poppet valve, and the difficulty of obtaining a solution is correspondingly greater. It is to be hoped that at least a partial solution can be found, since the inherent immunity to dirt of the jet-pipe and its comparatively small force requirements make it very attractive for service at low power levels. Even as a purely empirical device, it has found many useful applications and undoubtedly will find more.

[22] J. F. Dunn, Jr., "A Study of Some Characteristics of the Jet-Pipe Valve," Sc.D. Thesis, Department of Mechanical Engineering, Massachusetts Institute of Technology, Cambridge, Mass., 1957.

11

R. H. Frazier

Electromagnetic Actuators

11.1. VALVE ACTUATORS

Valve actuators require for satisfactory operation and application certain minimum or maximum achievements, principally with respect to force, stroke, speed of response, sensitivity, gross bulk, linearity, hysteresis, and power consumption. The relative importance of these various quantities depends on the application. For missile applications, speed of response, sensitivity, and small size generally have priority, whereas for industrial applications, substantial force and stroke may take precedence, while response speed and bulk are not primary concerns. Whether fast action is paramount or not, a certain minimum starting force is commonly specified to give reasonable assurance that stiction and valve stroking forces can be overcome at the neutral position, in which vicinity most action takes place. Dither sometimes is provided to reduce stiction. Power consumption in itself usually is not a primary concern because the power involved generally is but a small fraction of the power involved in the controlled system. In missile applications, however, power consumption is of indirect concern owing to its influence on bulk and speed of response.

In addition to the basic requirements relating to performance and size, especially in military applications, rigid specifications concerning immunity to vibration and shock, and to environmental conditions of atmosphere and temperature, may have to be met. As in all design problems, compromises must be made since optimization with respect to one desirable feature is generally incompatible with best design from another point of view. Furthermore, theoretically optimum design conditions may be too difficult or too costly to achieve from a manufacturing standpoint, or may be insufficiently immune to environmental conditions. Therefore, while complicated over-all "figures of merit" can be developed that involve the various quantities of interest, the significance of the figure in a specific instance is questionable because in different applications different quantities require different emphases and because practical problems of manufacture and immunity to environment may overrule. Rather, the designer should have at hand a few simple criteria emphasizing certain features of design, from which he can select, as a starting guide, in accordance with his objectives. A number of such criteria are indicated for various styles of motor in Sec. 11.25.

11.2. ELECTRIC ACTUATORS

Electric actuators are of the electromagnetic type involving either moving coils or moving iron. Action is controlled either through modulation of a polarizing field by means of a control current or through modulation of an alternating reference field by modulating an alternating control current of the same frequency as the reference field. The alternating-current devices are not considered in this treatment. An a-c torque motor has been made and operated successfully,[1] but experience with this type of device is limited. If a suitable means of feeding an a-c signal to it could be developed, such a device might have considerable advantages and, even though larger than a d-c motor suited to the same job, might achieve net space saving through conservation in auxiliary electronic equipment.

Electrostatic, piezoelectric, or magnetostriction devices theoretically may be considered but practically are relatively inefficient and inadequate owing to insufficient force and stroke for reasonable sizes and driving power. Crystals are rather difficult to handle, are subject to cracking, and are rather sensitive to temperature variation. The electromagnetic types of actuators are commonly known as force motors or torque motors because their primary function is to deliver a force or torque through a small displacement in response to a control signal. Usually the action of the torque motor is converted to translational mo-

[1] W. E. Sollecito, "An A-C Torque Motor—Analysis and Design." S.M. Thesis, Department of Electrical Engineering, Massachusetts Institute of Technology, Cambridge, Mass., 1953.

tion. An advantage of a torque motor is that it can be counterbalanced against rotation caused by high translational acceleration of the vehicle carrying it. The counterbalancing can be accomplished by splitting the valve, though difficulties may arise if slight bending of parts causes phase error of one half of the valve with respect to the other half. The force motor can be counterbalanced against motion due to translational acceleration by use of a rocker arm, but at the expense of excess mechanism and space, and decreased speed of response.

11.21. Equations of Electromagnetic Force

Equations of electromagnetic force or torque of polarized devices can be placed in the general form

$$f = K_{1f}i_c + K_{2f}x \tag{11.1}$$

$$\tau = K_{1t}i_c + K_{2t}\theta \tag{11.2}$$

provided that operation of devices that are inherently nonlinear is restricted to an essentially linear range. Here K_{1f} and K_{1t} are called electromechanical coupling constants, i_c represents the control current, K_{2f} and K_{2t} are called electromagnetic elastance coefficients, and x and θ are, respectively, linear and angular displacements from neutral positions. The electromagnetic elastance coefficient may be positive, negative, or zero in moving-iron devices but is zero in moving-coil devices. As used in Eqs. 11.1 and 11.2, positive electromagnetic elastance is opposite in effect to the mechanical elastance of a centering spring, as indicated by the static equations

$$K_{0f}x = K_{1f}i_c + K_{2f}x \tag{11.3}$$

$$K_{0t}\theta = K_{1t}i_c + K_{2t}\theta \tag{11.4}$$

from which the quantities

$$K_{ef} = K_{0f} - K_{2f}$$

$$K_{et} = K_{0t} - K_{2t}$$

are defined as effective or net elastances. Obviously, for positive K_2, an unstable situation exists if K_0 is not sufficiently large. The expressions for K_1 and K_2 in terms of physical quantities involved are different for each type of device, but for any one kind of device they have the same form. Hence, derivations are made only for the moving-coil device and for two kinds of moving-iron devices (for both translatory and rotary actions) with the thought that the derivations and results for other configurations of these three kinds of devices should be similar to the illustrations. The force (or torque) equations are given in terms of the air-gap polarization flux density, $\mathcal{B}_p$, which is the air-gap flux density with the armature centered and zero control current, i_c. This polarization flux density must be established either by polarizing magnetomotive forces or by permanent magnets. The equations in terms of polarizing magnetomotive forces appear different in form from the equations in terms of polarizing fluxes or flux densities, but the one form can be translated into the other. In the limits of constant finite polarizing magneto-

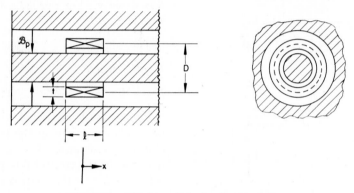

Fig. 11.1.   Moving-coil device for translation.

motive force with no associated series reluctance and constant polarizing flux with no associated shunt permeance, the equations differ in the electromagnetic elastance and nonlinear terms. Since the upper limit of pull at an air gap is determined practically by a reasonable upper limit of flux density in the adjacent magnetic material, equations that involve flux density are convenient for preliminary design exploration.

For the moving-coil device in translation, Fig. 11.1, the force equation is simply

$$f = \mathcal{B}_p \pi D N i_c \tag{11.5}$$

in which $\mathcal{B}_p$ is the radial flux density of the polarizing field in the air gap, D is the mean diameter of the coil, and N is the number of turns on the coil. Hence

$$K_{1f} = \mathcal{B}_p \pi D N \tag{11.6}$$

$$K_{2f} = 0 \tag{11.7}$$

and the device theoretically is linear if operation is restricted to the range of constant $\mathcal{B}_p$.

For the rotational counterpart, Fig. 11.2,

$$\tau = \mathcal{B}_p h D N i_c \tag{11.8}$$

$$K_{1t} = \mathcal{B}_p h D N \tag{11.9}$$

$$K_{2t} = 0 \tag{11.10}$$

The moving-iron device may have its motions either normal or parallel to the direction of the air-gap flux density. For normal action, Fig. 11.3,

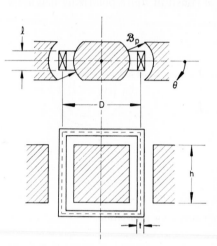

$$f = 4\mathcal{B}_p w N i_c \tag{11.11}$$

$$K_{1f} = 4\mathcal{B}_p w N \tag{11.12}$$

$$K_{2f} = 0 \tag{11.13}$$

if the motion is restricted within the central portions of the poles, and the reluctance of magnetic material in the control flux path is negligible. If the control magnetomotive force causes appreciable reluctance drop in magnetic material, K_{2t} is negative; the device then is self-centering, within a residual range determined by

Fig. 11.2. Moving-coil device for rotation.

magnetic hysteresis, just as if it had a mechanical centering spring attached to it. This action is discussed further in Sec. 11.25 in association with interpretation of over-all electromechanical performance of the devices.

For the rotational counterpart, Fig. 11.4,

$$\tau = 4\mathcal{B}_p w r N i_c \tag{11.14}$$

$$K_{1t} = 4\mathcal{B}_p w r N \tag{11.15}$$

$$K_{2t} = 0 \tag{11.16}$$

For parallel action of the moving-iron device, Fig. 11.5,

$$f = \frac{4wa}{[1 - (x/g_0)^2]^2}\left(\frac{\mathcal{B}_p N}{g_0} i_c + \frac{\mathcal{B}_p{}^2}{\mu_a g_0} x + \frac{\mu N^2}{g_0{}^3} i^2 x + \frac{\mathcal{B}_p N}{g_0{}^3} i x^2\right) \tag{11.17}$$

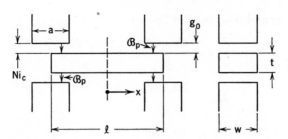

Fig. 11.3. Moving-iron device with translational motion normal to direction of air-gap flux density.

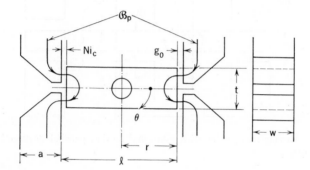

Fig. 11.4. Moving-iron device with rotational motion normal to direction of air-gap flux density.

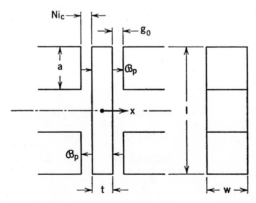

Fig. 11.5. Moving-iron device with translational motion parallel to direction of air-gap flux density.

and if $x \ll g_0$ and $Ni_c/4g_0 \ll \mathcal{B}_p$,

$$f \approx \frac{4\mathcal{B}_p waN}{g_0} i_c + \frac{4\mathcal{B}_p{}^2 wa}{\mu_a g_0} x \qquad (11.18)$$

$$K_{1f} = \frac{4\mathcal{B}_p waN}{g_0} \qquad (11.19)$$

$$K_{2f} = \frac{4\mathcal{B}_p{}^2 wa}{\mu_a g_0} \qquad (11.20)$$

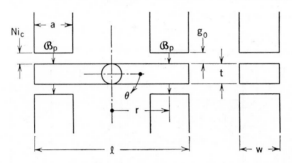

Fig. 11.6. Moving-iron device with rotational motion parallel to direction of air-gap flux density.

For the rotational counterpart, Fig. 11.6,

$$\tau = \frac{4wa}{[1 - (\theta/\theta_0)^2]^2} \left(\frac{\mathcal{B}_p N}{\theta_0} i_c + \frac{\mathcal{B}_p{}^2 r}{\mu_a \theta_0} \theta + \frac{\mu_a N^2}{\theta_0{}^2 r} i_c{}^2 \theta + \frac{\mathcal{B}_p N}{\theta_0{}^3} i_c \theta^2 \right) \quad (11.21)$$

in which

$$\frac{x}{r} = \theta \qquad (11.22)$$

$$\frac{g_0}{r} = \theta_0 \qquad (11.23)$$

or

$$\tau \approx \frac{4\mathcal{B}_p waN}{\theta_0} i_c + \frac{4\mathcal{B}_p{}^2 wa}{\mu_a \theta_0} \theta \qquad (11.24)$$

$$K_{1t} = \frac{4\mathcal{B}_p waN}{\theta_0} \qquad (11.25)$$

$$K_{2t} = \frac{4\mathcal{B}_p{}^2 war}{\mu_a \theta_0} \qquad (11.26)$$

Sometimes mechanical construction and separation of control flux paths and polarizing flux paths are facilitated, compactness is increased, or accessibility of moving parts is improved by use of a magnetic structure that contains a nonworking gap, for which the common E-I structure serves as an example. For Figs. 11.7a and 11.8a, the equations are

$$f = 2\mathcal{B}_p w N i_c - \frac{2\mu_a w N^2}{g_0 l_0 \left(\dfrac{\mathcal{R}_0}{2\mathcal{R}_n} + 1\right)} i_c^2 x \tag{11.27}$$

$$\tau = 2\mathcal{B}_p w r N i_c - \frac{2\mu_a w r^2 N^2}{g_0 l_0 \left(\dfrac{\mathcal{R}_0}{2\mathcal{R}_n} + 1\right)} i_c^2 \theta \tag{11.28}$$

in which

$$\mathcal{R}_0 = \frac{g_0}{\mu_a l_0 w} \tag{11.29}$$

and $\mathcal{R}_n$ is the reluctance of the nonworking gap. For Figs. 11.7b and 11.8b, the equations are

$$f = \frac{2\mathcal{B}_p w}{\left(\dfrac{2\mathcal{R}_n}{\mathcal{R}_0} + 1\right)} N i_c - \frac{2\mathcal{B}_p w g_0}{\mu_a l_0 \left(\dfrac{\mathcal{R}_0}{2\mathcal{R}_n} + 1\right)} x \tag{11.30}$$

$$\tau = \frac{2\mathcal{B}_p w r}{\left(\dfrac{2\mathcal{R}_n}{\mathcal{R}_0} + 1\right)} N i_c - \frac{2\mathcal{B}_p^2 w g_0 r^2}{\mu_a l_0 \left(\dfrac{\mathcal{R}_0}{2\mathcal{R}_n} + 1\right)} \theta \tag{11.31}$$

In terms of polarizing magnetomotive force, electromechanical coupling constants for the arrangements of Figs. 11.7a and 11.7b are the same, and those for the arrangements of Figs. 11.8a and 11.8b are the same. For Figs. 11.7a and 11.8a,

$$\mathcal{B}_p = \frac{\mu_a N I_0}{g_0 \left(\dfrac{2\mathcal{R}_n}{\mathcal{R}_0} + 1\right)} \tag{11.32}$$

and the control magnetomotive force is $N i_c$ across each gap for starting, while for Figs. 11.7b and 11.8b,

$$\mathcal{B}_p = \frac{\mu_a N I_0}{g_0} \tag{11.33}$$

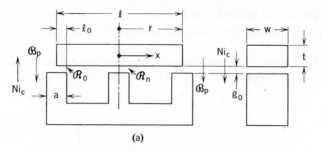

(a)

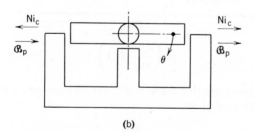

(b)

Fig. 11.7. E-I structure for translational motion normal to direction of air-gap flux density.

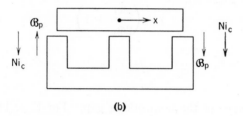

(a)

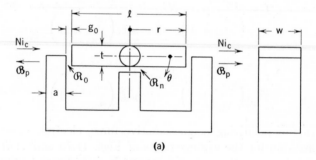

(b)

Fig. 11.8. E-I structure for rotational motion normal to direction of air-gap flux density.

and the control magnetomotive force is $Ni_c/(2\Re_n/\Re_0 + 1)$ across each gap for starting. Hence, for either arrangement,

$$K_{1f} = \frac{2\mu_a N^2 I_0 h}{g_0 \left(\dfrac{2\Re_n}{\Re_0} + 1\right)} \tag{11.34}$$

$$K_{1t} = \frac{2\mu_a N^2 I_0 h}{\theta_0 \left(\dfrac{2\Re_n}{\Re_0} + 1\right)} \tag{11.35}$$

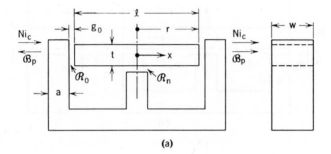

(a)

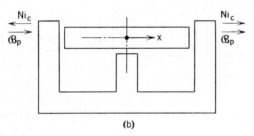

(b)

Fig. 11.9. E-I structure for translational motion parallel to direction of air-gap flux density.

For Figs. 11.9a and 11.10a,

$$f = \frac{2wa \left(\dfrac{2\Re_n}{\Re_0} + 1\right)}{\left[\dfrac{2\Re_n}{\Re_0} + 1 - \left(\dfrac{x}{g_0}\right)^2\right]^2} \left[\frac{\mathfrak{B}_p \left(\dfrac{2\Re_n}{\Re_0} + 1\right) Ni_c}{g_0} + \frac{\mathfrak{B}_p^2 \left(\dfrac{2\Re_n}{\Re_0} + 1\right) x}{\mu_a g_0} \right.$$
$$\left. + \frac{\mu_a N^2 i_c^2 x}{g_0^3} + \frac{\mathfrak{B}_p Ni_c x^2}{g_0^3} \right] \tag{11.36}$$

in which

$$\mathcal{R}_0 = \frac{g_0}{\mu_a w a} \tag{11.37}$$

or

$$f \approx \frac{2\mathcal{B}_p w a N}{g_0} i_c + \frac{2\mathcal{B}_p{}^2 w a}{\mu_a g_0} x \tag{11.38}$$

$$\tau \approx \frac{2\mathcal{B}_p w a N}{\theta_0} i_c + \frac{2\mathcal{B}_p{}^2 w a r}{\mu_a \theta_0} \theta \tag{11.39}$$

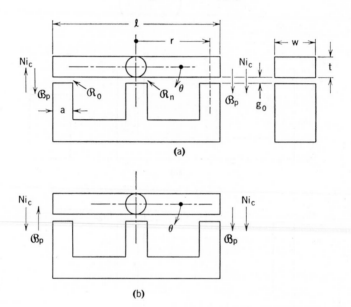

Fig. 11.10. E-I structure for rotational motion parallel to direction of air-gap flux density.

for the linear range. For Figs. 11.9b and 11.10b,

$$f = \frac{2wa}{\left[\dfrac{2\mathcal{R}_n}{\mathcal{R}_0} + 1 - \left(\dfrac{x}{g_0}\right)^2\right]^2} \left[\frac{\mathcal{B}_p\left(\dfrac{2\mathcal{R}_n}{\mathcal{R}_0} + 1\right)Ni_c}{g_0} + \frac{\mathcal{B}_p{}^2\left(\dfrac{2\mathcal{R}_n}{\mathcal{R}_0} + 1\right)x}{\mu_a g_0} \right.$$
$$\left. + \frac{\mu_a N^2 i_c{}^2 x}{g_0{}^3} + \frac{\mathcal{B}_p N i_c x^2}{g_0{}^3} \right] \tag{11.40}$$

or

$$f \approx \frac{2\mathfrak{B}_p waN}{g_0 \left(\dfrac{2\mathfrak{R}_n}{\mathfrak{R}_0} + 1\right)} i_c + \frac{2\mathfrak{B}_p{}^2 wa}{\mu_a g_0 \left(\dfrac{2\mathfrak{R}_n}{\mathfrak{R}_0} + 1\right)} x \qquad (11.41)$$

$$\tau \approx \frac{2\mathfrak{B}_p waN}{\theta_0 \left(\dfrac{2\mathfrak{R}_n}{\mathfrak{R}_0} + 1\right)} i_c + \frac{2\mathfrak{B}_p{}^2 war}{\mu_a \theta_0 \left(\dfrac{2\mathfrak{R}_n}{\mathfrak{R}_0} + 1\right)} \theta \qquad (11.42)$$

for the linear range.

In terms of polarizing magnetomotive force, the electromechanical coupling constants for the arrangements of Figs. 11.9a and 11.9b are the same, and those for the arrangements of Figs. 11.10a and 11.10b are the same. For Figs. 11.9a and 11.10a,

$$\mathfrak{B}_p = \frac{\mu_a NI_0}{g_0 \left(\dfrac{2\mathfrak{R}_n}{\mathfrak{R}_0} + 1\right)} \qquad (11.43)$$

and the control magnetomotive force is Ni_c across each gap for starting, while for Figs. 11.9b and 11.10b,

$$\mathfrak{B}_p = \frac{\mu_a NI_0}{g_0} \qquad (11.44)$$

and the control magnetomotive force is $Ni_c/(2\mathfrak{R}_n/\mathfrak{R}_0 + 1)$ across each gap for starting. Hence, for either arrangement,

$$K_{1f} = \frac{2\mu_a N^2 I_0 wa}{g_0{}^2 \left(\dfrac{2\mathfrak{R}_n}{\mathfrak{R}_0} + 1\right)} \qquad (11.45)$$

$$K_{1t} = \frac{2\mu_a N^2 I_0 wa}{r\theta_0{}^2 \left(\dfrac{2\mathfrak{R}_n}{\mathfrak{R}_0} + 1\right)} \qquad (11.46)$$

but the electromagnetic elastances and $i_c{}^2 x$ terms are different. If any of these E-I structures are made as double E-I four-pole structures, the forces (or torques) are doubled, and as $\mathfrak{R}_n$ approaches zero, the equations for the double E-I structures reduce to Eqs. 11.11, 11.14, 11.17, and 11.21, or the linearized versions of Eqs. 11.17 and 11.21.

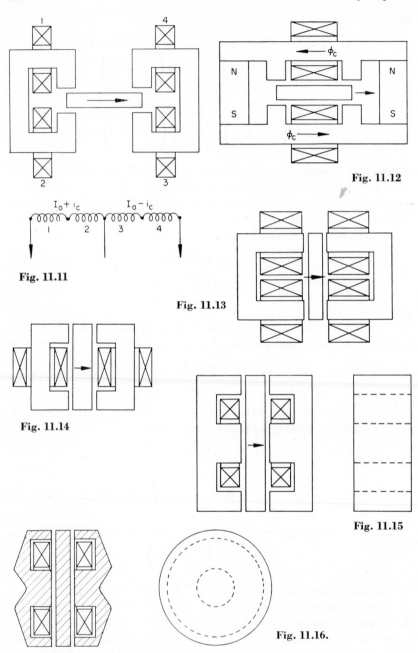

Fig. 11.12

$I_0 + i_c$ $I_0 - i_c$

Fig. 11.11

Fig. 11.13

Fig. 11.14

Fig. 11.15

Fig. 11.16.

Figs. 11.11–11.16   Various coil and magnetic-circuit arrangements.

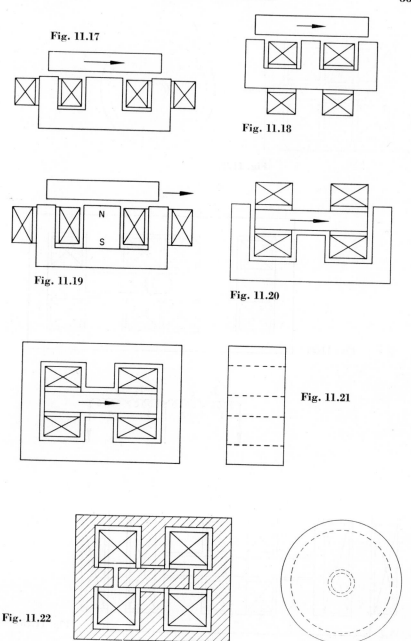

Fig. 11.17

Fig. 11.18

Fig. 11.19

Fig. 11.20

Fig. 11.21

Fig. 11.22

Figs. 11.17–11.22. Various coil and magnetic-circuit arrangements.

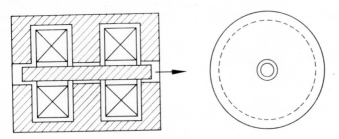

Fig. 11.23

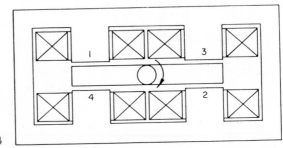

Fig. 11.24

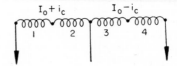

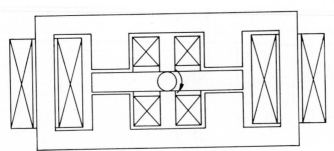

Fig. 11.25

Figs. 11.23–11.25. Various coil and magnetic-circuit arrangements.

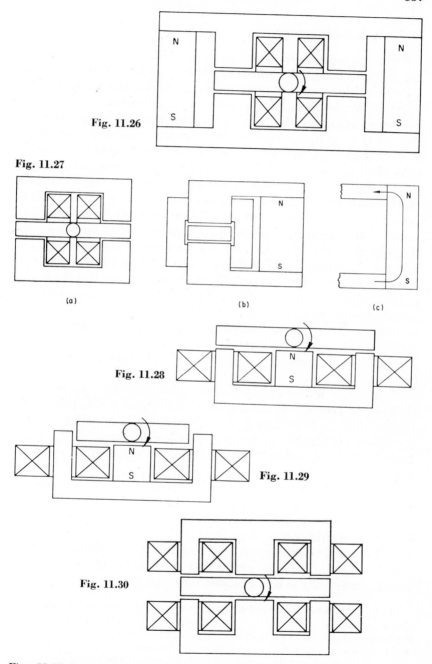

Fig. 11.26

Fig. 11.27

(a)   (b)   (c)

Fig. 11.28

Fig. 11.29

Fig. 11.30

Figs. 11.26–11.30. Various coil and magnetic-circuit arrangements.

11.22. Coil Arrangements

Coil arrangements for obtaining the combinations of polarizing and control fluxes indicated schematically in Figs. 11.1 to 11.10, often in association with permanent magnets, are numerous, frequently dictated by space requirements. Some arrangements, various ones of which achieve substantially the same operational results, are illustrated in Figs. 11.11 to 11.30. Though control magnetomotive force is indicated as Ni_c per gap, sometimes the same coils supply two control-flux paths in parallel, as shown in Figs. 11.26 and 11.27. In general, to reduce leakage fluxes, coils and magnets should be located as near to air gaps as feasible, but restrictions of space and shape or manufacturing simplifications often dominate. Usually coils are operated from a push-pull vacuum-tube amplifier. The quiescent currents may be used to supply the polarizing field, Fig. 11.11, but if a permanent magnet is used, the coils may be arranged so that the magnetic effects of the quiescent currents cancel. In either situation, the quiescent currents give the principal power loss in the device.* A moving-coil device must have two coils or must have a midtapped coil for push-pull operation. Such operation gives twice the maximum dissipation in coil heating, concentrated in half the winding, as does operation with control current only, to obtain a given force or torque, or requires a larger coil to obtain a given force or torque with a given temperature rise.

Moving-iron devices may have coil spaces that are very restricted through the necessity of making the movable part small, Figs. 11.12, 11.17 to 11.21, 11.23 to 11.30, or may have coil space limited only through the necessity of keeping the device as a whole within bounds, Figs. 11.11, 11.13, 11.15, 11.16, and 11.22. The coils may be operated push-pull and may supply both control and polarizing flux, Figs. 11.11, 11.13 to 11.18, 11.20 to 11.24, and 11.30; separate windings may be used to supply the respective fluxes, Figs. 11.11, 11.13, and 11.25; or windings may be used to supply the control flux, and permanent magnets may be used to supply the polarizing flux, Figs. 11.12, 11.19, 11.26 to 11.29. Whereas control flux and polarizing flux may be supplied by separate coils magnetically in series, a permanent magnet and control coil should not be placed magnetically in series because they have contrary functions, the

---

* In fact, the power dissipated in the driving amplifier and associated electronic circuits commonly far exceeds the power required by the motor. Bridge schemes can be devised that shunt the quiescent currents around the motor coils, but the net results are greater dissipation of heat elsewhere than elimination of heat from the motor and decrease in speed of response of the motor owing to reduction of impedance seen by looking out of the motor terminals.

one to hold flux constant and the other to provide a varying component, and the resulting control sensitivity is very low owing to the high incremental reluctance of the magnet.

If windings are kept thin enough that the difference between interior and surface temperature is not great, the size of the winding needed to keep the temperature rise within a tolerable limit may be estimated by allowing 0.005 to 0.01 watt per square inch of exposed coil surface per degree C of temperature rise in still air. Cooling, of course, can be facilitated by forced circulation of air, or by good thermal contact of the coil with metal parts having extensive surface. The nominal allowable temperatures of some magnet wire insulations are

Formex or Formvar	105°C	
Silicone enamel	180°C	
Ceroc single silicone	200°C	continuous
Teflon, Tetrox	225°C	
Ceroc single or heavy Teflon	250°C	

all of which give good space factors. Teflon has the advantage of being essentially inert chemically. These temperature limits may be exceeded for short periods, and of course current limits can be exceeded for short periods without exceeding temperature limits because of the temperature time lag introduced by the thermal capacity of the winding. Recently developed insulations, as yet not very thoroughly tried, are available for much higher temperatures than indicated in the list. For example, anodized aluminum wire may be used in the vicinity of 400°C, and ceramic-insulated nickel-plated copper wire may be used in the vicinity of 500 to 600°C. Coil space may be saved by casting coils in plastic to eliminate frames and bobbins.

The manner of apportioning control flux and polarizing flux is discussed under general design procedure, Sec. 11.26.

11.23. Permanent-Magnet Design

Permanent-magnet material and shape also are governed by apportionment of control flux and polarizing flux and by the air-gap reluctances. The relations

$$\mathfrak{B}_m A_m = 2\mathfrak{B}_p A \eta \tag{11.47}$$

$$\frac{\mathfrak{B}_m}{\mathfrak{K}_m} = \frac{l_m}{A_m} \mathcal{P}\eta \tag{11.48}$$

can be used as guides. Here $\mathfrak{B}_m$ is the flux density and $\mathfrak{K}_m$ is the magnetic-field intensity in the magnet, A_m is the cross-sectional area and

l_m is the length of the magnet, A is a polar area (for the device of Fig. 11.27, for example), $\mathcal{P}$ is the permeance seen by the magnet in the air gaps of the device, and η is a coefficient to account for leakage flux.

The demagnetization curve of Alnico V is shown in Fig. 11.31, and Eq. 11.48 is the slope of the permeance line seen by the magnet but transposed to the $\mathcal{B}_m$-$\mathcal{K}_m$ plane so that the intersection with the demagnetization curve locates a static operating point for the magnet. If as a trial $\mathcal{B}_m$ is taken to correspond to the maximum energy product,

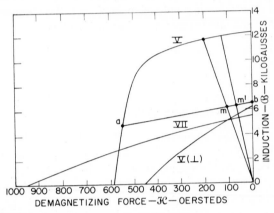

Fig. 11.31. Demagnetization curves for various Alnicos, $\mathcal{B}$ versus $\mathcal{K}$, with equivalent air-gap lines.

A_m can be computed from Eq. 11.47, $\mathcal{K}_m$ can be read from the curve, and l_m then can be computed from Eq. 11.48, provided that η has been estimated. Leakage may well be of the order of 100 per cent, so $\eta = 2$ can be used as a first trial, and η can be calculated later as closely as worth while by various methods.[2] However, the static operating point is not desirable, owing to the ready possibility of inadvertent demagnetization, and A_m and l_m may not be practical for the desired structure. In general, Alnico V tends to be appropriate where A_m needs to be small and l_m can be large, whereas a material such as Alnico VII tends to be appropriate where l_m needs to be small and A_m can be large. After a trial or two for obtaining reasonable A_m and l_m, they are adjusted further in accordance with the judgment of the designer to suit the structure and to permit improved stability of operation by demagnetizing the magnet somewhat so that the operating point is on line a-b, which is an idealization of a thin minor loop.

[2] H. C. Roters, *Electromagnetic Devices*, John Wiley and Sons, New York, 1941.

When the magnet is of irregular shape, analysis becomes very complex owing to nonuniformity of flux density; in fact, in a magnet of uniform cross-sectional area, a similar situation exists owing to leakage flux, and the magnet, therefore, sometimes is shaped to achieve more nearly uniform density by adding a "leakage shell." Furthermore, many magnets are "oriented" by cooling during the casting process in a magnetic field. The orienting should be done along lines corresponding to the desired magnetic path in the circuit of which the magnet is to be a part because magnetization at right angles to the direction of orientation is relatively difficult, as indicated by the lower curve of Fig. 11.31. For example, use of a bar magnet as in Fig. 11.27c requires portions of the magnetic path to go across the direction of orientation if the magnet is oriented lengthwise.

From the point of view of the device, the situation can be better appreciated by plotting the demagnetization curve in terms of flux and magnetomotive force, Fig. 11.32, derived from Fig. 11.31 by multiplying ordinates by A_m and abscissas by l_m. The operating point for the magnet is at m, but the operating point for the air gaps, armature centered, is at p, for 100 per cent leakage. If the rotor translates, Fig. 11.3, operation remains at m and p, but if the armature turns in either direction, Fig. 11.6 or 11.27, the operating points move to m' and p'. If leakage is ignored, the operating points are erroneously supposed to fall at q and q'. Likewise, for the translational device, the control magnetomotive force does not influence the polarizing flux, whereas for the rotational device,

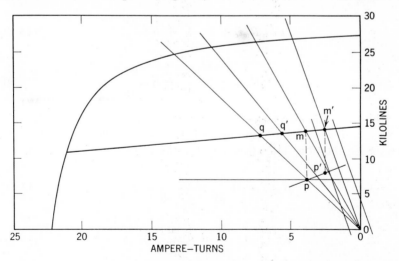

Fig. 11.32. Demagnetization curve for Alnico V, flux versus magnetomotive force, with actual air-gap lines, illustrating influence of leakage.

the displacement of the armature permits the magnet to see a fraction of the control magnetomotive force, which is represented on Fig. 11.32 by displacement of the air-gap line from the origin. This explanation illustrates the facts that leakage flux tends to emphasize the change of polarizing flux due to armature rotation and thus contributes to non-linearity, and that a large l_m/A_m ratio tends to reduce the change of polarizing flux due to armature rotation. Increased length within limited space sometimes can be achieved by use of a horseshoe magnet.

11.24. Dynamic Analysis

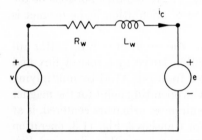

Fig. 11.33. Electric circuit of actuator.

Dynamic analysis of the devices can be simply generalized by means of linear theory, if operation of the nonlinear devices is restricted to the substantially linear range, and reduced to a single equivalent-circuit representation. For the electric-circuit equations, only the control current needs to be considered (except for the inclusion of polarizing current directly or indirectly in the expressions for certain parameters), though for computation of power losses, total current must be considered. For any of the devices,

$$v = R_w i_c + L_w \frac{di_c}{dt} + K_{1f}\frac{dx}{dt} \tag{11.49}$$

or

$$v = R_w i_c + L_w \frac{di_c}{dt} + K_{1t}\frac{d\theta}{dt} \tag{11.50}$$

in which v is the incremental terminal voltage, R_w is the winding resistance, and L_w is the winding self-inductance (armature centered). Either Eq. 11.49 or 11.50 describes the circuit of Fig. 11.33. Also,

$$K_{1f}i_c = M \frac{d^2x}{dt^2} + B_f \frac{dx}{dt} + K_{ef}x \tag{11.51}$$

or

$$K_{1t}i_c = J \frac{d^2\theta}{dt^2} + B_t \frac{d\theta}{dt} + K_{et}\theta \tag{11.52}$$

Here M represents the mass and J represents the moment of inertia of the moving parts, and B_f and B_t are viscous-friction coefficients. Use of

$$e = K_{1f} \frac{dx}{dt} \tag{11.53}$$

or

$$e = K_{1t} \frac{d\theta}{dt} \tag{11.54}$$

with Eq. 11.51 or 11.52, respectively, gives

$$i = \frac{M}{K_{1f}^2} \frac{de}{dt} + \frac{B_f}{K_{1f}^2} e + \frac{K_{ef}}{K_{1f}^2} \int e\, dt \tag{11.55}$$

or

$$i = \frac{J}{K_{1t}^2} \frac{de}{dt} + \frac{B_t}{K_{1t}^2} e + \frac{K_{et}}{K_{1t}^2} \int e\, dt \tag{11.56}$$

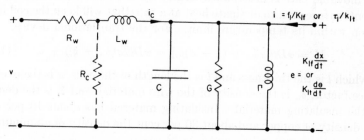

Fig. 11.34. Equivalent electromechanical circuit of actuator.

and leads to the all-electric equivalent circuit of Fig. 11.34, in which

$$C = \frac{M}{K_{1f}^2} \text{ or } \frac{J}{K_{1t}^2} \tag{11.57}$$

$$G = \frac{B_f}{K_{1f}^2} \text{ or } \frac{B_t}{K_{1t}^2} \tag{11.58}$$

$$\Gamma = \frac{K_{ef}}{K_{1f}^2} \text{ or } \frac{K_{et}}{K_{1t}^2} \tag{11.59}$$

The circuit may be made into a two-terminal pair by adding an impressed force, f_1, to the electromagnetic force, Eq. 11.51, and an impressed torque, τ_1, to the electromagnetic torque, Eq. 11.53, the equivalent of which is the impressed current, f_1/K_{1f} or τ_1/K_{1t}, whereas the equivalent of an impressed velocity is $K_{1f}(dx/dt)$ or $K_{1t}(d\theta/dt)$. Core losses may be represented approximately by resistance, R_c, Fig. 11.34, which is a function of frequency and flux density and can be estimated only roughly. It serves as an aid in visualizing the influence of core losses.

11.25. Figures of Merit

Figures of merit may be defined as aids in preliminary design procedure and in selection of motors. The ratio of starting force to mass or starting torque to inertia may be taken as a partial measure of speed of response. In computing these ratios, certain approximations are made for purposes of generality, since figures of merit generally need signify only substantial differences among various devices; if selection is very critical, accurate computation must be made for the specific devices or in specific design comparisons. For the moving-coil devices, the mass of the movable part is determined largely by the necessity of having sufficient surface area to dissipate heat at a rate that will keep the coil insulation within its temperature limit. For the translational device,

$$M = \pi Dtl[\sigma\delta_w + (1 - \sigma)\delta_i] = \pi Dtl[\sigma(\delta_w - \delta_i) + \delta_i] \quad (11.60)$$

in which t is the thickness and l is the length of the coils, σ is the over-all space factor, δ_w is the density of the wire material, and δ_i is the density of the insulating material. Insulating material has about 10 per cent the density of copper and about 30 per cent the density of aluminum:

$$\delta_e = \sigma(\delta_w - \delta_i) + \delta_i \approx (0.9\sigma + 0.1)\delta_c \quad (11.61)$$

for copper, and

$$\delta_e = \sigma(\delta_w - \delta_i) + \delta_i \approx (0.7\sigma + 0.3)\delta_a \quad (11.62)$$

for aluminum, and

$$M \approx \pi Dtl\delta_e \quad (11.63)$$

The current limit in the coil is determined by

$$i_c{}^2 R_w = i_c{}^2 \frac{\rho\pi DN^2}{\sigma lt} = k\pi DlT \quad (11.64)$$

if the coil is operated with control current only, and by

$$(2i_c)^2 \frac{R_w}{2} = (2i_c)^2 \frac{\rho\pi DN^2}{2\sigma lt} = k\pi D \frac{l}{2} T \quad (11.65)$$

if the winding is operated push-pull and hence carries quiescent current. Here ρ is the resistivity of the wire material, k is the dissipation factor in power per unit coil-surface area per degree of temperature rise, and

T is the temperature rise. The coil is supposed to be thin enough that the internal temperature does not differ greatly from the surface temperature. The current limit then is

$$i_c = \sqrt{(k\pi DlT)/R_w} = \frac{l}{N} \sqrt{(\sigma ktT)/\rho} \tag{11.66}$$

or

$$i_c = \frac{1}{2} \sqrt{(k\pi DlT)/R_w} = \frac{l}{2N} \sqrt{(\sigma ktT)/\rho} \tag{11.67}$$

Hence, for maximum allowable flux density $\mathscr{B}_p = \mathscr{B}_m$,

$$f = \pi \mathscr{B}_m \sqrt{(\sigma kT)/\rho}\, Dl\sqrt{t} \tag{11.68}$$

or

$$f = \frac{\pi}{2} \mathscr{B}_m \sqrt{(\sigma kT)/\rho}\, Dl\sqrt{t} \tag{11.69}$$

and

$$\frac{f}{M} = \frac{\mathscr{B}_m}{\delta_e} \sqrt{(\sigma kT)/(\rho t)} \tag{11.70}$$

or

$$\frac{f}{M} = \frac{\mathscr{B}_m}{2\delta_e} \sqrt{(\sigma kT)/(\rho t)} \tag{11.71}$$

For the rotational device,

$$\tau = \mathscr{B}_m \sqrt{(\sigma kT)/\rho}\, hDl\sqrt{t} \tag{11.72}$$

or

$$\tau = \frac{\mathscr{B}_m}{2} \sqrt{(\sigma kT)/\rho}\, hDl\sqrt{t} \tag{11.73}$$

and

$$J = \frac{tlD\delta_e}{2}\left(hD + \frac{D^2 + l^2}{3} \right) \tag{11.74}$$

$$\frac{\tau}{J} = \frac{2\mathscr{B}_p \sqrt{(\sigma kT)/(\rho t)}}{\delta_e \left(D + \dfrac{D^2 + l^2}{3h} \right)} \tag{11.75}$$

or

$$\frac{\tau}{J} = \frac{\mathscr{B}_p \sqrt{(\sigma kT)/(\rho t)}}{\delta_e \left(D + \dfrac{D^2 + l^2}{3h} \right)} \tag{11.76}$$

If the angular accelerations of Eqs. 11.75 and 11.76 are translated into linear accelerations at the coil side, for comparison with Eqs. 11.70 and 11.71, the equivalent force-to-mass ratios are

$$\left(\frac{f}{M}\right)_e = \frac{\mathcal{B}_m D \sqrt{(\sigma k T)/(\rho l)}}{\delta_e \left(D + \dfrac{D^2 + l^2}{3h}\right)} \tag{11.77}$$

or

$$\left(\frac{f}{M}\right)_e = \frac{\mathcal{B}_m D \sqrt{(\sigma k T)/(\rho l)}}{2\delta_e \left(D + \dfrac{D^2 + l^2}{3h}\right)} \tag{11.78}$$

Hence, the rotational device cannot have as high starting acceleration as the translational device and can approach the starting acceleration of the translational device only if its coil is relatively high, short, or thin. Whereas the maximum force or torque of moving-coil devices is limited by the practical flux-density limit of the magnetic material adjacent to the air gap in terms of polarizing flux, $\mathcal{B}_m = \mathcal{B}_p$, the moving-iron devices are so limited by the flux density of polarizing plus control flux, $\mathcal{B}_m = \mathcal{B}_p + \mathcal{B}_c$, and for a given upper limit of total flux density, the maximum force is achieved when the polarizing and control components are equal, $\mathcal{B}_m = 2\mathcal{B}_p$. Hence for the device of Fig. 11.3 the maximum force achievable is

$$f = \frac{\mathcal{B}_m{}^2 w g_0}{\mu_a} \tag{11.79}$$

and the minimum mass of the armature is

$$M = \frac{w a l \delta}{2} \tag{11.80}$$

so that

$$\frac{f}{M} = \frac{2\mathcal{B}_m{}^2 g_0}{\mu_a a l \delta} \tag{11.81}$$

Likewise, for the device of Fig. 11.4,

$$\tau = \frac{\mathcal{B}_m{}^2 w g_0 l}{2\mu_a} \tag{11.82}$$

$$J = \frac{w t l^3 \delta}{12} \tag{11.83}$$

This rotor must be twice as thick as necessary to carry the maximum flux because it must overlap each pole by $a/2$. It could be thinned in the center, but the decrease in J would not be great. Hence, somewhat pessimistically,

$$\frac{\tau}{J} = \frac{6\mathcal{B}_m{}^2 g_0}{\mu_a a l^2 \delta} \tag{11.84}$$

$$\left(\frac{f}{M}\right)_e = \frac{3\mathcal{B}_m{}^2 g_0}{\mu_a a l \delta} \tag{11.85}$$

for comparison with Eq. 11.76.

For the device of Fig. 11.5,

$$f = \frac{\mathcal{B}_m{}^2 w a}{\mu_a} \tag{11.86}$$

$$M = w t l \delta \tag{11.87}$$

$$\frac{f}{M} = \frac{\mathcal{B}_m{}^2 a}{\mu_a l t \delta} \tag{11.88}$$

and for the device of Fig. 11.6,

$$\tau = \frac{\mathcal{B}_m{}^2 w a (l - a)}{2\mu_a} \tag{11.89}$$

$$\frac{\tau}{J} = \frac{6\mathcal{B}_m{}^2 a (l - a)}{\mu_a l^3 t \delta} \tag{11.90}$$

$$\left(\frac{f}{M}\right)_e = \frac{3\mathcal{B}_m{}^2 a (l - a)}{\mu_a l^2 t \delta} \tag{11.91}$$

For normal action, the rotational device has $\frac{3}{2}$ the starting acceleration of the translational device, and for parallel action, the rotational device has $3(l - a)/l$ as much starting acceleration as the translational device. For the two types of translational devices, the ratio of starting forces is

$$\frac{f_n}{f_p} = \frac{g_{0n}}{a_p} \tag{11.92}$$

which means that for equal forces the normal-acting device must be much the larger. The ratio of starting accelerations is

$$\frac{(f/M)_n}{(f/M)_p} = \frac{2g_{0n}}{a_p} \tag{11.93}$$

which reflects the same information. For the two types of rotational devices, the ratio of starting torques is

$$\frac{\tau_n}{\tau_p} = \frac{g_{0n}l_n}{a_p(l_p - a_p)} \tag{11.94}$$

which means that for equal torques the normal acting device must be much the larger. The ratio of starting accelerations is

$$\frac{(\tau/J)_n}{(\tau/J)_p} = \frac{g_{0n}l_p}{a_p(l_p - a_p)} \tag{11.95}$$

Actually, the comparison can be made still more favorable towards the parallel-acting device. The armature dimensions in all cases have been determined on the assumption that the flux densities in the armature cannot exceed the maximum flux density stipulated for the pole faces. However, if for the parallel-acting devices the armature cross-sectional area is reduced so that in the structure of Fig. 11.27, for example, the armature approaches saturation as it turns or so that in the structure of Fig. 11.24 the armature is always saturated by the polarizing flux, the nonlinearity is much reduced and the moment of inertia is reduced, without influencing starting torque. Furthermore, by saturating the armature, Fig. 11.24, it can be made solid and still have little hysteresis loss in it, which is a manufacturing advantage. If saturation is introduced into the normal-acting devices, they acquire negative K_2 and become nonlinear.

This effect can be visualized, for example, by consideration of Fig. 11.11 for balanced currents and displaced armature. The magnetomotive force across the air gaps on the end that has been pulled further into the stator then is less than the magnetomotive force across the air gaps on the other end; the armature tends to return to center, but the centering force is not in direct proportion to displacement, since it depends on the degree of saturation and hysteresis of the magnetic material.

Of the moving-iron devices, the parallel-acting device has the greater force and starting acceleration, and the rotational device exceeds the translational device in starting acceleration. A comparison of the parallel-acting rotating device with the moving-coil device in translation therefore is of interest. The ratio of forces is

$$\frac{f_c}{f_i} = \frac{\pi \mu_a \sqrt{(\sigma k T)/\rho}}{\mathscr{B}_m} \times \frac{Dl_c \sqrt{t}}{w_i a_i} \tag{11.96}$$

if the moving coil does not carry quiescent currents, and is half the above ratio if the moving coil does carry quiescent currents. The ratio of starting accelerations is

$$\frac{(f/M)_c}{(f/M)_{ie}} = \frac{\mu_a \delta_i \sqrt{(\sigma k T)/\rho}}{3\delta_e \mathcal{B}_m} \times \frac{l_i^2 \sqrt{t}}{(l_i - a_i)a_i} \qquad (11.97)$$

if the moving coil does not carry quiescent currents, and is half the above ratio if the moving coil does carry quiescent currents.

These ratios each have been factored into a part that depends on assumed limits for physical quantities and a part that depends on dimensions. Two comparisons now are made. First, for a copper winding,

$$\sigma \approx 0.50$$

$$k \approx 10 \text{ w/m}^2 \text{ °C}$$

$$T \approx 200°C$$

$$\rho \approx 3.0 \times 10^{-8} \text{ ohm-m at } 200°C$$

$$\mu_a = 12.6 \times 10^{-7}$$

$$\mathcal{B}_m = 1.5 \text{ webers m}^{-2}$$

$$\frac{\delta_i}{\delta_e} = 1.6$$

which gives

$$\frac{\pi \mu_a \sqrt{(\sigma k T)/\rho}}{\mathcal{B}_m}$$

$$= \frac{3.14 \times 12.6 \times 10^{-7} \sqrt{(0.50 \times 10 \times 200)/(3.0 \times 10^{-8})}}{1.5} \approx 0.48$$

$$(11.98)$$

and

$$\frac{\mu_a \delta_i \sqrt{(\sigma k T)/\rho}}{3\delta_e \mathcal{B}_m}$$

$$= \frac{12.6 \times 10^{-7} \times 1.6 \sqrt{(0.050 \times 10 \times 200)/(3.0 \times 10^{-8})}}{3 \times 1.5} = 0.084$$

$$(11.99)$$

If the permissible temperature rise is reduced, or if the flux-density limit is raised, or both, the factors computed above are reduced.

Now for an edgewise strip-aluminum winding, using no bobbin or frame and relying on oxide insulation,

$$\sigma \approx 0.90$$

$$\rho \approx 5.0 \times 10^{-8} \text{ ohm-m at } 200°C$$

$$\frac{\delta_i}{\delta_e} = 3.2$$

and if other quantities remain the same,

$$\frac{\pi\mu_a\sqrt{(\sigma kT)/\rho}}{\mathcal{B}_m} = 0.50 \tag{11.100}$$

and

$$\frac{\mu_a\delta_i\sqrt{(\sigma kT)/\rho}}{3\delta_e\mathcal{B}_m} = 0.17 \tag{11.101}$$

which shows little gain in force ratio but double the acceleration ratio.

To complete the comparison, some assumptions must be made for dimensions. For devices approximately the same size

$$\frac{Dl_c}{w_ia_i} \approx 1 \tag{11.102}$$

and generally

$$t \ll l_c \tag{11.103}$$

Therefore

$$\frac{Dl_c}{w_ia_i}\sqrt{t} \approx 10^{-1} \rightarrow 10^{-2} \tag{11.104}$$

for t between 10^{-2} and 10^{-3} m. Also, if

$$l_i \approx 4a_i \tag{11.105}$$

then

$$\frac{l_i^2\sqrt{t}}{(l_i - a_i)a_i} \approx 0.2 \rightarrow 2 \tag{11.106}$$

Evidently for comparable sizes, the moving-coil device is by far the weaker and has considerably smaller starting accelerations when sizes are small but could be superior in starting acceleration for large sizes. Specifications sometimes are made in terms of starting force, maximum stroke, minimum resonant frequency (with or without specified load),

$$\omega_0^2 = \frac{K_{ef}}{M} = \frac{K_{et}}{J} \tag{11.107}$$

and control current. Specification of starting force and maximum stroke fixes effective elastance and, hence, fixes maximum mass or moment of inertia, which may fix too much. If the moving-coil translational device can develop the required force with a coil of a mass that is essentially the maximum tolerable, the moving-iron rotational device could develop

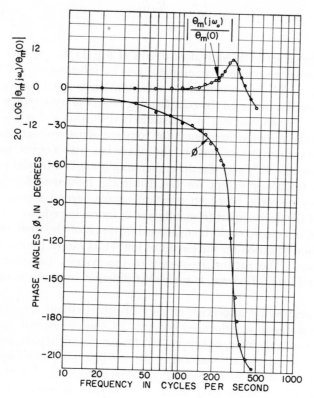

Fig. 11.35. Frequency-response characteristic of typical small torque motor.

the required force with an armature having much less than the tolerable moment of inertia, or in other words, could substantially exceed the specified resonant frequency. On the other hand, if the moving-iron rotational device can develop the required force with an armature having a moment of inertia that is essentially the required maximum, the moving-coil translational device could not develop the required force without using a coil of excessive mass.

Whereas resonant frequency may be taken as an indirect measure of speed of response, a better measure is phase lag. A frequency-response characteristic is plotted in Fig. 11.35.[3] For low frequencies, the phase lag can be computed from

$$\tan \theta = -\frac{\omega(L_w + L)}{R} \qquad (11.108)$$

in which

$$R = \frac{R_w R_c}{R_w + R_c} \qquad (11.109)$$

in accordance with the equivalent circuit of Fig. 11.34. In fact, the entire frequency-response characteristic can be interpreted qualitatively from Fig. 11.34, but quantitative computation is not feasible beyond low frequencies owing to the complicating effects of hysteresis and eddy currents. The current in Γ is a measure of deflection and is not a maximum for the resonant frequency, for constant-signal-voltage excitation. For constant-current excitation, the phase lag with respect to *current* passes through 90°, and the deflection amplitude is a maximum at the resonant frequency except for the disturbing influence of core losses. Frequency response based on constant source current gives a characteristic of the device apart from influence of source impedance. Dynamic behavior, of course, could be expressed in terms of the roots of the determinantal equation for the circuit of Fig. 11.34 except for the effects of core losses. The natural frequency of the system, including voltage source and load, of course, is somewhat different from the resonant frequency.

As is evident by inspection of the equivalent circuit, Fig. 11.34, the speed of response is influenced not only by the mechanical parameters and the available force, or torque, but by the electrical parameters also, since the electrical time constant of the winding is a prime factor in determining the speed with which force can be developed. Hence, whereas for a given total flux-density limit, equal division of polarizing and control flux densities gives maximum possible force for a moving-iron device, reduction of control magnetomotive force by using larger polarizing and smaller control flux densities may be desirable to reduce winding inductances, control current, armature thickness, or coil heating. Division into 0.6 polarizing flux density and 0.4 control flux density gives 25 per cent reduction in control current (and perhaps in armature thickness), 36 per cent reduction in heating, and only 4 per cent reduction in force. For a given control magnetomotive force, reduction of turns re-

[3] R. H. Frazier and R. D. Atchley, "A Permanent-Magnet-Type Electric Actuator for Servovalves," Dynamic Analysis and Control Laboratory Report No. 66, Cambridge, Mass., June 1, 1952, Fig. 14, p. 23.

duces winding self-inductance and increases control current (or vice versa), but self-inductance is a squared function of turns, whereas control current is an inverse function of turns.

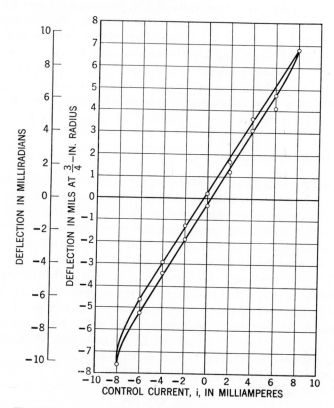

Fig. 11.36. Static characteristic of typical small torque motor.

The static sensitivity

$$\frac{\theta}{i} = \frac{K_1}{K_e} \tag{11.110}$$

(though θ/Ni is better as a characteristic of a given structure) is indicated by a static characteristic such as Fig. 11.36,[4] which gives an indication of linearity and of hysteresis, generally entirely of magnetic origin if centering springs are properly designed. To achieve very low hysteresis in moving-iron devices, one must select low-hysteresis materials with proper annealing, subject them to a minimum of machining or strain

[4] Frazier and Atchley, *op. cit.*, Fig. 12, p. 23.

thereafter, and operate at low flux-density levels and with relatively long air gaps. These requirements tend to make large size and large control current necessary. Since pressure for small size and high sensitivity tends to dominate in actuator design, hysteresis is rarely less than 2 per cent of maximum stroke, though the situation might be improved somewhat by better attention to annealing and by methods of fabrication that avoid machining after annealing and strains produced by mounting. Hysteresis effects in moving-coil devices readily can be negligible, since the control current imposes only a relatively small magnetomotive force on magnetic material which is already polarized well towards saturation.

11.26. Design Procedure

Design procedure should be systematized to prevent running computations round and round without converging on a conclusion that meets the specifications or without realizing what items in the specifications are in conflict or are unreasonable and, hence, prevent a sensible solution. Since the achievement of ideal features in all respects inevitably involves conflicts in most design problems, the manner in which various requirements influence the design should be clearly evident so that compromises can be made in a way that best preserves the most desirable features for the application. Procedures can be systematized in various ways. For example, if starting force, maximum static deflection, resonant frequency, and current sensitivity are specified, a procedure for the moving-coil translational device, Fig. 11.1, is as follows.

Combination of Eqs. 11.63, 11.68, and 11.107 gives

$$\sqrt{t} = \frac{\mathcal{B}_m \sqrt{(\sigma k T)/\rho}}{\omega_0{}^2 d \delta_e} \tag{11.111}$$

in which d is the maximum static deflection. Use of Eqs. 11.63 and 11.107 gives

$$Dl = \frac{f}{\pi \omega_0{}^2 t d \delta_e} \tag{11.112}$$

Hence, by assuming maximum air-gap flux density and quantities pertinent to winding design and heating which determine $\sqrt{(\sigma k T)/\rho}$ and δ_e, the coil thickness can be estimated from Eq. 11.111. (For a coil carrying quiescent current, Eq. 11.69 would be used instead of Eq. 11.68, which would halve the right-hand side of Eq. 11.111 and give t one-fourth as thick.) If the first result appears unreasonable, then a reasonable result must be achieved by modification of $\sqrt{(\sigma kT)/\rho}$, δ_e, specified ω_0^2, d, or all of them. The reasonableness of t is judged in part by the fact that it, plus clearances, determines the air-gap length across which $\mathfrak{B}_m$ must act.

When a preliminary figure for t has been determined, combinations of D and l can be tabulated from Eq. 11.112. If the valve mass is included in the specification of resonant frequency, it must be added to the right-hand side of Eq. 11.63, and Eqs. 11.111 and 11.112 modified accordingly. If acceptable combinations do not result, either t must be revised or specified ω_0 must be changed. When acceptable preliminary figures for D, l, and t have been obtained, then the magnetic circuit of the device, including permanent magnet or polarizing coil, can be designed, and the number of turns and wire size for the moving coil can be determined by use of Eq. 11.5 and the coil dimensions. Then if the preliminary design still appears practical, it can be developed in full detail.

If the same specifications are used for a parallel-acting moving-iron rotational device, Fig. 11.6, a possible procedure is as follows. Use of Eq. 11.89 gives

$$fxr = \frac{4\mathfrak{B}_m^2 z(1-z)wyr^2}{\mu_a} \tag{11.113}$$

in which

$$r' = xr \tag{11.114}$$

is the load radius (not to exceed $l/2$), and

$$a = yr \tag{11.115}$$

$$\mathfrak{B}_c = z\mathfrak{B}_m \tag{11.116}$$

$$\mathfrak{B}_p = (1-z)\mathfrak{B}_m \tag{11.117}$$

The armature thickness becomes

$$t = 2yzr \tag{11.118}$$

Use of Eqs. 11.107 and 11.83 gives

$$fxr = \omega_0^2 J\frac{d}{xr} = \frac{2yzwr^3(2+y)^3\delta d\omega_0^2}{12x} \tag{11.119}$$

and combination with Eq. 11.113 gives

$$r = \frac{24\mathfrak{B}_m^2(1-z)x}{(2+y)^3\delta d\omega_0^2\mu_a} \tag{11.120}$$

If the valve mass, M_v, is included in the specification of resonant frequency, and if the valve is split or counterbalanced, $2M_v(xr)^2$ must be added to the right-hand side of Eq. 11.83, and Eqs. 11.119 and 11.120 must be modified accordingly. Transposition of Eq. 11.113 gives

$$wyr = \frac{\mu_a f x}{4 \mathfrak{B}_m{}^2 z(1 - z)} \qquad (11.121)$$

The last two equations, Eqs. 11.120 and 11.121, with Eq. 11.117 may be taken as the preliminary design equations. By assuming systematically combinations of x, y, and z, recognizing that practically

$$1 < x < 1.5 \qquad (11.122)$$

$$0.3 < y < 1 \qquad (11.123)$$

$$0.2 < z < 0.5 \qquad (11.124)$$

the possibility of achieving reasonable dimensions can quickly be examined. If one starts, for example, with $x = 1$, $y = 0.5$, $z = 0.3$, preliminary values for r', r, w, a, and t are obtained which readily can be adjusted by proportion through juggle of x, y, and z. Whereas coil heating does not enter directly into these computations, as it does for the moving-coil device, it enters indirectly because r and a and the space needed for a shaft must be such that sufficient space is allowed for coils. The gap length will be of the order of three times the stroke unless a saturation effect is used to reduce nonlinearity. From this point, coil design (and magnet design, if permanent-magnet polarization is used) can proceed, using high-temperature insulation if necessary, and spring requirements can be investigated.

The electromagnetic elastance is

$$K_{2t} = \frac{(1 - z)f x r^2}{g_0 z} \qquad (11.125)$$

(when all reluctance is concentrated in air gaps), and the required spring elastance is, therefore,

$$K_{0t} = K_{et} + K_{2t} = \frac{f x^2 r^2}{d} + \frac{(1 - z)f x r^2}{g_0 z} = f x r^2 \left[\frac{x}{d} + \frac{(1 - z)}{g_0 z} \right] \qquad (11.126)$$

The spring torque for this type of device can conveniently be provided by the twist of its armature shaft. If a combination of x, y, and z cannot be found that gives acceptable r, w, a, and t, both from the point of view of over-all size and of practical coil and magnet design, specifications must be changed accordingly.

Procedures similar to the illustrations can be developed for any of the types of devices. The design of a motor using an E-I magnetic structure is given in Sec. 13.41. The type of device is often determined by mechanical adaptability to the application, shape of space available, or ease or expense of manufacture rather than by ideal analyses that optimize certain features. Furthermore, certain shapes, such as the cylindrical devices of Figs. 11.16, 11.22, and 11.23 and the magnetic circuits of the translational moving-coil device, are not readily laminated. The normal-acting moving-iron devices in translation may have high side forces owing to slight dissymmetries or unbalances, which may be quite troublesome if valves are rigidly attached to the armatures. Incidentally, any one of the actuators illustrated herein can operate as a velocity pickoff. If the actuator is driven mechanically, it can generate a voltage proportional to velocity.

NOMENCLATURE

Symbol	Definition
a	polar length, radial direction
A	polar area at air gap
A_m	cross-sectional area of magnet
B	viscous-damping factor
$\mathcal{B}_c$	control component of flux density
$\mathcal{B}_m$	maximum allowable flux density
$\mathcal{B}_n$	flux density at neutral section of magnet
$\mathcal{B}_p$	polarizing component of flux density
C	capacitance
D	diameter
e	electromotive force
f	force
g_0	air-gap length, rotor in neutral position
G	conductance
h	height
$\mathcal{H}_m$	magnetic-field intensity at neutral section of magnet
i_c	control component of current
I_0	polarizing component of current
J	moment of inertia
k	surface-heat-dissipation factor
K_0	mechanical elastance
K_1	electromechanical coupling constant
K_2	electromagnetic elastance
K_e	effective or net elastance
l	length

Symbol	Definition
l_0	rotor overlap with stator pole
l_m	length of magnet
L_w	self-inductance of winding
M	mass
N	number of turns
$\mathcal{P}$	permeance
r	radius
R	resistance
R_c	core-loss resistance
R_w	resistance of winding
$\mathcal{R}_0$	reluctance of working gap, rotor in neutral position
$\mathcal{R}_n$	reluctance of nonworking gap
t	thickness
T	temperature
v	voltage
w	width
x	linear displacement
x	auxiliary design parameter
y	auxiliary design parameter
z	auxiliary design parameter
Γ	reciprocal inductance
δ	density
η	magnetic-leakage coefficient
θ	angular displacement
θ_0	g_0/r
μ_a	permeability of air
ρ	resistivity
σ	space factor of coil
τ	torque
ω_0	resonant angular frequency

ADDITIONAL REFERENCES

Blitzer, S. D., "The Design of a Torque Motor for Hydraulic Servo Applications," S.M. Thesis, Department of Electrical Engineering, Massachusetts Institute of Technology, Cambridge, Mass., 1948.

Dunn, J. F., Jr., "A Study of Permanent-Magnet Torque Motors," S.M. Thesis, Department of Mechanical Engineering, Massachusetts Institute of Technology, Cambridge, Mass., 1953.

Frazier, R. H., and R. D. Atchley, "An Electric Valve Actuator for Hydraulic Servo-mechanism," Meteor Report No. 42, Dynamic Analysis and Control Laboratory, Massachusetts Institute of Technology, Cambridge, Mass., 1949.

Massachusetts Institute of Technology Electrical Engineering Staff, *Magnetic Circuits and Transformers*, John Wiley and Sons, New York, 1943.

12

J. F. Blackburn
J. L. Coakley
F. D. Ezekiel

Transient Forces and Valve Instability

12.1. INTRODUCTION

Instability in a fluid-power drive or servo may arise from causes of two types: instability of the servo loop as a whole and instability of a portion of the loop, usually the control valve. Servo-loop instability is discussed elsewhere, both in this book (for example, in Chapter 14) and in many others, and will not be considered further in the present chapter. Instability or rough operation is occasionally caused by resonant oscillations of some part of the system in response to pump pulsations or similar periodic disturbances, but once this possibility is realized, both diagnosis and cure are straightforward if sometimes laborious. Control-valve instability, however, is frequently encountered, seldom understood, and sometimes very difficult to cure. We are far from a complete understanding of this phenomenon—or group of phenomena—but we do know something about it, and this chapter will attempt to summarize that knowledge.

The whole question of instability and oscillation may be considered in several different ways. One which seems to be palatable to the average engineer is Barkhausen's criterion. This states that in order for a self-sustaining oscillation to occur the system must contain both a source of energy and a feedback loop which controls the flow of energy to the load, and that when an oscillation occurs with constant amplitude the gain around the loop will be unity and the frequency will be that for which the phase shift around the loop is 360°. If the gain is greater than unity, the amplitude of the oscillation will increase until, in physical systems, some kind of limiting occurs which reduces the gain to unity. If it is initially less than unity, the amplitude will steadily decrease.

The foregoing statement was phrased in the language of sinusoidal oscillations but can easily be generalized to include such cases as that of sharp pulses traveling along transmission lines.

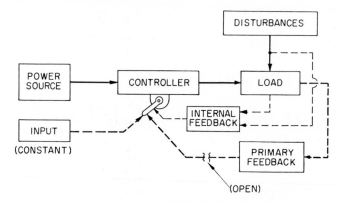

Fig. 12.1. System with internal feedback.

Consider the system shown symbolically in Fig. 12.1. It contains a source of power and a controller which determines the rate of flow of the power to a load. The load may also be affected by external disturbances. The flow of power is indicated by solid lines. The power produces some kind of effect as it is absorbed by the load, and this effect can be considered as information, whose flow is indicated by dashed lines. Another source of information is an input signal. If the system is a "controller," the input signal will probably have a constant value; if it is a "servo," the input will vary. For the purposes of this chapter, we shall assume that the primary feedback loop is opened and the input is held constant. If there were no other feedback paths, the system would then be stable since the gain around an opened loop is obviously zero. In practically all physical systems, however, at least one internal feedback path exists and is a potential cause of regeneration or oscillation. In the following sections we shall discuss certain possible feedback paths associated with control valves.

In a valve-controlled system the one element with by far the highest gain from input to output is almost always the valve itself, and therefore stray feedback around the valve is most likely to cause oscillation. Of course other elements such as the controller and the power supply may be affected directly by external disturbances, and the flow of information may be direct from disturbance source to controller input, or it may take place via some other path. Detailed enumeration of the possible configurations would be superfluous.

The actual effect on the controller setting which a stray feedback can produce depends both upon the strength of the fed-back signal (the attenuation in the feedback path) and upon the "stiffness" of the controller input at the point of entry of the feedback. In the case of a valve the input point is the spool, flapper, poppet, or other moving member which controls the valve opening. If this member is positioned by a very stiff device, such as a strong spring, a cam or dog, or a hydraulic positioner, even a strong feedback will have little effect and the system will probably be stable. On the other hand, if the positioning means is "soft," as is an electromagnetic or a pneumatic valve actuator, a comparatively small fed-back force will have a considerable effect on the valve position, and the chances of instability or oscillation will be very much greater. Valve oscillation has always been troublesome in both pneumatic and hydraulic systems, and with the recent trend to higher-gain and far more sensitive valves operating at high pressures and actuated by soft actuators, the problem has become even more serious.

The gain around the internal loop depends upon the stiffness of the controller input, the gain of the forward path, and the gain of the feedback path. Input stiffness usually depends upon the method of actuation and in many cases cannot be much changed. Forward gain usually must be kept high to obtain the desired system performance, and so the only remaining way to ensure controller stability is to decrease the internal feedback gain.

We shall confine our attention to feedback paths which travel from the flowing medium to the valve input, that is, with forces exerted upon the valve by the liquid or gaseous power-transmitting fluid. Chapter 10 has discussed certain steady-state forces and we shall now continue the discussion.

For expository purposes it is convenient to classify valve instabilities into four types, with the usual proviso that, like all classifications, this one is arbitrary and many intermediate or overlapping cases will arise. The four types are:

1. Instability of flow, which persists even if the moving member of the valve is rigidly clamped.

2. Instability caused by steady-state forces, due either to static pressure unbalance or to momentum effects.

3. Instability due to forces associated with transient-flow effects.

4. Instability dependent principally upon resonant or travel-time effects external to the valve itself.

Undoubtedly there are other types also, but little seems to be known about them at present.

12.2. FLOW INSTABILITY

The phenomenon of an unstable flow is very common, so much so that nearly all flows except those of highly viscous fluids are unstable at least in part. When a fluid flows past an obstacle, or even along a smooth wall,

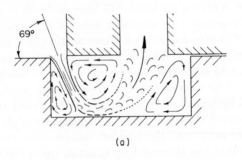

(a)

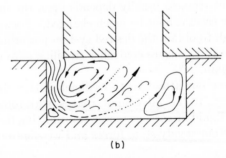

(b)

Fig. 12.2. "Normal" and "abnormal" positions of jet in liquid-filled chamber.

(a) "Normal" position as predicted by von Mises' theory.
(b) "Abnormal" position with jet clinging to wall.

at more than a certain minimum velocity, the flow is unstable, and this instability manifests itself by the formation of eddies and turbulence. When a jet exhausts from an orifice into a downstream chamber, as in Fig. 12.2a, it usually creates large eddies upstream of the point where the jet itself breaks up into a turbulent mass. These large eddies act in a sense as roller bearings upon which the jet rides, and serve to eliminate the discontinuity in velocity at the jet boundary which would occur with the hypothetical nonviscous fluid.

Depending upon the geometry of the upstream and downstream chambers, the properties of the fluid, and the conditions of flow, these eddies may or may not be stable and anchored in place. If they are not, the jet may become "nervous"; it may follow the path assigned to it by a simplified theory such as that of von Mises,[1] or it may cling to one or the other side wall [2] or take some other preferred position, or it may alternate between two or more preferred positions. The change from one flow configuration to another usually involves at least a transient change in flow rate or pressure drop. One of these "abnormal" flow patterns is sketched in Fig. 12.2b.

Rapidly varying flow configurations are common in hydraulic equipment, but in most cases the frequencies involved are high and the only effect of consequence is the generation of noise. This may be annoying but seldom affects system performance seriously. In rare cases, of course, the amplitude of the noise is high enough and its frequency low enough to degrade the performance. This is particularly true when the system contains resonant elements with fairly low damping, such as loops of pipe,[3] whose frequencies coincide with those of eddy formation. The case is much more serious when the eddy-induced oscillations are those of a control element such as a relief valve, which may amplify the noise to serious or even to destructive proportions.

[1] See Sec. 10.32, fn. 15. The "nervousness" of the jet was beautifully shown in the glass-sided model used in Lee's original work on valve forces; the jet could be made visible very easily by viewing between crossed Polaroids. This birefringence due to flow occurs with some oils such as Esso Univis P-48 but not with others such as the British DTD-545 or with water. If the oil contains dissolved air, the jet will often be made visible by a cloud of fine bubbles that forms almost instantly when the pressure is released as the liquid passes through the orifice. It is also possible to introduce bubbles upstream as a means of visualization. The same phenomenon can be observed with gases by the standard interferometric techniques.

[2] Markus Reiner, "The Teapot Effect . . . A Problem," *Phys. Today* (September 1956), pp. 16–20; J. B. Keller, "Teapot Effect," *J. Appl. Phys.*, Vol. 28 (August 1957), pp. 859–865.

[3] Hans Krug, *Das Flüssigkeitsgetreibe bei spanenden Werkzeugmaschinen*, Springer-Verlag, Berlin/Göttingen/Heidelberg, 1951, Erster Teil, Sec. C., "Die Schwingungen im Leitungssystem," pp. 30–35. This applies specifically to pipe vibrations excited by pump pulsations, but the same phenomenon can also occur with eddy-induced pulsations.

12.3. INSTABILITY CAUSED BY STEADY-STATE FORCES

Valve instabilities are not necessarily oscillatory, for in many cases a valve spool may be driven against its stops in one or the other direction by forces that persist even when the flow rate is constant. We may distinguish two types: static pressure unbalance, and instability due to steady-state flow forces.

12.31. Static Pressure Unbalance

In valves with this type of instability the pressure difference across the valve acts upon some equivalent piston area to produce an opening or closing force on the valve. In what follows we shall consider a valve force positive if it tends to close the valve. Sliding valves are often balanced against static pressure unbalance; seating valves usually are not, and indeed it is almost impossible to get very accurate pressure balance with such a valve. This static unbalance is not necessarily bad, since it may be used in various ways. Perhaps the commonest example of such a use is the check valve, which is so connected that a positive pressure forces the ball against the seat, sealing off the flow, while a negative pressure drives the ball away from the seat and the fluid flows freely. The common globe valve may be installed in either direction; domestic faucets and the like are usually installed so that the stem closes against the pressure. Any looseness or incomplete closure will then permit the valve to leak, but on the other hand with usual designs this installation keeps the full line pressure from the stem packing and allows the use of the very cheap construction ordinarily encountered. Industrial globe and angle valves are often installed in the opposite sense so that if the stem or sealing washer is slightly loose, the valve will fail safe since it will act as a check valve. Poppet valves in gasoline engines are always installed in this way so that they will seal tightly during the compression and explosion strokes. Flapper-and-nozzle valves are practically always installed the other way since they are not intended to close tightly.

A valve may be made intentionally unbalanced in order to obtain positive reversal or "toggle" action, as in Fig. 12.3. Here the valve will be in unstable equilibrium when it is exactly centered; any small displacement will unbalance the pressures acting on the end faces and will tend to move the spool even farther in the same direction. If this motion is not restrained, the spool will jump to and remain in one or the other extreme position at all times. This toggle action is convenient, for example, to reverse the direction of drive of a machine-tool table. With

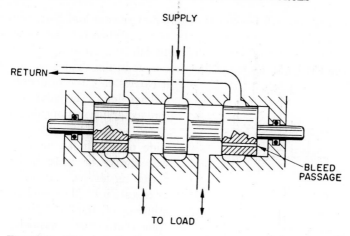

Fig. 12.3. "Toggle-action" valve, shown in unstable center position.

reasonable care in construction, it is possible to get positive reversal of the valve consistent from one stroke to the next within much less than 0.0001 in.[4]

12.32. Steady-Flow Instability

Even in cases where a valve is statically balanced for the no-flow condition, it may be unbalanced in the direction of instability when fluid flows through it.   Consider the case sketched in Fig. 12.4.   Here the

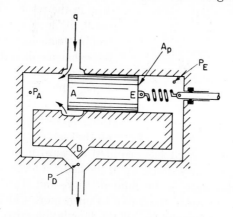

Fig. 12.4. Valve with static pressure balance.

[4] A. H. Dall, "Machine Hydraulics," *Machine Design*, Vol. 18 (October 1946), p. 87.

valve will be statically balanced at zero flow since A and E are connected and the pressures P_a and P_e are the same. When the valve opens and flow starts from A to D, however, there will be a pressure drop of $P_a - P_d$ in the left-hand discharge line, but none in the right-hand line since it carries only the negligibly small leakage flow. The valve will therefore experience an opening force of $A_p(P_a - P_d)$ and will tend to jump wide open. In an actual case it will always be restrained by its actuating means, and its stability or instability will depend upon the stiffness of the actuating means (plus that of the "Bernoulli" force discussed in Sec. 10.32). Since the flow in the conduit A–D will always be highly turbulent in practical cases, the pressure drop will increase roughly as the square of the flow rate, q, and if the conduit is excessively long, narrow, or tortuous, this steady-state drop may be sufficient to overpower a soft actuator.

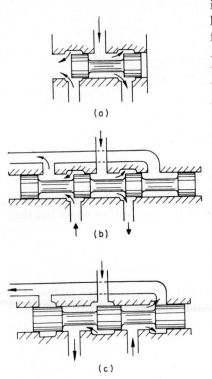

Fig. 12.5. Use of sealing lands.

(a) Two-land valve, unbalanced against "skyrocket effect."

(b) Four-land valve, balanced.

(c) Three-land valve, balanced.

Even if the steady-flow pressure drop in the return line is negligibly small, it is still possible to have trouble from instability when the jet exhausts directly into a large chamber. In the valve of Fig. 12.5a, this is the case and we shall assume that there is no frictional pressure drop downstream of the valve. As the fluid leaves the left end of the bore, it has acquired leftward momentum from the piston, and consequently exerts upon the latter a force equal to $q^2\rho/A_p$, where q is the flow rate, ρ the fluid density, and A_p the piston area. This force tends to cause the valve to open more widely, and thus to increase q and cause instability. If the flow is constricted by having to exhaust through a narrow downstream passage, the force is correspondingly increased and may be serious. In the original Model 1 valves for the DACL flight table we used some existing ⁵⁄₁₆-in. two-land pilot valves and connected them to the torque motors with rather long and compliant push wires. The resulting instability was sometimes almost catastrophic. This effect can also be serious with gaseous media; in a torpedo control mechanism we tried to use a sliding-plate two-land valve with a 500-psi air supply. Both ends of the valve exhausted directly into free air. When the supply was first turned on the valve plate happened to be centered, but as soon as it was slightly displaced it shot out of the block and across the room. The remedy for this type of instability (and for all others that depend upon the action of a pressure upon the valve as a piston) is obvious; simply add sealing lands to the ends of the spool and take the discharge radially out through the block, as in Fig. 12.5b or 12.5c.

This "skyrocket effect" type of instability is frequently encountered. One homely example is the chattering float valve in a tank-type toilet. Here the soft sealing washer of the poppet valve initially has a shape something like the sketch of Fig. 12.6a, and while the turbulent flow into the tank may be somewhat noisy, the valve will not be unstable. After months or years of sitting pressed against its seat, however, it may cold flow into a shape something like Fig. 12.6b. Here the water leaving the valve has made a turn of nearly 180° and the resulting momentum force on the valve has greatly increased. With the contour shown the force may change very rapidly with flow and poppet position

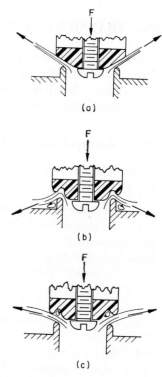

Fig. 12.6. Effect of deformation of soft poppet.

(a) New washer, reaction force small.

(b) Deformed washer, valve partly closed, force large.

(c) Deformed washer, valve wide open, force moderate.

for certain openings, and this high rate of change may so increase the feedback that a most annoying and even destructive oscillation results. This condition can be cured by partially closing the supply-line stop valve, which adds damping to the system. Naturally a better cure is to replace the damaged washer with a new one. Essentially the same type of instability sometimes occurs with old faucets, where wear of the stem and threads combines with the spring of the stem packing and the shape of the washer to promote oscillation.

12.4. TRANSIENT-FLOW INSTABILITY

The types of instability just discussed are fairly easy to visualize since they depend either upon static pressure unbalances or upon steady-flow phenomena. Another class is somewhat less obvious since it depends upon rate of change of flow. Two examples will be given.

12.41. Return-Line Acceleration Instability

Consider the valve of Fig. 12.7, which does not have sealing lands. The piston is acted upon by the following forces:

F_1, the static pressure unbalance, which is zero for this construction.

F_2, the pressure unbalance due to the steady-state pressure drop in the downstream conduit, $A-D$. Since the flow in this conduit is highly turbulent, this drop is proportional to q^2.

F_3, the steady-state "Bernoulli" force discussed in Sec. 10.32.

F_4, the force required to accelerate the fluid in $A-D$.

F_5, the force required to accelerate the piston.

F_6, the force required to accelerate the fluid in the balance conduit, $E-D$.

F_7, the force required to overcome the viscous friction in the clearance space between piston and bore.

F_8, the force exerted on the piston by its external actuator through its spring of stiffness, k_8.

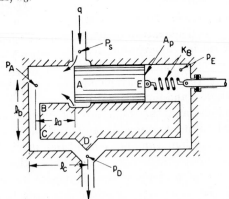

Fig. 12.7. Statically balanced, dynamically unbalanced valve.

A complete solution of the problem would involve writing down the expressions for all these forces, adding and equating to zero, solving for the value of the valve opening, x, at the equilibrium point, and then evaluating the significant parameters at this operating point, x_o. At the moment, however, we are interested only in *dynamic* stability and may therefore assume that the valve will be statically stable at x_o. We therefore omit the determination of x_o and investigate the coefficients of the dynamic equation.

We assume further:

1. The fluid is incompressible.
2. The supply pressure, P_s, is constant.
3. The maximum pressure in the downstream valve chamber is small compared to P_s so that the flow rate, q, depends only on P_s and on the valve opening, x.
4. The leakage flow past the piston is negligibly small.
5. The length of the downstream chamber is so great that the jet has room to break down into a turbulent mass at a uniform pressure, P_A, which will be taken as "the" pressure of the portion A–B of the downstream conduit.

Forces F_2, F_3, and F_8 are functions of x and therefore represent springs. The first acts as a negative spring whose stiffness is approximately proportional to x, and the last two as positive springs. We have assumed static stability, and so the positive springs must be the stronger. We shall assume that the resultant spring has an effective stiffness of K_o at the operating point.

Forces F_5 and F_6 are inertial forces which are proportional to $(d^2x)/(dt^2)$. We assume an equivalent mass, M_e, which is equal to the actual mass of the piston and attached mechanical parts, plus the effective mass of the fluid in the passage E–D.

The two remaining forces represent damping terms. The damping contributed by F_7 is positive and therefore stabilizing, being due to friction. That of F_4 is negative, as can be shown by the following argument.

Suppose that the downstream conduit is represented as a series of n sections, each of length l_j and of constant area A_j, where j is the number of the section. Consider an elementary length, dl, of the j'th section. The mass of fluid enclosed in this elementary volume will be $\rho A_j dl$ and it will be moving at the velocity q/A_j. If the flow varies at the rate dq/dt, the acceleration of the mass is $(1/A_j)(dq/dt)$ and this acceleration requires a force $\rho dl(dq/dt)$. This force is exerted by a pressure dp acting between the faces of the elementary slice of fluid over the area A_j of the cross section.   Integrating over section j of the conduit gives

$$\Delta p_j = \rho \frac{l_j}{A_j} \frac{dq}{dt} \tag{12.1}$$

and summing over all the n sections of the conduit from A to D gives

$$\Delta p = \rho \frac{dq}{dt} \sum_{j=1}^{n} \frac{l_j}{A_j} = \frac{dq}{dt} \rho \Sigma_1 \tag{12.2}$$

where Σ_1 is a factor which depends only on the geometry of the conduit.

We assume that the steady-state orifice equation holds for the metering orifice up to rates of change of flow considerably higher than any we shall encounter.   If this is true, we can differentiate it to get

$$\frac{dq}{dt} = \frac{d}{dt} (C_d w x \sqrt{2P_s/\rho}) = C_d w \sqrt{2P_s/\rho}\, \frac{dx}{dt} \tag{12.3}$$

where C_d = discharge coefficient

w = peripheral width of metering orifice, in.

P_s = supply pressure, psi

ρ = fluid density, lb $\sec^2/\text{in.}^4$

x = axial dimension of metering orifice, in.

Combining Eqs. 12.2 and 12.3 and multiplying by the piston area, A_p, to get the force on the piston, we have

$$F_4 = -A_p \Sigma_1 C_d w \sqrt{2\rho P_s}\, \frac{dx}{dt} = k_4 \frac{dx}{dt} \tag{12.4}$$

The positive damping term will be

$$F_7 = \frac{\mu A_s}{h} \frac{dx}{dt} = k_7 \frac{dx}{dt} \tag{12.5}$$

where μ = fluid viscosity, lb $\sec/\text{in.}^2$

A_s = surface area of clearance space, in.^{2}

h = radial clearance, in. (assumed uniform)

We can now combine the several terms to get the final force equation

$$M_e \frac{d^2x}{dt^2} + B_e \frac{dx}{dt} + K_e x = 0 \tag{12.6}$$

where $B_e = k_4 + k_7$.   This is the standard form of a second-order system, which will be dynamically stable or unstable accordingly as B_e is positive or negative.   Since k_4 is inherently negative because the chamber pressure rises and tends to open the valve when the flow is increased, the system will be unstable if its absolute magnitude is greater than that of k_7.

To illustrate the relative magnitudes of these two terms we may consider the case of a typical small control valve. We assume the parameters to be:

$$w = 1 \text{ in.}$$

$$A_p = 0.08 \text{ in.}^2$$

$$C_d = 0.625$$

$$P_s = 1000 \text{ psi}$$

$$\rho = 8 \times 10^{-5} \text{ lb sec}^2/\text{in.}^4$$

$$\mu = 1.6 \times 10^{-6} \text{ lb sec/in.}^2$$

$$\frac{dx}{dt} = 10 \text{ in./sec}$$

$$A_s = 0.5 \text{ in.}^2$$

$$h = 0.0001 \text{ in.}$$

We assume that the downstream conduit has a diameter of $\frac{1}{4}$ in. and a length of 3 in. so that $\Sigma_1 = 60$ in.$^{-1}$ Substituting these numbers into Eqs. 12.4 and 12.5, we find that $F_7 = 0.08$ lb and $F_4 = 12.4$ lb, or that F_4 is over 150 times larger than F_7. The valve is therefore violently unstable unless some effective additional positive damping is provided, or the construction is changed. The easiest way to provide the damping is to introduce a restriction into the balance conduit, though this makes the valve considerably slower acting. It is far better, if possible, to use sealing lands as in Fig. 12.5.

12.42. Damping Length and Instability

Even when sealing lands are used and the fluid does not have access to the ends of the valve piston, the valve may still be unstable. This may be shown as follows.[5]

[5] S.-Y. Lee and J. F. Blackburn, "Contributions to Hydraulic Control 2: Transient-Flow Forces and Valve Instability," *Trans. ASME*, Vol. 74 (August 1952), pp. 1013–1016. Note that the convention for the sign of the damping length adopted in this book is opposite to that used in the article so that here a *positive* damping length gives *positive* damping and a stable valve.

Consider the case of Fig. 12.8, in which a length of pipe is so supported by flexible couplings that it can move in an axial direction. As long as the flow remains constant, the horizontal component of the force F_1 which halts the upward movement of the fluid and directs it to the right is balanced by the horizontal component of F_2 which bends the flow downward. (We assume for simplicity that the flows in the legs are parallel.) There is no net horizontal force on A–B.

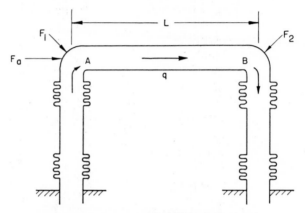

Fig. 12.8. Forces caused by accelerated flow.

F_1, F_2, steady-state forces.

F_a, acceleration force.

Suppose now that the flow is increasing at the rate dq/dt. The mass of fluid between A and B must be accelerated in a horizontal direction and therefore exerts a force on the wall of the tube at A. The tube can only be kept stationary by an equal force, F_a, pushing it toward the right. It can be shown by using the momentum principle that this force is given by

$$F_a = \rho L \frac{dq}{dt} \qquad (12.7)$$

where L is the horizontal distance between the centers of the incoming and outgoing flows. This derivation is very general; it assumes only that the fluid is incompressible and that the flow-velocity profiles across the inlet and outlet change only in magnitude and not in shape with changes in flow rate. The force is completely independent of the shape of the container or of the nature of the flow pattern within it.

When the derivation given in the Appendix of the reference in fn. 5

is applied to the valve chamber of Fig. 12.9, the resulting force equation is

$$F = -\rho q u \cos \theta + \rho L \frac{dq}{dt} \qquad (12.8)$$

where u is the average velocity of the flow through the metering orifice. If we assume that the orifice equation holds and that the pressure drop is constant,

$$\frac{dq}{dt} = C_d w \sqrt{2p/\rho} \frac{dx}{dt} \qquad (12.9)$$

where p is the drop across the orifice and the other symbols are as in the previous section

But also $u = q/wx$ and $q^2 = 2C_d{}^2 w^2 x^2 p/\rho$, and when these expressions are substituted in Eq. 12.8,

$$F = -(2C_d{}^2 w p \cos \theta)x - (C_d w \sqrt{2p\rho} \, L) \frac{dx}{dt} \qquad (12.10)$$

Thus the effect of the oil flowing past the corner of the piston land is that of a linear spring plus a linear damping term. The spring always

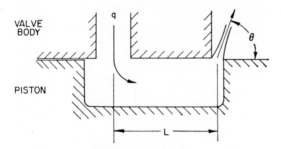

VALVE BODY

PISTON

Fig. 12.9. Damping length of a valve chamber.

tries to close the orifice, and therefore tends to make the valve *statically* stable. The damping will be positive and stabilizing, or negative and destabilizing, depending on the sign of L. With oil flowing *outwards* through the metering orifice, as shown, L is *positive*, the damping is *positive*, and the valve is *dynamically stable*. Reversing the direction of flow reverses the sign of L and the damping, and the valve tends to be *dynamically* unstable even though the sign of the spring is unchanged and the valve is still *statically* stable.

In a conventional four-way valve there are two orifices in series, and with most constructions the fluid flows outward through one and inward

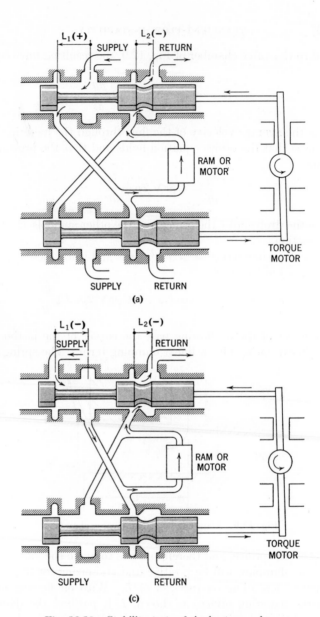

Fig. 12.10. Stability tests of single-stage valve.

Config-uration	L_1, in.	L_2, in.	Net L, in.	Stability	Critical Pressure, psi	Steady-State Balance
(a)	+0.33	−0.08	+0.25	stable	...	yes
(b)	−0.33	+0.08	−0.25	unstable	1100	no
(c)	−0.33	−0.08	−0.41	unstable	500	yes
(d)	+0.33	+0.08	+0.41	stable	...	no

374

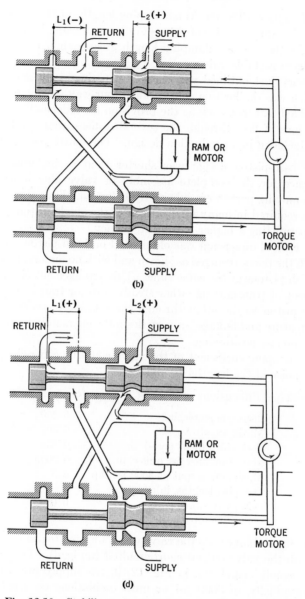

Fig. 12.10. Stability tests of single-stage valve (*continued*).

through the other. The effective damping length for the whole valve is the algebraic sum of the L's of the separate orifices, and it is very desirable to make the positive damping length greater than the negative in order to have a stable valve.

The theory just outlined was checked experimentally with a special valve which was similar to a conventional four-way servovalve but had certain added features, such as the provision for independent control of the directions of flow through the several orifices, that would not normally be included in a production design. These features were:

1. The twin-piston design was shorter than a single-piston valve, saving space for the complete assembly of valve plus torque motor.

2. The moving parts were inherently balanced, eliminating the counterweight required in many applications of the single-piston design and thereby permitting a higher natural frequency.

3. One critical dimension was eliminated on both piston and sleeve.

4. With the ports arranged as shown and with no underlap, fluid never passed both pistons at the same time. This arrangement eliminated the requirement of transmitting either steady-state or transient flow forces from one piston to the other through the relatively compliant torquemotor armature and linkage—a very desirable characteristic for a device requiring precise positioning.

5. The damping coefficient of the valve could be controlled easily by suitable choice of the lengths L_1 and L_2.

The principal dimensions of interest were as follows:

Piston diameter, in.	0.250
Maximum stroke, in.	0.007
Port width (360° ports), in.	0.785
Maximum area per orifice, in.^{2}	0.0055
Maximum output at 3000 psi, hp	3.5
Damping lengths, in.:	
Negative-force ports	0.08
Positive-force ports	0.33

During the stability tests the valve was connected to a standard torque motor with the valve output shortcircuited (no load). For each of the four flow conditions shown in Fig. 12.10, the supply pressure was increased gradually to 3000 psi or until oscillations occurred. It was found that for the oscillatory configurations the onset of oscillations was fairly consistent, the critical pressure varying over a range of perhaps 5 to 10 per cent. The test results showed the following:

1. Configuration a had a small negative damping in the downstream port and a larger positive damping upstream, since L_1 was purposely

made larger than L_2. This was the normal configuration for the valve; the steady-state force was compensated (this compensation has already been described in Sec. 10.321) and the valve was stable for all pressures up to 3000 psi.

2. For configuration b the supply and return lines were interchanged, reversing all flows through the valve and load. The negative-force port now acted as an ordinary square-land port and there was no steady-state force compensation. The small positive damping only partly compensated the larger negative damping of L_1; the net negative damping overcame the viscous damping, and the valve oscillated violently for all pressures above 1100 psi.

3. Configuration c was similar to a except for the reversal of flow through the left-hand port. The steady-state force was compensated but both damping terms were negative, and the valve oscillated for all pressures above 500 psi.

4. The supply and return lines were again interchanged, giving configuration d. This was the most stable of the four, since both damping terms were positive, but the steady-state compensation was destroyed by reversing the flow through the upstream port.

The viscous damping forces were not known and it was therefore impossible to predict the values of the critical pressures above which the valve would oscillate. The inverse process of calculating the effective piston-to-sleeve clearance from the observed critical pressures gave clearance values that were within the expected range, so the agreement of experiment and theory was at least semiquantitative. The qualitative agreement with the theoretical predictions was excellent.

12.5. RESONANT INSTABILITY

Even in the absence of all of the types of instability which have already been discussed, a valve—or more accurately the system of which the valve is a part—will often oscillate. Instability of this type (or types) is often very difficult to diagnose since it appears and disappears erratically, and one is always left with the worry that an apparent cure of the particular system may not be permanent. The following sections may help in the diagnosis of such cases, and may also increase the sense of security of the practitioner of the hydraulic art after he has diagnosed and cured one of them.

Oscillation of a valve which should be stable according to the criteria of the previous sections has led to investigations at DACL and elsewhere. Before describing these works, however, it will be instructive to consider in some detail the characteristics of the feedback mechanism which produces the oscillations.

The argument of the next two sections rests upon the fact that it is often convenient to consider many systems—including the ones with which we are now concerned—as being made up of two parts, a generator and a load. In the hydraulic case the generator may include the power supply and the valve, and the transducer (hydraulic motor or ram) and its attached mechanical load, plus the fluid in the various connecting pipes, plus the load disturbances, if there are any, make up the load. The pressure difference between the two connecting pipes is obviously the same for the generator as for the load, and the flow from the generator is equal to the flow into the load for each pipe separately. The flows are not necessarily the same for the two pipes since an additional component of flow may be required to change the pressure of a trapped volume on one or both lines. In this connection, it must be remembered that in drawing equivalent circuits, the capacitance which represents a charged volume must always have one terminal grounded, just as must the capacitance which represents a mass in the Firestone mobility analogue of mechanical quantities. When a difference due to charging flow does exist between the flow in the two pipes, however, it can always be taken into account by adding a suitable volume to one or both, as required, and considering this volume to be part of the generator or the load.

The method to be followed in the analysis can be applied to oscillators of many different types. It has been applied to the case of a poppet valve terminating a resonant compressed-air line,[6] but in what follows we shall confine our attention to the case of a square-land sliding valve with rectangular ports. This case is very important in practice, and the laws of flow and force are so well established for this type of valve that the results to be presented should inspire some confidence. Experimental verification of these results is somewhat meager, but what is available does agree with the theory.

12.51. The Equivalent Generator—D-C Characteristics

The system to be considered is shown schematically in Fig. 12.11. To avoid obscuring the argument we assume that there are no charging flows and that the system is a simple series circuit with the flow rate everywhere the same. In an actual case the cut between generator and

---

[6] E. Frederiksen, "Some Problems of Unstable Air Flow," *Ingenjøren*, Nr. 2, 12 Januar 1957, pp. 74–83; P. Hansens Bogtrykkeri, København.

load can be taken at any convenient place, all admittance to the left of the cut A–A being charged to the generator and all to the right being part of the load. (Sometimes it is more convenient to consider impedances than admittances.)

Initially, consider only the d-c case, where all variables are changing so slowly with time that inertial and viscous forces can be neglected. The

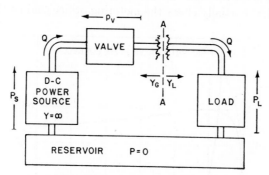

Fig. 12.11. Generator and load.

force equation for the valve becomes

$$Ep_vx + k(x - x_0) = 0 \qquad (12.11)$$

where $E = 2C_dw \cos \theta = 0.45w$ for the given valve, in.
C_d = discharge coefficient $\doteq 0.625$ for square lands
w = peripheral width of valve ports, in.
p_v = pressure drop across metering orifice, psi
x = axial displacement of valve stem, in.
x_0 = initial displacement when $p_v = 0$, in.
k = spring constant of actuator spring, lb/in.

E is the coefficient of the Bernoulli force.

The flow equation for a square-land valve is

$$q = Gx\sqrt{p_v} \qquad (12.12)$$

where q = flow rate, in.^{3}/sec
$G = C_dw\sqrt{2/\rho}$
ρ = fluid density, lb sec^2/in.^{4}

G is a fluid conductance coefficient for the valve orifice.

It will be convenient to define a dimensionless quantity, the feedback factor, which is physically the ratio of the Bernoulli spring rate to the mechanical spring rate. It is given by

$$\gamma = \frac{E p_v}{k} \tag{12.13}$$

When fluid flows through the metering orifice, it exerts a force on the valve spool that partially closes the metering orifice against the restraint

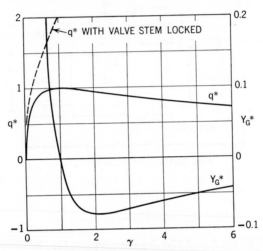

Fig. 12.12. Normalized d-c flow and conductance of valve.

of the actuator spring. The degree of closure depends upon γ, according to the expression

$$x = \frac{x_0}{1 + \gamma} \tag{12.14}$$

As γ increases with increasing pressure, the decrease in x decreases the flow rate below what it would be in the absence of the Bernoulli force, and the resulting flow rate can be found by combining the last three equations to get

$$q = G x_0 \sqrt{k/E} \left(\frac{\sqrt{\gamma}}{1 + \gamma} \right) \tag{12.15}$$

which is plotted in normalized form in Fig. 12.12. Here q^* is a dimensionless flow, obtained by dividing the actual flow by the peak flow, which occurs when $\gamma = 1$. As can be seen, the flow rate rises rapidly at first with increasing γ, goes through a flat maximum at $\gamma = 1$, and thereafter falls and approaches the γ-axis asymptotically.

We now define the hydraulic incremental conductance of the valve (the reciprocal of the hydraulic incremental resistance) as the derivative of flow rate with respect to valve pressure drop. This can be found by differentiating Eq. 12.15 and is

$$Y_G = \frac{Gx_0}{2} \sqrt{E/k} \left[\frac{(1 - \gamma)}{\sqrt{\gamma}(1 + \gamma)^2} \right] \qquad (12.16)$$

This can be normalized to a nondimensional conductance, Y_G^*, by dividing by the quantity outside the parentheses; Y_G^* also is plotted in Fig. 12.12. It starts at positive infinity, crosses the axis at $\gamma = 1$, and remains negative thereafter. It attains its maximum negative value when $\gamma = 1 + (2/\sqrt{3}) = 2.154$, at which time $Y_G^* = -0.0792$.

So far we have not specified how the pressure is established and therefore cannot define the value of γ. We now assume that the valve and load are supplied in series by a constant-pressure (infinite-admittance) source, as in Fig. 12.11. We may now draw the generator flow curve and superpose upon it the flow-pressure curve of the load, but with the pressure scale reversed and having its origin at P_s. This load curve will intersect the generator curve at one or more points. Each of these intersections represents a point of equilibrium, but the stability of the equilibrium is given by the following rules:

1. An intersection to the left of the maximum ($\gamma < 1$) is stable.

2. An intersection to the right of the maximum is stable if the slope of the load curve at the intersection is greater than the slope of the generator curve, that is, if $-Y_L > Y_G$. If the two curves are tangent, the system is metastable; if $-Y_L < Y_G$, the equilibrium is unstable and the system will jump to an adjacent stable point. Actually the slope and curvature of the descending leg are so small that this type of instability is very unlikely.

12.52. The Equivalent Generator—A-C Characteristics

It was assumed in the previous section that p_v, q, and x would change very slowly. In an actual system this is rarely the case, and both inertial and viscous forces are very important. The complete force equation for the valve then becomes

$$mD^2x + bDx + Ep_vx + k(x - x_0) + \rho LDq + F_{ext} = 0 \qquad (12.17)$$

where m = mass of valve spool and attachments, lb sec^2/in.

D = derivative with respect to time, $\dfrac{d}{dt}$, sec^{-1}

b = viscous-friction coefficient, lb sec/in.

L = damping length of valve per Eq. 12.10, in.

F_{ext} = externally applied force, lb

and the other symbols are as before.

We assume that both the steady-state force equation

$$F_B = (2C_d w \cos \theta)x p_v = E x p_v \tag{12.18}$$

and the flow equation (12.12) hold up to very high frequencies. Obviously this cannot be true beyond a certain point, but what experimental evidence we have indicates that the equations are valid at least up to several hundred cycles per second. It might be possible to develop a more general theory analogous to the extended theory of vacuum-tube oscillators which takes into account the electron transit time, but the practical value of such a theory appears to be small at present.

We now apply the perturbation method, allowing all the variables of Eqs. 12.12, 12.17, and 12.18 to undergo small changes, the increments in the original values being symbolized by Δ. For the stability study we assume that F_{ext} and x_0 remain constant. After a little rearrangement the equations in incremental form become

$$\frac{\Delta q}{Q} = \frac{\Delta x}{X} + \frac{\Delta p_v}{2P_v} \tag{12.19}$$

and

$$(mD^2 + bD + k + EP_v)\,\Delta x + EX\,\Delta p_v + \rho L D\,\Delta q = 0 \tag{12.20}$$

where Q, X, and P_v represent the constant (d-c) components of q, x, and p_v.

These two equations can be combined to eliminate x. After this has been done and the terms are rearranged,

$$Y_G = \frac{\Delta q}{\Delta p} = \frac{Q}{2P_v} Y_G{}^* = \frac{Q}{2P_v} \cdot \frac{D^2 + \left(\dfrac{b}{m}\right)D + \left(\dfrac{k}{m} - \dfrac{EP_v}{m}\right)}{D^2 + \left(\dfrac{b}{m} + \dfrac{\rho L Q}{mX}\right)D + \left(\dfrac{k}{m} + \dfrac{EP_v}{m}\right)} \tag{12.21}$$

Since we are considering small steady-state oscillations in a linearized system, we may replace the operational symbol D by $j\omega$, where j is the imaginary operator and ω the angular frequency. We now define four dimensionless quantities:

$$Y_G{}^* = \frac{2P_v}{Q} Y_G = \text{normalized complex generator admittance} \quad (12.22)$$

$$\eta_G = \frac{\omega}{\sqrt{k/m}} = \frac{\text{frequency}}{\text{undamped mechanical natural frequency of valve}}$$

$$(12.23)$$

$$\alpha = \frac{\rho L Q}{bX} = \frac{\text{fluid-acceleration damping}}{\text{viscous damping}} \quad (12.24)$$

$$\zeta = \frac{b}{2\sqrt{km}} = \text{viscous damping constant} \quad (12.25)$$

Substituting these four equations plus Eq. 12.13 into Eq. 12.21 and rearranging give

$$Y_G{}^* = \frac{(1 - \eta^2 - \gamma) + j2\zeta\eta}{(1 - \eta^2 + \gamma) + j2\zeta\eta(1 + \alpha)} \quad (12.26)$$

Rationalizing this equation gives us the final expression for the normalized complex admittance of the generator:

$$Y_G{}^* = \frac{(1 - \eta^2)^2 + 4\eta^2\zeta^2(1 + \alpha) - \gamma^2}{(1 - \eta^2)^2 + 2\gamma(1 - \eta^2) + 4\eta^2\zeta^2(1 + \alpha)^2 + \gamma^2}$$

$$+ j \frac{2\zeta\eta[\gamma(2 + \alpha) - \alpha(1 - \eta^2)]}{(1 - \eta^2)^2 + 2\gamma(1 - \eta^2) + 4\eta^2\zeta^2(1 + \alpha)^2 + \gamma^2}$$

$$= G_G{}^* + jB_G{}^* \quad (12.27)$$

Of the quantities involved in this equation, η and ζ are always positive. The feedback factor, γ, is usually positive, though it may be negative for an overcompensated valve, as shown by the curves of Fig. 10.16. The acceleration damping factor, α, may have either sign, though in a well-designed valve it will be zero or positive. If negative, it may cause valve oscillations even if γ is zero, as was shown in Sec. 12.42. The denominator of Eq. 12.27 is always positive. The imaginary part, $B_G{}^*$, of the impedance may have either sign, but this is not significant in determining the stability of the system. The important quantity is γ, which will make $G_G{}^*$ negative if it is large enough.

The effect of γ on the likelihood of oscillation can be illustrated as follows. Assuming that there are no other sources of positive feedback such as a negative α, and assuming also that the (passive) load is always such as to make oscillation most likely, an expression can be derived for the value of γ which is just sufficient to cause oscillation. It is

$$\gamma_c{}^2 = \eta_c{}^4 + [4\zeta^2(1 + \alpha) - 2]\eta_c{}^2 + 1 \qquad (12.28)$$

and is obtained from Eq. 12.27 by setting $G_G{}^* = 0$, which is one condi-

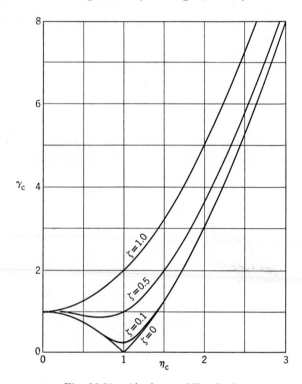

Fig. 12.13. Absolute-stability limits.

tion for the onset of oscillation. The subscript c denotes this condition of incipient oscillation.

Equation 12.28 is plotted in Fig. 12.13 for several values of the quadratic damping factor, ζ. Here, as throughout the rest of the section, α is taken as zero to reduce the number of parameters. This is reasonable since the effect of α is not large, and to a first approximation it can be considered as merely increasing ζ somewhat. It varies with p_v, but only as the square root rather than as the first power, like γ. It can be seen from Fig. 12.13 that for large damping γ_c increases continuously with frequency, but that for a valve with small damping it dips to a minimum near $\eta = 1$ and rises rapidly thereafter. This behavior is obvious from physical considerations, since a sharply tuned system oscillates easily but only in the neighborhood of resonance.

The curves of Fig. 12.13 represent boundaries of the region of absolute stability; oscillation is impossible in the region below the curve for any passive load whatever. Above the curve, oscillation may or may not occur depending upon the parameters of the load. Similar boundary curves can be drawn for given combinations of generator and load, as will be seen later.

From Eqs. 12.13 and 12.18,

$$\gamma = \frac{2C_d w \cos \theta}{k} p_v = 0.45 \frac{w}{k} P_v \qquad (12.29)$$

For a given valve and actuator spring, all these quantities except P_v (the average pressure drop across the valve orifice) are fixed, so that γ is proportional to P_v. This fact explains the frequent observation that a valve which is stable at one supply pressure may be very unstable at a somewhat higher pressure. This condition occurs not only with spool valves, but also with other types,[7,8] and the mechanism in these cases is presumably much the same. As a matter of fact, most loads are more or less linear so that the drop across the load increases roughly as the flow. The valve, however, is nonlinear, and as the supply pressure is increased, other things being equal, it will take a larger and larger fraction of the total pressure, and γ will increase faster than linearly with P_s.

γ is inversely proportional to the actuator stiffness, k, and as has been stated, one of the most effective ways of stabilizing a valve is to make this stiffness as great as possible. This is one more argument, and a potent one, for the use of electrohydraulic actuation as described in Chapter 13.

This statement can be supported by a numerical example. Assume that a 360°-port, 5/16-in.-diameter valve is to be driven by a typical torque motor with a spring constant of 325 lb/in. Then according to Eq. 12.29, $\gamma = 1$ for $p_v = 740$ psi. This condition is made more serious by higher supply pressures, by operation in such a way that the drop across the valve can sometimes be considerably greater than the supply pressure, and by the decrease in effective stiffness that is characteristic of many electromagnetic actuators near the ends of their strokes.

[7] G. Reethof, "On the Dynamics of Pressure-Controlled Hydraulic Systems," Paper No. 54-SA-7, presented at ASME meeting, Pittsburgh, June, 1954, and also as Secs. 17.3 through 17.37 of this book.

[8] See Frederiksen, fn. 6.

Roughly speaking, a γ of 1 will often be reached; a γ of 10 should be very rare.

The viscous damping coefficient, ζ, is usually small and depends greatly upon temperature in hydraulic valves, since the viscosities of liquids vary rapidly with temperature. With gases (if the viscous friction is truly due to the viscosity of the gas and not to oil-film friction) ζ is extremely small. The valve friction is often masked by actuator friction

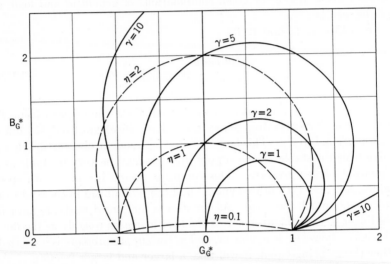

Fig. 12.14. Complex generator admittance for critical damping ($\zeta = 1$).

in either case. So long as it is small (unless α is very large, as is unusual), ζ has only a small effect on the real component, $G_G{}^*$, of $Y_G{}^*$ but is primarily important in determining the imaginary component, $B_G{}^*$.

The effects of these various parameters on $Y_G{}^*$ are not easily visualized from the equations. They may be somewhat more easily seen from Fig. 12.14, in which $G_G{}^*$ is plotted against $B_G{}^*$ for several values of η and γ. Unfortunately this method of plotting does not bring out the great effect of frequency, particularly for the usual case of low damping. This effect is illustrated in Figs. 12.15 and 12.16, in which the two components of $Y_G{}^*$ are plotted separately against the normalized frequency, η.

In Figs. 12.14 and 12.15 the valve is assumed to be critically damped ($\zeta = 1$). This assumption is perhaps unrealistic, but it does bring out the fact that values of γ much above 1 can cause instability even in heavily damped valves. In Fig. 12.16, $\zeta = 0.1$, which is more typical of actual spool valves. Here the peaks of both $G_G{}^*$ and $B_G{}^*$ are very much sharper and higher, showing that oscillation is possible even with

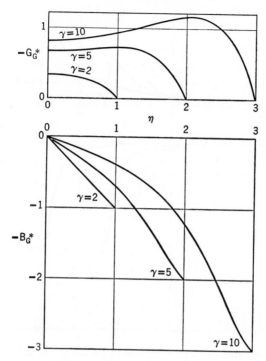

Fig. 12.15. Generator negative conductance, $G_G{}^*$, and susceptance, $B_G{}^*$, versus normalized frequency, η, for critical damping ($\zeta = 1$).

fairly highly damped loads and at small values of γ, but only over very narrow ranges of frequency.

Inspection of the last four figures leads to several conclusions as to the possibility of oscillation with various types of loads:

1. At very low frequencies $\gamma_c \cong 1$, and ζ has little effect. The valve can be unstable only if $|G_L| < |G_G|$, which is just what was deduced in Sec. 12.51 for the d-c case.

2. For values of ζ above about 0.5 the minimum γ_c never gets below 0.9 and the valve is essentially nonresonant.

3. For small values of ζ, γ_c also becomes small and the chance of oscillation increases, but only for a small range of frequencies around the undamped natural frequency of the valve. This fact suggests that added damping would be an effective stabilizing agent in such cases. This is true, but the remedy is often difficult to apply if the valve must respond rapidly to changes in the input signal. If the valve is driven by an electromagnetic torque motor and a valve-position pickoff is used to

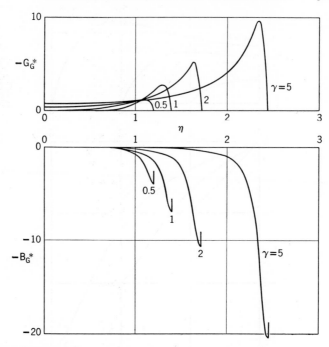

Fig. 12.16. Generator negative conductance, G_G^*, and susceptance, B_G^*, versus normalized frequency, η, for lightly damped valve ($\zeta = 0.1$).

close a tight position loop around it, both natural frequency and damping can be controlled easily. The added complication involved in this scheme is a disadvantage, but the potential advantages often outweigh it.

4. For large values of η, γ_c is very large and oscillation is correspondingly unlikely. This is the low-pass filter effect of any resonant mechanical system. For most practical purposes we can confine our attention to values of η less than 2.

5. For positive values of α the curves may enter the third quadrant but not the fourth. For $\alpha = 0$ the valve is never inductive since the curves do not extend below the real axis. In either case oscillation is impossible with a capacitive load; as seen by the generator the load must look like a mass and not a spring.

These general conclusions are useful, but it is often necessary to determine either whether the valve will be stable for a given set of conditions or else at what value of applied pressure (or other condition) it will become unstable. These questions demand a more detailed investigation of the interaction between generator and load than has been given so far, and this can be done in four ways:

1. The obvious method is to set up the system and try it out. This is very expensive and usually uninformative. It does tell when oscillation occurs, but only under the actual conditions of the experiment; there is no guarantee of stability if the conditions are slightly altered.

2. The characteristic curves of the load admittance can be superposed on the generator curves (with due attention to scale factors), and the stability margins can be determined from the resulting intersections. This is informative but frequently laborious because of the number of parameters that must be considered and the computational labor involved. The method seems potentially useful but requires further development. As an aid to the qualitative understanding of the phenomena it is excellent.

3. The system can be attacked analytically by equating the load admittance to the negative of the generator admittance and solving for the values of the two unknowns, which are usually η_c and γ_c. This is a very powerful and general method, but the high order of the resulting equations often makes it impractical. With sufficient ingenuity in selecting approximations the equations can sometimes be reduced to tractable form.

4. The whole system can be set up on an analogue computer and the parameters varied at will to obtain families of solutions. This is the most general method of all since it places no limitations upon the nature of the system or of its elements, but it does require a computer of considerable versatility and accuracy and it does demand great care in the translation to the language of the machine and still more in the retranslation from the language of the machine to that of the original inquirer, the fluid-power engineer.

The remainder of this chapter will be devoted to brief summaries of applications of the last two of the four methods just listed.

12.53. The Resonant Line—Analytical Method

Even a rather brief attempt to devise a mechanical load that will cause a valve to oscillate serves to demonstrate that such loads must be almost nonexistent in practice. This is because nearly always the resonant frequency of the load is much smaller than that of the valve, and it is very difficult to satisfy both the phase-shift and the gain conditions of the Barkhausen condition simultaneously. There is one class of system elements, however, to which the valve is closely coupled and which may have not one but several resonances in the critical region. This is the fluid line, which connects the valve to the power source, the load, and the tank. Under a fairly wide range of conditions, fluid lines can act as fairly sharply tuned resonators, and their admittances as seen by the valve can easily be such as to satisfy the requirements for oscillation.

When experiments are made on valve instability, it is often found that the condition of instability seems to correlate rather more closely with changes in the configuration of the fluid lines than with most other variables. In one experiment at DACL, for example, the valve was connected to the accumulator by a composite conduit made up of a long and a short piece of hose coupled together. When the flow passed from accumulator to valve in one particular direction, the valve oscillated persistently; when the hose was reversed so as to put the other section next to the valve, oscillation would not occur under any condition. Results such as these have indicated to various people the necessity of studying the interaction of valve and line, and several such studies have been made. In Denmark, Frederiksen[9, 10] has done extensive work, both theoretical and experimental, on air lines. At the Minneapolis-Honeywell Regulator Company, Ainsworth has done some work on oil lines,[11] using the Nyquist method in the form commonly employed by the communications engineer. At DACL, Friedensohn made a preliminary and only partially successful investigation [12] of the subject, which was followed up in much greater detail by Ezekiel.[13] This last work, which was carried on concurrently with but independently of that of Ainsworth, will be summarized in Sec. 12.54. Ezekiel's thesis includes an analytical treatment which is much like that which follows immediately, though with different nomenclature, but its principal contribution is the use of an analogue computer to simulate the valve-conduit system. The following analysis is based upon unpublished work by J. L. Coakley of DACL.

[9] Frederiksen, fn. 6.

[10] E. Frederiksen, "Pulsating Air Flow in Pipe Systems" (in Danish), I Kommission hos Teknisk Forlag, København, 1954. This excellent monograph does not discuss the oscillating valve, but gives an exhaustive analysis of forced oscillations in pipe systems of many different configurations.

[11] F. W. Ainsworth, "The Effect of Oil-Column Acoustic Resonance on Hydraulic Valve Squeal," *Trans. ASME*, Vol. 78 (May 1956), pp. 773–778.

[12] G. Friedensohn, "Stability of Control Valves," S.M. Thesis, Department of Mechanical Engineering, Massachusetts Institute of Technology, Cambridge, Mass., 1953.

[13] F. D. Ezekiel, "Effect of a Hydraulic Conduit with Distributed Parameters on Control Valve Stability," Sc.D. Thesis, Department of Mechanical Engineering, Massachusetts Institute of Technology, Cambridge, Mass., 1954; "The Effect of Conduit Dynamics on Control-Valve Stability," *Trans. ASME*, Vol. 80 (May 1958), pp. 904–908.

In the first attempt made at DACL by Friedensohn to analyze the combination of valve and line, it was assumed that the compressibility of the oil could be neglected so that the load seen by the valve would be essentially the mass of the oil in the line resonated by the equivalent spring of the accumulator into which the upstream end of the line opened. The experimental arrangement is shown in Fig. 12.17. (It may be

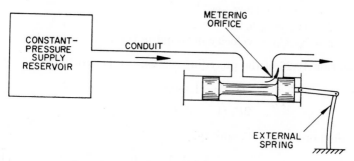

Fig. 12.17.   Diagram of valve and line.

desirable to point out that for the purposes of the analysis it is immaterial whether the line is upstream or downstream of the valve.) It was found, however, that the measured critical frequencies agreed very poorly with those predicted by the theory, and it became evident that it is necessary to take into account the distributed nature of the line parameters.

The following assumptions are made with respect to the line of Fig. 12.17:

1. There are no losses in the line or by radiation from the far end.

2. The termination at the accumulator offers no resistance to flow; it acts as a short circuit.

3. Conditions and fluid properties are uniform across every transverse section of the line.

4. For a section of the line of length dx, in the momentum equation, $\partial V/\partial t \gg V(\partial V/\partial x)$.

5. For the same section, in the continuity equation, $\rho(\partial V/\partial x) \gg V(\partial \rho/\partial x)$.

6. The valve is connected directly to the downstream end of the line, without either series or shunt admittance.

Under these conditions the admittance of the line as seen by the valve is

$$Y_L = \frac{\Delta p\ (j\omega)}{\Delta q\ (j\omega)} = \frac{-jl}{Z_0 \tan \theta} \qquad (12.30)$$

where $Z_0 = \dfrac{\sqrt{\rho\beta}}{A} = $ characteristic impedance of the line, lb sec/in.[5]

$\rho = $ density of fluid, lb sec^2/in.[4]

$\beta = $ bulk modulus of fluid, lb/in.[2]

$\theta = \dfrac{\omega l}{c} = $ angle of the line

$\omega = 2\pi \cdot$frequency, sec^{-1}

$l = $ line length, in.

$c = \sqrt{\beta/\rho} = $ velocity of sound in the fluid, in./sec

Equation 12.26 of the preceding section can be written

$$Y_G = \frac{Q}{2P_v} \frac{(1 - \eta^2 - \gamma) + j2\zeta\eta}{(1 - \eta^2 + \gamma) + j2\zeta\eta(1 + \alpha)} \qquad (12.31)$$

At the border line of stability,

$$Y_G + Y_L = 0 \qquad (12.32)$$

After substituting Eqs. 12.30 and 12.31 in Eq. 12.32, separating the real and imaginary parts, and rearranging, we have

$$(1 - \eta^2 - \gamma_c)Z_0^* \tan \theta + 2\zeta\eta(1 + \alpha) = 0 \qquad (12.33)$$

and

$$(1 - \eta^2 + \gamma_c) - 2\zeta\eta Z_0^* \tan \theta = 0 \qquad (12.34)$$

where $Z_0^* = \dfrac{Q}{2P_v} Z_0$.

Combining these two equations so as to eliminate $Z_0^* \tan \theta$ yields a quadratic equation in η^2 whose solution is

$$\eta^2 = 1 - 2S \pm \sqrt{S^2 - 2S + \gamma_c^2} \qquad (12.35)$$

where $S = 2\zeta^2(1 + \alpha)$.

Substitution of Eq. 12.35 in either Eq. 12.33 or Eq. 12.34 will give the value of $Z_0^* \tan \theta$, from which θ can be determined. From Eq. 12.34, for example,

$$\theta = \frac{1}{Z_0^*} \tan^{-1}\left(\frac{1 - \eta^2 + \gamma_c}{2\zeta\eta}\right) \qquad (12.36)$$

Since η has been eliminated between Eqs. 12.35 and 12.36, it is now possible to plot θ as a function of γ, the other parameters being assumed to remain constant. The plot can be made more generally applicable, however, by noting that

$$\frac{\omega l}{c} = \frac{\pi}{2} \eta \left(\frac{\lambda_L}{\lambda_n}\right) = \frac{\pi}{2} \eta L^* \tag{12.37}$$

where λ_L = wave length of sound at fundamental resonant frequency of the line

and

       λ_n = wave length at undamped natural frequency of the valve

whence

       L^* = length of line in quarter wave lengths at the natural frequency of the valve

Figure 12.18 shows a plot of L^* against γ_c for the case $\alpha = 0$, $\zeta = 0.1$. The boundary is cyclic because of the cyclic nature of the tangent func-

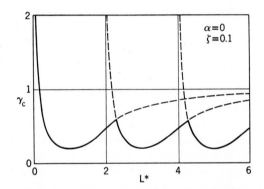

Fig. 12.18. Stability margin for system of Fig. 12.17.

tion. The critical value of γ_c is least when L^* is an odd integer, at which times the natural frequency of the valve coincides with one of the resonant frequencies of the line.

12.54. The Resonant Line—Simulation Method

One potential disadvantage of the analytical method which was used in the preceding three sections is the doubt as to the validity of the linearization of the characteristic equations. When the necessary facilities are available, the simulation method is a powerful tool which does not suffer from this handicap. The problem of the preceding section was simulated on an analogue computer [14] and the approximate boundary

[14] Ezekiel, fn. 13.

of the region of stability was determined by observing the computer response.

In applying this method to the system of Fig. 12.17, one of the principal problems is the simulation of the distributed-parameter conduit. In the present investigation this was accomplished by the method discussed in Sec. 5.3. The actual simulation of the conduit can be arrived at as follows.

It is assumed that the kinetic energy of mass transport and the energy losses of the fluid in the conduit can be neglected. If this is done, it can be shown that

$$2p_s(t - T) = p_v(t) + p_v(t - 2T) + Z_0 q_v(t) - Z_0 q_v(t - 2T) \quad (12.38)$$

where p_s = instantaneous pressure at upstream end of conduit, psi

$\quad p_v$ = instantaneous pressure at downstream end of conduit, psi

$\quad T = \dfrac{l}{c}$ = travel time of a pressure wave along the conduit, sec

$\quad Z_0 = \dfrac{\sqrt{\rho\beta}}{A}$ = characteristic impedance of conduit, lb sec/in.[5]

$\quad q_v$ = instantaneous flow at downstream end of conduit, in.³/sec

Now assume that there is no energy loss accompanying the flow of fluid from the reservoir into the upstream end of the conduit. This implies that the impedance of the reservoir as seen by the conduit is zero, and thus that for a.c. the upstream termination of the conduit is a short circuit. Then p_s is equal to the constant supply pressure, P_s, and is independent of time, and the left side of Eq. 12.38 becomes simply $2P_s$.

It will be convenient to express this equation in normalized form. We now write

$$p = \frac{p_v}{P_s} \quad (12.39)$$

and

$$q = \frac{q_v}{Q_i} \quad (12.40)$$

where Q_i is the constant steady-state value of the flow. Equation 12.38 now becomes

$$2 = p(t) + p(t - 2T) + \frac{2Z_0 Q_i}{2P_s} q(t) - \frac{2Z_0 Q_i}{2P_s} q(t - 2T) \quad (12.41)$$

We define a nondimensional characteristic impedance as

$$Z_0{}^* = \frac{Q_i}{2P_s} Z_0 \quad (12.42)$$

which is identical with the Z_0^* of Eqs. 12.33 and 12.34 since we here assume no pressure drop in the conduit. Substituting this quantity in Eq. 12.39 and rearranging, we get

$$p(t) = 2 - p(t - 2T) - 2Z_0^*q(t) + 2Z_0^*q(t - 2T) \quad (12.43)$$

This equation can be mechanized easily. The two summing units symbolized in Fig. 12.19 by the pentagons and the coefficient unit symbolized by the circle with the value of the coefficient enclosed are stand-

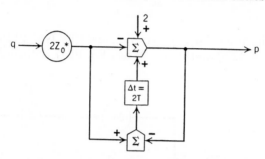

Fig. 12.19. Computer-setup diagram for distributed-parameter conduit.

ard components which are provided in any analogue computer. The delay unit is not a common piece of equipment; in the present study one was available which had been developed at DACL for the Delay-Line Synthesizer.[15] It proved to be very convenient for the purpose.

There are two basic equations to be mechanized for the valve, the force equation and the flow equation. As in Sec. 12.52, we start with Eq. 12.17, which is

$$mD^2x + bDx + Ep_vx + k(x - x_0) + \rho LDq_v + F_{ext} = 0 \quad (12.17)$$

In the steady state, and when $F_{ext} = 0$, this equation reduces to the form

$$k(x_i - x_0) + EP_ix_i = 0 \quad (12.11)$$

Therefore

$$x_0 = x_i\left(\frac{EP_i}{k} + 1\right) \quad (12.44)$$

We define the nondimensional valve displacement, X, as

$$X = \frac{x}{x_i} \quad (12.45)$$

[15] J. B. Reswick, "The Design and Application of the Delay Line Synthesizer," Sc.D. Thesis, Department of Mechanical Engineering, Massachusetts Institute of Technology, Cambridge, Mass., 1954.

and the nondimensional external force as

$$f = \frac{F_{\text{ext}}}{kx_i} \tag{12.46}$$

The undamped natural frequency of the valve is defined in the usual way by the equation

$$\omega_n{}^2 = \frac{k}{m} \tag{12.47}$$

and the nondimensional parameters γ, α, and ζ by Eqs. 12.13, 12.24, and 12.25 from Secs. 12.51 and 12.52. When Eq. 12.17 is divided by kx_i and

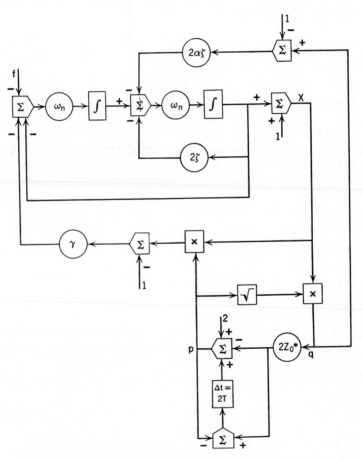

Fig. 12.20. Computer-setup diagram for system of Fig. 12.17.

the substitutions just indicated are made, we get the normalized force
equation

$$\frac{1}{\omega_n{}^2} D^2 X + \frac{2\zeta}{\omega_n} DX + \gamma(pX - 1) + X - 1 + \frac{2\zeta\alpha}{\omega_n} Dq + f = 0 \quad (12.48)$$

The flow equation is the same as that used previously, or

$$q_v = Gx\sqrt{p_v} \quad (12.49)$$

After nondimensionalizing, if $G = Q_i/x_i\sqrt{P_s}$, this equation becomes

$$q = X\sqrt{p} \quad (12.50)$$

The three equations 12.43, 12.48, and 12.50 represent the system of
Fig. 12.17 completely. The block diagram for the simulation of the system is shown in Fig. 12.20 and the results of a typical series of experiments in Fig. 12.21. The boundary between the stable and unstable
regions was determined point by point by injecting small impulses into
the simulator at point f and observing the response at point X on an
oscilloscope. Preliminary experiments showed that, as expected, the
parameters γ_c and L^* were the most important in determining the stability of the system. Fig. 12.21 differs from Fig. 12.18 in detail partly
because of a different choice of system parameters and partly because of
the departures from ideality of some of the computer components, which

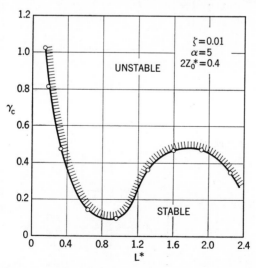

Fig. 12.21. Typical stability-margin plot for system of Fig. 12.17.

made it difficult to determine the boundary of stability with great accuracy. The resemblance between the two curves is more striking than the differences, however, and constitutes a pretty good proof that the linearization used in the analytical treatment is legitimate.

From various experiments with actual valves it became apparent that the effect of the valve chamber at the downstream end of the line cannot be neglected in the analysis. This case was also simulated on the computer, on the assumption that the flow through the downstream end of the line equals the flow through the valve plus the flow required to

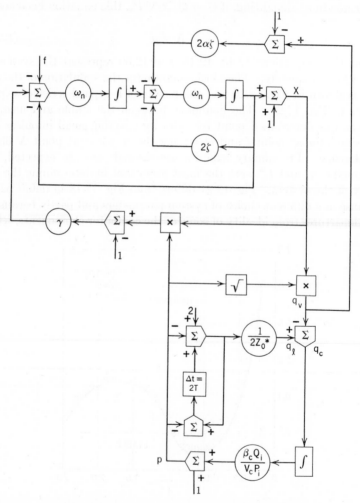

Fig. 12.22. Computer-setup diagram for system of Fig. 12.17 plus valve chamber.

change the pressure in the chamber. The additional equations required to do this are

$$q_l = q_v + q_c \tag{12.51}$$

and

$$q_c = \frac{V_c P_i}{\beta_c Q_i} \frac{dp}{dt} \tag{12.52}$$

where V_c is the chamber volume and β_c the effective bulk modulus of the oil in the chamber. The modified computer diagram is shown in Fig. 12.22. (It should be noted that the simulation of the conduit is performed somewhat differently from the previous case; here it was more convenient to use pressure as an input, obtaining flow as an output.)

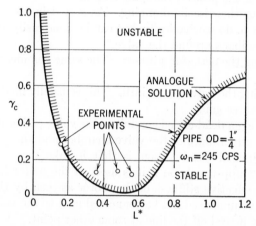

Fig. 12.23. Comparison of experimental and simulator results for system of Fig. 12.22.

The results of a typical run with this modified setup are shown in Fig. 12.23. As can be seen, the addition of the chamber has effectively lengthened the line. The figure also shows the results of some experiments with an actual valve and line. The experimental evidence is unfortunately rather meager, but what is available agrees fairly well with the theory. The three points in the region of the minimum are higher than the simulator solution because the depth of the minimum is greatly affected by damping in the system, and the actual damping was somewhat greater than that used in the computer. The two points on the ascending legs of the curve agree very well, since these portions of the curve are less affected by the damping.

From both the theoretical analysis and the experimental work it can be concluded that the phenomenon of resonance in the conduit is one important cause of valve instability. Thus in attempting to stabilize an unstable valve one very useful remedy is to change the length of the conduit. If this is impossible, one solution is to replace the simple conduit with a composite one made up of two sections of different diameters. The additional reflection from the junction of the two sections will help greatly in detuning the conduit. This method is very effective if the lengths and diameters are properly chosen. For example, in a certain automotive steering gear the hose which joins the pump to the valve is intentionally made in this way to minimize the generation and transmission of noise from the pump.

Other methods for stabilizing a valve include the following:

1. Increasing the positive damping length.

2. Increasing the viscous damping. As stated in an earlier section, this is difficult to do mechanically, but can be done easily and effectively by position feedback around the valve.

3. Increasing the actuator stiffness. The same comments apply as to item 2.

4. Decreasing the valve port width. This is usually inadmissible since it affects the valve gain.

5. Connecting a dead-ended conduit to the valve end of the line. This combines the advantages of the valve chamber and the composite conduit. Its principal disadvantage is that it (like the valve chamber) increases the trapped volume if it is used on the motor lines, and thus decreases the effective stiffness of the motor as seen by the load. The shunt line or volume need not be connected next to the valve; it may be more effective if teed off the line at some other point.

6. Making the pipe as short and of as large a diameter as possible.

7. Making the pipe conical, with the small end toward the valve. This decreases the reflection coefficient at the upstream end by improving the match from conduit to the large volume of the accumulator. A tapered pipe is hard to make, but a suitably stepped one is a good substitute.

The theory of resonant valve oscillations which has been outlined in the preceding four sections is capable of very wide application and should be investigated further. It is hoped that such investigations, both theoretical and experimental, can be carried out and reported in the near future.

13

S.-Y. Lee

J. L. Shearer

Electrohydraulic Actuation

13.1. INTRODUCTION

In the three preceding chapters we have seen (1) that the operating member of a valve must be moved against various forces of considerable magnitude when it is in service, (2) that there are serious limitations both as to the amount of force that can be obtained from an electromechanical actuator and also as to its stiffness, and (3) that insufficient actuator stiffness may permit sufficient motion under the influence of transient valve forces to make the whole system unstable. Thus, it is often desirable and sometimes imperative to obtain an actuator that pushes hard and is very stiff. The obvious solution is to use fluid power for actuating the main valve, for exactly the reasons that dictated its choice for the final task of the system, namely, its strength and its stiffness. Our valve now becomes a two-stage device, with some kind of a hydraulic (or pneumatic) pilot stage controlling the main valve.

Even if we neglect the dynamics and consider only steady-state conditions, it is not enough merely to supplement the torque motor with a pilot valve and a ram because the valve-ram combination is an integrator and the torque motor alone is not. In Fig. 13.1, the electromagnetic portion of the torque motor merely converts from current to force (for simplicity we are assuming that the elements of the system are

401

linear) and multiplies by a constant, while the restoring spring converts
from force to displacement and multiplies by another constant. Thus,
the final result is that the output displacement is proportional to the

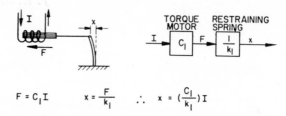

Fig. 13.1. Spring-restrained electromagnetic actuator.

input current. In Fig. 13.2 we have added a valve and a ram, which will
give us the desired greater force and stiffness. The valve changes displacement to *flow rate*, and the ram changes this rate to output *velocity*,

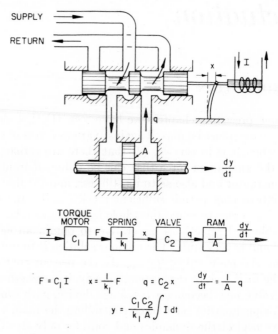

Fig. 13.2. Torque-motor–valve–ram system.

so the output *position* is proportional to the *time integral* of the input
current. If the assembly of Fig. 13.2 is to be equivalent to that of Fig.
13.1, we must add some element that will destroy this integration.

This added element will be feedback from the output. The feedback may represent the output position as force, as shown in Fig. 13.3, or as position directly, or as current, but so long as the quantity fed back is proportional to output position, the integration will be destroyed so far as the output is concerned, and the output will be proportional to the input. (The integration remains in the forward loop, and this is a valuable feature since it ensures that the error must become zero after sufficient time.) We also gain the usual benefit of any feedback scheme;

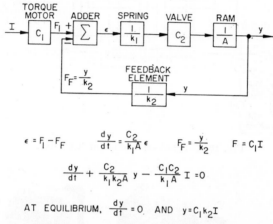

Fig. 13.3. Torque-motor–valve–ram system with feedback.

since the feedback device can be made highly linear, if the forward gain from summing point to output can be made high without causing instability, the gain from input to output becomes dependent only on the feedback element and is therefore highly linear also. Thus, our system consisting of torque motor, valve, and ram has become in effect a very fast, powerful, and stiff valve actuator, which is what we set out to obtain.

13.2. PRACTICAL TWO-STAGE VALVES

Although many different types of two-stage valves have been built, for our purposes it will be sufficient to describe only two of them.

The first was originally designed for an application in which the response had to be very rapid and added electronic equipment was not objectionable. Accordingly, the position of the power piston (second-stage valve spool) was picked off by a linear differential transformer, and

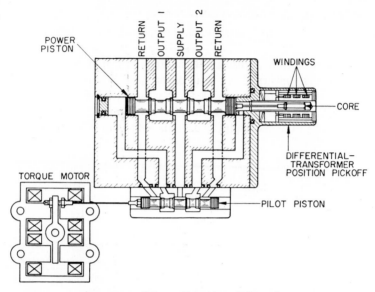

Fig. 13.4. Schematic section of XA valve.

Fig. 13.5. Exploded view of XA valve.

the resulting a-c position signal was demodulated and filtered before combining with the d-c command signal. A schematic cross section of this valve, a photograph, and a block diagram of the system are shown in Figs. 13.4 through 13.6. About thirty of these valves were built in two sizes at DACL for various applications, and modified versions are now commercially available.* Typical performance data are given in

* From the Hydraulic Controls Co., 217 California St., Newton 58, Mass., among others. Similar valves not derived from the DACL model described here are available from other manufacturers.

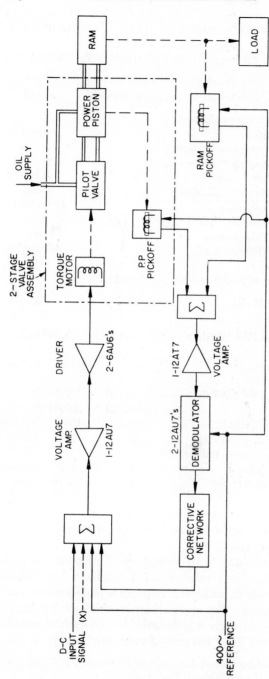

NOTE : IF A D-C RAM PICKOFF IS USED, ITS OUTPUT CAN BE RETURNED TO POINT (X).

Fig. 13.6.   Block diagram of system employing XA valve.

Table 13.1. Characteristics of DACL Two-Stage Valves, Models MX and XA

(The Two Models Are Identical except for Port Width)

Total Weight, with model 5 torque motor	3.9 lb
Total Volume, with torque motor, approximately	25 in.³

DACL Model 5 Torque Motor:

Displacement	±0.005 in.
Force (with armature centered), max	5.6 lb
Resonant Frequency, approximately	400 cps
Quiescent Power, total	0.9 watt
Quiescent Current, each side	10 ma
Resistance, each side	4500 ohms

Power-Piston Position Pickoff:

Excitation Voltage, max	10 v
Excitation Frequency, min	400 cps
Sensitivity	0.9 v/in.

Characteristics with 3000-psi Supply Pressure:

Model	MX	XA	
Pilot Valve:			
Peripheral Port Width	0.156	0.064	in.
Stroke	±0.005	±0.005	in.
Flow Sensitivity	240	100	in.²/sec
Power Valve:			
Peripheral Port Width	0.220	0.100	in.
Stroke	±0.040	±0.040	in.
Flow Sensitivity	350	150	in.²/sec
Max Stroking Velocity	27	11.2	in./sec
Max Flow Rate (into short-circuited load)	34	15.5	in.³/sec
Peak Output Power	5	2.3	hp

Table 13.1 and in the curves of Fig. 13.7. This performance is representative of modern fast-response hydraulic servomechanisms in the integral-horsepower class.

The use of a spool valve as the first stage of a high-performance two-stage valve is expensive because a small, accurate, low-leakage spool valve is very difficult to make and requires a high-performance torque motor for its actuation. For applications in which a fairly high leakage flow is permissible and where it is desirable to avoid additional electronic elements, the Moog valve * has proved very satisfactory. It uses an in-

* Originally designed by Cornell Aeronautical Laboratory, and now produced by Moog Valve Co., Inc., Proner Airport, East Aurora, N.Y.

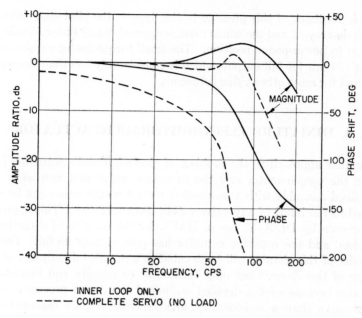

Fig. 13.7. Typical performance of DACL two-stage-type XA valve.

expensive torque motor and a twin flapper-and-nozzle valve as a pilot stage. The power piston is restrained by heavy springs, as shown in Fig. 13.8. Since these springs are linear, the deflection of the power piston will be proportional to the effective ram pressure (as in most two-stage valves, the power piston and the ram are the same piece), and because of the

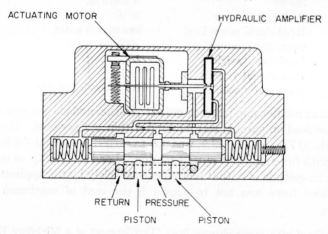

Fig. 13.8. Schematic section of Moog 2000-series valve.

built-in feedback of this pressure to the flapper, the pilot-stage integration is destroyed, and the whole valve is approximately linear from input current to power-piston position. The small torque motor requires only about 125 mw for full output, which is considerably less than the power required for most other valve actuators.

13.3. A MINIATURE ELECTROHYDRAULIC ACTUATOR

Another approach to the problem of electrohydraulic actuation is to make the torque motor and the first-stage valve and ram as a self-contained assembly with a mechanical output member that can be connected to the input of any valve within its capabilities. This approach was chosen by Dr. S.-Y. Lee of DACL for the solution of a particular problem, and the resulting actuator has proved very useful. The remainder of this chapter will be devoted to a detailed description of the design of this device,[1] not only because of its novelty and importance but also because such a detailed outline of the design process is more informative than a considerably greater bulk of pure description of various valve assemblies.

13.31. Design Requirements

The actuator to be discussed was designed specifically for stroking a 5-hp four-way valve. The design requirements are as follows:

Stroke	±0.015 in.
Peak output force	50 lb
Input-signal power level	less than 5 watts
Operating pressure	2000 psi
Load mass (weight)	1 lb
Peak output velocity	18 ips
Speed of response	90° phase lag at 200 cps

A linear relationship between output position and electric-input signal (with no load on output) was required within ±1 per cent. Hysteresis, the ratio h/Y_{max} in Fig. 13.9, was not to exceed 1 per cent during a test made with full-amplitude output motion. Load sensitivity as measured by the magnitude of output motion caused solely by the application of a 35-lb load force was not to exceed 5 per cent of maximum output motion.

[1] Reprinted with minor changes from "Development of a Miniature Electrohydraulic Actuator," by S.-Y. Lee and J. L. Shearer, *Trans. ASME*, Vol. 77 (October 1955), pp. 1077–1086, by permission of the ASME.

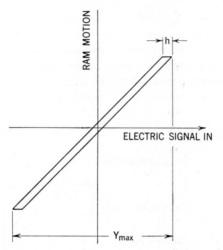

Fig. 13.9. Plot of output position versus input signal at very low frequency.

Of utmost importance was the need for reliability of operation, with emphasis on simplicity of design and ease of application. Operation with hydraulic fluid slightly contaminated by foreign particles was essential, but incorporation of a very small filter in the actuator was permissible.

13.32. Description of Miniature Actuator

Figure 13.10 is a schematic diagram of the complete actuator system. A small four-way control valve with a hole-slot-and-plug construction

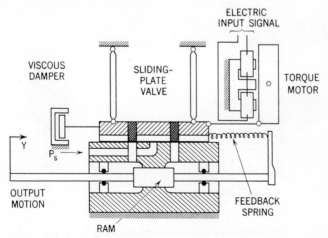

Fig. 13.10. Schematic diagram of miniature electrohydraulic actuator.

(described in Sec. 9.54) is used to modulate the flow of hydraulic fluid from a constant-pressure source to the chambers of a double-acting ram which delivers the power required to produce output motion. The control valve moves in response to various forces acting on it. These forces consist of the electromagnetic force generated by the flow of electric current (input signal) through the torque-motor coils and the force

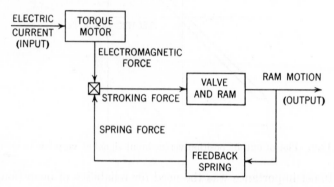

Fig. 13.11. Functional block diagram of miniature electrohydraulic actuator.

exerted by the feedback spring, the right end of which is attached to the ram. The feedback is negative. Initially, the torque-motor current is steady, the valve is centered, and the ram is motionless at a position such that the net force due to deflection of the feedback spring just balances the electromagnetic force generated by the flow of current in the torque-motor coils. A change in torque-motor current may cause leftward valve

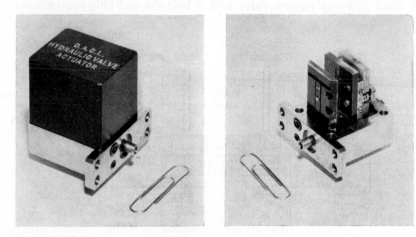

Fig. 13.12. Miniature actuator compared to ordinary paper clip.

motion which produces a flow tending to move the ram to the right, thereby increasing the spring deflection and creating a force acting in opposition to that generated by the torque motor; hence, if the system is dynamically stable, the valve eventually will be returned to its center position, and the ram will assume a new position which is proportional to the magnitude of the torque-motor current. Figure 13.11 is a functional block diagram of the system showing the sequence of events just described.

Although the series of events in the preceding description starts with operation of the torque motor and although the valve and torque-motor designs are intricately interrelated, the control valve is in many respects the heart of the system.

A photograph of the complete actuator is shown in Fig. 13.12.

13.33. Design Characteristics of the Components

13.331. Control Valve. Because of the decision to use a low-power-level torque motor with a small stroke, it is necessary to employ a valve which will provide the maximum required ram velocity with small valve displacement. In order to minimize quiescent leakage and meet the load-sensitivity requirement, a closed-center valve (zero overlap) with small clearance between fixed and moving parts is required. Such factors make it necessary to achieve very accurate relative location of the corners of the metering orifices in the fixed and moving parts of the valve. Friction forces in the valve must be minimized in order to keep torque-motor size down. From an economic point of view, ease of manufacture is an important factor, since cost inevitably must play an important role in the acceptance of the actuator for general use.

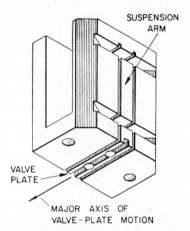

Fig. 13.13. Isometric view of valve suspension unit.

Figure 13.13 is an isometric drawing of the flexure-suspended valve-plate construction which was developed for the actuator. This unit measures approximately ½ in. x ¾ in. x ¾ in., and it may be machined from a solid piece of steel or fabricated from simple parts. The thin portions of the vertical suspension arms serve as flexure pivots that enable the valve plate to move freely along only its major axis while the upper ends of the arms are held motionless by the rest of the structure, which is firmly fastened at the bottom to a block serving as the fixed part of the valve. A similar type of suspension employing a construction with the suspension arms in tension rather than in compression proved to be more difficult to manufacture and adjust.

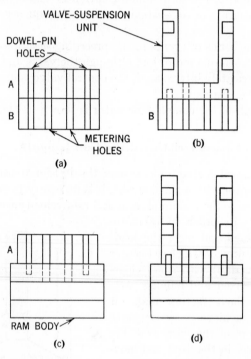

Fig. 13.14.   Use of drilling and reaming jigs in making suspension valve.

A nominal clearance of 0.0002 in. between the valve plate and valve body is attained during final grinding of the bottom surface of the suspension unit. Very accurate alignment of the metering orifices can be attained by drilling and lapping the metering holes of the suspension unit and the fixed part of the valve when they are clamped together. However, for quantity production, it has been found advantageous to machine the two parts separately by using a pair of drilling and reaming jig plates as shown in Fig. 13.14. This technique ensures that the holes match very accurately at the mating surfaces. Parts thus made are interchangeable, an important factor in quantity production. The accuracy of valves made in this way can be determined best from a displacement-flow curve; a typical one is shown in Fig. 13.15.

The upward force due to hydraulic pressure acting on the bottom face of the valve plate is the cause of two significant effects. (1) It causes the suspension arms to deflect under compression. This makes metal-to-metal contact between the valve plate and valve body impossible, thereby minimizing valve friction, but it also increases the valve clearance and therefore the quiescent leakage of the valve and the load sensitivity of the actuator. Actual measurements show that the significant factors in suspension-arm deflection are direct compression in the thin

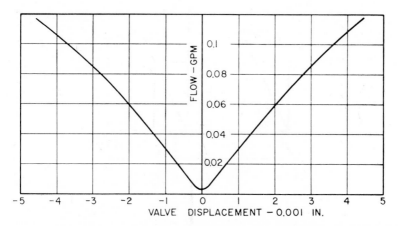

Fig. 13.15. Supply flow versus valve displacement for suspension valve with no pressure drop across the load. Univis 40 hydraulic fluid, port width = 0.040 in., P_s = 1750 psi.

sections and column-type bending in the thick sections. The valve was so designed that the total deflection under maximum pressure force does not exceed 0.0001 in. (2) The upward pressure force can cause the valve to tend to open in one direction or the other, depending on how much the valve plate is deflected relative to its reed pivots, as shown in Fig. 13.16. When the pressure force is constant, any unbalance due to initial deflection relative to the pivots when the valve is centered can be canceled by an adjustment somewhere in the system or by biasing the torque-motor current so that the opening effect simply acts like a negative spring of constant stiffness acting on the valve plate. When the pressure varies, as, for example, because of supply-pressure variation, the opening force will vary with pressure and the error appearing in ram position will be proportional to the pressure variation. Only proper initial alignment of the centered valve plate relative to its reed pivots will make possible a steady ram position that is insensitive to supply-pressure variations.

Sticking of the valve due to operation with dirt-contaminated hydraulic fluid has been minimized by incorporating a small felt or sintered-bronze filter in the fluid supply line.

A valve motion of ± 0.004 in. proved to work well with the torque-motor design and the flow requirements of the valve. At 2000-psi drop across the ram, a net ram area of 0.025 in.[2] is required to produce a force of 50 lb. With a maximum desired no-load ram velocity of 18 ips, a maximum flow of 0.5 in.[3]/sec is required through the valve. When

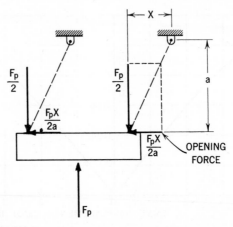

Fig. 13.16. Sketch of valve plate, showing how an opening force results from pressure acting on bottom.

the pressure drop across each orifice is 1000 psi, the maximum valve orifice opening is found from

$$A_0 = \frac{Q}{C_d} \sqrt{\rho/(2P)} = 1.6 \times 10^{-4} \text{ in.}^2 \qquad (13.1)$$

where the quantities are defined in Sec. 13.34. In order to produce this area with a stroke of 0.004 in., a port width of 0.04 in. is required. The total quiescent leakage flow to the valve when it is supplied with hydraulic oil at 2000 psi is 0.03 in.[3]/sec, a value which represents a quiescent power drain of 0.009 hp. The maximum power which the valve can deliver to the ram is 0.15 hp.

13.332. Torque Motor. In providing a torque motor for the actuator it was necessary to employ a design simple to make and maintain, compact, powerful enough to overcome any sticking effects in the valve, large enough to provide the required valve motion, and easily driven from a vacuum-tube amplifier. Figure 13.17 is an isometric view of the torque motor which was designed for use in the actuator. A detailed analysis of the torque-motor design is given in Sec. 13.4. This design is very nearly insensitive to linear accelerations because of its nearly symmetrical construction. The armature is supported by a single flexure pivot which is mounted on the center leg of an E-shaped field-magnet structure. Two coils each consisting of 2050 turns of No. 44 copper wire

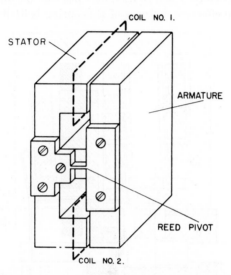

Fig. 13.17. Isometric sketch of electromagnetic torque motor.

are energized electrically to establish electromagnetic forces in the air gaps at the outer legs of the field magnet. Each coil is connected in series with the plate of a vacuum tube, and the vacuum tubes are driven in push-pull fashion so that the current in one coil is increased as that in the other is decreased, thereby increasing the force at one air gap and decreasing the force at the other. The net effect is a couple or torque acting on the armature proportional to the difference of the coil currents. Each of the working air gaps is 0.0045 in. thick and the maximum torque that can be generated with the armature in its center position is 0.19 in. lb. This corresponds to a valve-stroking force of 0.5 lb. The input impedance of the torque motor may be approximated by 500 ohms resistance in series with an inductance of 0.615 henry. The maximum current for each coil is 40 ma. The over-all dimensions of the torque motor are approximately $\frac{3}{8}$ in. x $\frac{5}{8}$ in. x $1\frac{1}{8}$ in.

13.333. Other Components. The ram consists of a single piece of hardened steel with a piston diameter of $\frac{7}{32}$ in. and a shaft diameter of

$\frac{5}{32}$ in. The ram shaft is sealed at each end by O-rings as shown in Fig. 13.18.

The feedback mechanism is a cantilever-type spring rigidly attached to the ram. Since the deflection of this spring is equal to the difference between the ram motion and the valve motion, the force exerted by the spring on the valve plate is proportional to ram motion only when the valve plate has been returned to its center position (as during any steady condition). The effective stiffness of this spring is 31 lb/in.

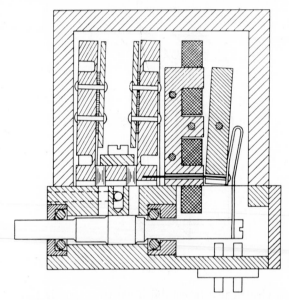

Fig. 13.18. Assembly drawing of miniature actuator.

13.34. Dynamic Analysis of Actuator System

In order to reveal a number of important facts about the complete system, a dynamic analysis of the actuator system has been made for small changes of all system variables. This analysis, which consists of carefully characterizing each part of the system mathematically and then combining the various component equations into a single mathematical expression, is given in the following paragraphs.

The following nomenclature is used:

a = length of valve-plate suspension arms, in. (0.750)

$\left. \begin{array}{c} a_0 \\ a_1 \\ a_2 \\ a_3 \end{array} \right\}$ = coefficients of characteristic equation of actuator

A = ram area, in.^{2} (0.025)

A_0 = orifice area, in.^{2}

b = viscous-damping coefficient, lb sec/in.

b_1 = height of outer legs of torque-motor field magnet, in. (0.125)

c = half-height of center leg, in. (0.063)

C_1 = valve coefficient, in.^{5}/lb sec

C_2 = ram leakage coefficient, in.^{5}/lb sec

C_d = orifice discharge coefficient (0.625)

d = width of all magnet legs, in. (0.250)

D = derivative with respect to time, d/dt

F_1 = force at air gap No. 1, lb

F_2 = force at air gap No. 2, lb

F_f = friction force on valve plate, lb

F_l = external load force, lb

F_p = pressure force on bottom of valve plate, lb (20)

F_{rs} = friction force at ram seals, lb

F_s = feedback-spring force, lb

F_{sf} = steady-flow force, lb

F_t = torque-motor force, lb

F_{us} = unsteady-flow force, lb

F_{vd} = viscous-damper force, lb

F_{vs} = valve-spring force, lb

g = thickness of center-leg air gap of torque motor, in. (0.0015)

g_1 = thickness of air gap No. 1, in.

g_2 = thickness of air gap No. 2, in.

g_0 = thickness of g_1 and g_2 when they are equal, in. (0.0045)

I_1 = electric current in coil No. 1, amp

I_2 = electric current in coil No. 2, amp

k = composite spring stiffness associated with torque motor and suspension valve including stiffness of feedback spring, lb/in. (73.7)

k_1 = valve coefficient, in.^{2}/sec (168.0)

k_2 = $C_1 + C_2$, in.^{5}/lb sec (4×10^{-5})

k_3 = fluid-compliance coefficient, in.^{5}/lb

k_4 = electromagnetic constant, psi/weber^2

k_5 = electromagnetic constant, webers/amp-turn-in.

k_a = torque-motor coefficient, lb/amp (11.1)

k_b = torque-motor coefficient, lb/in. (20.7)

k_{rp} = effective stiffness of torque-motor-reed pivot, lb/in. (43.0)

k_s = stiffness of feedback spring, lb/in. (29.6)

k_{sf} = steady-flow-force coefficient, lb/in. (80.0)

k_{us} = unsteady-flow-force coefficient, lb sec/in.

k_v = stiffness of valve suspension, lb/in. (11.8)

m = net effective mass of valve plate and moving parts of torque motor, lb $\sec^2$/in. (6×10^{-6})

m_1 = mass load attached to ram, lb $\sec^2$/in. (1.29×10^{-3})

n = number of turns per coil (2000)

P = pressure drop across orifice, psi

P_m = pressure difference across ram, psi

P_s = supply pressure, psi (2000)

Q = flow rate, in.^{3}/sec

Q_l = leakage flow rate past ram, in.^{3}/sec

r = radius of torque-motor armature, in. (0.375)

$\mho$ = volume of fluid under compression in one end of ram, in.^{3} (0.0025)

X = valve position, in.

Y = ram position, in.

β = bulk modulus of hydraulic fluid, psi (2.5×10^5)

Δ = small change

ρ = mass density of fluid, lb $\sec^2$/in.^{4} (8×10^{-5})

τ = time constant, sec (0.0002)

ϕ_1 = magnetic flux in air gap No. 1, webers

ϕ_2 = magnetic flux in air gap No. 2, webers

ω_n = natural frequency, radians/sec (3500)

ω_{ns} = natural frequency of ram and mass load, radians/sec (8870)

i = subscript denoting initial condition

ss = subscript denoting steady state

OUTPUT FORCE OF TORQUE MOTOR. From the detailed analysis in Sec. 13.4, the force developed on the valve plate by the torque motor is found to be a function both of the net difference of the coil currents and of the position of the armature. With the assumption that only small changes occur, the following equation may be used for the torque motor:

$$\Delta F_t = k_a(\Delta I_1 - \Delta I_2) - k_b(\Delta X) \qquad (13.2)$$

where Δ indicates a small change.

FEEDBACK-SPRING FORCE. The feedback spring exerts a force which tends to oppose the force from the torque motor, as shown in Fig. 13.19. This force is given by

$$\Delta F_s = k_s(\Delta Y + \Delta X) \qquad (13.3)$$

Also acting on the suspension valve plate are: (1) an upward pressure force, F_p, which is directly related to the system supply pressure; (2) a spring force, F_{vs}, due to the stiffness of the flexure pivots (represented by an equivalent coil spring of stiffness k); (3) a viscous-damping force,

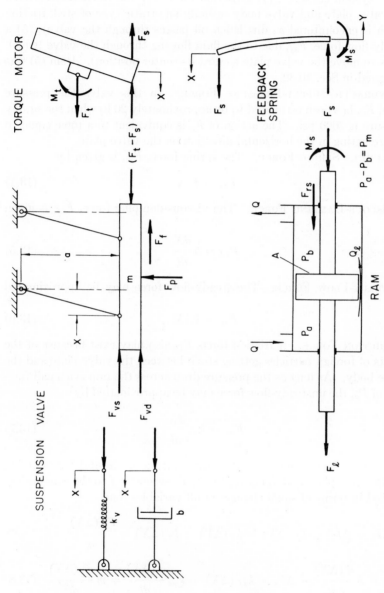

Fig. 13.19. Free-body diagrams of various parts of actuator system.

F_{vd}, due to a damping device which may be incorporated to provide adequate system stability; (4) a friction force, F_f, due to friction between the valve plate and valve body (usually an erratic type of stick friction which is encountered as dirt-laden oil passes through the valve); (5) a steady-flow force, F_{sf}, due to the fluid flowing through the valve (which tends to move the valve plate toward its center position). Item (5) was discussed in Sec. 10.32.

HYDRAULIC-SUPPLY-PRESSURE FORCE. In this valve the pressure force, F_p, has been estimated to be approximately 20 lb when the supply pressure is 2000 psi. The action of F_p is equivalent to a force equal to F_pX/a acting in the horizontal direction on the valve plate.

FLEXURE SPRING FORCE. The spring force, F_{vs}, is given by

$$F_{vs} = k_v X \tag{13.4}$$

VISCOUS-DAMPING FORCE. The viscous-damping force, F_{vd}, is given by

$$F_{vd} = b \frac{dX}{dt} \tag{13.5}$$

STEADY-FLOW FORCE. The steady-flow force may be approximated by

$$F_{sf} = k_{sf} X \tag{13.6}$$

FRICTION FORCE. A friction force, F_f, also may exist because of the effects of foreign particles getting stuck between the valve plate and the valve body. As long as the pressure drop across the ram is a small fraction of P_s, the unsteady-flow force may be approximated by

$$F_{us} = k_{us} \frac{dX}{dt} \tag{13.7}$$

SUMMATION OF FORCES. After summing the horizontal forces acting on the valve plate and applying Newton's second law to the summation, we find in terms of small changes of all variables that

$$k_a(\Delta I_1 - \Delta I_2) - k_b \, (\Delta X) - k_s \, (\Delta Y) - k_s \, (\Delta X) - \frac{F_p \, (\Delta X)}{a} - k_v \, (\Delta X)$$

$$- b \frac{d \, (\Delta X)}{dt} - \Delta F_f - k_{sf} \, (\Delta X) + k_{us} \frac{d \, (\Delta X)}{dt} = m \frac{d^2 \, (\Delta X)}{dt^2} \tag{13.8}$$

where m is the effective mass of the moving parts of the torque motor and the valve.

When terms are collected,

$$m \frac{d^2 (\Delta X)}{dt^2} + (b - k_{us}) \frac{d (\Delta X)}{dt} + k (\Delta X) = k_a(\Delta I_1 - \Delta I_2)$$
$$- k_s (\Delta Y) - \Delta F_f \quad (13.9)$$

where

$$k = k_b + k_s + k_r + k_{sf} - \frac{F_p}{A} \quad (13.10)$$

PRESSURE-FLOW CHARACTERISTICS. The pressure-flow characteristics of the valve are similar to those discussed in Sec. 7.31. For small changes, the flow rate, Q, to the ram may be expressed by

$$\Delta Q = k_1 (\Delta X) - C_1 (\Delta P_m) \quad (13.11)$$

where $P_m = P_a - P_b$, the pressure difference across the ram.

An estimate based on the work reported in Chapter 15 indicates that the natural frequency of the ram system driving a mass of $\frac{1}{2}$ lb is more than 1000 cps. This means that the effects of load mass and fluid compressibility are probably negligible in this system. Therefore, the flows to and from the valve are exactly equal to the displacement of the ram plus leakage past the ram; therefore,

$$A \frac{d (\Delta Y)}{dt} = \Delta Q - \Delta Q_l \quad (13.12)$$

For small pressure difference across the ram, the leakage flow past the ram is viscous (laminar) and is given by

$$\Delta Q_l = C_2 (\Delta P_m) \quad (13.13)$$

RAM FORCES. The forces acting on the ram include the pressure forces from the hydraulic fluid, the spring force, F_s, a friction force, F_{rs}, and an external load force, F_l. Although the exact characteristics of F_{rs} are not known in detail, this friction force, which is caused by rubbing action in the ram seals, seems to act very much like Coulomb friction. After summing the horizontal forces acting on the ram and ignoring the small mass load, we have

$$(\Delta P_m) A = k_s(\Delta Y + \Delta X) + \Delta F_{rs} + \Delta F_l \quad (13.14)$$

FINAL DYNAMIC EQUATION. The information contained in Eqs. 13.9 through 13.14 completely describes the system for small changes. The set may be assembled in a number of ways in order to examine the static and dynamic behavior of the actuator.

A block diagram consisting of integrations, coefficients, and summations can be made up to represent a system of this kind. Figure 13.20 is a block diagram of this type for the actuator.

When Eqs. 13.9 through 13.14 are combined into a single differential equation,

$$
\left[\frac{[mD^2 + (b - k_{us})D + k]\left(AD + \dfrac{k_2 k_s}{A}\right)}{k_1 - \dfrac{k_2 k_s}{A}} + k_s \right] (\Delta Y)
$$

$$
= k_a(\Delta I_1 - \Delta I_2) - \Delta F_f
$$

$$
+ \frac{[mD^2 + (b - k_{us})D + k]k_2(\Delta F_{rs} + \Delta F_l)}{\left(k_1 - \dfrac{k_2 k_s}{A}\right)A}
\qquad (13.15)
$$

where D represents d/dt, and $k_2 = C_1 + C_2$.

Of primary interest in assessing the merits and demerits of the actuator are steady-state characteristics including errors in ram position due to changes of external load, ΔF_l, ram friction, ΔF_{rs}, and valve-plate friction, ΔF_f, and dynamic characteristics including degree of system damping and speed of response.

ANALYSIS. Since, during any steady-state condition, following a step- or pulse-type disturbance, all the derivatives in Eq. 13.14 are zero,

$$
\left(\frac{kk_2}{k_1 A - k_2 k_s} + 1 \right) k_s (\Delta Y) = k_a(\Delta I_1 - \Delta I_2)
$$

$$
- \Delta F_f - \frac{kk_2}{k_1 A - k_2 k_s} (\Delta F_{rs} + \Delta F_l) \qquad (13.16)
$$

If the coefficient k_2 could be made to vanish, the actuator output motion would be insensitive to changes in ram friction force, ΔF_{rs}, or external load force, ΔF_l. In order to make k_2 equal to zero, it would be necessary to build a valve which would be perfectly closed at its center position and to provide a seal around the ram piston which would allow no leakage past the ram. Because of valve clearance, imperfect location of all of the metering orifices, and rounding of the corners of the metering orifices, a truly closed-center valve cannot be built. The suspension valve employed here nevertheless very nearly approaches being a truly closed-center valve. Leakage past the ram can be reduced to practically zero when a packing is used, but space limitations in this design made it necessary to try to achieve a reasonably good seal around the ram piston by minimizing the clearance between piston and cylinder walls.

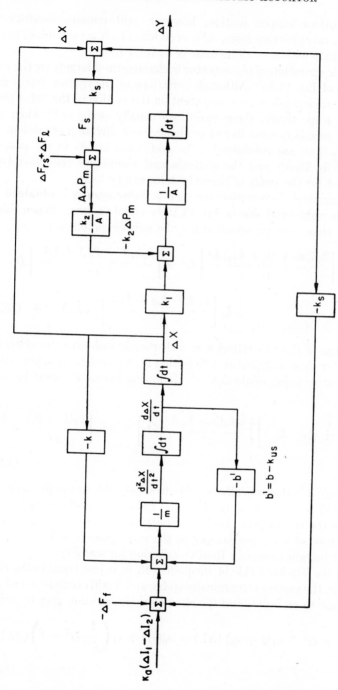

Fig. 13.20.   Analogue block diagram of basic differential equations of actuator.

The actuator output motion, however, still remains sensitive to changes in valve friction force, ΔF_f, even when k_2 is zero, tending to decrease slightly with a small increase of k_2 from zero.

Dynamic operation of the actuator is characterized largely by the left-hand side of Eq. 13.15. Although variations of the friction forces ΔF_f and ΔF_{rs} are dependent to some extent on the motion of the valve plate and ram, respectively, these variations usually occur only when the respective members come to rest or change their direction of motion. If these variations are considered to be additional inputs to the system, Eq. 13.15 is linear, and the methods and techniques that have been worked out for the study of linear systems can be used.

The characteristic equation for such a linear system is obtained by letting the right-hand side of Eq. 13.15 be equal to zero. When this is done, the characteristic equation may be written as follows:

$$\left\{ mD^3 + \left[\frac{k_2 k_s m + (b - k_{us})A^2}{A^2} \right] D^2 + \left[\frac{k_2 k_s (b - k_{us}) + kA^2}{A^2} \right] D \right.$$
$$\left. + k_s \left[\frac{k_2(k - k_s) + k_1 A}{A^2} \right] \right\} (\Delta Y) = 0 \quad (13.17)$$

Application of Routh's criterion reveals that in order for the system to have better than marginal stability (that is, respond to an input with a damped or decaying oscillation) the following inequality must be satisfied:

$$\left[\frac{k_2 k_s m}{A^2} + (b - k_{us}) \right] \left[\frac{k_2 k_s (b - k_{us})}{A^2} + k \right] > mk_s \left[\frac{k_2(k - k_s)}{A^2} + \frac{k_1}{A} \right]$$

$$(13.18)$$

Although it is theoretically possible to satisfy this inequality when the quantity $(b - k_{us})$ is zero, the fact that k_2 has been made very small to minimize the steady-state load sensitivity of the actuator suggests the possible need of a viscous damper on the valve plate in order to provide sufficient margin to satisfy Routh's criterion for stability.

When the left-hand side of Inequality 13.18 is just equal to the right-hand side, the system is marginally stable and it will oscillate unendingly in response to an input, and the characteristic equation may be written

$$(a_3 D^3 + a_2 D^2 + a_1 D + a_0)(\Delta Y) = a_0(\tau D + 1)\left(\frac{1}{\omega_n^2} D^2 + 1 \right)(\Delta Y) = 0$$

$$(13.19)$$

where

$$a_0 = \frac{k_s}{A^2}[k_2(k - k_s) + k_1 A] \qquad a_1 = \frac{1}{A^2}[k_2 k_s(b - k_{us}) + kA^2] \quad (13.20)$$

$$a_2 = \frac{1}{A^2}[k_2 k_s m + (b - k_{us})A^2] \qquad a_3 = m \qquad\qquad (13.21)$$

and τ and ω_n are the time constant and natural frequency, respectively, of the system. The values of τ and ω_n are useful in making an estimate of the system's speed of response. By equating the coefficients of terms with like powers of D in Eq. 13.19, the expressions for τ and ω_n are found to be

$$\tau = \frac{k_2 k_s(b - k_{us}) + kA^2}{[k_2(k - k_s) + k_1 A]k_s} \qquad (13.22)$$

$$\omega_n = \sqrt{\frac{[k_2(k - k_s) + k_1 A]k_s}{k_2 k_s m + (b - k_{us})A^2}} \qquad (13.23)$$

Because it is usually desirable to have a system which responds with a rapidly decaying oscillation rather than with an unending oscillation, the time constant and natural frequency of such a system will differ somewhat from τ and ω_n given in Eqs. 13.20 and 13.21, but τ and ω_n are useful if only a rough estimate of speed of response is required.

13.35. Experimental Work

DYNAMIC CHARACTERISTICS. When the actuator was first assembled, the viscous damper on the valve plate was omitted, and the system was found to oscillate with increasing amplitude when subjected to a small disturbance at any point in the system. In order to provide adequate system stability, viscous damping was employed on the valve plate. This damping action was accomplished by attaching a small plate to each of the vertical suspension arms in such a way that it covered, but did not touch, the adjacent solid vertical supports as shown in Fig. 13.18. The space between the damping plates and vertical supports is filled with oil which exerts a damping force tending to oppose valve-plate motion as the thickness of the space is changed by valve-plate motion.

Because of the experimental difficulties anticipated in making useful measurements of the characteristics of this damping device, its final design was achieved by trial and error. Satisfactory system stability was obtained with a total effective plate area of 0.25 sq in. with a space

0.004 in. thick with the valve plate in its center position. The fluid used in the dampers is the same as that used to drive the ram, namely, Univis 40.

Figure 13.21 shows the amplitude and phase-lag characteristics of the complete actuator system that were obtained from frequency-response measurements after the viscous-damping plates were installed. A set of calculations, given in Sec. 13.4, shows that the values of τ and ω_n for this system are 0.0004 sec and 3500 radians/sec, respectively, when the

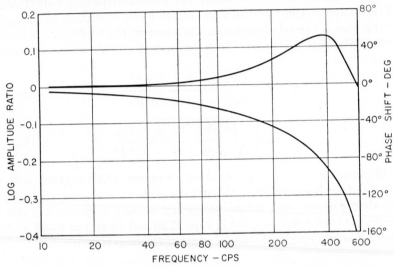

Fig. 13.21. Amplitude and phase-shift characteristics of miniature actuator.

system is marginally stable. Figure 13.21 shows that the real (well-damped) system has a somewhat lower natural frequency than the analysis of a marginally stable system indicates. The graphed frequency-response characteristics include the dynamic characteristics of the electronic amplifier and torque motor. Frequency-response measurements relating torque-motor current to amplifier input signal revealed very little lag or attenuation up to 400 cps.

Since the speed of response of the actuator is more than adequate to meet the requirements, it was not necessary to seek an optimum system speed of response. A thorough study of the fundamental equations together with an analogue study based on the block diagram of Fig. 13.20 would be useful to the designer who wished to arrive at a design with optimum speed of response.

STATIC CHARACTERISTICS. The relationship between actuator output position and input current during very-low-frequency variations of the input is closely linked to the steady-state characteristics discussed in the analysis section. All of the derivatives in Eq. 13.15 are small in this case, and the steady-state equation, 13.16, may be employed to gain an insight into the factors which cause a hysteresis effect to exist between the input and the output of the actuator. The friction forces ΔF_f and ΔF_{rs} may be thought of as consisting of Coulomb friction. As such, they are nearly independent of velocity of relative motion and always act to oppose motion so that they change rapidly from a negative value to a

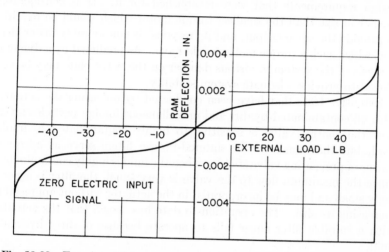

Fig. 13.22. Experimental measurements of ram deflection versus external load.

positive value as relative velocity changes from a positive value to a negative value. Thus, during a test at very low frequency, these forces change only at the extreme positions of the valve and ram, respectively. Because of the very low force levels involved, these forces have not been measured directly. The experimental data that have been derived from low-frequency response tests, however, do indicate that this concept of Coulomb friction together with Eq. 13.16 does describe the important effects which take place. As an example, decreasing the net spring stiffness, k, of the torque motor and valve relative to the feedback-spring stiffness, k_s, resulted in a proportional decrease in hysteresis as indicated by Eq. 13.16 so that it was possible to reduce the hysteresis to a value less than 1 per cent.

Another important factor in the operation of the actuator is its sensitivity to an external load force, ΔF_l. Figure 13.22 is a graph showing how ram position is related to external load force as determined by steady-state measurements. The slope of this curve at any point is given by $\Delta Y/\Delta F_l$, which may also be obtained from Eq. 13.16. Of primary concern is this slope at the origin where the stiffness is smallest. The slope measured from the curve is 0.00018 in./lb, while that computed in Sec. 13.4 from the system constants is 0.00019 in./lb.

13.36. Conclusions

The miniature actuator which has been developed does satisfy the design requirements that were established for it. It is considerably smaller in size than the smallest torque motor which would be required to provide the same output, and its response is considerably faster than that which such a torque motor would have. As predicted by a dynamic analysis of the system, a viscous damper on the valve plate may be employed to provide adequate system stability.

Extensive tests that have been conducted by operating the actuator with oil contaminated by dirt and miscellaneous foreign particles reveal that the control valve is sensitive to highly contaminated hydraulic fluid, but that small felt or sintered-bronze filters serve as adequate means of preventing valve sticking when the fluid is highly contaminated. Since the maximum flow to the valve is very small, the filter is accordingly small and may be incorporated in the actuator without appreciably increasing its size. The operation to date has shown that the actuator with a suitable filter never fails to operate because of dirty hydraulic fluid.

Preliminary tests with the actuator operating on compressed air at 1000 psi instead of hydraulic oil indicate that development of an electropneumatic actuator with speed of response nearly as fast as the electrohydraulic actuator is feasible.

13.4. APPENDIX

13.41. Torque-Motor Analysis

Figure 13.23 is a drawing of the torque-motor configuration employed in the actuator. The electromagnetic force exerted on the armature at each of the active gaps is given by

$$F_1 = \frac{k_4}{b_1 d} (\phi_1)^2 \tag{13.24}$$

$$F_2 = \frac{k_4}{b_1 d} (\phi_2)^2 \tag{13.25}$$

In order to consider small changes from a set of initial conditions, Eqs. 13.24 and 13.25 may be written

$$F_{1i} + \Delta F_1 = \frac{k_4}{b_1 d}(\phi_{1i} + \Delta\phi_1)^2 \quad (13.26)$$

$$F_{2i} + \Delta F_2 = \frac{k_4}{b_1 d}(\phi_{2i} + \Delta\phi_2)^2 \quad (13.27)$$

so that when Eq. 13.24 is combined with 13.26, and 13.25 is combined with 13.27,

$$\Delta F_1 = \frac{k_4}{b_1 d}[2\phi_{1i}\,\Delta\phi_1 + (\Delta\phi_1)^2] \quad (13.28)$$

$$\Delta F_2 = \frac{k_4}{b_1 d}[2\phi_{2i}\,\Delta\phi_2 + (\Delta\phi_2)^2] \quad (13.29)$$

When $\Delta\phi_1$ and $\Delta\phi_2$ are only a few per cent of ϕ_{1i} and ϕ_{2i}, the terms $(\Delta\phi_1)^2$ and $(\Delta\phi_2)^2$ may be neglected so that

$$\Delta F_1 = \frac{2k_4}{b_1 d}\phi_{1i}\,\Delta\phi_1 \quad (13.30)$$

$$\Delta F_2 = \frac{2k_4}{b_1 d}\phi_{2i}\,\Delta\phi_2 \quad (13.31)$$

The flux in each air gap is a function of the magnetomotive force generated by the current flowing in each coil and the total reluctance of each magnetic cir-

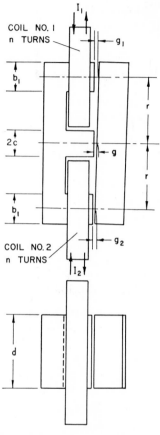

Fig. 13.23. Essential dimensions of torque motor.

cuit. As a first approximation, the reluctance of the magnetic material in the field magnet and armature may be neglected so that the reluctance of each magnetic circuit is due to its air gaps. Half the area of the air gap at the center leg may be considered to contain ϕ_1 and the other half to contain ϕ_2. The following equations then may be used:

$$\phi_1 = k_5 \frac{nI_1}{\dfrac{g_1}{b_1 d} + \dfrac{g}{cd}} \quad (13.32)$$

$$\phi_2 = k_5 \frac{nI_2}{\dfrac{g_2}{b_1 d} + \dfrac{g}{cd}} \quad (13.33)$$

The values of g_1 and g_2 are the same and equal to g_0 when the armature is centered. When the armature is away from center,

$$g_1 = g_0 + X \tag{13.34}$$

$$g_2 = g_0 - X \tag{13.35}$$

Combining Eqs. 13.32 with 13.34 and Eqs. 13.33 with 13.35 and using logarithmic differentials gives

$$\Delta\phi_1 = \phi_{1i}\left(\frac{\Delta I_1}{I_{1i}} - \frac{\Delta X}{g_0 + X_i + \dfrac{b_1}{c}g}\right) \tag{13.36}$$

$$\Delta\phi_2 = \phi_{2i}\left(\frac{\Delta I_2}{I_{2i}} + \frac{\Delta X}{g_0 - X_i + \dfrac{b_1}{c}g}\right) \tag{13.37}$$

The changes in ϕ_1 and ϕ_2 may be eliminated by combining Eq. 13.30 with 13.36 and Eq. 13.31 with 13.37 to give

$$\Delta F_1 = \frac{2k_4}{b_1 d}(\phi_{1i})^2\left(\frac{\Delta I_1}{I_{1i}} - \frac{\Delta X}{g_0 + X_i + \dfrac{b_1}{c}g}\right) \tag{13.38}$$

$$\Delta F_2 = \frac{2k_4}{b_1 d}(\phi_{2i})^2\left(\frac{\Delta I_2}{I_{2i}} + \frac{\Delta X}{g_0 - X_i + \dfrac{b_1}{c}g}\right) \tag{13.39}$$

Summing the forces acting on the torque-motor armature gives

$$\Delta F_t = \Delta F_2 - \Delta F_1 - k_{rp}(\Delta X) \tag{13.40}$$

When the initial conditions are $I_{1i} = I_{2i} = I_i$, $X_i = 0$, and $F_{ti} = 0$,

$$\Delta F_t = k_a(\Delta I_1 - \Delta I_2) - k_b(\Delta X) \tag{13.41}$$

where

$$k_a = \frac{2k_4 k_5{}^2 n^2 I_i b_1 d}{\left(g_0 + \dfrac{b_1}{c}g\right)^2} \tag{13.42}$$

$$k_b = k_{rp} - \frac{4k_4 k_5{}^2 n^2 I_i{}^2 b_1 d}{\left(g_0 + \dfrac{b_1}{c}g\right)^3} \tag{13.43}$$

For the miniature actuator, k_a and k_b may be calculated by use of the equation

$$k_4 k_5^2 = 1.4 \times 10^{-7} \text{ lb/(amp turn)}^2$$

and the values of the constants given in the list of nomenclature:

$$k_a = \frac{2(1.4 \times 10^{-7})(2000)^2(0.02)(0.125)(0.25)}{0.0045 + \dfrac{(0.125)(0.0015)^2}{(0.063)}} = 11 \text{ lb/amp} \quad (13.44)$$

$$k_b = (43.0) - \frac{4(1.4 \times 10^{-7})(2000)^2(0.02)^2(0.125)(0.25)}{(0.0075)^3}$$

$$= 43.0 - 63.7 = -20.7 \text{ lb/in.} \quad (13.45)$$

The maximum force developed by the torque motor in its center position may be determined by using Eqs. 13.22 and 13.32 because the current in one coil is a maximum of 40 ma and the current in the other coil is zero.

$$F_t = \frac{k_4 k_5^2 n^2 I_1^2 b_1 d}{\left(g_0 + \dfrac{b_1 g}{c}\right)^2} \quad (13.46)$$

or

$$F_t = \frac{(1.4 \times 10^{-7})(2000)^2(0.04)^2(0.125)(0.25)}{(0.0075)^2} = 0.5 \text{ lb} \quad (13.47)$$

13.42. Natural Frequency of Ram and Mass Load

With the constant k_2 very small, the natural frequency of the ram and mass load may be approximated as follows:

$$\omega_{ns} = \sqrt{(2A^2\beta)/(\mathcal{U}m_l)} \quad (13.48)$$

Calculating ω_{ns} when m_l weighs 0.5 lb gives

$$\omega_{ns} = \sqrt{\frac{(2)(0.025)^2(2.5 \times 10^5)(368)}{(0.0025)(0.5)}} = 8870 \text{ radians/sec} \quad (13.49)$$

13.43. Calculation of τ and ω_n

For unending oscillation, Routh's criterion gives

$$[k_2 k_s m + (b - k_{us})A^2][k_2 k_s(b - k_{us}) + kA^2]$$

$$= m k_s A^2 [k_2(k - k_s) + k_1 A] \quad (13.50)$$

Assuming k_2 is zero we have

$$(b - k_{us}) = \frac{k_1 k_s}{k} \frac{m}{A} \tag{13.51}$$

and

$$\tau = \frac{kA}{k_1 k_2} \tag{13.52}$$

Solving for τ gives

$$\tau = \frac{(73.7)(0.025)}{(168)(29.6)} = 0.0004 \text{ sec} \tag{13.53}$$

Continuing with the assumption that $k_2 = 0$, we find that

$$\omega_n = \sqrt{k/m} = \sqrt{(73.7)/(6 \times 10^6)} = 3500 \text{ radians/sec} \tag{13.54}$$

A more exact expression for ω_n is given by Eq. 13.23 or by its counterpart

$$\omega_n = \sqrt{\frac{k}{m}\left[1 + \frac{k_s k_2 (b - k_{us})}{kA^2}\right]} \tag{13.55}$$

The assumption that k_2 is zero leads to a somewhat higher natural frequency than when k_2 is not zero, but the term $(k_s k_2)/kA^2(b - k_{us})$ was found to be approximately 0.0007 in this case. Therefore the foregoing assumption regarding k_2 is justified. In any event, this assumption yields a conservative value of ω_n.

13.44. Calculation of Load Sensitivity

The steady-state load sensitivity is given by

$$\left.\frac{\Delta Y}{\Delta F_l}\right|_{ss} = \frac{-kk_2}{k_s(k_1 A - k_2 k_s)\left(\dfrac{kk_2}{k_1 A - k_2 k_s} + 1\right)} = \frac{-1}{k_s\left(1 + \dfrac{k_1 A}{k_2 k} - \dfrac{k_s}{k}\right)} \tag{13.56}$$

The value of k_1/k_2 at the center position of the valve was found experimentally by blocking the ram and finding load pressure as a function of valve position. These measurements showed k_1/k_2 could be as low as 5.3×10^5 lb/in.³ Calculation of the load sensitivity gives

$$\left.\frac{\Delta Y}{\Delta F_l}\right|_{ss} = \frac{-1}{(29.6)\left(1 + \dfrac{(5.3 \times 10^5)(0.025)}{(73.7)} - \dfrac{(29.6)}{(73.7)}\right)}$$

$$= -1.88 \times 10^{-4} \text{ in./lb} \tag{13.57}$$

14

J. A. Hrones

The Analysis of
the Dynamic Performance
of Physical Systems

14.1. INTRODUCTION

The systematic determination of the behavior of a physical system
in the presence of known disturbances is based upon a well-developed
body of mathematics. It is the purpose of this chapter to present certain
selected portions of this material which are most essential to the full
understanding and use of the contents of this book. While a complete
coverage could only be achieved in a much greater space, the tools herein
presented are powerful ones. These tools of analysis can be brought into
use after a mathematical formulation of the actual system under consideration has been made. The creation of such a model involves important decisions of what is of first rank importance and of what is of
low-order importance and hence can be neglected. The skillful building
of relatively simple models of complex systems is the work of the top-
notch engineer, and the powerful methods of analysis presented below
can be used only when the mathematical description of the model is in
the relatively simple forms that permit handling in a reasonable time.

433

14.2. LINEAR ELEMENTS

The largest available useful body of analysis has been developed for systems composed entirely of "linear elements." A linear element is one in which an effect (output) is di-

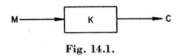

Fig. 14.1.

rectly proportional to its cause (input). The element may be represented by the block shown in Fig. 14.1. The output, C, is proportional to the input, M. There are many examples of such elements. See Table 14.1.

Table 14.1. EXAMPLES OF LINEAR ELEMENTS

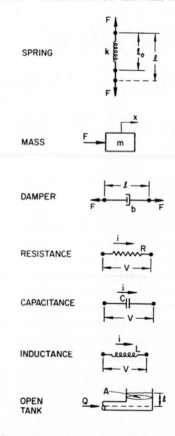

14.3. DEFINITIONS

A *system* is a collection of matter within prescribed boundaries whose behavior can be expressed in terms of quantities which can be measured by an observer positioned outside the boundaries of the system.

A system *quantity* is any characteristic of a system which can be observed or measured from outside the boundaries of the system.

An *input quantity* or an *input* is a characteristic of a system whose value is determined by conditions outside the boundaries of the system.

An *output quantity* or an *output* is a characteristic of a system whose value depends upon the behavior of the system and the values of the inputs to the system.

A *constant quantity* or a *constant* is a characteristic of a system whose value does not change with time over the interval under observation.

A *block diagram* is a simple schematic representation of a system. (See Fig. 14.2.) The inputs to the system are represented by the arrows

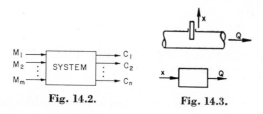

Fig. 14.2.   Fig. 14.3.

$M_1, M_2, \cdots M_m$. The outputs of the system are shown by the arrows $C_1, C_2, \cdots C_n$. The arrows represent a unidirectional cause-effect relationship. For example, the valve shown in Fig. 14.3 controlling the flow, Q, can be represented by the block diagram of Fig. 14.1 where the valve position (x) is the input, M, and the flow, Q, is the output, C. Changes in valve position, x, will cause changes in the flow, Q. However, changes in the flow, Q, will not produce significant changes in the valve setting, x, in most cases. There exists, therefore, a unidirectional cause-effect relationship between x and Q.

The *transfer characteristic* of a system relates a change in system output to the change in input which causes the output change. Where a system has a number of outputs and inputs, a number of transfer characteristics exists.

To illustrate the meaning of these important definitions, consider the system shown in Fig. 14.4, consisting of a disk of moment of inertia J rotating about the axis O–O with a speed θ under the action of the driving

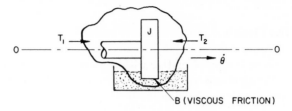

Fig. 14.4.

torque, T_1, and the load torque, T_2, in the presence of viscous friction. The jagged line is the arbitrarily selected boundary of the system, which

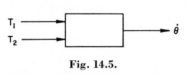

Fig. 14.5.

can be represented by the *block dia-gram* of Fig. 14.5. The torques, T_1 and T_2, are set from outside the boundaries of the system and are *input* quantities. The speed, $\dot{\theta}$, is an *output*, being de-termined by the system *constants:* the moment of inertia, J, and the damping constant, B; and the variable *inputs* T_1 and T_2.

14.4. DETERMINATION OF SYSTEM TRANSFER CHARACTERISTICS

The relationships between output and input quantities can, in general, be established by:

1. Calculations based upon previously determined physical laws.
2. Measurement of output changes responding to known input changes.
3. A combination of items 1 and 2.

In this case the application of Newton's laws will yield the desired result. Summing up the torques acting on the rotating disk in the posi-tive direction of motion gives

$$T_1 - T_2 - B\dot{\theta} = J\ddot{\theta} \qquad (14.1)$$

While Eq. 14.1 is adequate, it is usually desirable to describe the be-havior of the system in terms of changes in system quantities from an initial state rather than in terms of the absolute values of the system quantities. Assume the system of Fig. 14.4 is initially in equilibrium with its various quantities having the values listed as follows.

At $t < 0$:

$$T_1 = T_{1i} \qquad T_2 = T_{2i}$$
$$\dot{\theta} = \dot{\theta}_i \qquad \ddot{\theta} = 0$$

(14.2)

Then at any time later, t, these quantities may be expressed as the initial values plus the changes in value occurring in the interval t.

At $t = t$:

$$T_1 = T_{1i} + \Delta T_1 \qquad T_2 = T_{2i} + \Delta T_2$$
$$\dot{\theta} = \dot{\theta}_i + \Delta\dot{\theta} \qquad \ddot{\theta} = \ddot{\theta}_i + \Delta\ddot{\theta} \qquad \ddot{\theta}_i = 0$$

(14.3)

Substitution of Eqs. 14.2 and 14.3 in Eq. 14.1 yields:

$$T_{1i} - T_{2i} = B\dot{\theta}_i$$

(14.4)

$$(T_{1i} + \Delta T_1) - (T_{2i} + \Delta T_2) = J(\Delta\ddot{\theta}) + B(\dot{\theta}_i + \Delta\dot{\theta})$$

(14.5)

Subtraction of Eq. 14.4 from 14.5 will give Eq. 14.6:

$$\Delta T_1 - \Delta T_2 = J(\Delta\ddot{\theta}) + B(\Delta\dot{\theta})$$

(14.6)

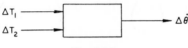

Fig. 14.6.

The *transfer characteristics* of the system are obtained directly from Eq. 14.6, by using the differential operator, $D = d/dt$, and the system may also be represented by the block diagram shown in Fig. 14.6.

$$G_{11} = (\text{transfer characteristic})_{11} = \frac{(\text{output})_1}{(\text{input})_1}$$

(14.7)

$$G_{12} = (\text{transfer characteristic})_{12} = \frac{(\text{output})_1}{(\text{input})_2}$$

(14.8)

$$G_{11} = \left[\frac{\Delta\dot{\theta}}{\Delta T_1}\right]_{\Delta T_2 = 0}$$

(14.9)

$$G_{11} = \frac{1}{JD + B}$$

(14.10)

$$G_{12} = \left[\frac{\Delta\dot{\theta}}{\Delta T_2}\right]_{\Delta T_1 = 0} = -\frac{1}{JD + B}$$

(14.11)

Thus, for the system under discussion, the behavior of the output speed is known in terms of the input torques and the system constants.

$$\Delta \dot{\theta} = \frac{\Delta T_1}{JD + B} - \frac{\Delta T_2}{JD + B} \qquad (14.12)$$

or

$$\Delta \dot{\theta} = G_{11} \, \Delta T_1 + G_{12} \, \Delta T_2 \qquad (14.13)$$

14.5. NONDIMENSIONALIZATION

It is often helpful to write all relationships in nondimensional form. Results then have a greater generality of application, and the actual numbers encountered are usually more easily handled. The method of nondimensionalizing is often arbitrary, and the best way may vary from case to case. However, in many instances convenient reference bases are the normal operating values of the system quantities. In the case of the rotating disk, let the initial values of the system quantities be the normal operating values of those quantities.

$$\frac{\Delta T_1}{T_{1i}} - \left(\frac{T_{2i}}{T_{1i}}\right) \frac{\Delta T_2}{T_{2i}} = \frac{B\theta_i}{T_{1i}} \left(\frac{J}{B} D + 1\right) \frac{\Delta \dot{\theta}}{\dot{\theta}_i} \qquad (14.14)$$

Let

$$m_1 = \frac{\Delta T_1}{T_{1i}} \qquad m_2 = \frac{\Delta T_2}{T_{2i}}$$

$$c = \frac{\Delta \dot{\theta}}{\dot{\theta}_i} \qquad k_{11} = \frac{T_{1i}}{B\dot{\theta}_i}$$

$$\tau = \frac{J}{B} \qquad k_{12} = \frac{T_{2i}}{B\dot{\theta}_i} \qquad (14.15)$$

Then

$$m_1 - \frac{k_{12}}{k_{11}} m_2 = \frac{1}{k_{11}} (\tau D + 1)c \qquad (14.16)$$

Each term appearing in Eq. 14.16 is dimensionless. Block-diagram

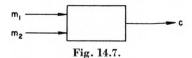

Fig. 14.7.

representation of the original system in terms of the nondimensional quantities of Eq. 14.16 is shown in Fig. 14.7.

The nondimensional transfer characteristics of the system are:

$$g_{11} = \left[\frac{(\text{output})_1}{(\text{input})_1} \right]_{(\text{input})_2=0} \tag{14.17}$$

$$g_{12} = \left[\frac{(\text{output})_1}{(\text{input})_2} \right]_{(\text{input})_1=0} \tag{14.18}$$

$$g_{11} = \frac{k_{11}}{(\tau D + 1)} \tag{14.19}$$

$$g_{12} = \frac{-k_{12}}{(\tau D + 1)} \tag{14.20}$$

14.6. DYNAMIC RESPONSE OF A SYSTEM TO DISTURBANCES

A change in input quantity is frequently referred to as a *disturbance*. The resulting changes in output quantities with time describe the dynamic behavior of the system. The following types of disturbances are often used, both analytically and experimentally:

1. A step change in an input. See Fig. 14.8. The output response as a function of time for this type of disturbance is often called the *step response* of the system. This response is also called the transient re-

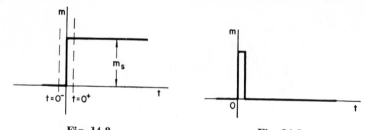

Fig. 14.8.   **Fig. 14.9.**

sponse, although the term "transient" here is somewhat misleading because the response to a step input is often said to consist of two parts: (1) a transient part, and (2) a steady-state part.

2. A pulse change in an input. See Fig. 14.9.

3. A sinusoidal input. See Fig. 14.10. In this test the amplitude m_a is maintained constant and at any given frequency, ω_m, the amplitude c_a and the phase angle, ϕ, of the output are measured after the transients

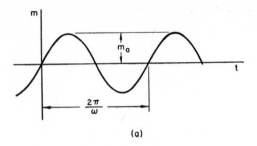

(a)

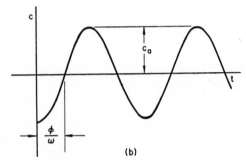

(b)

Fig. 14.10.

have disappeared. The test is repeated a number of times with the same amplitude m_a but with various values of frequency, ω.

4. A random input disturbance. See Fig. 14.11.

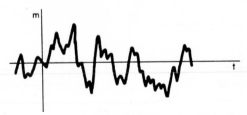

Fig. 14.11.

Because of space limitations only disturbances of the types 1 and 3 will be discussed here.

A person who seeks to gain a thorough understanding of the dynamic performance of physical systems must rely heavily on the accumulated experiences of daily life. The change in temperature of a well-stirred pan of water when subjected to the step input of being placed over a heat source, the change in temperature of a house when subjected to a step change in thermostat setting, the motion of a car moving over a "washboard" road, the characteristic behavior of a weight suspended by a spring, and countless other common everyday experiences give us a wealth of "know-how" to use in dealing with other problems in dynamics which may be less familiar to us.

14.7. THE SINGLE-CAPACITY SYSTEM [1]

The rotating disk and the pan of well-stirred water are systems in which all or nearly all of the energy contained within the boundaries of the system is stored in a single reservoir and hence can be precisely stated if one and only one output quantity is known. Thus in the case of the rotating disk the angular speed completely determines the amount of energy stored, and in the case of the well-stirred pan of water, the temperature of the water defines explicitly the energy content of that system. Such systems are known as single-capacity systems, and if linear, their dynamic behavior can be expressed by a linear differential equation of the first order.[1] See Eq. 14.16.

14.8. THE SOLUTION OF THE FIRST-ORDER DIFFERENTIAL EQUATION

Solving the first-order differential equation, like solving many more difficult problems, rests upon our ability to estimate at the very outset what the answer is. In other words, we can solve such equations if we know the form of the answer beforehand. This situation is not as absurd as it may appear if we draw upon the resources of experience mentioned above.

Consider the rotating disk (Figs. 14.4 and 14.7), whose behavior is given by Eqs. 14.1 and 14.16. This statement of Newton's law establishes the relationship that the difference between the driving and load torques is equal to the viscous friction or damping torque if the system is operating at constant speed. If the load torque is held constant and the driving torque is subjected to a step change in its value (see Fig. 14.8), a new value of speed must result if the system is again to operate at constant speed.

[1] J. A. Hrones and J. B. Reswick, "Nondimensional Study of Proportional-Plus-Reset Control of a Single-Capacity System," *Trans. ASME*, Vol. 73 (July 1951), pp. 511–517.

Letting $m_2 = 0$ in Eq. 14.16, we have

$$(\tau D + 1)c = k_{11}m_1 \qquad (14.21)$$

or

$$\tau \frac{dc}{dt} + c = k_{11}m_1 \qquad (14.22)$$

Thus, if the system does reach a new steady state (new level of constant-speed operation), the acceleration at that time will be zero,

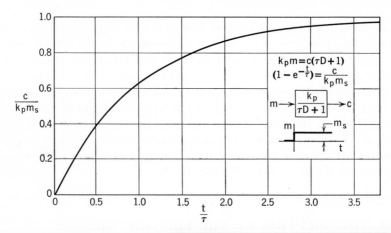

Fig. 14.12.

and the nondimensional speed change which the system will experience will be given by: As $t \to \infty$,

$$c_{ss} = k_{11}m_{1s} \qquad (14.23)$$

and the constant k_{11} is sometimes referred to as the steady-state gain of the system.

However, immediately after the application of the step change in torque, there can be no speed change, because any speed change requires the existence of an acceleration for a finite length of time. Therefore: At $t = 0^+$, $c = 0$ (see Fig. 14.12),

$$\left.\frac{dc}{dt}\right|_{t=0} = \frac{k_{11}m_{1s}}{\tau} \qquad (14.24)$$

Equation 14.24 establishes the initial slope of the speed change versus time plot, and Eq. 14.23 sets the final value of that plot. Intermediate values can be arrived at by continuing this line of reasoning. The existence of an acceleration for a finite time produces a speed change, thus changing the damping torque. The net torque available for acceleration is thereby decreased, and hence the acceleration decreases to zero in the steady state. Thus, the curve shown in Fig. 14.12 may be a reasonable estimate of system performance. A possible mathematical means for expressing the change in speed is:

$$c_t = c - c_{ss} = Ke^{\lambda t} \qquad (14.25)$$

The quantity $(c - c_{ss})$ is the transient part of the solution.

Equation 14.22 may be rewritten as:

$$(\tau D + 1)c = 0 + k_{11}m_1 \qquad (14.26)$$

and for the special case when a step change in m_1 occurs, Eq. 14.26 in turn may be re-expressed as two equations:

$$(\tau D + 1)c_t = 0 \qquad (14.27a)$$

$$(\tau D + 1)c_{ss} = k_{11}m_{1s} \qquad (14.27b)$$

The sum of the *solutions* of Eqs. 14.27a and b gives the general solution to Eq. 14.26. Equation 14.27a is known as the characteristic equation of the system and describes the fundamental nature of the system independent of the kind of disturbance to which it is subjected. The solution of this equation is the transient part of the complete solution.

The solution for Eq. 14.27b is the steady-state solution for a particular disturbance, and it is already given by Eq. 14.23. Substitution of the assumed transient part of the solution from Eq. 14.25 in Eq. 14.27a gives

$$(\tau\lambda + 1)Ke^{\lambda t} = 0 \qquad (14.28)$$

When $\lambda = -(1/\tau)$, the equation is satisfied, and the solution to Eq. 14.27a is

$$c_t = Ke^{-t/\tau} \qquad (14.29)$$

Hence adding Eqs. 14.29 and 14.23, we get

$$c = c_t + c_{ss}$$
$$c = Ke^{-t/\tau} + k_{11}m_{1s} \qquad (14.30)$$

Imposing the initial condition that at $t = 0^+$, $c = 0$, we get

$$c = k_{11}m_{1s}(1 - e^{-t/\tau}) \qquad (14.31)$$

Equation 14.30 is a general expression for the dynamic response of a single-capacity system to a step disturbance in an input. The time constant τ is a measure of how fast the system approaches a new steady state following a disturbance.

In summary the solution of the first-order linear differential equation,

$$\tau \frac{dc}{dt} + c = k_p m \tag{14.32}$$

or

$$(\tau D + 1)c = k_p m$$

for a step input is of the form

$$c = K_1 + K_2 e^{-t/\tau} \tag{14.33}$$

The values of the coefficients K_1 and K_2 are determined from known conditions, often the initial and final conditions. The time constant τ is a measure of the speed of response of the system, 95 per cent of the total output change being realized in a time interval of about three times the time constant τ. See Fig. 14.12 for a graphical presentation of the solution. A single-capacity system of this kind is often called a first-order lag.

14.9. THE OPERATIONAL BLOCK DIAGRAM

It is often convenient and helpful to break the system block diagram into a number of simpler operational blocks. Equation 14.32 can thus be schematically represented by a detailed operational block diagram as shown in Fig. 14.13.

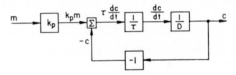

Fig. 14.13.

It will be noted that three types of operational blocks are used:

1. *An adder* (Σ) capable of adding a number of quantities.
2. *An integrator* $(1/D)$ capable of continuously producing an output which is the time integral of the input.
3. *A coefficient* capable of multiplying an input by a constant. *Note:* in order not to confuse this operation with that of multiplication of two variables, this block will be called a "coefficient block," and will be labeled with the name of the coefficient.

14.10. THE GENERALITY OF THE SINGLE-CAPACITY SYSTEM

While the solution of a first-order differential equation has been derived with reference to a specific mechanical system, the same solution results from considering many diverse situations in nature,[2] some of which are covered below.

I. AN ELECTRICAL SYSTEM.   (Fig. 14.14.)

$$V = Ri + \frac{1}{C} \int i \, dt$$

$$V = \left(R + \frac{1}{CD} \right) i$$
(14.34)

$$CDV = (RCD + 1)i$$
(14.35)

The operator, D, was manipulated above as an algebraic quantity and may be handled in this fashion as long as the equations are those of linear systems.   The equation in its classical form is directly recoverable from Eq. 14.35 and is reproduced below.

$$C \frac{dV}{dt} = RC \frac{di}{dt} + i$$
(14.36)

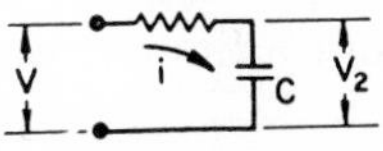

Fig. 14.14.

RC is the time constant τ.

The solution for the current i produced by a step change in V is from Eq. 14.33:

$$i = K_1 + K_2 e^{-t/\tau}$$
(14.37)

The known conditions are:
At $t = 0^+$

$$DV = 0$$

$$V = V_s \qquad V_2 = 0 \qquad i = \frac{V_s}{R}$$

At $t = \infty$

$$DV_2 = 0$$

$$i = 0 \qquad V_2 = V_s$$

[2] J. D. Trimmer, *Response of Physical Systems*, John Wiley and Sons, New York, 1950.

Thus at time, t, equal to infinity Eq. 14.37 yields, $i = K_1 + 0 = 0$, so that

$$K_1 = 0$$

And at time, t, equal to 0^+ (14.38)

$$i = 0 + K_2 = \frac{V_s}{R}$$

Thus the complete solution is given by

$$i = \frac{V_s}{R} e^{-t/\tau} \tag{14.39}$$

II. THE FLOW OF LIQUIDS. (Fig. 14.15.) Consider the change of the height of the liquid as a function of time in the tank of area A when a

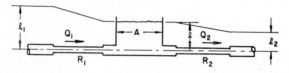

Fig. 14.15.

step change is made in the entrance pressure head, l_1. Using the concept of continuity, we obtain

$$Q_1 - Q_2 = A\, Dl \tag{14.40}$$

In terms of changes from an initial steady-flow condition, Eq. 14.40 becomes

$$\Delta Q_1 - \Delta Q_2 = A\, D(\Delta l) \tag{14.41}$$

If the flow resistances are linear, the change in flow rate ΔQ_1 is

$$\Delta Q_1 = \frac{\Delta l_1 - \Delta l}{R_1} \tag{14.42}$$

and the change in flow rate ΔQ_2 is

$$\Delta Q_2 = \frac{\Delta l - \Delta l_2}{R_2} \tag{14.43}$$

Let the exit pressure head, l_2, be constant and combine Eqs. 14.41, 14.42, and 14.43 to obtain

$$\left(AD + \frac{1}{R_1} + \frac{1}{R_2} \right) \Delta l = \frac{\Delta l_1}{R_1}$$

or

$$\left(\frac{R_1}{R_1} + \frac{R_1}{R_2}\right)\left[\frac{A}{(1/R_1) + (1/R_2)}D + 1\right]\Delta l = \Delta l_1$$

$$(\tau D + 1)\,\Delta l = \left[\frac{1}{1 + (R_1/R_2)}\right]\Delta l_1 \qquad (14.44)$$

where $\tau = \dfrac{A}{(1/R_1) + (1/R_2)}$

Again it is found that the same basic relationship governs the dynamic behavior of the system. The time solution when Δl_1 is a step change is given by

$$\Delta l = K_1 + K_2 e^{-t/\tau} \qquad (14.45)$$

From Eq. 14.44, when $t = \infty$, the system will again operate in a steady-flow condition, hence the steady-state value of Δl will be

$$\Delta l_{ss} = \left[\frac{1}{1 + (R_1/R_2)}\right]\Delta l_{1s} \qquad (14.46)$$

By analogy to Eq. 14.31 the full solution is

$$\Delta l = \left[\frac{1}{1 + (R_1/R_2)}\right]\Delta l_{1s}\,[1 - e^{-t/\tau}] \qquad (14.47)$$

III. The Flow of Heat.   Figure 14.16 illustrates a common situation in the process industries. A liquid is being heated by an immersed electric coil. A mixer is used to keep the entire liquid at essentially uniform temperature (T_m). In order to design the system properly the response of changes in temperature, T_m, to changes in the input flow of energy must be known. If the mixer work is assumed negligible, the container walls have negligible capacity to store energy, and the bath temperature is uniform, the following relationships hold:

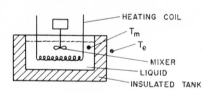

Fig. 14.16.

$$Q_i - h_o(T_m - T_e) = C_h\rho\mathtt{v}\frac{dT_m}{dt} \qquad (14.48)$$

and therefore

$$(C_h\rho\mathtt{v}D + h_o)T_m = Q_i + h_oT_e \qquad (14.49)$$

where Q_i = rate of heat flow into process, in. lb/sec

h_o = over-all heat-transfer coefficient, in. lb/sec °R

C_h = specific heat of fluid, in.^{2}/sec^2 °R

$\mathcal{V}$ = volume of fluid, in.^{3}

ρ = density of fluid, lb sec^2/in.^{4}

T_e = temperature of surroundings, °R

In terms of changes from some initial steady-state temperatures, T_{ei} and T_{mi},

$$\Delta Q_i + h_o \, \Delta T_e = h_o \left(\frac{C_h \rho \mathcal{V}}{h_o} D + 1 \right) \Delta T_m \qquad (14.50)$$

The time constant, τ, of this system is $(C_h \rho \mathcal{V})/h_o$. It, again, is the first-order linear differential equation characteristic of a single-capacity system. If we assume that the change in environmental temperature is zero, a step change in rate of flow of heat input results in a bath temperature change given by

$$\Delta T_m = \frac{\Delta Q_{is}}{h_o} (1 - e^{-t/\tau}) \qquad (14.51)$$

While in actual fact few systems are *exactly* represented as single-capacity linear systems, a great many systems can be so represented with results that fall well within the bounds of acceptability. Nevertheless, there are systems where appropriate mathematical modeling requires higher-order differential equations.

14.11. THE SECOND-ORDER DIFFERENTIAL EQUATION

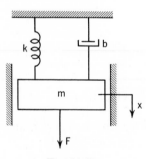

Fig. 14.17.

One of the best-known physical systems whose behavior is described by the second-order linear differential equation is the mass-spring-dashpot system shown schematically in Fig. 14.17. Such a system is capable of storing energy in two elements of the system. By compressing the spring, energy is stored, which can be recovered upon removal of the compression. Energy is also stored in the mass whenever it possesses a velocity. The flow of energy from one storage capacity (the spring) to a second storage capacity (the mass) and the reversal of such flow causes the interesting, and often undesirable, phenomenon of oscillation. The presence of the dashpot provides an element of the system capable of dissipating energy to the environs. In Fig. 14.18 the mass is shown in a displaced position and all forces are indicated.

Summing the forces in the direction of positive displacement according to Newton's law yields the following result:

$$F - kx - b\dot{x} = m\ddot{x}$$

or

$$(mD^2 + bD + k)x = F \qquad (14.52)$$

The determination of the displacement, x, with time following a step change in the force, F, rests upon sufficient prior knowledge to permit an estimate of the answer. If, for example, we say that k is small enough so that the term kx is negligible, then the system becomes a single-capacity one for which we already know the answer. In this instance we know from Eq. 14.33 that the answer is

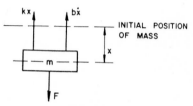

Fig. 14.18.

$$\dot{x} = K_1 + K_2 e^{-t/\tau} \qquad (14.53)$$

If, on the other hand, we assume that the energy-dissipating term $b\dot{x}$ is negligibly small, we have the case of a mass-supported spring. From previous experience we know that such a system subjected to a disturbance will oscillate endlessly at some frequency, ω. Such behavior can be mathematically portrayed by the following relationship:

$$x = K_1 + K_3 \sin \omega t + K_4 \cos \omega t \qquad (14.54)$$

14.111. Complex Numbers

While Eqs. 14.53 and 14.54 appear in different forms, they may be shown to be of the same general form. In this connection the concept of a complex number is useful. Consider the plane shown in Fig. 14.19. The location of any point in the plane can be designated in the following manner:

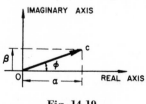

Fig. 14.19.

$$C = \alpha + j\beta$$

where j indicates $\sqrt{-1}$. All numbers which contain the $\sqrt{-1}$ are plotted in the "imaginary" direction. All other numbers are plotted in the "real" direction. This representation can also indicate the vector OC. For example, the length of the vector is $\sqrt{\alpha^2 + \beta^2}$ and its direction is given by

$$\tan \phi = \frac{\beta}{\alpha}$$

Now consider a vector of unit length plotted at an angle, ϕ, to the positive real axis. See Fig. 14.20.

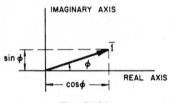

Fig. 14.20.

By use of the arbitrary definition of the complex plane established above, the following relation results:

$$\bar{1} = \cos \phi + j \sin \phi \qquad (14.55)$$

Both $\cos \phi$ and $\sin \phi$ can be written in the form of series expansions:

$$\cos \phi = 1 - \frac{\phi^2}{2!} + \frac{\phi^4}{4!} - \frac{\phi^6}{6!} + \cdots \qquad (14.56)$$

$$\sin \phi = \phi - \frac{\phi^3}{3!} + \frac{\phi^5}{5!} - \frac{\phi^7}{7!} + \cdots \qquad (14.57)$$

$$\cos \phi + j \sin \phi = 1 + j\phi - \frac{\phi^2}{2!} - j\frac{\phi^3}{3!} + \frac{\phi^4}{4!} + j\frac{\phi^5}{5!} - \cdots \qquad (14.58)$$

The exponential $e^{j\phi}$ can also be expanded in series form:

$$e^{j\phi} = 1 + j\phi + \frac{(j\phi)^2}{2!} + \frac{(j\phi)^3}{3!} + \frac{(j\phi)^4}{4!} + \frac{(j\phi)^5}{5!} + \cdots \qquad (14.59)$$

$$= 1 + j\phi - \frac{\phi^2}{2!} - j\frac{\phi^3}{3!} + \frac{\phi^4}{4!} + j\frac{\phi^5}{5!} - \cdots \qquad (14.60)$$

Equations 14.58 and 14.60 are identical. Therefore,

$$e^{j\phi} = \cos \phi + j \sin \phi \qquad e^{j\phi} + e^{-j\phi} = 2 \cos \phi \qquad e^{j\phi} - e^{-j\phi} = 2j \sin \phi$$

$$(14.61)$$

Hence $e^{j\phi}$ indicates a unit vector in the complex plane located at an angle, ϕ, to the real axis. Therefore, the transient part of the solution given in Eq. 14.54 may also be written

$$K_3 \sin \omega t + K_4 \cos \omega t = x_t = K_a e^{j\omega t} + K_b e^{-j\omega t} \qquad (14.62)$$

where $K_a = \dfrac{-K_3 j}{2} + \dfrac{K_4}{2} \qquad K_b = \dfrac{K_3 j}{2} + \dfrac{K_4}{2} \qquad |K_a| = |K_b|$

It is important to recognize that the right-hand side of Eq. 14.62 represents counter-rotating vectors of length $|K_a| = |K_b|$, each rotating at an angular speed of ω radians per unit time, as in Fig. 14.21.

Returning to the general solution of Eq. 14.52, for a step change in F, assume the transient part of the solution to be of the same form as that assumed for the single-capacity system.

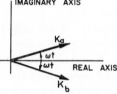

Fig. 14.21.

then

$$x = Ke^{\lambda t} \qquad (14.63a)$$

$$Dx = K\lambda e^{\lambda t} \qquad (14.63b)$$

$$D^2 x = K\lambda^2 e^{\lambda t} \qquad (14.63c)$$

In order to find values of λ which describe the possible transient parts of the solution to Eq. 14.52, the right-hand side of Eq. 14.52 is set equal to zero in order to form the characteristic equation, as in the previous case of a first-order system. Then employing Eqs. 14.63a, b, and c yields

$$(m\lambda^2 + b\lambda + k)Ke^{\lambda t} = 0 \qquad (14.64)$$

If k is not to be zero and λ is not to be equal to minus infinity, then

$$\lambda = -\frac{b}{2m} \pm \sqrt{(b/2m)^2 - (k/m)}$$

or

$$\lambda_a = -\frac{b}{2m} + \sqrt{(b/2m)^2 - (k/m)} \qquad (14.65a)$$

and

$$\lambda_b = -\frac{b}{2m} - \sqrt{(b/2m)^2 - (k/m)} \qquad (14.65b)$$

The values λ_a and λ_b are often referred to as the *roots* of the characteristic equation.

There are thus two possible terms in the transient part of the solution as indicated previously for the special case when $b = 0$. The following substitutions are now made:

$$\frac{k}{m} = \omega_n{}^2 \tag{14.66}$$

$$\frac{b}{2m} = \zeta\omega_n \tag{14.67}$$

The symbol ω_n denotes the undamped natural frequency of the system (the frequency of the oscillation that results from a step input when there is negligible energy dissipated in the system so that the oscillation decays infinitely slowly). The symbol ζ is defined as the damping ratio of the system and represents the ratio of the value of the viscous damping coefficient to the value that would be required to produce critical damping (that is, roots of the characteristic equation are negative, real, and equal). Thus only two constants, ζ and ω_n, are required to specify the *dynamic performance* of the system instead of the three system constants, k, b, and m. Note, however, that k is then needed to describe the *steady-state response* of the system. (See Eq. 14.74.) The two roots λ_a and λ_b may also be used to describe the *dynamic performance* of the system because they are functions only of ζ and ω_n (or vice versa). It is also possible to express the left-hand side of Eq. 14.52 in terms of ζ and ω_n because of the following identity:

$$(mD^2 + bD + k)x \equiv \left(\frac{D^2}{\omega_n{}^2} + \frac{2\zeta D}{\omega_n} + 1\right)kx \tag{14.68}$$

From Eq. 14.63a the two terms of the transient part of the solution for x are given by

$$x_t = K_a e^{\lambda_a t} + K_b e^{\lambda_b t} \tag{14.69}$$

or

$$x_t = K_a \exp\left[-\omega_n(\zeta + \sqrt{\zeta^2 - 1})t\right] + K_b \exp\left[-\omega_n(\zeta - \sqrt{\zeta^2 - 1})t\right] \tag{14.70}$$

or

$$x_t = e^{-\zeta\omega_n t}(K_a e^{j\omega t} + K_b e^{-j\omega t})$$

where $\omega = \omega_n\sqrt{\zeta^2 - 1}$

and it is seen that when $\zeta = b = 0$, the solution agrees with Eq. 14.62.

In general, three types of performance exist:

1. When $\zeta > 1$, the roots λ_a and λ_b of the characteristic equation are negative and real. The transient response is nonoscillatory and the system is said to be overdamped.

2. When $\zeta = 1$, both roots are equal and negative real numbers. The performance of the system is on the boundary between oscillation and nonoscillation and is referred to as critically damped.

3. When $\zeta < 1$, the roots are complex numbers. The performance of the system is oscillatory in character and is referred to as an underdamped system.

For the case of equal roots, Eq. 14.71 must be used instead of Eq. 14.69

$$x_t = (K_c + K_d t)e^{-\zeta \omega_n t} \qquad (14.71)$$

For the case where the roots are complex ($\zeta < 1$)

$$\lambda_a, \lambda_b = -\omega_n(\zeta \pm j\sqrt{1 - \zeta^2}) \qquad (14.72)$$

Equation 14.69 becomes

$$x_t = K_t e^{-\zeta \omega_n t} \cos \left[(\omega_n \sqrt{1 - \zeta^2})t - \psi\right] \qquad (14.73)$$

The transient solutions obtained above do not involve the form of input or disturbance which acts on the system. To obtain a complete solution, the steady-state response to a particular disturbance or to a particular group of disturbances must be added to the transient solution.

Let us assume that the system shown in Fig. 14.17 was initially at rest. At time $t = 0$ a step change in the force ΔF_s is made, as shown in Fig. 14.22. What is x as a function of time?

The solution is the sum of the transient and steady-state solutions, where the transient solution is the general time solution of the differential equation in the absence of disturbances (the solution to Eq. 14.64), and the steady-state solution is the solution at $t = \infty$ for the applied disturbance.

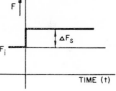

Fig. 14.22.

The steady-state solution is obtained by setting $D = 0$ in Eq. 14.52 because after sufficient time from the application of the step disturbance shown in Fig. 14.22, all time derivatives are zero. Therefore

$$x_{ss} = \frac{\Delta F_s}{k} \qquad (14.74)$$

The complete solution is

$$x = x_t + x_{ss}$$

$$x = K_t e^{-\zeta \omega_n t} \cos (\omega t - \psi) + \frac{\Delta F_s}{k} \qquad (14.75)$$

where $\omega = \omega_n \sqrt{1 - \zeta^2}$

If the system is initially at rest, the conditions at $t = 0^+$ are

$$x = 0 \qquad \dot{x} = 0$$

Therefore

$$0 = K_t \cos (-\psi) + \frac{\Delta F_s}{k} \qquad (14.76)$$

$$0 = -\zeta \omega_n K_t \cos (-\psi) - K_t \omega \sin (-\psi) \qquad (14.77)$$

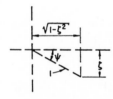

$$\tan (-\psi) = -\frac{\zeta \omega_n}{\omega} = -\frac{\zeta}{\sqrt{1 - \zeta^2}}$$

From Fig. 14.23 $\cos (-\psi) = \sqrt{1 - \zeta^2}$.

From Eq. 14.76 $K_t = -(1/\sqrt{1 - \zeta^2})(\Delta F_s/k)$.

Fig. 14.23.   Substitution in Eq. 14.75 gives

$$x = \frac{\Delta F_s}{k} \left[1 - \frac{1}{\sqrt{1 - \zeta^2}} e^{-\zeta \omega_n t} \cos (\omega t - \psi) \right] \qquad (14.78)$$

where $\psi = \tan^{-1} \dfrac{\zeta}{\sqrt{1 - \zeta^2}}$

or

$$\psi = \cos^{-1} \sqrt{1 - \zeta^2}$$

$$\zeta = \frac{b}{2\sqrt{km}}$$

$$\omega = \omega_n \sqrt{1 - \zeta^2} = \sqrt{\frac{k(1 - \zeta^2)}{m}}$$

Figure 14.24 is a plot of Eq. 14.78 for various values of the damping ratio, ζ. It is plotted to the nondimensional quantities c/c_{ss} and $\omega_n t$.
From the curves of Fig. 14.24 it can be seen that:

1. ω is the angular frequency in radians per unit time of the damped motion.

2. ω_n is the angular frequency in radians per unit time for the special case when no damping is present ($\zeta = 0$). This is called the undamped

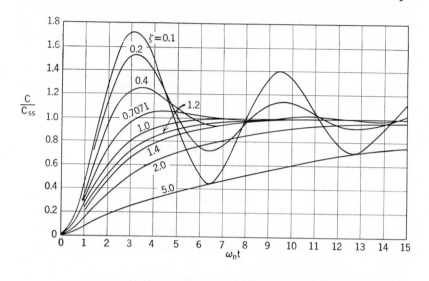

RESPONSE OF A SECOND-ORDER SYSTEM TO A STEP INPUT

Fig. 14.24.

natural frequency of the system. It is a measure of how fast the system responds to an input.

3. ζ is a measure of the rate of decay of the oscillation in an underdamped system. In an overdamped system, higher values of ζ result in slower responses.

While the second-order differential equation has been discussed in connection with the mass-spring-dashpot system, it is characteristic of many well-known and often-encountered physical situations. Thus, complete solutions are available without further work where the disturbances and initial conditions are the same as those for which the plots were developed.

The importance of complete familiarity with the behavior of the single- and two-capacity system cannot be overstressed. Such knowledge gives the physical insight essential to the handling of more complex situations.

14.12. THE SOLUTION OF HIGHER-ORDER DIFFERENTIAL EQUATIONS

It is not the intent to cover in detail the general solution of differential equations. There are many excellent texts available.[3-6] However, it is possible to draw from our experiences with the simple first- and second-order systems and make valuable projections concerning the behavior of more complicated systems.

The results already obtained are repeated below.

First Order.

$$(\tau D + 1)C = k_p M \tag{14.79}$$

$$\frac{C}{k_p M} = \frac{1}{\tau D + 1} = \frac{1}{\tau(D - \lambda)} \tag{14.80}$$

and for a step input

$$\frac{C}{C_{ss}} = \frac{C}{k_p M_s} = K_1 + K_2 e^{\lambda t} \tag{14.81}$$

Second Order.

$$(a_2 D^2 + a_1 D + 1)C = \left[\left(\frac{D}{\omega_n} \right)^2 + 2\zeta \left(\frac{D}{\omega_n} \right) + 1 \right] C = k_p M \tag{14.82}$$

---

[3] Trimmer, *op. cit.*

[4] C. S. Draper, W. McKay, and S. Lees, *Instrument Engineering*, Vols. I, 1952, II, 1953, and III, 1955, McGraw-Hill Book Co., New York.

[5] A. W. Porter, *Introduction to Servomechanisms*, 2nd ed., John Wiley and Sons, New York, 1953.

[6] M. F. Gardner and J. L. Barnes, *Transients in Linear Systems*, Vol. 1, John Wiley and Sons, New York, 1942.

$$\frac{C}{k_p M} = \frac{1}{(D/\omega_n)^2 + 2\zeta(D/\omega_n) + 1} = \frac{\omega_n{}^2}{(D - \lambda_a)(D - \lambda_b)} \qquad (14.83)$$

and for a step input

$$\frac{C}{C_{ss}} = \frac{C}{k_p M_s} = K_1 + K_a e^{\lambda_a t} + K_b e^{\lambda_b t} \qquad (14.84)$$

THIRD ORDER.

$$(a_3 D^3 + a_2 D^2 + a_1 D + 1)C = \left[\left(\frac{D}{\omega_n}\right)^2 + \frac{2\zeta D}{\omega_n} + 1\right](\tau D + 1)C = k_p M$$

$$(14.85)$$

$$\frac{C}{k_p M} = \frac{1}{[(D/\omega_n)^2 + 2\zeta(D/\omega_n) + 1](\tau D + 1)}$$

$$= \frac{1}{a_3(D - \lambda_a)(D - \lambda_b)(D - \lambda_c)} \qquad (14.86)$$

Projecting the results of the first- and second-order systems for a step input gives

$$\frac{C}{C_{ss}} = \frac{C}{k_p M_s} = K_1 + K_a e^{\lambda_a t} + K_b e^{\lambda_b t} + K_c e^{\lambda_c t} \qquad (14.87)$$

14.121. Nature of the Roots of a Third-Order System

Three important possibilities exist:

1. At least one root is a positive real number or a complex number with a positive real part. In such instances, the magnitude will increase without limit as time proceeds.

2. All roots are negative real roots. The system will reach a new steady-state condition without oscillation.

3. Two roots are conjugate complex with negative real parts, and the third root is negative real. The system will reach a new steady-state condition following a nonrepetitive disturbance but will exhibit oscillation in doing so.

Thus, it is seen that the roots of the characteristic equations yield a great deal of information, and once in hand, they permit a rapid evaluation of the dynamic behavior of a system.

14.122. Third-Order System with All Negative Real Roots

The three-capacity system is represented in Fig. 14.25 and in equivalent form in Fig. 14.26, when all roots are real and negative.

By use of the results of the single-capacity system previously obtained, the response of the three-capacity system to a step input can readily be estimated. The sketches in Fig. 14.26 are shown for the case where τ_1, τ_2, and τ_3 are approximately the same. A system with a large number of equal time lags gives a response to a step input nearly the equivalent of a delay in cascade with a time lag as shown in Fig. 14.27.

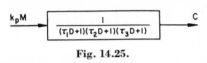

Fig. 14.25.

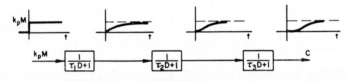

Fig. 14.26.

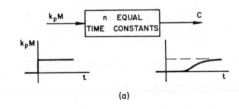

(a)

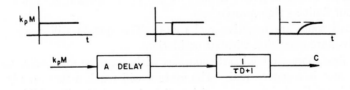

(b)

Fig. 14.27.

14.123. Third-Order System with One Negative Real Root and Two Complex Roots

In Fig. 14.28, the system is shown as a single-capacity system in cascade with a two-capacity system, and from the knowledge of each of these systems the passage of a step input through the three-capacity system can be estimated.

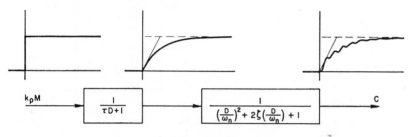

Fig. 14.28.

The foregoing treatment of the solution of linear differential equations of low order yields results of importance which can be used in a straightforward and simple fashion, particularly when the order is no greater than 2.   For higher-order equations, which occur often, the classical methods become cumbersome and long, and in such cases use is made of the analogue-computer and root-locus techniques.

The rapid growth of the communications field in the 1930's provided a well-developed foundation for the steady-state analysis of systems subjected to sinusoidal inputs.   During the 1940's this was further developed and brought to bear on the solution of many control problems.   A brief presentation of this analysis is made in the following pages.

14.13. THE STEADY-STATE FREQUENCY RESPONSE OF PHYSICAL SYSTEMS

Consider the linear system shown in Fig. 14.29.   If the input, M, is sinusoidal and sufficient time is allowed for the initial transient to die out, the output, C, will also be sinusoidal with the same frequency as the input but out of phase with it.   See Figs. 14.10a and 14.10b. If the amplitude ratio, $C_a/k_p M_a$, and the phase angle, ϕ, are recorded and the test is repeated at a number of frequencies, the accumulated results may be plotted as

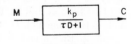

Fig. 14.29.

shown in Fig. 14.30. The results shown in Fig. 14.30 are for a single-capacity system. They can be obtained by experiment on an actual system or they can be determined mathematically.[7-11]

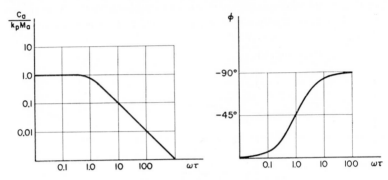

Fig. 14.30.

14.131. Mathematical Determination of System Frequency Response of Single-Capacity System

Consider the single-capacity system whose transfer characteristic is

$$\frac{C}{k_p M} = \frac{1}{\tau D + 1} \tag{14.88}$$

The input, M, is

$$M = M_a \sin \omega t$$

which is often represented by   (14.89)

$$= M_a e^{j\omega t}$$

Assume the output to be represented by

$$C = A e^{j\omega t} \tag{14.90}$$

then

$$DC = j\omega A e^{j\omega t} \tag{14.91}$$

[7] A. C. Hall, *The Analysis and Synthesis of Linear Servomechanisms*, Technology Press, Cambridge, Mass., 1943.

[8] G. S. Brown and D. P. Campbell, *Principles of Servomechanisms*, John Wiley and Sons, New York, 1948.

[9] H. Chestnut and R. W. Mayer, *Servomechanisms and Regulating System Design*, Vol. I, 1951 and Vol. II, 1955, John Wiley and Sons, New York.

[10] G. S. Brown and A. C. Hall, "Dynamic Behavior and Design of Servomechanisms," *Trans. ASME*, Vol. 68 (July 1946), pp. 503–524.

[11] R. Oldenburger, "Frequency-Response Data Presentation, Standards and Design Criteria," *Trans. ASME*, Vol. 76 (November 1954), pp. 1155–1169.

Substitution in Eq. 14.88 yields

$$(j\omega\tau + 1)Ae^{j\omega t} = k_p M_a e^{j\omega t} \tag{14.92}$$

$$\left(\frac{C}{k_p M}\right)_{j\omega} = \frac{A}{k_p M_a} = \frac{1}{j\omega\tau + 1} \tag{14.93}$$

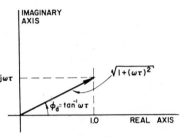

Fig. 14.31.

where the *subscript jω* signifies that the quantity to which it applies is a function of $j\omega$.

It will be observed from Eq. 14.93 that the steady-state output for a sinusoidal input can be obtained directly from Eq. 14.88 by the mere substitution of $j\omega$ for D. Hence, the solution is obtained directly without the need of determining the roots of the equation. For higher-order systems this is a distinct advantage. To obtain a somewhat more usable form of Eq. 14.93, the denominator of the right-hand side may be plotted as a vector on the complex plane. See Fig. 14.31.

The quantity $(j\omega\tau + 1)$ may thus be expressed in terms of a vector of amplitude $\sqrt{1 + (\omega\tau)^2}$ at an angle ϕ_d. Thus

$$\left|\frac{k_p M_a}{A}\right| = \frac{k_p M_a}{C_a} = \sqrt{1 + (\omega\tau)^2}$$

$$\measuredangle \frac{k_p M_a}{A} = \phi_d = \tan^{-1}\omega\tau$$

and

$$A = C_a e^{-j\phi_d}$$

Similarly, the quantity $1/(j\omega\tau + 1)$ may be expressed by a vector at an angle ϕ. Its amplitude is given by

$$\left|\frac{A}{k_p M_a}\right| = \frac{C_a}{k_p M_a} = \left|\frac{1}{j\omega\tau + 1}\right| = \frac{1}{\sqrt{1 + (\omega\tau)^2}} \tag{14.94}$$

and its phase angle, ϕ, is given by

$$\phi = \measuredangle \left(\frac{A}{k_p M_a}\right) = -\measuredangle \left(\frac{k_p M_a}{A}\right) = -\tan^{-1}(\omega\tau) \tag{14.95}$$

and

$$A = C_a e^{j\phi}$$

The functions of $(\omega\tau)$ expressed by Eqs. 14.94 and 14.95 are illustrated numerically in Table 14.2.

Table 14.2

Dimensionless Frequency, $\omega\tau$	$\sqrt{1 + (\omega\tau)^2}$	$\dfrac{C_a}{k_p M_a}$	ϕ
0	1.000	1.000	0
0.1	1.005	0.995	$-5.7°$
0.5	1.118	0.894	$-26.6°$
1	1.414	0.707	$-45°$
5	5.10	0.196	$-78.7°$
10	10.05	0.100	$-84.3°$
∞	∞	0	$-90°$

At low values of $\omega\tau$ [values of frequency ω that are small relative to the reciprocal of the time constant $(1/\tau)$], the amplitude and phase of the input are faithfully reproduced at the output of the system; at large values of $\omega\tau$, the output is markedly attenuated and lags the input by a substantial angle. At very high values of $\omega\tau$ the output for any input is essentially zero and whatever output exists lags the input by 90°.

14.132. The Log Plot of Single-Capacity Systems

The frequency response of the single-capacity system may be expressed by

$$\left(\frac{C}{k_p M}\right)_{j\omega} = \left(\frac{C_a}{k_p M_a}\right)_{j\omega} e^{j\phi} = \frac{1}{j\omega\tau + 1} = \frac{e^{j\phi}}{\sqrt{(\omega\tau)^2 + 1}} \qquad (14.96)$$

Taking the logarithm of the magnitude of both sides gives

$$\log\left(\frac{C_a}{k_p M_a}\right)_{j\omega} = 0 - \frac{1}{2}\log\left[(\omega\tau)^2 + 1\right] \qquad (14.97)$$

For very small values of $\omega\tau$, $(\omega\tau \ll 1)$

$$\log\left(\frac{C_a}{k_p M_a}\right)_{j\omega} \approx 0$$

For very large values of $\omega\tau$, $(\omega\tau \gg 1)$

$$\log\left(\frac{C_a}{k_p M_a}\right)_{j\omega} \approx -\log \omega\tau$$

Therefore, for large values of $\omega\tau$ the plot of $(C_a/k_p M_a)_{j\omega}$ versus $\omega\tau$ on log-log paper is a straight line of slope -1, passing through the $(1, 1)$ point. Hence, the amplitude-versus-frequency plot can be approximately represented by two straight lines. See Fig. 14.32. It is conventional practice to construct the amplitude plot using log-log scales and to use a linear-log plot for phase angle versus frequency.

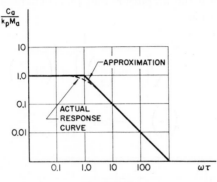

Fig. 14.32.

It will be noted that the intersection of the straight lines occurs at a frequency of $\omega = 1/\tau$. This frequency is often referred to as the "break frequency." At this point the true value of $(C_a/k_p M_a)_{j\omega} = 0.707$, and the straight-line approximation is least accurate.

14.133. The Two-Capacity System

Referring again to the mass-spring-dashpot system of Fig. 14.17 but this time considering the disturbance to be a sinusoidal change of the force, F, rather than a step change, we can write

$$(mD^2 + bD + k)x = F \tag{14.98}$$

then

$$\frac{x}{F/k} = \frac{1}{(D/\omega_n)^2 + 2\zeta(D/\omega_n) + 1} \tag{14.99}$$

To obtain the expression for the frequency response, $j\omega$ is substituted for D, giving

$$\left(\frac{x}{x_s}\right)_{j\omega} = \frac{1}{-(\omega/\omega_n)^2 + 2\zeta j(\omega/\omega_n) + 1} \tag{14.100}$$

where $x_s = (F/k) =$ amplitude of x at zero frequency

By analogy to the previous case of a single-capacity system, we may write

$$\left(\frac{x}{x_s}\right)_{j\omega} = \frac{e^{j\phi}}{\sqrt{\left[1 - \left(\dfrac{\omega}{\omega_n}\right)^2\right]^2 + \left(2\zeta\dfrac{\omega}{\omega_n}\right)^2}} \qquad (14.101)$$

or we may separate the amplitude and phase characteristics as before and use for amplitude

$$\left|\frac{x}{x_s}\right|_{j\omega} = \frac{1}{\sqrt{\left[1 - \left(\dfrac{\omega}{\omega_n}\right)^2\right]^2 + \left(2\zeta\dfrac{\omega}{\omega_n}\right)^2}} \qquad (14.102)$$

and for phase

$$\phi = \measuredangle\left(\frac{x}{x_s}\right)_{j\omega} = -\tan^{-1}\frac{2\zeta(\omega/\omega_n)}{1 - (\omega/\omega_n)^2} \qquad (14.103)$$

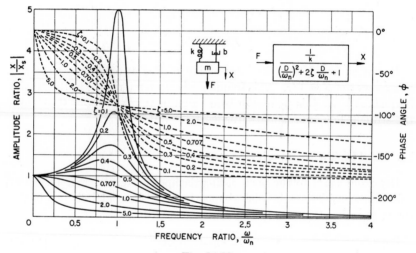

Fig. 14.33.

Figure 14.33 shows Eqs. 14.102 and 14.103 plotted for various values of the damping ratio, ζ. For the purposes of generality, the reader may consider the results shown in Fig. 14.33 to apply to any system having the same form of second-order differential equation.

14.134. The Log Plot of Two-Capacity Systems

$$\log\left|\frac{x}{x_s}\right|_{j\omega} = -\frac{1}{2}\log\left\{\left[1 - \left(\frac{\omega}{\omega_n}\right)^2\right]^2 + \left(2\zeta\frac{\omega}{\omega_n}\right)^2\right\} \qquad (14.104)$$

At $\omega/\omega_n \ll 1$,

$$\left|\frac{x}{x_s}\right|_{j\omega} \approx 1 \qquad (14.105a)$$

At $\omega/\omega_n \gg 1$,

$$\log\left|\frac{x}{x_s}\right|_{j\omega} \approx -2\log\left(\frac{\omega}{\omega_n}\right) \qquad (14.105b)$$

Again the amplitude ratio $|x/x_s|_{j\omega}$ may be represented approximately by two straight lines, as shown in Fig. 14.34. As indicated in Eq. 14.105b the slope at values of $\omega/\omega_n \gg 1$ is -2. Figure 14.34 also shows that the straight-line approximation is again less acceptable in the vicinity of $\omega/\omega_n = 1$. The break (or resonant) frequency occurs at $\omega/\omega_n = 1$.

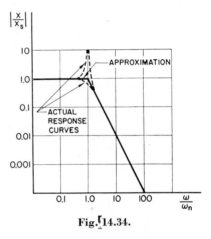

Fig. 14.34.

The results obtained above are borne out by experience. If we apply a sinusoidal force at frequencies low with respect to ω_n, the mass readily follows. At the frequency $\omega = \omega_n$ (the natural undamped frequency of the system) the amplitude ratio may become larger than 1; if no damping were present and no limiting constraints existed, the amplitude ratio would be infinite. At frequencies large with respect to ω_n the mass can no longer follow, its phase increases, and its amplitude decreases, reaching limits of 180° phase shift and a zero amplitude ratio.

14.14. RESPONSE TO A STEP FUNCTION—
ITS RELATION TO FREQUENCY RESPONSE

By applying a step change to the input of a system and measuring the output continuously as a function of time, the dynamic properties of the system may be determined. For a single-capacity system the time constant, τ, can be determined by measuring or computing the initial slope (or the time required to achieve a certain percentage of the final steady-state change). In a two-capacity system the damping ratio and undamped natural frequency can be determined by observing the rate of decay of oscillation and the time per cycle. In many instances it is difficult to apply a step and to measure the resulting transient. If the system is of high order, it may be difficult to determine its roots analytically.

The frequency-response method is based on a steady-state situation; hence, it is often easier to instrument. It is easy to handle analytically. In the case of linear systems it provides the same information as can be obtained from a transient analysis. The break point indicates ω_n and the peak amplitude ratio is a measure of the damping ratio, ζ. While there are more general and fundamental mathematical transformations for shifting from the time domain to the frequency domain, a familiarity with the first- and second-order linear systems as pictured to this point will give the reader powerful tools to deal with many complex problems in dynamics and control. Then, knowing the frequency response of a system, one may accurately predict its transient response and vice versa.

14.15. FREQUENCY RESPONSE OF CASCADED ELEMENTS

In many circumstances a complicated system is composed of a number of relatively simple elements in cascade as shown in Fig. 14.35.

The representation of elements in cascade as shown in Fig. 14.35 implies existence of a unidirectional movement of cause to effect from

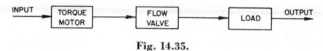

Fig. 14.35.

left to right. This may not be the case in numerous situations. For example, a change in the speed of the output might demand a change in the flow through the valve. Such a change in flow could produce changes in the forces on the valve of sufficient magnitude to change the valve opening independently of the input signal to the valve.

Consider the cascaded elements shown in Fig. 14.36.

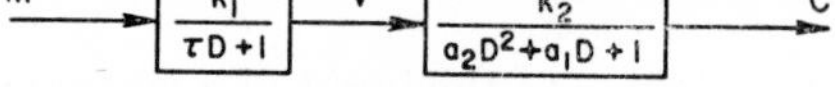

$$M \xrightarrow{\qquad} \boxed{\dfrac{k_1}{\tau D + 1}} \xrightarrow{\quad V \quad} \boxed{\dfrac{k_2}{a_2 D^2 + a_1 D + 1}} \xrightarrow{\qquad} C$$

Fig. 14.36.

It is desired to obtain the frequency response of the over-all system. The transfer characteristic of the system is

$$\frac{C}{M} = \left(\frac{k_1}{\tau D + 1}\right)\left(\frac{k_2}{a_2 D^2 + a_1 D + 1}\right) \tag{14.106}$$

The logarithmic relationship gives

$$\log\left(\frac{C_a}{M_a}\right)_{j\omega} = \log k_1 k_2 - \log (j\omega\tau + 1) - \log (-a_2\omega^2 + ja_1\omega + 1)$$

$$(14.107)$$

Equation 14.107 is written for the case of steady-state sinusoidal response. The log of the over-all system output-input performance is merely the sum of the logs of the component elements. It is convenient

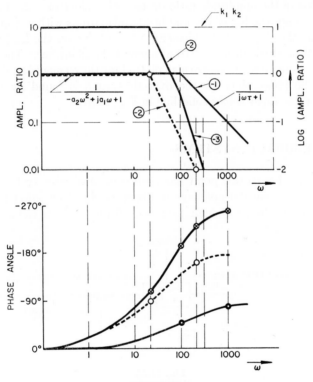

Fig. 14.37.

to collect all steady-state gains into a single term (k_1k_2), leaving only frequency-dependent factors in the other terms.

The straight-line approximations for each of the terms on the right-hand side of Eq. 14.107 may be readily plotted. (Fig. 14.37.)

The first term, $\log k_1 k_2$, is the log of 10 which is 1 and is plotted as shown. For the second term the break frequency occurs at $1/\tau = 1/0.01 = 100$ radians/sec. The slope at frequencies above this is -1, and at frequencies below this the slope is 0. See the solid line of Fig. 14.37.

For the quadratic term the break frequency occurs at $\omega_n = \sqrt{1/a_2} = \sqrt{400} = 20$ radians/sec. The slope at low frequencies is 0 and at frequencies above 20 radians/sec is -2. See the dotted line of Fig. 14.37. The plot which describes the over-all system frequency response is obtained by adding the plots of the elements together in accordance with Eq. 14.107. The result is the plot shown in the heavy solid line. From 0 to 20 radians/sec the amplitude ratio is 10. From 20 to 100 radians/sec, the log of the amplitude ratio drops with a slope of -2. From 100 radians/sec up, the log of the amplitude ratio drops with a slope of -3 (note that the slopes of the elements are additive on the log-log plot). The amplitude ratio drops to 0.01 at about 400 radians/sec. The output-input phase-angle relations are also plotted for the elements on linear-log plots. The phase shift for the over-all system is found by addition and gives the upper heavy solid line in Fig. 14.37.

14.16. FEEDBACK SYSTEMS

The discussion to this point has been concerned with systems where the over-all system did not have feedback of signal or power from one element to another, although feedback within a given element could

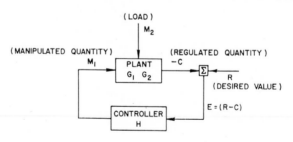

Fig. 14.38.

exist. As demonstrated in other chapters, automatic control of an output of a system is frequently obtained by measuring the output, comparing it to the desired value of that output, and using the difference as an input to a controller which manipulates the input to the controlled system (plant) in such a manner as to reduce the difference between actual and desired output quantity. Such a system is shown in the block diagrams of Figs. 14.38 and 14.39, where G_1, G_2 are the transfer characteristics of the plant to M_1 and M_2, and H is the transfer characteristic of the controller.   The relation between the error, E, and the inputs to the system: M_2, the load, and R, the desired value of the regulated quantity, can readily be determined.

$$E = R - C$$

$$C = G_1 M_1 + G_2 M_2$$

$$M_1 = HE$$

therefore

$$E = \frac{R - G_2 M_2}{1 + HG_1} \tag{14.108}$$

Equation 14.108 is the basic relationship describing the behavior of a single-loop feedback system consisting of linear components; it can

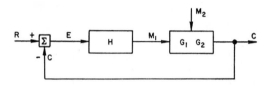

Fig. 14.39.

always be simply developed from the block diagram.   Immediate consideration of Eq. 14.108 yields two important facts:

1. Zero errors in the presence of load disturbances (M_2) can only be obtained if the open-loop transfer characteristic ($G_1 H$) can reach a value of infinity.

2. Infinite values of error (instability) result if the open-loop transfer characteristic ever has a value of -1.

14.17. CRITERIA OF PERFORMANCE OF CLOSED-LOOP CONTROL SYSTEMS

It is essential to establish criteria of performance toward which the design of the system is aimed.   While such criteria may vary depending upon circumstances, the objectives of control can be generally stated in the following manner.

1. The error or deviation should be reduced to acceptable values after the application of disturbances to the system, or more briefly: The steady-state error must not exceed acceptable limits.   For example, in the speed-governing problem satisfactory operation in certain applica-

tions may require that the steady-state error in speed shall be less than 10 per cent of the rated speed when the load varies from zero to rated load.

2. The error must reach its steady-state value rapidly.
3. The system must be stable. It must not oscillate excessively.

It should be observed that while the above objectives are common to all control systems, it is not possible to give specific numbers which will

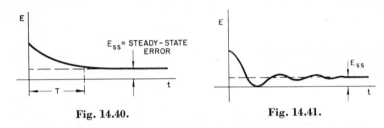

Fig. 14.40.   Fig. 14.41.

hold for all cases. For example, in the temperature regulation of a home, a steady-state error of $\pm 2°\text{F}$ is probably acceptable, a time of 30 minutes in reaching the steady state after a disturbance (sudden drop in outside temperature) would be considered

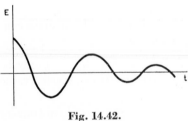

Fig. 14.42.

adequate, minor oscillation would not be regarded as disastrous. In comparison, the specifications for certain controlled-atmosphere chambers limit steady-state errors to an order of $0.1°\text{F}$, with a speed of response of a few seconds and with zero or very small overshoot. Thus system performance is always measured relative to the requirements which the system must meet to fulfill its function.

In Figs. 14.40, 14.41, and 14.42 three responses to step changes in desired value for a closed-loop system are shown. In Fig. 14.40 the response is nonoscillatory in character with a steady-state error of E_{ss}. The time required to reach essentially the steady state is approximately T. In Fig. 14.41 a similar response is shown; however, in this case some oscillation is present, and a considerably shorter time is required for the error to pass through the steady-state position. While there is some oscillation in this case, the oscillation dies out quite rapidly, and in many applications such oscillation would not be regarded as harmful. In Fig. 14.42 a response is shown in which the damping is considerably lower; the oscillation is of large amplitude and requires considerable time to damp out. A response of this kind would be considered too oscillatory or unstable for almost all applications. In this case the response is shown oscillating about the zero steady-state error position, indicating that when the system does steady down, the steady-state error will be zero.

In many cases steps which are taken to ensure small errors and fast response reduce the stability of operation. The final design is usually a compromise yielding the lowest errors and fastest response consistent with adequate stability.

14.18. THE ANALYSIS OF THE PERFORMANCE OF FEEDBACK SYSTEMS

The techniques which have been outlined for obtaining the transient and frequency responses of systems are directly applicable to the systems where feedback is present. Consider the system shown in Fig. 14.39, where the transfer characteristic of the plant is

$$G_1 = \frac{k_p}{a_2 D^2 + a_1 D + 1}$$

A proportional controller is to be used which produces a change in the manipulated variable (M_1) proportional to the error, E. It is desired to determine the value of the controller gain, k_c, where

$$H = k_c$$

Substitution in Eq. 14.108 gives

$$E = \frac{R - G_2 M_2}{1 + \dfrac{k_p k_c}{a_2 D^2 + a_1 D + 1}} \tag{14.109}$$

14.181. Determination of Steady-State Error Following a Step Input

If a system is stable, it will reach a new steady state, given sufficient time following a step input. In such a steady state all time derivatives of the error are zero. Hence, the steady-state error may be found merely by allowing all D's to be zero. Therefore, from Eq. 14.109 the steady-state error is

$$E_{ss} = \frac{R - G_2 M_2}{1 + k_c k_p} \tag{14.110}$$

If the input is a step change in the input R (see Fig. 14.43), the steady-state error is

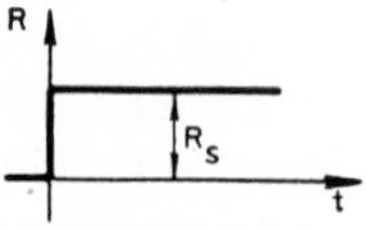

Fig. 14.43.

$$E_{ss} = \frac{R_s}{1 + k_c k_p} \qquad (14.111)$$

Consequently, large values of gain are necessary if small errors are to be realized.

14.19. DETERMINATION OF DEGREE OF STABILITY

The stability of a linear system depends upon the characteristics of the system and is independent of the nature of the disturbances. The degree of stability can be determined from an analysis of either the transient behavior or the response to sinusoidal inputs. Both approaches will be discussed.

14.191. Stability: Transient Analysis

In general, the characteristic equation of the closed loop is, from Eq. 14.108,

$$(1 + HG)E = 0 \qquad (14.112)$$

When H and G are inserted, Eq. 14.112 is of the form

$$(a_n D^n + a_{n-1} D^{n-1} + \cdots + a_1 D^1 + a_0)E = 0 \qquad (14.113)$$

The general solution of Eq. 14.113 for a step input is:

$$E = A_0 + A_1 e^{\lambda_1 t} + A_2 e^{\lambda_2 t} + \cdots + A_n e^{\lambda_n t} \qquad (14.114)$$

where $\lambda_1, \lambda_2, \cdots \lambda_n$ are the roots of Eq. 14.113. If any one of the roots is positive and real or contains a positive real part, one of the terms in Eq. 14.114 will become infinitely large: the system will not reach a new finite steady state; hence instability will exist.

Therefore: A necessary and sufficient condition of stability is that all roots shall have their real parts negative.

In some cases the roots can be determined by inspection. Usually, however, their numerical evaluation is laborious, and in such cases one very useful procedure is the use of *Routh's criterion of stability.* It is established as follows:

1. Write the characteristic equation in order of descending powers of

D, with the coefficient of the highest power positive:

$$[a_n D^n + a_{n-1} D^{n-1} + a_{n-2} D^{n-2} + \cdots + a_1 D + a_0]x = 0$$

2. Write the coefficients alternately in two rows:

$$a_n \qquad a_{n-2} \quad a_{n-4} \cdots 0$$
$$a_{n-1} \quad a_{n-3} \quad a_{n-5} \cdots 0$$

3. Form the determinant of columns 1 and 2, and divide by $-a_{n-1}$ to get a term b_1:

$$b_1 = -\frac{1}{a_{n-1}} \begin{vmatrix} a_n & a_{n-2} \\ a_{n-1} & a_{n-3} \end{vmatrix} = \frac{a_{n-1}a_{n-2} - a_n a_{n-3}}{a_{n-1}}$$

4. Similarly, form a term b_2 from the first and third columns:

$$b_2 = -\frac{1}{a_{n-1}} \begin{vmatrix} a_n & a_{n-4} \\ a_{n-1} & a_{n-5} \end{vmatrix} = \frac{a_{n-1}a_{n-4} - a_n a_{n-5}}{a_{n-1}}$$

5. Continue the process to form additional b's until further terms are all zero.

6. Write the b's in order to form a third row of the array:

$$a_n \qquad a_{n-2} \quad a_{n-4} \cdots$$
$$a_{n-1} \quad a_{n-3} \quad a_{n-5} \cdots$$
$$b_1 \qquad b_2 \qquad b_3 \cdots$$

7. Derive a term c_1 from the first two columns of rows 2 and 3, as before:

$$c_1 = -\frac{1}{b_1} \begin{vmatrix} a_{n-1} & a_{n-3} \\ b_1 & b_2 \end{vmatrix} = \frac{b_1 a_{n-3} - b_2 a_{n-1}}{b_1}$$

8. In a like manner, derive the other c's as in steps 4 and 5, and write the c's in order to form a fourth row of the array.

9. Similarly form rows d, e, and so on, until all further terms are zero.

10. Routh's criterion is now evident by inspection of the first column of the complete array:

$$a_n$$
$$a_{n-1}$$
$$b_1$$
$$c_1$$
$$d_1$$
$$\cdot$$
$$\cdot$$
$$\cdot$$

etc.

The number of roots of the characteristic equation which have positive real parts will be equal to the number of changes of sign (from + to − and from − to +) in the first column. If the system is to be stable, all the coefficients in the column must be positive, since a_n is positive by hypothesis.

11. Occasionally the process of forming the array cannot be completed because certain coefficients will be zero. For the procedure in such exceptional cases, see any good text on dynamics, such as Gardner and Barnes.[12]

Routh's criterion leads to the following results for the four types of systems most often encountered:

For the linear equation, $a_1 D + a_0 = 0$, and for the quadratic equation, $a_2 D^2 + a_1 D + a_0 = 0$, all the a's must be positive.

For the cubic equation, $a_3 D^3 + a_2 D^2 + a_1 D + a_0 = 0$,

$$a_1 a_2 > a_3 a_0$$

and

$$a_0 > 0$$

For the quartic equation, $a_4 D^4 + a_3 D^3 + a_2 D^2 + a_1 D + a_0 = 0$,

$$a_3 a_2 > a_4 a_1$$
$$a_3 a_2 a_1 > a_3{}^2 a_0 + a_4 a_1{}^2$$

and

$$a_0 > 0$$

Similar sets of conditions can be written for equations of higher degree.

The Routh criterion of stability is simple to apply and readily establishes the conditions for stable operation, but it cannot be used directly to determine the *degree* of stability. It is convenient to refer to a system whose characteristics are such that it is on the boundary between stable and unstable operation as one having "boundary stability" or "marginal stability."

The degree of stability in the case of a system characterized by a quadratic equation is directly given by the parameter damping ratio, ζ. If ζ is zero, the system has boundary stability and will oscillate without attenuation when disturbed. When ζ is 1, the system is critically damped and will not oscillate following a disturbance. For values of ζ greater than 1 the system is more heavily damped. It is general practice to design feedback control systems with a damping ratio between 0.4 and 1.

[12] M. F. Gardner and J. L. Barnes, *Transients in Linear Systems*, Vol. 1, John Wiley and Sons, New York, 1942, pp. 197–201.

As an illustration, consider the system whose error is given by Eq. 14.109. The value of the open-loop parameters are

$$a_1 = 0.6 \text{ sec}$$

$$a_2 = 0.05 \text{ sec}^2$$

$$k_p = 2$$

It is desired to determine a value of the proportional-controller gain, k_c, which will give a steady-state error not greater than 5 per cent of a step change in the input, R, and a damping ratio not less than 0.5.

The characteristic equation is found from Eq. 14.109 by setting all inputs equal to zero.

$$\left(1 + \frac{k_c k_p}{a_2 D^2 + a_1 D + 1} \right) E = 0 \qquad (14.115)$$

$$a_2 \lambda^2 + a_1 \lambda + 1 + k_c k_p = 0$$

$$0.05 \lambda^2 + 0.6 \lambda + 1 + 2k_c = 0 \qquad (14.116)$$

$$\frac{0.05}{1 + 2k_c} \lambda^2 + \frac{0.6}{1 + 2k_c} \lambda + 1 = 0$$

This last equation is in the form

$$\left(\frac{\lambda}{\omega_n} \right)^2 + 2\zeta \left(\frac{\lambda}{\omega_n} \right) + 1 = 0$$

where $\omega_n{}^2 = (1 + 2k_c)/0.05$, and $(2\zeta/\omega_n) = 0.6/(1 + 2k_c)$.

And if ζ is to be 0.5, we may solve for ω_n from

$$1 + 2k_c = 0.6\omega_n$$

and

$$1 + 2k_c = 0.05\omega_n{}^2$$

$$0.6\omega_n = 0.05\omega_n{}^2$$

whence

$$\omega_n = 0 \text{ or } 12$$

Using the value $\omega_n = 12$ (0 is a trivial solution), we may now solve for k_c:

$$1 + 2k_c = 0.6\omega_n$$

therefore

$$k_c = 3.1$$

The steady-state error, E_{ss}, is then found from Eq. 14.111.

$$\frac{E_{ss}}{R} = \frac{1}{1 + k_c k_p}$$

By employing the given value of k_p and the value of k_c found to give a ζ of 0.5

$$\frac{E_{ss}}{R_s} = \frac{1}{1 + 3.1(2)} = 0.14$$

The steady-state error is 14 per cent rather than the desired 5 per cent.

If the 5 per cent limitation is imposed upon the design, the system will not be sufficiently stable (that is, its damping ratio ζ will be too small), as may be verified by the following computations:

$$\frac{E_{ss}}{R_s} = 0.05 = \frac{1}{1 + k_c k_p}$$

$$1 + 2k_c = 20$$

$$k_c = 9.5$$

Therefore

$$\omega_n{}^2 = \frac{1 + 2k_c}{0.05} = 400$$

$$\omega_n = 20 \text{ radians/sec}$$

$$\frac{1 + 2k_c}{0.6} = \frac{\omega_n}{2\zeta}$$

and

$$\zeta = 0.3$$

which is too low.

With the simple controller employed, the system cannot be designed to meet both the specified steady-state error and the stability criteria.

k_c	ζ	ω_n	E_{ss}/R_s
3.1	0.5	12	0.14
9.5	0.3	20	0.05

From the above table it can be seen that increasing values of gain produce higher speeds of response, as indicated by higher values of ω_n, and lower steady-state errors, but decreased stability. The decreased stability with increasing gain occurs in many systems but this is not universally true for all systems.

The design of higher-order systems for certain specified degrees of stability by the use of transient analysis will be discussed later in Sec. 14.20, Root-Locus Method.

14.192. Stability: Frequency Response Analysis

The basic expression for the behavior of a linear feedback system is

$$E = \frac{R - G_2 M_2}{1 + HG} \qquad (14.108)$$

where $G \overset{\Delta}{=} G_1$.

As previously indicated in Sec. 14.16, closed-loop instability occurs when

$$HG = -1 \qquad (14.117)$$

This relationship may be divided into two separate statements.

$$|HG| = 1$$
$$\measuredangle HG = 180° \qquad (14.118)$$

The conditions for boundary stability of the closed-loop (feedback) system are, therefore, expressed in terms of the simpler loop transfer characteristic, HG. Because of the ease with which HG can be determined for the case of a sinusoidal input, wide use is made of Eq. 14.118 in the form

$$|HG|_{j\omega} = 1$$
$$\measuredangle (HG)_{j\omega} = 180° \qquad (14.119)$$

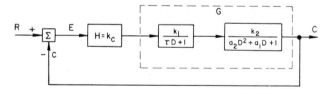

Fig. 14.44.

Consider the closed-loop system shown in Fig. 14.44. The system constants are

$$\tau = 0.01 \text{ sec} \qquad a_1 = 0.05 \text{ sec} \qquad a_2 = 0.0025 \text{ sec}^2$$
$$k_1 = 2 \qquad \qquad k_2 = 5$$

The plant transfer characteristic, evaluated for the steady-state response to a sinusoidal input, is

$$G_{j\omega} = (k_1 k_2) \left(\frac{1}{j\omega\tau + 1} \right) \left(\frac{1}{-a_2\omega^2 + ja_1\omega + 1} \right)$$

Let $G_{j\omega}' = \left(\frac{1}{j\omega\tau + 1} \right) \left[\frac{1}{ja_1\omega + (1 - a_2\omega^2)} \right]$. From Fig. 14.37, plot $G_{j\omega}'$ as shown in Fig. 14.45. Since

$$(HG)_{j\omega} = (k_c k_1 k_2)G_{j\omega}' = K(G_{j\omega}')$$

Fig. 14.45 presents the open-loop amplitude-ratio plot for $K = 1$. The phase plot does not depend upon K.

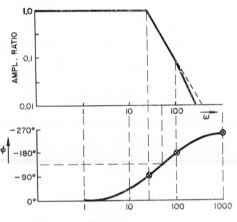

Fig. 14.45.

When the phase angle, ϕ, is $-180°$, the amplitude ratio is about 0.1 and therefore $k_c k_1 k_2 = 10$ gives the boundary-stability condition of Eq. 14.119 for the closed-loop system.

14.193. Gain and Phase Margin

For satisfactory operation some margin of stability must be provided, since the condition of boundary stability is unacceptable. Values of gain from 25 per cent to 40 per cent of the value which produces boundary stability are often used.[7-11]

The *gain margin* is defined as shown in Fig. 14.46, or by

$$\text{Gain margin} = \frac{\text{gain for boundary stability}}{\text{actual gain}}$$

If the gain margin of 4 is used in the example

$$K = 10 \text{ for boundary stability}$$

$$K_{\text{actual}} = {}^{10}\!/_{4} = 2.5$$

and therefore

$$k_c = 0.25$$

Another approach is to establish a value of over-all system gain such that the phase angle when the amplitude ratio is 1 does not exceed a certain value (usually 90° to 150°). The difference between 180° and this phase angle is called the *phase margin*. A design range often used as indicated above is a phase margin of 90° to 30°.

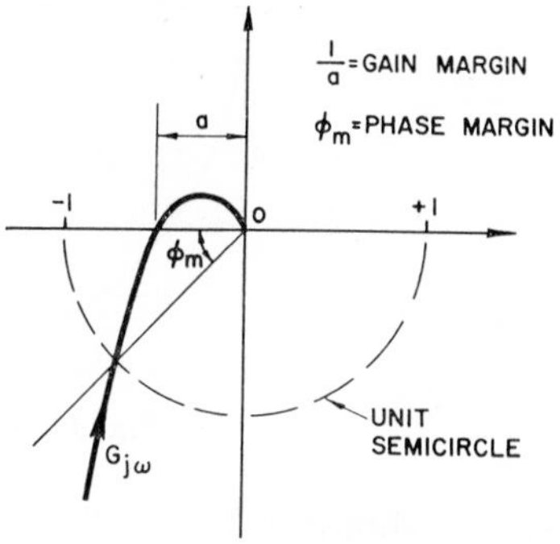

Using as a design criterion a phase margin of 45°, Fig. 14.45 indicates an amplitude ratio, $|G_{j\omega}'| = \frac{1}{3}$ (at $\phi = -135°$). Since $K|G_{j\omega}'| = 1$ is permitted at $\phi = -135°$,

$$K = 3$$

$$k_c = 0.3$$

Fig. 14.46.

14.194. Polar Plots

The information contained in the amplitude ratio and phase angle versus frequency plots is often combined in a single *polar plot* such as that shown shown in Fig. 14.46. It is constructed by representing the

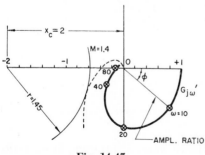

amplitude ratio and phase angle at any one frequency as a vector whose length is determined by the amplitude ratio and whose direction is determined by the phase angle. The construction of a polar plot from the information given in Fig. 14.45 is shown in Fig. 14.47. The terminal points of the vectors are plotted and joined by a smooth curve to give the total picture. Each

Fig. 14.47.

point is marked with the frequency for which it applies. Figures 14.48 to 14.50, inclusive, show such plots for a number of systems frequently encountered.

The polar plot for a system consisting of a number of elements in cascade can be constructed by multiplying the amplitude ratios and adding the phase angles for each frequency. As an illustration consider the

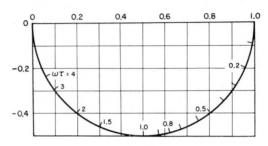

Fig. 14.48. Plot of $1/(j\omega\tau + 1)$.

system shown in Fig. 14.51. Figure 14.48 shows that the polar plot of the simple lag $1/(\tau D + 1)$ is a semicircle. The polar plot for two equal lags in cascade is obtained by the vector multiplication procedure just

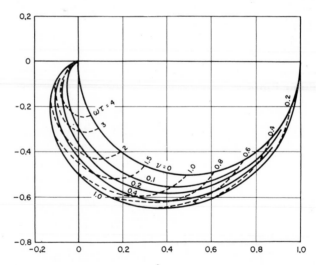

Fig. 14.49. Plots of $\dfrac{1}{(j\omega\tau + 1)(\nu j\omega\tau + 1)}$ for several values of ν.

described and shown in Fig. 14.49. Hence, at $\omega\tau = 1$ the amplitude ratio for each lag is 0.707 and the phase angle is 45°. The amplitude ratio of the two lags in cascade at $\omega\tau = 1$ is $0.707 \times 0.707 = 0.5$. The corresponding phase angle is $45° + 45°$ or 90°. If this process is carried out for a number of frequencies, the full polar plot for the two lags will be

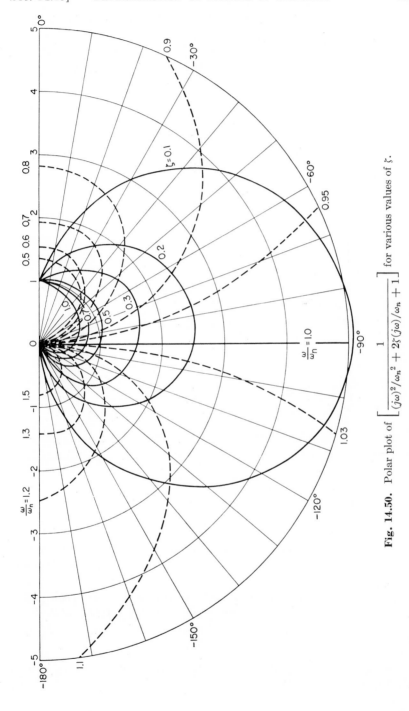

Fig. 14.50. Polar plot of $\left[\dfrac{1}{(j\omega)^2/\omega_n{}^2 + 2\zeta(j\omega)/\omega_n + 1}\right]$ for various values of ζ.

completed as shown in Fig. 14.49. For the given τ/τ_i ratio in Fig. 14.51 the two component plots of Figs. 14.49 and 14.52 can be combined to give the complete HG polar plot of Fig. 14.53. Both gain margin and phase margin can be directly obtained from this polar plot.

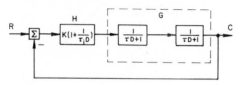

Fig. 14.51.

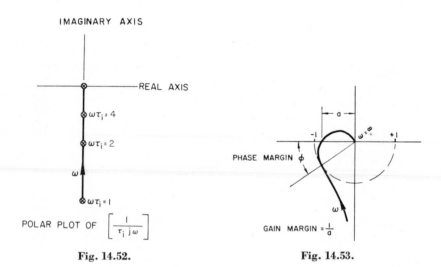

Fig. 14.52.   **Fig. 14.53.**

14.195. Nyquist Stability Criterion [13]

Nyquist developed the necessary conditions which the polar plot must satisfy in order to represent a stable system. For almost all cases that are encountered in automatic-control work the following simple statement is valid.

A closed-loop system is stable if in traversing the open-loop polar plot in the direction of increasing positive frequencies the −1 point lies at the left of the point of closest approach, and it is unstable if the −1 point is on the right.

Some examples are shown in Fig. 14.54.

[13] H. Nyquist, "Regeneration Theory," *Bell System Technical Journal*, Vol. II (January 1932), pp. 126–147.

The relationship between error, E, and the reference input, R, is given by

$$\frac{E}{R} = \frac{1}{1 + HG}$$

For the case of a sinusoidal input

$$\left[\frac{E}{R}\right]_{j\omega} = \frac{1}{1 + (HG)_{j\omega}} \tag{14.120}$$

At any particular frequency (see Fig. 14.54a) the vectors 1 and $(HG)_{j\omega}$ are directly available, and the vector $[1 + (HG)_{j\omega}]_{\omega=\omega_1}$ is obtained

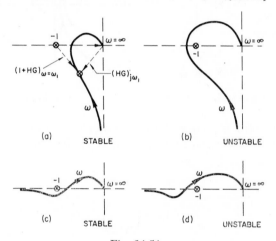

(a)   STABLE   (b)   UNSTABLE

(c)   STABLE   (d)   UNSTABLE

Fig. 14.54.

by vector addition. If small errors are to be realized, $[1 + (HG)_{j\omega}]$ must be large with respect to 1. This is generally true at low frequencies relative to the system response speed, but the error increases with frequency, and finally becomes equal to the reference input at high frequency.

For design purposes it is general practice to use the ratio of output, C, to reference input, R. For the perfect system this would be unity. In general the ratio of C to R is given by

$$\left(\frac{C}{R}\right)_{j\omega} = \frac{(HG)_{j\omega}}{1 + (HG)_{j\omega}} \tag{14.121}$$

14.196. Use of Polar Plot for Synthesis

The polar plot can be conveniently used in the determination of desirable system parameters. Often a maximum closed-loop amplitude ratio (M) is specified. A system with $M = 1.4$ has acceptable stability for many applications.[7-11]

It can be shown that the locus of

$$\left|\frac{C}{R}\right|_{j\omega} = M = \left|\frac{(HG)_{j\omega}}{1 + (HG)_{j\omega}}\right| = \text{constant}$$

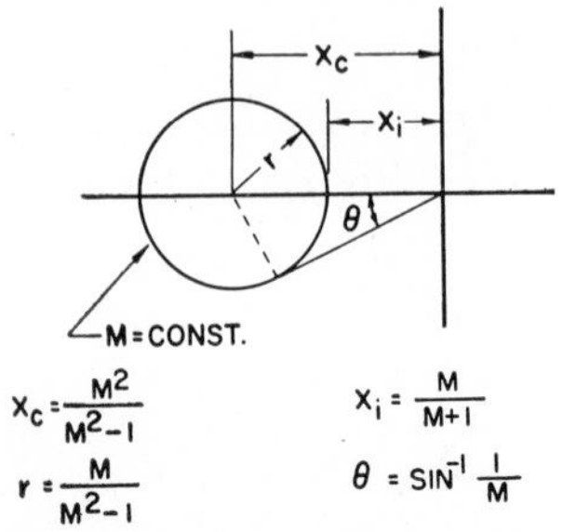

$$x_c = \frac{M^2}{M^2-1}$$

$$x_i = \frac{M}{M+1}$$

$$r = \frac{M}{M^2-1}$$

$$\theta = \sin^{-1}\frac{1}{M}$$

Fig. 14.55. 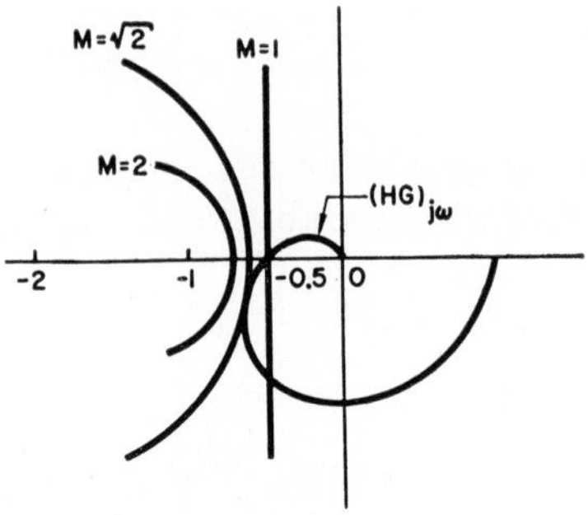   **Fig. 14.56.**

for $M = \text{constant}$ is a circle with its center on the real axis. See Fig. 14.55. This construction may be used to design for the specified M. For $M = \sqrt{2}$

$$x_c = 2 \qquad r = \sqrt{2}$$

$$\theta = \sin^{-1}\left(\frac{1}{\sqrt{2}}\right) = 45°$$

Consider the design of the system shown in Fig. 14.44. The open-loop frequency response of Fig. 14.47 is shown for an over-all system gain, $K = k_c k_1 k_2 = 1$. It is desired to determine the value of K which will give a maximum amplitude ratio of $M = 1.4$.

Construct the $M = 1.4$ circle as shown in Fig. 14.47. In Fig. 14.56 loci for a number of values of amplitude ratio, M, are shown. For $M = 1$ the locus is a straight line perpendicular to the real axis through the -0.5 point. As the value of M increases, the radius of the M circle decreases, with the center moving toward the -1 point. At $M = \infty$ the radius is zero, with the center located at the -1 point. The system gain must be designed so that $(HG)_{j\omega}$ is tangent to the required M circle, in this case $M = 1.4$. In reference to the $G_{j\omega}'$ plot of Fig. 14.47, no intersection of $G_{j\omega}'$ with the $M = 1.4$ circle occurs. If we recall that

$$(HG)_{j\omega} = K(G_{j\omega}')$$

we can select values of K which, when multiplied with $G_{j\omega}'$, will produce an $(HG)_{j\omega}$ plot just tangent to the $M = 1.4$ circle. See the dotted curve in Fig. 14.47. While this procedure can be used, it involves a laborious trial-and-error process. It is shorter and more convenient to deal with changing the size of the circle than with changing the size of the open-loop frequency response. This may be readily done because a change in scale does not destroy the geometry of the problem. From Fig. 14.55, one geometric condition is that the tangent from the origin shall make an angle with the real axis such that its sine is $1/M$. This figure also indicates necessary ratios of circle radius to location of the center of the circle. Hence, by decreasing the scale to which the M circle is drawn in Fig. 14.47 we can find a scale which will make the circle tangent to the $G_{j\omega}'$ plot. This is easily done graphically with the aid of a compass.

1. Select a suitable scale and plot $G_{j\omega}'$.

2. Using the design value of M, draw a straight line from origin making an angle θ with the real axis such that

$$\sin^{-1} \theta = \frac{1}{M}$$

3. By trial and error draw a circle with its center on the negative real axis which is tangent both to the straight line and to $G_{j\omega}'$.

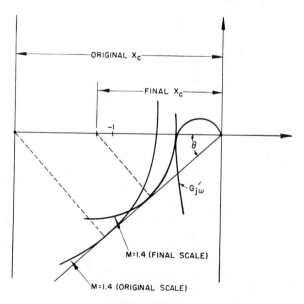

Fig. 14.57.

4. The ratio of the original scale of the M-circle to the final scale of the M-circle is the desired gain, K.

5. From Fig. 14.57 for the example chosen the value of K is

$$K = \frac{\text{original } x_c}{\text{final } x_c}$$

14.20. ROOT-LOCUS METHOD

It was pointed out previously that the mathematics of transient analysis was complicated by the necessity of obtaining the roots of the characteristic equation of the closed-loop system. This process becomes time-consuming for higher-order equations. However, in many cases, the roots of the open-loop characteristic equation are directly available, particularly in those instances where the systems consist of elements in cascade. Evans,[14] Yeh,[15, 16] Chu,[17] and others [18] have developed a background of literature which adequately presents the method and techniques.

Again the basic tool is the equation for closed-loop performance.

$$\frac{E}{R} = \frac{1}{1 + HG} \tag{14.108}$$

Hence, the characteristic equation is

$$(1 + HG)E = 0 \tag{14.112}$$

which may be written

$$|HG| = 1$$
$$\angle HG = 180° \tag{14.118}$$

[14] W. R. Evans, *Control System Dynamics*, McGraw-Hill Book Co., New York, 1954.

[15] V. C.-M. Yeh, "The Generalized Root Locus Method," Sc.D. Thesis, Department of Mechanical Engineering, Massachusetts Institute of Technology, Cambridge, Mass., 1952.

[16] V. C.-M. Yeh, "The Study of Transients in Linear Feedback Systems by Conformal Mapping and the Root Locus Method," *Trans. ASME*, Vol. 76 (April 1954), pp. 349–361.

[17] Y. Chu and V. C.-M. Yeh, "Study of Cubic Characteristic Equation by Root Locus Method," *Trans. ASME*, Vol. 76 (April 1954), pp. 343–348.

[18] Akira Nomoto (Chuo University, Tokyo), "Contribution to the Root-Locus Analysis of the Feedback Control System," *Proceedings of the Second Japan National Congress for Applied Mechanics*, 1952, pp. 359–362.

Thus, the conditions for closed-loop operation are expressed in terms of the open-loop transfer characteristic, HG.

Consider the system shown in Fig. 14.58.

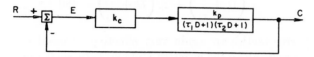

Fig. 14.58.

The open-loop transfer characteristic, HG, is

$$HG = \frac{k_c k_p}{(\tau_1 D + 1)(\tau_2 D + 1)} = \frac{C}{E} \tag{14.122}$$

Substitution in Eq. 14.118 yields the closed-loop relationships

$$\frac{k_c k_p}{\tau_1 \tau_2} = \left(D + \frac{1}{\tau_1}\right)\left(D + \frac{1}{\tau_2}\right)$$

$$0 - \measuredangle\left(D + \frac{1}{\tau_1}\right) - \measuredangle\left(D + \frac{1}{\tau_2}\right) = \pm 180° \tag{14.123}$$

When the loop is open ($k_c = 0$), the open-loop characteristic equation is recovered with the roots $-1/\tau_1$ and $-1/\tau_2$.

$$(D - \lambda_1)(D - \lambda_2) = \left(D + \frac{1}{\tau_1}\right)\left(D + \frac{1}{\tau_2}\right) = 0 \tag{14.124}$$

In the root-locus technique employed here the symbol λ will be used to represent a root of the characteristic equation (that is, a value of D which satisfies Eq. 14.123). The roots may be plotted on the complex plane. See Fig. 14.59. It is now desired to obtain the values of closed-loop roots which satisfy Eq. 14.123, as the over-all system gain ($K =$

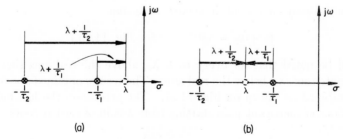

(a)   (b)

Fig. 14.59.

$k_c k_p / \tau_1 \tau_2)$ is increased. The locus of such roots can readily be found by using the angle relationship of Eq. 14.123.

$$\not\subset \left(D + \frac{1}{\tau_1} \right) + \not\subset \left(D + \frac{1}{\tau_2} \right) = 180° \qquad (14.123)$$

Trial-and-error tactics are again very effective. Consider the possible existence of roots along the negative real axis between 0 and $-1/\tau_1$ such as λ in Fig. 14.59a.

Equation 14.123 is then

$$\not\subset \left(+\lambda + \frac{1}{\tau_1} \right) + \not\subset \left(+\lambda + \frac{1}{\tau_2} \right) = 180° \qquad (14.125)$$

$(+\lambda + 1/\tau_1)$ and $(+\lambda + 1/\tau_2)$ are vectors as plotted in Fig. 14.59a. The angle which both vectors make with the positive real axis is zero. Therefore, no points lying to the right of $-1/\tau_1$ can satisfy Eq. 14.125. Therefore, no closed-loop roots can exist to the right of $-1/\tau_1$. Consider a point lying between $-1/\tau_1$ and $-1/\tau_2$. Substitution in Eq. 14.125 of the new vector angles gives:

$$180 + 0 = 180°$$

Therefore, points lying between $-1/\tau_1$ and $-1/\tau_2$ do satisfy Eq. 14.125.

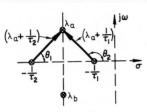

Fig. 14.60.

A trial of points to the left of $-1/\tau_2$ indicates that this axis segment cannot contain closed-loop roots. Hence, on the real axis, only roots lying between $-1/\tau_1$ and $-1/\tau_2$ satisfy the equations which describe closed-loop performance. For a complete determination it is necessary to examine the possibility of complex roots. In this case, it is clear that roots lying on the perpendicular bisector of the line from $-1/\tau_2$ to $-1/\tau_1$ satisfy the angle equation. All points on this line result in vectors such that

$$\theta_1 + \theta_2 = 180° \qquad \text{(See Fig. 14.60.)}$$

It will be noted that for every root, λ_a, above the real axis there is a conjugate root, λ_b, below the real axis. The results obtained above may be combined as shown on Fig. 14.61. The arrows on the loci indicate directions of increasing gain starting from a gain of zero at roots $-1/\tau_1$ and $-1/\tau_2$.

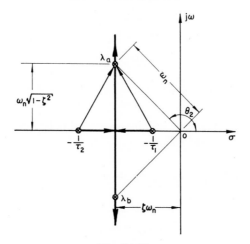

Fig. 14.61.

It is conventional to write the complex roots as follows:

$$\lambda_a = -\zeta\omega_n + j\omega_n\sqrt{1 - \zeta^2}$$

$$\lambda_b = -\zeta\omega_n - j\omega_m\sqrt{1 - \zeta^2}$$

From Fig. 14.61

$$-\cos\theta_2 = \frac{\zeta\omega_n}{\omega_n} = \zeta$$

$$|\lambda_a^2| = (+\zeta^2 + 1 - \zeta^2)\omega_n^2$$

$$|\lambda_a| = \omega_n$$

Thus it is seen that radial lines from the origin are lines of constant damping ratio, ζ, and circles with their centers at the origin are loci of constant undamped natural frequency ω_n. See Fig. 14.62.

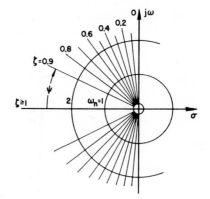

Fig. 14.62.

Consider the system of Fig. 14.63. The open-loop transfer function for the complete system of Fig. 14.63 is

$$\frac{k_c k_p \tau_c}{\tau_1 \tau_2}\left[\frac{D + \dfrac{1}{\tau_c}}{\left(D + \dfrac{1}{\tau_1}\right)\left(D + \dfrac{1}{\tau_2}\right)}\right] = HG$$

It is desired to determine the values of τ_c and k_c, the controller constants, to give satisfactory performance.

$$\measuredangle HG = 180$$

$$\measuredangle \left(D + \frac{1}{\tau_c} \right) - \measuredangle \left(D + \frac{1}{\tau_1} \right) - \measuredangle \left(D + \frac{1}{\tau_2} \right) = 180° \quad (14.125)$$

Values of D which make HG infinite are called poles whereas values of D which make HG zero are called zeros. Therefore $-1/\tau_c$ is called a zero

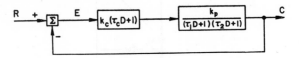

Fig. 14.63.

of the system. Plot the zeros and poles of Eq. 14.125 on the complex plane. See Fig. 14.64. A trial value of $1/\tau_c$ lying between $-1/\tau_1$ and zero is first used. The root locus is determined as before, by first moving along the real axis. As the gain is increased, one root of the closed loop approaches $-1/\tau_c$ at $K = \infty$. The other root approaches ∞ at $K = \infty$. Trial indicates that no roots lying off the real axis satisfy the angle equation; hence the complete locus lies along the real axis.

When $|\tau_1| > |\tau_c| > |\tau_2|$, the root locus is as shown in Fig. 14.65.

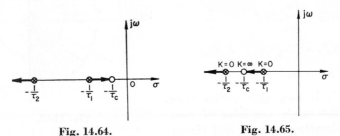

Fig. 14.64.   **Fig. 14.65.**

In general the closed-loop dynamic response is controlled by the poles and zeros near the complex-plane origin, as may be shown by mathematical solution of the differential equation. If this two-capacity system has a proportional controller only, then the root-locus plot is as shown in Fig. 14.61. Increase of loop gain, K, results in higher natural frequency but smaller damping ratio although theoretically the system never becomes unstable, even as $K \to \infty$. The addition of the derivative control produces the open-loop zero shown in Figs. 14.64 through 14.67, and results in more highly damped closed-loop systems in each case.

If $\tau_c = \tau_1$, then the controller effectively cancels the "slow" lag of the system, and a faster first-order response occurs, the speed of which is determined by the time constant, τ_2, and gain K. Even if the cancellation is not exact (Figs. 14.64 and 14.65) this result is still approximated. Note that in both cases the response is nonoscillatory since no complex roots exist for any K.

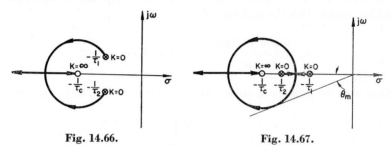

Fig. 14.66.   Fig. 14.67.

As the derivative "gain," τ_c, decreases further, the zero at $-1/\tau_c$ moves to the left away from the system poles. For the limiting case $\tau_c = 0$, the zero is at infinity and the pure proportional control case considered above is recovered. For intermediate values $[0 < |\tau_c| < |\tau_2|]$ the derivative-control zero causes the root loci from the open-loop poles to follow a circular path around the zero and to return finally to the real axis, thus retaining a stable system (cf. Figs. 14.66 and 14.67), although there is probably no greater relative gain in speed of response. Note

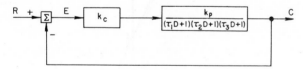

Fig. 14.68.

also that stability passes through a minimum ($\zeta = \cos \theta_m$) at some K value.

THE THREE-CAPACITY SYSTEM. Consider the system shown in Fig. 14.68.

The open-loop roots $-1/\tau_1$, $-1/\tau_2$, and $-1/\tau_3$ are plotted in Fig. 14.69. Use of the equation

$$\angle HG = 180°$$

establishes quickly the portion of the root locus lying on the real axis. Two simple facts are generally useful in developing the root locus for complex roots.

1. Complex roots appear in conjugate pairs.

$$\lambda_a = \sigma + j\omega$$

$$\lambda_b = \sigma - j\omega$$

2. When the roots become infinite, the loci approach n asymptotes making angles with the real axis of

$$\theta_a = \frac{180 + k360}{n} \qquad k = 0, 1 \cdots, n-1$$

where n = the number of poles. In this case $\theta_a = 60°$, $180°$, and $300°$.

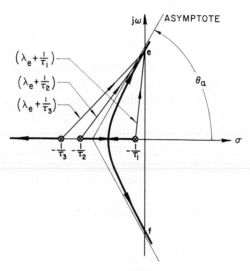

Fig. 14.69.

It can be shown that these asymptotes pass through the point

$$-\frac{1}{3}\left(\frac{1}{\tau_1} + \frac{1}{\tau_2} + \frac{1}{\tau_3}\right)$$

or in general

$$+\frac{1}{n}\,(\text{sum of the open-loop roots})$$

From the above, the root locus can be completed as shown in Fig. 14.69. While a number of helpful rules can be used, it is important to point out that the only statements really needed are

$$\angle HG(j\omega) = 180°$$

$$|HG| = 1 \tag{14.118}$$

Trial-and-error graphical methods will give solutions which rapidly converge to the desired root locus. Reference to Fig. 14.69 indicates the root locus crossing the imaginary axis at points e and f. Hence, the gain producing these roots yields an unstable system. This value of gain can readily be obtained from Fig. 14.69.

$$|HG| = 1$$

$$\frac{K}{\tau_1\tau_2\tau_3}\left|\frac{1}{\left(D + \dfrac{1}{\tau_1}\right)\left(D + \dfrac{1}{\tau_2}\right)\left(D + \dfrac{1}{\tau_3}\right)}\right| = 1 \tag{14.126}$$

Measure the vectors $(\lambda_e + 1/\tau_1)$, $(\lambda_e + 1/\tau_2)$, and $(\lambda_e + 1/\tau_3)$ to scale. Place the values obtained in Eq. 14.126 and solve for K.

The value of K needed to establish a desired damping ratio of the oscillatory factor can also be rapidly determined. Let a damping factor of $\zeta = 0.7$ be required. The root locus of Fig. 14.69 is reproduced in Fig. 14.70.

Construct radial lines such that $\cos\phi = 0.7$. The intersection of these lines with the locus establishes the closed-loop roots λ_a and λ_b. By using Eq. 14.126 and scaling off the vectors $(\lambda_a + 1/\tau_1)$, $(\lambda_a + 1/\tau_2)$, and $(\lambda_a + 1/\tau_3)$, the proper value of K can be found.

The application to more difficult cases is similar. One further example is discussed with an illustration. The system shown in Fig. 14.71 is the plant of Fig. 14.68 with a proportional-plus-derivative controller.

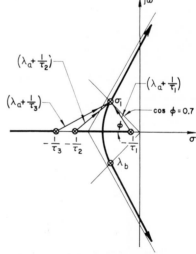

Fig. 14.70.

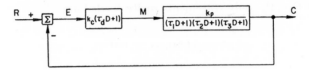

Fig. 14.71.

The open-loop transfer characteristic is

$$HG = \frac{K(\tau_d D + 1)}{(\tau_1 D + 1)(\tau_2 D + 1)(\tau_3 D + 1)}$$

$$HG = \frac{K\tau_d}{\tau_1 \tau_2 \tau_3} \frac{\left(D + \dfrac{1}{\tau_d}\right)}{\left(D + \dfrac{1}{\tau_1}\right)\left(D + \dfrac{1}{\tau_2}\right)\left(D + \dfrac{1}{\tau_3}\right)}$$

The poles and zeros are plotted. See Fig. 14.72. As τ_d is an unknown design factor, some assumption as to its value must be made. Its various possible values can be bracketed into four categories:

1. $0 < |\tau_d| < |\tau_3|$ Limits:
    $\tau_d = 0$—no derivative influence
    $\tau_d = \tau_3$—the smallest lag is compensated
        for

2. $|\tau_3| < |\tau_d| < |\tau_2|$ Limit: $\tau_d = \tau_2$—the intermediate lag is
        compensated for

3. $|\tau_2| < |\tau_d| < |\tau_1|$ Limit: $\tau_d = \tau_1$—the largest lag is compensated for

4. $|\tau_1| < |\tau_d| < |0|$

Cancellation of the largest lag as indicated in items 3 above appears desirable and would yield the result shown in Fig. 14.72. The case where $\tau_d = \tau_3$, the shortest time constant, is shown in Fig. 14.73. Note the

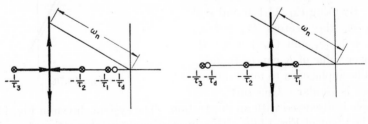

Fig. 14.72.   Fig. 14.73.

greater natural frequency of the system shown in Fig. 14.72 at the same damping ratio as shown in Fig. 14.73. It will be noted that at high values of the roots the locus approaches a 90° asymptote.

$$\measuredangle \text{ of asymptote} = \frac{180 + k360}{n - m} \qquad k = 0, 1 \cdots, n - m - 1$$

where n = number of poles
m = number of zeros

The asymptotes intersect the real axis at an intercept point given by

$$\frac{(\text{sum of values of open-loop poles}) - (\text{sum of values of open-loop zeros})}{(\text{number of open-loop poles}) - (\text{number of open-loop zeros})}$$

The shape of the root locus for intermediate values of τ_d can be determined by the application of the basic angle equation. Suitable values of gains can be determined as in the previous example.

There is great value to this technique of evaluating performance at obvious design limits. First, the computations involved are usually simple, and second, limits of attainable performance are established. If necessary, more complicated intermediate computation can then be undertaken with greater understanding.

14.21. SUMMARY OF TECHNIQUES CONSIDERED

1. TRANSFER CHARACTERISTICS—BLOCK DIAGRAMS. Separate components and complete systems all have transfer characteristics. These transfer characteristics are derived from the differential equations relating input and output of the component or system, or from measured input-output data. Block diagrams are convenient pictorial flow expressions of these relations.

Linear systems only are considered. Many important systems may be represented as first- or second-order linear transfer characteristics with sufficient accuracy.

2. SYSTEM RESPONSE CHARACTERISTICS. *Transient Response.* Every system has a characteristic dynamic response to a change in the nature of the input. The behavior of the system during the interval bounded by the end of one steady state and the beginning of the succeeding steady state is called the *transient response.* Common inputs to investigate this phenomenon are (1) pulses, (2) step inputs, and (3) random

signals. In stable systems the transient response dies out with time. The mathematical solutions for first-, second-, and higher-order systems have been found. It is noted that oscillatory response becomes possible with second- or higher-order equations.

Frequency Response. Among steady forcing functions, the sinusoidal input is widely used to investigate system response. Replacing D by $j\omega$ in the transfer characteristic permits simple evaluation of the amplitude ratio and phase angles as functions of the applied frequency, ω. Plots of these functions are considerably simplified if logarithmic coordinates are used for amplitude ratio and frequency, and, in particular, straight-line approximations may be quickly obtained for complicated systems by addition of the systems' component plots.

3. CLOSED-LOOP CONTROL SYSTEMS. In order to control a system, the output is compared with the reference input, and the error signal produces corrective action if the output is not at the desired value. In addition, it is possible to nullify the effect of undesired disturbance loads elsewhere in the system.

From the block diagram, the closed-loop relations between input, disturbances, and output can be quickly found.

The design problem requires system response to be rapid but stable—generally these are conflicting requirements.

Controllers are added to the system to effect the required design.

4. STABILITY. *Routh's Criterion.* This determines the absolute stability of any polynomial transfer characteristics.

Polar Plot—Nyquist Diagram. Similarly, the absolute stability is determined from the Nyquist diagram. This polar plot is obtained from the amplitude and phase plots of the frequency-response tests (or directly from the transfer characteristic) and has the advantage that only the *open-loop* characteristic is needed.

Gain and Phase Margins. These quantities are alternative design settings permitting the required relative stability to be obtained by adjustment of the polar plot. This adjustment is effected by varying some system parameter (for example, gain).

Constant-M Circles. This technique similarly uses the polar plot. Circles of constant-amplitude ratio, M, are employed, and again either analysis or synthesis is possible.

5. ROOT-LOCUS PLOTS. The root loci trace the closed-loop roots (poles and zeros) across the complex plane as the over-all loop gain is varied. These loci start from the open-loop poles, and therefore approximate sketches are quickly made. In addition, extra geometric short cuts facilitate this construction. The closed-loop behavior is characterized uniquely by its poles and zeros, and hence system per-

formance changes are quickly evaluated qualitatively. Actual quantitative results are obtainable but the method for obtaining them is more time-consuming.

CONCLUSION. The concepts and the techniques of analysis herein presented are powerful tools for the solution of system problems. The treatment has been necessarily brief but can be filled out by resorting to one or more of the references. However, a thorough understanding of the contents of this chapter will permit the reader to tackle successfully complex linear systems, or nonlinear systems which can be treated as linear systems.

ADDITIONAL REFERENCE

R. H. MacMillan, *An Introduction to the Theory of Control in Mechanical Engineering*, Cambridge University Press, 1951, pp. 176–180.

15

J. L. Shearer

═ *Hydraulic Drives*

15.1. INTRODUCTION

Some of the major advantages of hydraulic power control over other forms of power control are the relatively fast response, light weight, and small volume of the hydraulic control components. Hydraulic power control is hampered in many instances by the need for system cooling, careful filtration of dirt particles, prevention of foaming and air entrainment in the fluid and leakage of valuable fluid from high-pressure lines, and by the lack of ability to transmit power over long distances due to high cost of piping and relatively poor fluid dynamic characteristics of long lines. In addition, hydraulic control components usually require precision manufacture of intricate mechanisms and their cost is relatively high.

There are basically two ways to control the flow of fluid power to a load, namely: (1) by so varying some characteristic of the pump which generates fluid power from a prime power source that the rate at which fluid energy is generated is varied or (2) by so varying some characteristic of a valve, the motor, or the load that this change varies the rate at which fluid energy is converted to useful energy at the load. This chapter presents an analysis of one of each of these two basic types of control. An attempt is made to delineate the most important factors which enhance or limit performance in each case, without making a relative evaluation of either system with respect to the many other configurations which may be employed to accomplish the same end results. Both of the systems analyzed are hydraulic servomotor drives of the type often employed in high-performance servomechanisms. In each case the working fluid is assumed to be pure hydraulic fluid with no entrained air present in either the lines or the working chambers of the system. The load consists of mass, viscous friction, and an external load force.

Experience has shown that when the mass in the load is sufficiently large, the effects of fluid compressibility and fluid line elasticity play an important role in system performance. As industrial and military requirements have created demands for faster, stable systems driving massive loads, the problems associated with the compliant nature of the coupling between fluid-power modulator and load have grown more acute. Evidently it is important to understand thoroughly the characteristics of the servomotor and load in order to incorporate the hydraulic drive successfully into more complex systems. Block diagrams are employed as conceptual aids in providing insight into dynamic system performance.

15.2. PUMP-DISPLACEMENT-CONTROLLED SERVOMOTOR

Consider a hydraulic servomotor system such as that shown in Fig. 15.1 consisting of a variable-displacement hydraulic pump driven by a constant-speed power source, relatively short transmission lines, a fixed-displacement hydraulic motor, and mass plus viscous damping load. This is the type of drive that is often employed in machine-tool control systems, tension control systems, gun-turret drives, antenna drives, and

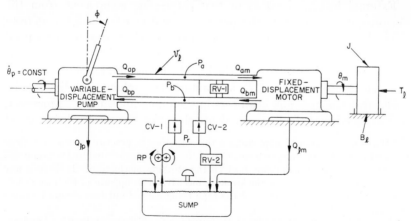

Fig. 15.1. Pump-displacement-controlled servomotor.

ship steering systems. Because the variable-displacement pump tends to generate fluid power only when it is needed to drive the load, this system wastes considerably less power than one which depends on a control valve or brake to control the flow of power to the load by varying the dissipation of large amounts of power. However, relatively large forces are required to vary the displacement of commercially available hydraulic pumps, and, therefore, it is usually necessary to provide a separate stroking mechanism with a maximum output power level that is 2 to 10 per cent of the maximum power output of the main drive. This stroking mechanism often employs some form of the valve-controlled servomotor, which will be discussed later in this chapter.

The purpose of this analysis is to determine how the output shaft position, θ_m, is related to variations in pump stroke, ϕ, and load torque, T_l. Thus, it is assumed that a stroking mechanism has been provided to vary ϕ at will by means completely external to the system and that changes which may occur within the system will have no effect on ϕ.*

15.21. Major Simplifying Assumptions

In order to make the following analysis as brief as possible without making it unrealistic, the following simplifying assumptions are made:

1. Internal leakage in the pump and internal leakage in the motor are each a linear function of only the difference between the pressure in line a and the pressure in line b. (*Note*: Internal leakage is sometimes exponentially related to this pressure difference, and in a few cases this leakage is also a function of pump or motor speed.) [1]

External leakage consists of two parts—one component from the low-pressure chambers to the pump case and another component from the high-pressure chambers to the pump case. Case pressure is assumed negligible. It is also assumed that cavitation does not exist so that the pump and motor displace only the liquid working fluid at all times.

2. The pressures P_a and P_b are uniform throughout each line and its connecting passages and chambers in the pump and motor. Moreover, it is assumed that the lines are identical so that the total volume, $\mathcal{U}_l$, of

---

* It is only possible to approximate this situation in practice because the force required to stroke the pump is a function of the pressure generated by the pump and this may represent a significant load on the stroking mechanism. If there is any question about the ability of the stroking mechanism to operate in the face of this load, a suitable system analysis of the type being carried out here should be made with this effect included.

[1] Donald Crockett, "A Study of Performance Coefficients of Positive-Displacement Hydraulic Pumps and Motors," S.M. Thesis, Department of Mechanical Engineering, Massachusetts Institute of Technology, Cambridge, Mass., 1952.

each line and its connecting passages is the same for each line at a given pressure. This assumption renders inertia effects in the fluid negligible, and is good only for relatively short lines. (See Chapter 5, Fluid-Power Transmission.)

3. The pressures P_a and P_b never exceed the setting of the relief valve RV-1 which incidentally must be a double-acting relief valve since either line may be at high pressure at various times during the operation of the system.

4. The replenishing system, consisting of a replenishing pump, RP, running at constant speed, a relief valve, RV-2, and check valves, CV-1 and CV-2, acts instantaneously to prevent either line pressure from falling below the replenishing pressure, P_r (which is set by RV-2), whenever the net flow into either line from the pump and motor falls below zero. Thus, in normal operation one line or the other will usually be at a constant pressure of P_r. However, in some cases it may not be reasonable to assume instantaneous response of the check valves.[2] For example, leaky check valves, sluggish check valves, or a sluggish relief valve, RV-2, might make this assumption invalid. It is interesting to speculate on the possibility of eliminating the need for a replenishing system by pressuring pump and motor cases at half the maximum system pressure. Obviously this could not be done with presently available pumps and motors since their cases and shaft seals cannot withstand high pressures, but units could be designed to have an inner pressurized region for this purpose. This scheme would tend to establish a high quiescent pressure in both transmission lines when there was zero torque load on the motor. Then as the system underwent dynamic changes, the pressure would increase in one line and decrease in the other in a manner similar to the operation of a valve-controlled hydraulic servomotor. Some auxiliary means of cooling the pump and motor might have to be provided to make up for the loss of beneficial cooling provided by the leakage flow through the replenishing system. Having both line pressures differentially variable would tend to decrease the sensitivity of this system to steady loads.

5. Friction torque in the motor consists of a component that is proportional to $(P_a + P_b)$ and a component that is proportional to shaft speed, $\dot{\theta}_m$. This assumption is not good for all motors. Friction torque has been observed to decrease as the speeds of some motors reach intermediate values. Also, in some motors there is a component proportional to $(P_a - P_b)$.

6. The shaft connecting the moving members of the motor to the load is very stiff so that its twist is negligible. Motor and load inertia may

[2] G. C. Newton, Jr., "Hydraulic Variable-Speed Transmissions as Servomotors," *J. Franklin Inst.*, Vol. 243 (June 1947), pp. 439–469.

thus be lumped together. An interesting case where this was not true has been reported by Stallard.[3]

15.22. Dynamic Analysis

In analyzing the dynamic behavior of this system it is necessary to assimilate the characteristics of each of its components into a comprehensive characterization of the system. In this case, mathematical equations and analogue block diagrams will be used to represent the complete system.

15.221. Pump Flow Equations. The various flows associated with the pump may be expressed as follows:

$$Q_{ap} = D_p\dot{\theta}_p - C_{ip}(P_a - P_b) - C_{ep}P_a \tag{15.1}$$

$$Q_{bp} = D_p\dot{\theta}_p - C_{ip}(P_a - P_b) + C_{ep}P_b \tag{15.2}$$

$$D_p = k_p\phi \tag{15.3}$$

$$Q_{lp} = C_{ep}(P_a + P_b) \tag{15.4}$$

where C_{ip} = internal-leakage coefficient for pump, in.^{5}/lb sec
C_{ep} = external-leakage coefficient for pump, in.^{5}/lb sec
D_p = pump displacement, in.^{3}/radian
k_p = displacement coefficient for pump, in.^{3}/radian deg
P_a = pressure in line a, lb/in.^{2}
P_b = pressure in line b, lb/in.^{2}
Q_{ap} = rate of flow from pump to line a, in.^{3}/sec
Q_{bp} = rate of flow from line b to pump, in.^{3}/sec
Q_{lp} = total external-leakage rate of flow, in.^{3}/sec
ϕ = pump stroke angle, deg
$\dot{\theta}_p$ = pump shaft speed, radians/sec

15.222. Transmission Lines (including chambers and passages in pump and motor). The pressure in each of the lines is related to the pump and motor flows as follows (see Sec. 3.351, Unsteady Flow with Negligible Momentum Effects):

$$\left(k_e + \frac{\upsilon_l}{\beta}\right)\frac{dP_a}{dt} = Q_{ap} - Q_{am} \tag{15.5}$$

$$\left(k_e + \frac{\upsilon_l}{\beta}\right)\frac{dP_b}{dt} = Q_{bm} - Q_{bp} \tag{15.6}$$

---

[3] D. V. Stallard, "Analysis and Performance of a Valve-Controlled Hydraulic Servomechanism," *Trans. AIEE*, Vol. 78, Part 2 (February 1956), pp. 75–83.

where k_e = elasticity coefficient of pipe, in.^{5}/lb
$\quad Q_{am}$ = rate of flow from line a to motor, in.^{3}/sec
$\quad Q_{bm}$ = rate of flow from motor to line b, in.^{3}/sec
$\quad \mathcal{V}_l$ = total volume of each line, in.^{3}
$\quad \beta$ = bulk modulus of fluid, lb/in.^{2}

15.223. Motor Flow Equations. The various flows associated with the motor may be expressed as follows:

$$Q_{am} = D_m\dot{\theta}_m + C_{im}(P_a - P_b) + C_{em}P_a \tag{15.7}$$

$$Q_{bm} = D_m\dot{\theta}_m + C_{im}(P_a - P_b) - C_{em}P_b \tag{15.8}$$

$$Q_{lm} = C_{em}(P_a + P_b) \tag{15.9}$$

where C_{im} = internal-leakage coefficient for motor, in.^{5}/lb sec
$\quad C_{em}$ = external-leakage coefficient for motor, in.^{5}/lb sec
$\quad D_m$ = motor displacement, in.^{3}/radian
$\quad \dot{\theta}_m$ = motor shaft speed, radians/sec
$\quad Q_{lm}$ = total external-leakage rate of flow, in.^{3}/sec

15.224. Motor Torque Equation. The motor torque is related to the pressures P_a and P_b and the speed $\dot{\theta}_m$ as follows:

$$T_m = D_m(P_a - P_b) - \left(\frac{\dot{\theta}_m}{|\dot{\theta}_m|}\right)(P_a + P_b)C_f'D_m - B_m\dot{\theta} \tag{15.10}$$

where C_f' = coefficient for friction due to pressure
$\quad B_m$ = viscous-friction coefficient, in. lb sec/radian

15.225. Load Equation. For the inertia and viscous damping load we find that

$$T_m - T_l = J\frac{d^2\theta_m}{dt^2} + B_l\frac{d\theta_m}{dt} \tag{15.11}$$

where J = polar moment of inertia of load mass, in. lb sec^2/radian
$\quad B_l$ = viscous-damping coefficient of load, in. lb sec/radian

Equations 15.1 through 15.11 give a complete description of the system within the limitations of the simplifying assumptions. It is now convenient to start bringing these equations together by making the following combinations.

Equations 15.1, 15.3, 15.5, and 15.7 give

$$2k_3\frac{dP_a}{dt} = k_p\dot{\theta}_p\phi - C_i(P_a - P_b) - C_eP_a - D_m\frac{d\dot{\theta}_m}{dt} \tag{15.12}$$

where $k_3 = \dfrac{1}{2}\left(k_e + \dfrac{\mathcal{v}_l}{\beta}\right)$

$\quad C_i = (C_{ip} + C_{im})$

$\quad C_e = (C_{ep} + C_{em})$

Equations 15.2, 15.3, 15.6, and 15.8 give

$$2k_3 \frac{dP_b}{dt} = -k_p\dot\theta_p\phi + C_i(P_a - P_b) - C_eP_b + D_m \frac{d\theta_m}{dt} \quad (15.13)$$

Note: It is important to remember that a restriction must be placed on the pressures P_a and P_b in Eqs. 15.12 and 15.13 as well as in the earlier equations: namely, neither P_a nor P_b can fall below the replenishing pressure, P_r. For the special case when both P_a and P_b are varying simultaneously, Eq. 15.13 may be subtracted from Eq. 15.12 to give

$$2k_3 \frac{d}{dt}(P_a - P_b) = 2k_p\dot\theta_p\phi - (2C_i + C_e)(P_a - P_b) - 2D_m \frac{d\theta_m}{dt} \quad (15.14)$$

However, P_a and P_b can vary simultaneously only during parts of a dynamic response when a previously lower pressure starts to increase from P_r before the previously higher pressure has fallen to P_r. During the period when only P_a is varying, $P_b = P_r$, $dP_b/dt = 0$, and we have

$$2k_3 \frac{d}{dt}(P_a - P_r) = 2k_3 \frac{d}{dt}P_a = k_p\dot\theta_p\phi - C_i(P_a - P_r) - C_eP_a - D_m \frac{d\theta_m}{dt}$$

$$(15.15)$$

and during periods when only P_b is varying, $P_a = P_r$, $dP_a/dt = 0$, and we have, from Eq. 15.13,

$$2k_3 \frac{d}{dt}(P_a - P_b) = -2k_3 \frac{d}{dt}P_b$$

$$= +k_p\dot\theta_p\phi - C_i(P_r - P_b) + C_eP_b - D_m \frac{d\theta_m}{dt} \quad (15.16)$$

Equations 15.10 and 15.11 give

$$D_m(P_a - P_b) - B_m \frac{d\theta_m}{dt} - \left(\frac{\dot\theta_m}{|\dot\theta_m|}\right)C_f'D_m(P_a + P_b) - T_l$$

$$= J\frac{d^2\theta_m}{dt^2} + B_l \frac{d\theta_m}{dt} \quad (15.17)$$

It is obvious that a simple mathematical solution to these differential equations is not possible. It is likely that a numerical solution with the

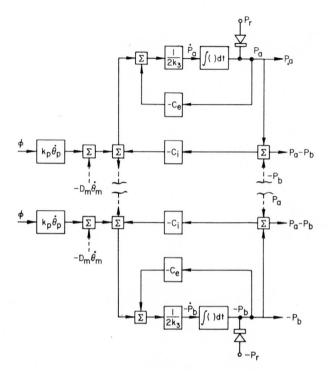

(a) Block diagrams of pump and lines a and b.

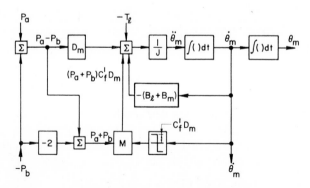

(b) Block diagram of motor and load.

Fig. 15.2. Functional block diagrams for pump-stroke-controlled hydraulic drive.

aid of a hand calculator or digital computer would be satisfactory in many instances, especially where a check solution is required, but for design purposes where it is desired to determine the effects of varying system parameters, an analogue solution would be much more satisfactory.[4,5]

The functional block diagram shown in Fig. 15.2 represents this system as described by the various equations which have just been derived.

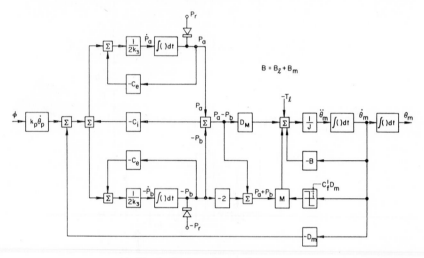

(c) Block diagram of complete system.

Fig. 15.2. (Continued.)

This diagram can also serve as the analogue block diagram for an analogue computer study if the various scale factors [6] are employed to make it possible to represent all of the system variables with variable voltage signals.

Except for the lower limits imposed on P_a and P_b, and the nonlinear friction characteristics in the motor, it would be possible to linearize the equations for this system in a manner similar to what is done in the following pages in the analysis of a valve-controlled hydraulic servomotor. The linearized analysis given by Newton (fn. 2) is only qualitatively applicable to most systems of this kind.

[4] T. M. Korn and G. R. Korn, *Electronic Analog Computers*, McGraw-Hill Book Co., New York, 1952.

[5] H. M. Paynter, ed., *A Palimpsest on the Electronic Analog Art*, G. A. Philbrick Researches, Inc., Boston, 1955.

[6] J. B. Reswick, "Scale Factors for Analog Computers," *A Palimpsest on the Electronic Analog Art*, G. A. Philbrick Researches, Inc., Boston, 1955, pp. 84–88.

15.23. Linearized Analysis

However, even a qualitative picture is often very useful for initial design purposes. Except for the effects of nonlinear friction characteristics in the motor, there are two possible situations that can be simply analyzed and which give results that may be potentially useful in the initial stages of dynamic system design. The first situation exists when only one line pressure varies by any appreciable amount. This can happen, for instance, in a system heavily loaded and running in one direction when there is an appreciable amount of air in the system— enough air so that bubbles exist in the low-pressure line, but not so much air that the high-pressure line is excessively compliant. (If there is so much air in the system that an appreciable amount of bubbles exists in the high-pressure line, then the effective bulk modulus of the air-liquid mixture is much lower than that of the liquid alone and in fact it becomes hard to define the bulk modulus.)

15.231. One Line Pressure Constant. Combining Eqs. 15.15 and 15.17, considering small changes of all variables from a set of initial conditions with motor speed initially zero, ignoring C_f', assuming that P_r is negligible compared to the actual line pressure, and noting that $\Delta(d\theta_m/dt) = D(\Delta\theta_m)$ when the motor speed is initially zero, we obtain

$$D_m D(\Delta\theta_m) - k_p \dot{\theta}_p (\Delta\phi)$$

$$= -[(C_i + C_e) + 2k_3 D]\left[\frac{(JD + B)D(\Delta\theta_m) + \Delta T_l}{D_m}\right]$$

or

$$\left[\frac{2k_3 J}{(C_i + C_e)B + D_m{}^2}D^2 + \frac{(C_i + C_e)J + 2k_3 B}{(C_i + C_e)B + D_m{}^2}D + 1\right]D(\Delta\theta_m)$$

$$= \frac{k_p \dot{\theta}_p(\Delta\phi) - [(C_i + C_e) + 2k_3 D](\Delta T_l)}{(C_i + C_e)B} \quad (15.18)$$

where D denotes derivative with respect to time, d/dt, Δ denotes a small change of the variable following it, and $B = B_m + B_l$.

15.232. Both Line Pressures Varying Simultaneously. The second situation is that which exists when both line pressures are varying, and it is likely that this situation exists more often than the first situation, discussed above, especially during the early part of dynamic responses that result from suddenly decreasing the pump stroke when the system has been driving a heavy load.

Combining Eqs. 15.14 and 15.17, considering small changes of all variables from a set of initial conditions with motor speed initially zero, noting that $\Delta(d\theta_m/dt) = D(\Delta\theta_m)$ when the motor speed is initially zero, and ignoring C_f', we obtain

$$\left[\frac{2k_3 J}{(2C_i + C_e)B + 2D_m^2} D^2 + \frac{(2C_i + C_e)J + 2k_3 B}{(2C_i + C_e)B + 2D_m^2} D + 1 \right] D(\Delta\theta_m)$$

$$= \frac{k_1 \dot{\theta}_p (\Delta\phi) - [(2C_i + C_e) + 2k_3 D](\Delta T_l)}{(2C_i + C_e)B + 2D_m^2} \quad (15.19)$$

15.233. Comparison of the Two Cases. The undamped natural frequency, ω_{ns}, and the damping ratio, ζ_s, of each case are given in Table 15.1.

Table 15.1

	First Case—Eq. 15.18	Second Case—Eq. 15.19
ω_{ns}	$\sqrt{\dfrac{(C_i + C_e)B + D_m^2}{2k_3 J}}$	$\sqrt{\dfrac{(C_i + C_e/2)B + D_m^2}{k_3 J}}$
ζ_s	$\dfrac{(C_i + C_e)J + 2k_3 B}{2\sqrt{2k_3 J[(C_i + C_e)B + D_m^2]}}$	$\dfrac{(C_i + C_e/2)J + k_3 B}{2\sqrt{k_3 J[(C_i + C_e/2)B + D_m^2]}}$

When the system is lightly damped, C_i, C_e, and B are negligible in the expressions for ω_{ns}, and it may readily be seen that the natural frequency of the second case is then about 40 per cent higher than that of the first case.

Although linearized analyses of this system are subject to question, it is possible to see from the block diagrams of Fig. 15.2 that the speed of response of the system tends to increase with increasing values of $1/k_3$, D_m, and $1/J$, while system damping, which is usually due to closure of inner loops, is most effectively provided by increasing values of C_e, C_i, and B.

15.3. VALVE-CONTROLLED SERVOMOTOR

Consider a hydraulic servomotor system such as that shown in Fig. 15.3, consisting of a four-way control valve supplied with hydraulic fluid at a constant supply pressure, P_s, a double-acting hydraulic ram having a net working area A, and a mass plus viscous damping load. This type of drive is often employed in hydraulic governors for prime movers, variable-displacement-pump stroking mechanisms, power steering mechanisms, machine-tool contour followers, aircraft control-surface actuators, and process-control-valve positioners. The analysis which follows is taken from an ASME paper [7] by the author and is based on a

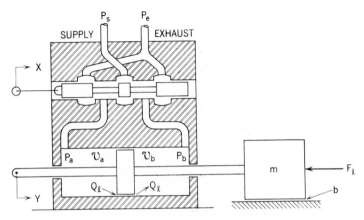

Fig. 15.3. Valve-controlled hydraulic servomotor with load.

number of simplifying assumptions that have proved to be reasonable in many instances.

15.31. Major Assumptions

Using the four-way control valve detailed in Fig. 15.4, the major assumptions to be made are as follows: (1) radial clearance between the valve spool and valve body is taken to be zero, and the metering edges are assumed to be perfectly sharp and perfectly oriented; (2) the flow through each orifice is assumed to be simple orifice-type flow with negligible viscous effects, and the flow rate is assumed to change instantaneously with change in either pressure drop across the orifice or with change in orifice area; (3) all of the connecting passages are short enough and large enough to eliminate the effects of fluid mass on the flows through them; (4) friction losses in the lines and passages are negligible, and leakage flow past the ram is assumed to be laminar; (5) supply pressure is constant, and the bulk modulus of the fluid is taken to be constant within the range of pressures employed (that is, the fluid is a pure liquid); (6) exhaust pressure, P_e, is zero.

[7] J. L. Shearer, "Dynamic Characteristics of Valve-Controlled Hydraulic Servomotors," *Trans. ASME*, Vol. 76 (August 1954), pp. 895–903.

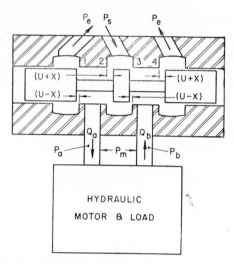

Fig. 15.4. Typical four-way control valve (U = valve underlap).

15.32. Dynamic Analysis

When the motion of the ram is restricted to small displacements from its center position, then $\mathcal{V}_a \approx \mathcal{V}_b$, and it can be shown [8] that the following relationship exists between ram velocity, dY/dt, valve position, X, leakage flow, Q_l, and pressure across the ram, P_m ($P_m = P_a - P_b$):

$$A\,\Delta(DY) = k_1\,(\Delta X) - C_1\,(\Delta P_m) - \Delta Q_{lr} - \frac{1}{2}\left(k_e + \frac{\mathcal{V}_i}{\beta}\right) D\,(\Delta P_m) \quad (15.20)$$

where Δ indicates the small change of each variable that has taken place from an initially steady value and D denotes the derivative with respect to time (d/dt). The coefficients are defined as follows:

$$A = \text{ram area}$$

$$k_1 = \left|\frac{\partial Q_m}{\partial X}\right|_{P_m=\text{const}}$$

$$C_1 = \left|\frac{\partial Q_m}{\partial P_m}\right|_{X=\text{const}}$$

$$\Delta Q_{lr} = \text{leakage flow past ram}$$

$$k_e = \text{coefficient of elasticity of each line}$$
$$\text{connecting ram to valve}$$

[8] Shearer, *op. cit.*, Appendix I.

$\mathcal{U}_i$ = volume of fluid between one side of ram
and valve when ram is centered

β = bulk modulus of elasticity of fluid

It is significant that when $\mathcal{U}_a \approx \mathcal{U}_b$ the flow rate Q_a is always equal
to the flow rate Q_b so that it is possible to speak of a single motor flow
$Q_m = Q_a = Q_b$.

Figures 15.5a and 15.5b show the pressure-flow curves which were obtained experimentally from a four-way valve for open-center and closed-
center operation, respectively. These curves were obtained only for the
quadrants in which fluid power was being delivered to the motor because
of the experimental difficulties involved in performing tests in which
fluid power is taken from the motor. They reveal imperfections in the
valve which may be impossible to predict or analyze. In this case, the
flow passages seem to be limiting the flow at large valve openings.

Graphical interpretations of k_1 and C_1 for small changes of Q_m, P_m, and
X are also given in Fig. 15.5a. The coefficient k_1 is a measure of the
amount by which the steady-state load flow, Q_m, varies when only a
change of X occurs, and C_1 is a measure of the amount by which Q_m
changes when only a change in P_m occurs. It is significant that the
pressure-flow characteristics of an open-center four-way valve yield
nearly constant values of k_1 and C_1 for widely different initial conditions.
This means that the analysis given in Appendix 1 of Shearer's article
(fn. 7) may be expected to hold for fairly large variations of Q_m, X, and
P_m when an open-center valve is being used. It is interesting to note
that C_1 is zero when operating about the origin of the pressure-flow
characteristics of a closed-center four-way valve, and that both k_1 and
C_1 vary more from point to point than with an open-center valve. The
flow sensitivity (k_1) of an open-center valve when $X < U$ should be
twice the value it is when $X > U$ or when U is zero. (See Chapter 7.)

It should be emphasized that the use of the four-way valve steady-
state pressure-flow curves can be justified rigorously only when $\mathcal{U}_a \approx \mathcal{U}_b$
(that is, when the ram is near its center position). When the ram is
near one end of the cylinder, the analysis is complicated by the lack of
symmetry in the system and strictly speaking it is no longer possible to
use a single motor flow rate, Q_m, since the flow rates Q_a and Q_b are equal
only in the steady state. For an exact analysis when $\mathcal{U}_a \neq \mathcal{U}_b$, it is necessary to consider each end of the ram to be controlled by a three-way
valve, each three-way valve consisting of two of the four orifices of the
four-way valve. This section will deal only with the case when $\mathcal{U}_a \approx \mathcal{U}_b$.
Experience has shown that the results of this analysis can usually be
used also to describe the performance of the system for small motions
about positions near the end of the ram travel.

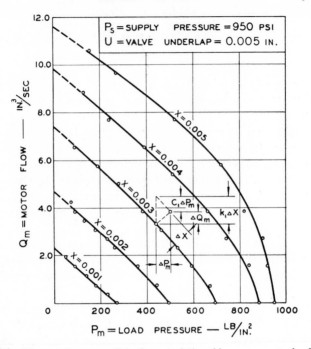

Fig. 15.5(a). Measured pressure-flow characteristics of four-way control valve—open-center operation.

For laminar leakage flow past the ram

$$\Delta Q_{lr} = C_2 \, (\Delta P_m)$$

where C_2 = laminar-flow coefficient for the leakage path. By defining new coefficients,

$$k_2 = C_1 + C_2$$

and

$$k_3 = \frac{1}{2}\left(k_e + \frac{\mathcal{V}_i}{\beta}\right) \tag{15.21}$$

Eq. 15.20 becomes

$$A \, \Delta(DY) = k_1 \, (\Delta X) - k_2 \, (\Delta P_m) - k_3 D \, (\Delta P_m) \tag{15.22}$$

Equation 15.22 is the fundamental equation which may be used to describe the behavior of the fluid part of the servomotor for small changes of all variables.

Turning now to the mechanical part of the system, consisting of the

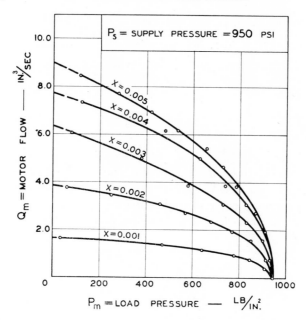

Fig. 15.5(b). Measured pressure-flow characteristics of four-way control valve—closed-center operation.

ram and mass plus viscous damping load, the requirement of force balance yields

$$(\Delta P_m) A = m D\Delta(DY) + b \Delta(DY) + \Delta F_l \qquad (15.23)$$

where F_l = external load force, lb

The relationship between ΔY, ΔX, and ΔF_l may be obtained by combining Eqs. 15.22 and 15.23 and noting that $\Delta(DY) = D\Delta Y$ when the initial value of DY is zero.

$$A D(\Delta Y) = k_1 (\Delta X) - (k_2 + k_3 D) \left[\frac{m D^2(\Delta Y) + bD (\Delta Y) + \Delta F_l}{A} \right] (15.24)$$

or

$$\left[\left(\frac{k_3 m}{k_2 b + A^2} \right) D^2 + \left(\frac{k_2 m + k_3 b}{k_2 b + A^2} \right) D + 1 \right] D(\Delta Y)$$

$$= \frac{k_1 A (\Delta X) - (k_2 + k_3 D) (\Delta F_l)}{k_2 b + A^2} \qquad (15.25)$$

which is a second-order differential equation.

The damping ratio, ζ_s, and the undamped natural frequency, ω_{ns}, of this system are given by the following equations:

$$\zeta_s = \frac{k_2 m + k_3 b}{2\sqrt{k_3 m (k_2 b + A^2)}} \tag{15.26}$$

and

$$\omega_{ns} = \sqrt{\frac{k_2 b + A^2}{k_3 m}} \tag{15.27}$$

Figure 15.6 illustrates the nature of the transient response of $D\Delta Y$ to a unit step change in ΔX for various values of ζ_s when $\Delta F_l = 0$. The values of ζ_s and ω_{ns} completely describe the dynamic characteristics of this system but are usually used only for underdamped cases ($\zeta_s < 1$).

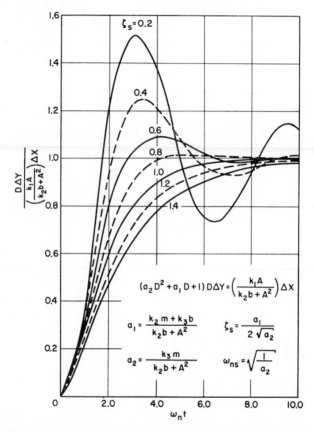

Fig. 15.6. Response of a second-order system to a step input.

Thus the expression

$$\left[\left(\frac{k_3 m}{k_2 b + A^2}\right) D^2 + \left(\frac{k_2 m + k_3 b}{k_2 b + A^2}\right) D + 1\right] D \Delta Y \qquad (15.28)$$

may be written

$$\left[\left(\frac{1}{\omega_{ns}}\right)^2 D^2 + \frac{2\zeta_s}{\omega_{ns}} D + 1\right] D \Delta Y \qquad (15.29)$$

When the system is overdamped ($\zeta_s > 1$), it is usually more convenient to describe the system in terms of its real time constants, τ_1 and τ_2, which are given by the following equation:

$$\tau_1, \tau_2 = \frac{k_2 m + k_3 b}{2(k_2 b + A^2)}\left[1 \pm \sqrt{1 - \frac{4 k_3 m (k_2 b + A^2)}{(k_2 m + k_3 b)^2}}\right] \qquad (15.30)$$

and which satisfy the following identity:

$$\left(\frac{k_3 m}{k_2 b + A^2}\right) D^2 + \left(\frac{k_2 m + k_3 b}{k_2 b + A^2}\right) D + 1 \equiv (\tau_1 D + 1)(\tau_2 D + 1) \qquad (15.31)$$

Figure 15.7 shows graphically how ζ_s is related to the dimensionless parameters π_a and π_b where

$$\pi_a = \frac{k_2}{A} \sqrt{m/k_3} \qquad (15.32)$$

and

$$\pi_b = \frac{b}{A} \sqrt{k_3/m} \qquad (15.33)$$

In addition to providing a means of rapidly determining ζ_s this figure demonstrates clearly how to alter k_2 and/or b to change the damping ratio, ζ_s, if it is not initially satisfactory.

For an analogue study, this system may be represented by the functional block diagram shown in Fig. 15.8, which shows clearly the complementary roles played by k_2 and b and by k_3 and m in determining over-all system performance. Analogue studies made during initial stages of design have proved to be a very useful means of arriving at optimum design parameters for systems of this kind.

Figures 15.7 and 15.8 may be used for the results of the two linear cases analyzed in Sec. 15.23 if appropriate substitutions are made for k_1, k_2, m, and b in each case.

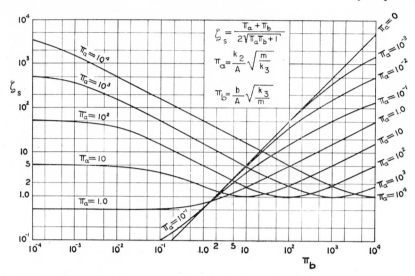

Fig. 15.7. ζ_s versus π_a and π_b for a valve-controlled hydraulic servomotor.

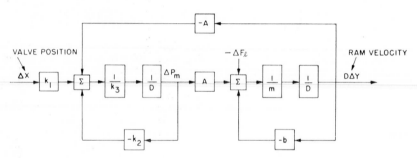

Fig. 15.8. Block diagram for valve-controlled servomotor.

15.4. CONCLUSIONS

15.41. Speed of Response

The most significant information relating to speed of response is contained in the expression for ω_{ns} if the servomotor is underdamped or in τ_1 if the servomotor is overdamped. Equation 15.27 shows that in order to obtain a high undamped natural frequency in an underdamped system, k_3m must be kept small in relation to $(k_2b + A^2)$; similarly for a small value of τ_1 in an overdamped system, Eq. 15.29 shows that $(k_2m + k_3b)$ must be kept small in relation to $k_2b + A^2$. Thus the smallest attainable values of k_3 and m will yield the fastest servomotor, in either case. The value of m is usually set by other considerations, but the designer has much latitude in varying k_3. The value of k_3 can be minimized by using rigid fluid lines (to reduce k_e) and by using the shortest possible ram cylinder and fluid lines (to reduce $\mathcal{U}_i$). In addition to the direct effect of ram area shown in Eqs. 15.27 and 15.29, the volume of fluid, $\mathcal{U}_i$, and hence k_3, is related to A, and if the volume of fluid in the lines is negligible, then $\mathcal{U}_i$ is proportional to A for a given ram stroke.

Although increasing A may give a system with higher ω_{ns} or lower τ_1 in some cases, the effect of varying A is usually weak and space limitations may prohibit the larger ram together with the larger valve, fluid lines, and power supply required to meet the same ram-velocity specifications.

15.42. Degree of Damping

From Eq. 15.26 it is seen that the servomotor damping ratio, ζ_s, is a function of all of the system constants except k_1; namely, k_2, k_3, A, m, and b. The values of k_3 and m are usually set by considerations such as those discussed in the foregoing section. The ram area often is determined by such practical considerations as maximum load and/or acceleration requirements to be met with a given supply pressure or range of supply pressures.

Once k_3, m, and A have been determined, it is necessary to investigate how a sufficient degree of damping may be attained. If k_2 and b are both zero, ζ_s is zero and the system will be unsatisfactory for most uses. An exception would be the case when an oscillatory motion at a fixed frequency is desired; here it would be desirable to keep k_2 and b small and select values of k_3, m, and A to produce the desired ω_{ns}.

For many applications the value of ζ_s must be greater than 0.7. Figure 15.7 illustrates clearly the effects of varying k_2 and b when k_3, m, and A are held constant. It is significant that appreciable values of k_2 and/or b may be required to obtain a value of ζ_s greater than 0.8. When k_2 is obtained by the use of an open-center valve, the stand-by power dissipation may be excessive if the servomotor is being used to control the position of the mass. If a closed-center valve is used, then C_1 is small and k_2 may be obtained by using a sufficiently large value of the laminar-leakage coefficient, C_2. In this case, power dissipation occurs whenever an appreciable pressure appears across the ram. In some instances a combination of valve underlap and leakage across the ram may produce the desired results. In either case, appreciable displacement of the valve from its center position may be required to hold a steady ram load.

A transient-flow stabilizer (Chapter 16) may be employed to gain stability, without sensitivity to steady-state loads. The large tanks necessary in a pneumatic servomotor can be replaced in the hydraulic case with relatively small accumulators; in fact, a single accumulator is sufficient if connected in series with a flow resistance (Fig. 15.9).

The value of b associated with the bearings on the output shaft may be sufficient in itself to produce the required value of ζ_s, but in some instances it may be desirable to add a viscous damper. The transfer of an

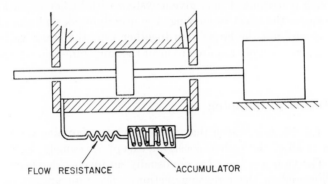

FLOW RESISTANCE   ACCUMULATOR

Fig. 15.9. Transient-flow stabilizer for valve-controlled hydraulic servomotor.

appreciable amount of heat, generated in the viscous damper, may become necessary at high output shaft speeds.

A number of nonlinear effects which exist in many real systems have not been considered here and the conclusions reached in this chapter may not be quantitatively applicable to such systems. On the other hand, the material covered gives a qualitative insight into the behavior of systems containing some nonlinearities. For instance, by determining the linearized dynamic characteristics in several regions of the valve characteristics, the variation of system performance may be estimated for a wide range of operating conditions. The block diagram and analogue techniques become essential to a thorough study of many nonlinear systems. The study of problems associated with the performance of hydraulic servomotors with various types of nonlinearities could be the basis of further work in this field.

Chapter 19, Comparison of Hydraulic and Pneumatic Servomechanisms, discusses a typical servomotor design and shows how the material in this chapter may be applied to a practical problem.

NOMENCLATURE

Symbol	Definition	Unit
A	ram area	in.^{2}
b	damping coefficient, associated with linear motion of load	lb sec/in.
B_l	damping coefficient, associated with angular motion of load	in. lb sec/radian
B_m	viscous-friction coefficient of motor	in. lb sec/radian
B	$B_l + B_m$ = total viscous-damping effect on output shaft	in. lb sec/radian
C_1	valve characteristic—partial derivative of flow with respect to pressure	in.^{5}/lb sec
C_2	laminar-leakage-flow coefficient	in.^{5}/lb sec
C_f	coefficient for friction due to pressure	
C_{ep}	external-leakage coefficient for pump	in.^{5}/lb sec
C_{ip}	internal-leakage coefficient for pump	in.^{5}/lb sec
C_{em}	external-leakage coefficient for motor	in.^{5}/lb sec
C_{im}	internal-leakage coefficient for motor	in.^{5}/lb sec
C_e	$C_{ep} + C_{em}$	
C_i	$C_{ip} + C_{im}$	
D	differential operator denoting derivative with respect to time, d/dt	1/sec
D_m	motor displacement	in.^{3}/radian
D_p	pump displacement	in.^{3}/radian
F_l	external load force	lb
J	polar moment of inertia of load mass	in. lb sec^2/radian
k_1	valve characteristic—partial derivative of flow with respect to valve position	in.^{2}/sec
k_2	$C_1 + C_2$	in.^{5}/lb sec
k_3	$\dfrac{1}{2}\left(k_e + \dfrac{v_i}{\beta}\right)$	in.^{5}/lb
k_e	coefficient of elasticity of each fluid line connecting ram to valve	in.^{5}/lb
k_p	displacement coefficient for pump	in.^{3}/radian deg
m	load mass	lb sec^2/in.
P_a	fluid pressure on left side of ram	lb/in.^{2}
P_b	fluid pressure on right side of ram	lb/in.^{2}
P_e	exhaust pressure	lb/in.^{2}
P_r	replenishing pressure	lb/in.^{2}
P_s	supply pressure	lb/in.^{2}
Q_1	rate of flow through orifice No. 1	in.^{3}/sec
Q_a	volume rate of flow of fluid into chamber on left side of ram	in.^{3}/sec

Symbol	*Definition*	*Unit*
Q_b	volume rate of flow of fluid into chamber on right side of ram	in.^{3}/sec
Q_{am}	rate of flow from line a to motor	in.^{3}/sec
Q_{bm}	rate of flow from motor to line a	in.^{3}/sec
Q_{ap}	rate of flow from pump to line a	in.^{3}/sec
Q_{bp}	rate of flow from line b to pump	in.^{3}/sec
Q_{lp}	total external-leakage rate of flow	in.^{3}/sec
Q_{lr}	leakage flow past ram	in.^{3}/sec
T_l	load torque	in. lb
T_m	motor torque	in. lb
$\mathcal{V}_a$	volume of fluid between left side of ram and valve	in.^{3}
$\mathcal{V}_b$	volume of fluid between right side of ram and valve	in.^{3}
$\mathcal{V}_i$	initial value of $\mathcal{V}_a$ and $\mathcal{V}_b$	in.^{3}
$\mathcal{V}_l$	total volume of each line	in.^{3}
X	displacement of valve from center position	in.
Y	position of ram from arbitrary datum	in.
β	bulk modulus of elasticity of fluid	psi
Δ	prefix indicating a small change	
ζ_s	servomotor damping ratio	
θ_m	motor-shaft angular displacement	radians
$\dot{\theta}_p$	pump-shaft angular velocity	radians/sec
π_a	dimensionless parameter	
π_b	dimensionless parameter	
τ_1	larger time constant of overdamped servomotor	sec
τ_2	smaller time constant of overdamped servomotor	sec
ϕ	pump stroke angle	deg
ω_{ns}	undamped natural frequency of servomotor	radians/sec

ADDITIONAL REFERENCES

J. L. Bower and F. B. Tuteur, "Dynamic Operation of a Force-Compensated Hydraulic Throttling Valve," *Trans. ASME*, Vol. 75 (October 1953), pp. 1395–1406.

S. Z. Dushkes and S. L. Cahn, "Analysis of Some Hydraulic Components Used in Regulators and Servomechanisms," *Trans. ASME*, Vol. 74 (May 1952), pp. 595–601.

H. E. Gold, W. Otto, and V. L. Ransom, "An Analysis of the Dynamics of Hydraulic Servomotors under Inertia Loads and the Application to Design," *Trans. ASME*, Vol. 75 (October 1953), pp. 1383–1394.

R. Hadekel, "Hydraulic Servos," Conference on Hydraulic Servo-mechanisms, published by the Institution of Mechanical Engineers, London, England, February 13, 1953.

N. F. Harpur, "Some Design Considerations of Hydraulic Servos of Jack Type," Conference on Hydraulic Servo-mechanisms, published by the Institution of Mechanical Engineers, London, England, February 13, 1953.

16

J. L. Shearer

═ *Pneumatic Drives*

16.1. INTRODUCTION

The flow of fluid under pressure is a widely used means of transmitting power from an energy source to the point of power utilization. Hydraulic systems have undergone a high degree of development, and they have been used effectively for the transmission and control of power in applications ranging from giant hydroelectric-power installations to compact hydraulic servomechanisms. Compressed gases have served successfully as the working fluid in systems as diversified as air brakes and steam and gas-turbine power plants.

Except for the low-pressure (10 to 20 psig) pneumatic control systems developed for industrial process controls, compressed gases seldom have been applied to the continuous control of motion required in many servomechanisms and automatic control systems. Use has been limited mainly to simple on-off control functions in systems where mechanical stops provide the required positive positioning action. The continuous-control systems developed for operation with compressed gas (usually air) as the working fluid are largely low-pressure instrument systems where speed of response is not a critical factor and the power controlled is usually a small fraction of 1 hp.

With a few possible exceptions, no rational engineering approach has been made to the problem of providing continuous pneumatic control of motion of a member having a significant mass when subjected to external load forces. Such existing systems seem to have evolved from the application of a fortunate combination of trial-and-error techniques and intuition.

In contrast to hydraulic drives, pneumatic drives are almost exclusively controlled with valves; variable-displacement air compressors are not available nor do they appear to be a promising means of modulating the flow of pneumatic power, largely because of the extra penalty of the more compliant working fluid and the difficulties involved in trying to locate the compressor close to the load.

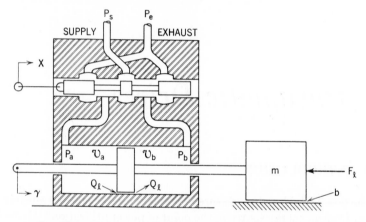

Fig. 16.1. Valve-controlled pneumatic servomotor with load.

The purpose of this chapter is to try to gain a thorough understanding of the fundamental factors governing the performance of a valve-controlled pneumatic servomotor in the presence of external load forces in order to evolve a rational design procedure that will facilitate the design and/or evaluation of systems of this kind in a straightforward manner. The material in this section is taken from two papers [1] by the author which were published in the *Transactions of the American Society of Mechanical Engineers.*

Figure 16.1 shows schematically the basic system under consideration. Factors deemed to be of outstanding importance in the evaluation of this system are speed of response, degree of dynamic stability, load sensitivity, and efficiency. Practical design considerations such as the strengths of structural members, means of fastening and sealing, arrangement of components, and selection of materials are not to be dealt with here.

[1] J. L. Shearer, "Study of Pneumatic Processes in the Continuous Control of Motion with Compressed Air," *Trans. ASME*, Vol. 78 (February 1956), pp. 233–242, 243–249.

16.2. DYNAMIC ANALYSIS OF RAM CHAMBERS

Figure 16.2 shows a valve-controlled ram schematically. The weight rate of flow of gas to chamber a is W_a, and the weight rate of flow of gas

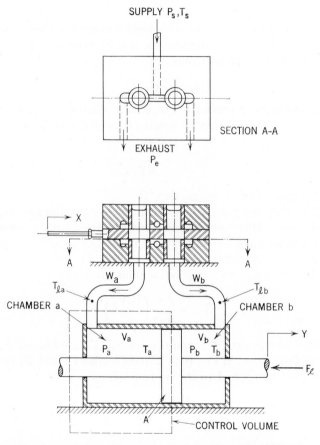

Fig. 16.2. Schematic cross section of valve-controlled ram.

to chamber b is W_b, with stagnation temperatures designated by T_{la} and T_{lb}, respectively. The ram chambers a and b have volumes of V_a and V_b, respectively. As a first approximation, an assumption of perfect mixing in chambers a and b makes possible the use of a single pressure and a single temperature to describe the state of the fluid in each chamber. The quantities P_a and T_a then will be the pressure and temperature in chamber a, and the quantities P_b and T_b the pressure and temperature in chamber b. The effective area of the ram is A, and Y is the displacement of the ram from its center position.

From the continuity equation of the control volume drawn around chamber a

$$W_a = g \frac{d}{dt} (\rho_a V_a) \tag{16.1}$$

Since the gas used here may be considered a pure substance as long as it never starts to liquefy,[2] its density, ρ_a, may be expressed in terms of its pressure, P_a, and its temperature, T_a.

$$\rho_a = f(P_a, T_a)$$

In many engineering problems (pneumatic control systems included) the perfect-gas equation may be used as the equation of state

$$\rho_a = \frac{P_a}{RT_a} \tag{16.2}$$

where R = gas constant, sq in./sec^2 °R

Equation 16.2 may not be used when the state of the gas approaches its liquid state. However, Eq. 16.2 may be used with reasonable accuracy (fn. 2) for air when its temperature is above 450°R (-10°F) and its pressure is below 4000 psi.

Equations 16.1 and 16.2 may be combined to give

$$W_a = \frac{g}{R} \frac{d}{dt} \left(\frac{P_a V_a}{T_a} \right) \tag{16.3}$$

Even when the flow rate, W_a, and the volume, V_a, are known, P_a and T_a still are undetermined in Eq. 16.3. To determine P_a and T_a the energy equation must be applied to the control volume.

$$\frac{c_p}{g} W_a T_{la} - P_a \frac{dV_a}{dt} + \frac{dQ_h}{dt} = \frac{d}{dt} (c_v \rho_a V_a T_a) \tag{16.4}$$

From Eq. 16.2 for the perfect gas,

$$W_a T_{la} - \frac{gP_a}{c_p} \frac{dV_a}{dt} + \frac{g}{c_p} \frac{dQ_h}{dt} = \frac{g}{kR} \frac{d}{dt} (P_a V_a) \tag{16.5}$$

In general, the rate of heat flow, dQ_h/dt, is difficult to evaluate by either analytical or experimental means. In practice, the value of dQ_h/dt lies

[2] J. L. Shearer, "Continuous Control of Motion with Compressed Air," Sc.D. Thesis, Department of Mechanical Engineering, Massachusetts Institute of Technology, Cambridge, Mass., 1954.

somewhere between two possible extremes:

1. Zero, when the ram is perfectly insulated.
2. Infinity, when sufficient heat is transferred to hold T_a constant regardless of the rate at which compression or expansion effects take place in the chamber.

Even without special provisions to insulate the ram chamber, heat-transfer rate is negligible in most fast-acting systems, but it may be appreciable in slower-acting systems.

Skinner and Wagner [3] investigated the basic processes of charging and discharging constant-volume containers; their experiments were conducted with cylindrical tanks surrounded with still air at atmospheric conditions. The results of these experiments may be employed to estimate the significance of heat-transfer effects in this type of system. Present experience indicates that heat-transfer effects are seldom important in fast-acting pneumatic control systems.

Here an attempt will be made to evaluate servomotor performance for the two extreme cases, namely, isothermal changes and adiabatic changes in the ram chambers. These two cases bracket most engineering applications of this work. An exception would be when a strong heat or energy source is present and acts as another disturbance to the system.

16.21. Isothermal Changes

When the pressure, P_a, changes isothermally, T_a is constant and Eq. 16.3 may be simplified to give

$$W_a = \frac{g}{RT_a} \frac{d}{dt} (P_a V_a) \tag{16.6}$$

or by integration with respect to time

$$P_a V_a = \frac{RT_a}{g} \int W_a \, dt + (P_a V_a)_{\text{initial}} \tag{16.7}$$

Similarly, for chamber b

$$W_b = \frac{g}{RT_b} \frac{d}{dt} (P_b V_b) \tag{16.8}$$

or

$$P_b V_b = \frac{RT_b}{g} \int W_b \, dt + (P_b V_b)_{\text{initial}} \tag{16.9}$$

[3] C. K. Skinner and F. D. Wagner, "A Study of the Processes of Charging and Discharging Constant Volume Tanks with Air," S.B. Thesis, Department of Mechanical Engineering, Massachusetts Institute of Technology, Cambridge, Mass., 1954.

When only small changes of P_a, P_b, V_a, and V_b occur from initially steady values, Eqs. 16.6 and 16.8 may be written as follows:

$$\Delta W_a = \frac{g}{RT_a} \left[P_{ai} \frac{d(\Delta V_a)}{dt} + V_{bi} \frac{d(\Delta P_a)}{dt} \right] \tag{16.10}$$

$$\Delta W_b = \frac{g}{RT_b} \left[P_{bi} \frac{d(\Delta V_b)}{dt} + V_{bi} \frac{d(\Delta P_b)}{dt} \right] \tag{16.11}$$

where the subscript i designates the initially steady value of each variable. If the ram moves small distances near its center position and the temperatures and pressures are initially the same on both sides of the ram, the following equation may be obtained by subtracting Eq. 16.11 from Eq. 16.10:

$$(\Delta W_a - \Delta W_b) = \frac{g}{RT_i} \left[2P_i A \frac{d(\Delta Y)}{dt} + V_i \frac{d}{dt} (\Delta P_a - \Delta P_b) \right] \tag{16.12}$$

Although Eq. 16.12 is valid only when P_a, P_b, and Y change by a small percentage of their initial values, it gives a qualitative picture of the relationship between pressures in the ram chambers, flow from the valve, and ram motion. Chamber pressure tends to vary directly with ram position but varies as the time integral of the valve flow. If a change in pressure is to result from a change in valve flow, sufficient time must elapse for this integrating action to occur before the desired pressure change can take place.

The equations just derived for small changes of all variables resemble the equations which apply to a similar hydraulic system.[4] Since such hydraulic systems operate very nearly isothermally, a complete analogy exists for the two cases.

16.22. Adiabatic Changes

When the changes in each chamber occur adiabatically, both Eq. 16.3 and Eq. 16.5 should be used. The term dQ_h/dt is zero, but T_a is now a variable. In most instances, the temperature, T_a, is of little direct interest and in a fast system it would be very difficult to measure. Therefore, for the time being, it should be eliminated. In Eq. 16.5, T_a does not appear directly as a variable. As long as the flow W_a remains positive, T_a can be ignored completely. However, when the flow W_a is negative, the temperature T_{la} becomes T_a.

---

[4] J. L. Shearer, "Dynamic Characteristics of Valve-Controlled Servomotors," *Trans. ASME*, Vol. 76 (August 1954), pp. 895–903.

The results from a thesis investigation (fn. 2) show that in most cases the temperature T_{la} does not vary appreciably from its value when W_a is positive. Therefore, Eq. 16.5 alone describes the relationship between pressure, flow, and ram motion for each of the ram chambers.

Accordingly, the equations describing the adiabatic changes in chambers a and b are

$$W_a T_{la} - \frac{g P_a}{c_p} \frac{dV_a}{dt} = \frac{g}{kR} \frac{d}{dt} (P_a V_a) \qquad (16.13)$$

$$W_b T_{lb} - \frac{g P_b}{c_p} \frac{dV_b}{dt} = \frac{g}{kR} \frac{d}{dt} (P_b V_b) \qquad (16.14)$$

When all variables undergo small changes from initially steady values, the equations may be written

$$T_{la}(\Delta W_a) - \frac{g P_{ai}}{c_p} \frac{d(\Delta V_a)}{dt} = \frac{g}{kR} \left[P_{ai} \frac{d(\Delta V_a)}{dt} + V_{ai} \frac{d(\Delta P_a)}{dt} \right] \qquad (16.15)$$

$$T_{lb}(\Delta W_b) - \frac{g P_{bi}}{c_p} \frac{d(\Delta V_b)}{dt} = \frac{g}{kR} \left[P_{bi} \frac{d(\Delta V_b)}{dt} + V_{bi} \frac{d(\Delta P_b)}{dt} \right] \qquad (16.16)$$

In addition, if the ram moves small distances near its center position, if the pressures are initially the same on both sides of the ram, and if $T_{la} = T_{lb} = T_i$, then the following equation may be obtained by subtracting Eq. 16.16 from Eq. 16.15 and using $1/c_p + 1/kR = 1/R$:

$$(\Delta W_a - \Delta W_b) = \frac{g}{RT_i} \left[2P_i A \frac{d(\Delta Y)}{dt} + \frac{V_i}{k} \frac{d}{dt} (\Delta P_a - \Delta P_b) \right] \qquad (16.17)$$

It is significant that the equations just derived for small adiabatic changes, namely, Eqs. 16.15, 16.16, and 16.17, are very nearly like Eqs. 16.10, 16.11, and 16.12, which were derived for small isothermal changes, the essential difference being the factor k that appears in the terms of the right-hand sides of Eqs. 16.15, 16.16, and 16.17. For air which has a value for k of 1.4, the adiabatic pressure changes are 1.4 times as large as the isothermal pressure changes.

16.3. CONTROL-VALVE CHARACTERISTICS

Experience gained in the development of fluid-power control systems indicates that a slide-type variable orifice is one of the best flow resistances to use in control valves. Although the flow characteristics are nonlinear, a variable orifice provides a means of obtaining a large change in resistance to flow with a very small valve motion. The valves shown as part of the servomotors in Figs. 16.1 and 16.2 are examples of slide-type valves, Fig. 16.1 showing a sliding-spool-type valve, and Fig. 16.2 showing a sliding-plate-type valve using the hole-slot-and-plug construction developed by Lee.[5] Although the clearance spaces between the moving parts must be kept at a minimum to reduce unwanted leakage effects, the forces required to stroke such valves are relatively small, and the valves require little space.

To control the flow of fluid from a single constant-pressure source of power to the ends of a double-acting ram, four flow resistances are required, one or more of which may be variable. It is usually important to keep the flow of fluid from the power source at a minimum when the ram is not moving. Since this quiescent leakage flow is related inversely to the values of the four resistances when the valve is in its neutral or centered position, the valve must be constructed so that the resistances are large when it is centered. Such a valve is termed a closed-center valve because the metering orifices comprising the flow resistances are very nearly closed when the valve is near its center position. Although a closed-center valve may be advantageous from other considerations, its small quiescent power loss due to leakage is usually its most important characteristic except when the fluid-power supply is overabundant or is generated in such a way that it must be consumed at a high rate at all times. This quiescent leakage is doubly important when the working fluid is a compressed gas because as the gas leaks through the valve, it expands to many times its initial volume without doing any useful work, whereas a large amount of work was required to compress the gas to its initial pressure.

When quiescent leakage flow is not a critical factor, the valve may be designed to have its orifices open when it is in the centered position. Such a valve is termed an open-center valve or underlapped valve. A four-orifice valve of the type illustrated in Figs. 16.1 and 16.2 is called a four-way valve.

A large-scale, two-dimensional model study of the flow of air through a single slide-type orifice has been made by Stenning.[6] This work shows

[5] S.-Y. Lee, "Contributions to Hydraulic Control—6, New Valve Configurations for High-Performance Hydraulic and Pneumatic Systems," *Trans. ASME*, Vol. 76 (August 1954), pp. 905–910.

[6] A. H. Stenning, "An Experimental Study of Two-Dimensional Gas Flow Through Valve-Type Orifices," ASME Paper No. 54-A-45, presented at the Annual Meeting of the ASME, New York, November 28–December 3, 1954.

that there are two flow conditions possible in such a valve orifice: one with a discharge coefficient of approximately 0.8 and the other with a discharge coefficient of approximately 1.0. Either flow condition may be obtained with greater than critical pressure drop across the orifice (that is, choked flow), but the higher coefficient is obtained more frequently at large-pressure-drop conditions. Furthermore, the discharge

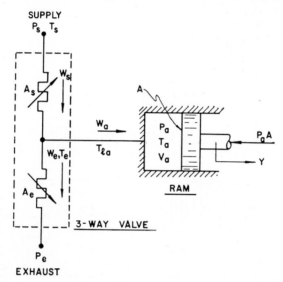

Fig. 16.3. Schematic diagram of single-acting ram controlled by three-way valve.

coefficient for an orifice of this type [7] operating with hydraulic fluid is only about 0.63.

The four orifices comprising a four-way valve may be considered as two pairs of orifices, one pair controlling the flow to or from one ram chamber, the other pair controlling the flow to or from the other ram chamber. Each pair of orifices may be considered as a three-way valve of the type often used to control the motion of a single-acting ram. Thus a study of three-way-valve characteristics also yields information that is directly useful in the design of single-acting systems, such as that shown in Fig. 16.3, where A_s and A_e designate the area of the opening of the supply and exhaust orifices and W_s and W_e are the weight rate of flow through A_s and A_e, respectively.

---

[7] S.-Y. Lee and J. F. Blackburn, "Contributions to Hydraulic Control—1, Steady-State Axial Forces on Control-Valve Pistons," *Trans. ASME*, Vol. 74 (August 1952), pp. 1005–1009.

A sliding-plate valve of the type shown in Fig. 16.2 may be used to illustrate the following analysis. One end of this valve contains the two orifices that make up a three-way valve as shown in detail in Fig. 16.4. The areas A_s and A_e are, respectively, $2w_s(U + X)$ and $2w_e(U - X)$ if the clearance between the valve plate and valve blocks is negligible. The quantity U, called the valve underlap, is a measure of the opening

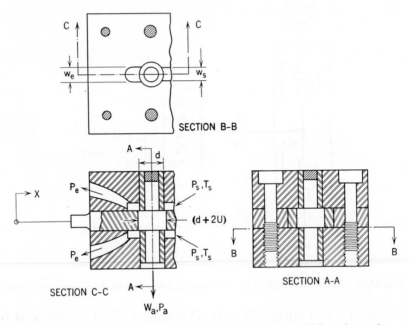

Fig. 16.4. Cross-sectional drawing of one end of four-way sliding-plate valve.

of each orifice when the valve is centered. Although valve underlap is undesirable when the quiescent leakage flow must be kept at a minimum, it is included in this discussion for generality.

For the time being, side leakage to atmosphere in the clearance spaces is assumed to be negligible. Then the flow to the ram, W_a, is equal to the difference between the supply flow, W_s, and the exhaust flow, W_e.

$$W_a = W_s - W_e \qquad (16.18)$$

Usually, the equation describing the flow of a compressible fluid through an orifice is of the form

$$W = C_d A_o f\left(P_u, T_u, \frac{P_d}{P_u}\right) \qquad (16.19)$$

The function $f(P_u, T_u, P_d/P_u)$ can be expressed as follows: [8]

$$f\left(P_u, T_u, \frac{P_d}{P_u}\right) = C \frac{P_u}{\sqrt{T_u}} f_1\left(\frac{P_d}{P_u}\right) \qquad (16.20)$$

where $C = g \sqrt{\dfrac{k}{R\left(\dfrac{k+1}{2}\right)^{(k+1)/(k-1)}}} \quad (\sqrt{°R}/\text{sec}) \qquad (16.21)$

Gases such as air having a value of $k = 1.4$ then have a value of $C =$

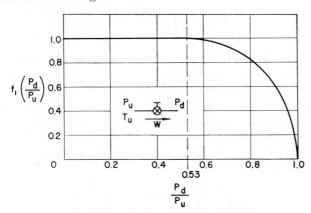

Fig. 16.5. Graphical representation of $f_1(P_d/P_u)$.

$0.54\sqrt{°R}/\text{sec}$. The function $f_1(P_d/P_u)$ for air is shown graphically in Fig. 16.5.

When the gas is air, the flow W_a then may be expressed as

$$W_a = C_{ds}(2w_s)(U + X)(0.54)\left(\frac{P_s}{\sqrt{T_s}}\right) f_1\left(\frac{P_a}{P_s}\right)$$

$$- C_{de}(2w_e)(U - X)(0.54)\left(\frac{P_a}{\sqrt{T_e}}\right) f_1\left(\frac{P_e}{P_a}\right) \qquad (16.22)$$

or when terms are collected, as

$$W_a = \frac{1.08 C_{ds} w_s U P_s}{\sqrt{T_s}} \left[\left(1 + \frac{X}{U}\right) f_1\left(\frac{P_a}{P_s}\right)\right.$$

$$\left. - \left(1 - \frac{X}{U}\right) \frac{C_{de} w_e P_a}{C_{ds} w_s P_s} \sqrt{T_s/T_e}\, f_1\left(\frac{P_e}{P_a}\right)\right] \qquad (16.23)$$

[8] A. H. Shapiro, "Compressible Fluid Flow," Ronald Press Co., New York, 1953, pp. 73–93.

The quiescent leakage flow, W_q, is the flow through A_s and A_e when $W_a = 0$ and $X = 0$, and is given by

$$W_q = \frac{1.08 C_{ds} w_s U P_s}{\sqrt{T_s}} f_1 \left(\frac{P_{aq}}{P_s} \right) \tag{16.24}$$

and an equation for W_a/W_q may be written as

$$\frac{W_a}{W_q} = \left(1 + \frac{X}{U} \right) \frac{f_1(P_a/P_s)}{f_1(P_{aq}/P_s)}$$

$$- \left(1 - \frac{X}{U} \right) \frac{C_{de} w_e P_a}{C_{ds} w_s P_s} \sqrt{T_s/T_e} \frac{f_1(P_e/P_a)}{f_1(P_{aq}/P_s)} \tag{16.25}$$

The quantities $(1 + X/U)$ and $(1 - X/U)$ can never become negative in a real valve. When the magnitude of X/U exceeds unity, one of the orifices is closed and either $(1 - X/U)$ or $(1 + X/U)$ is merely zero. In most applications the temperature T_e upstream of A_e may be assumed to be the same as T_s. Even when T_e does vary appreciably from T_s, the effect of this variation appears only as the square root of T_e.

The value of P_{aq} may be determined from Eq. 16.23 by setting $W_a = X/U = 0$ and $T_e = T_s$.

$$f_1 \left(\frac{P_{aq}}{P_s} \right) = \frac{C_{de} w_e P_{aq}}{C_{ds} w_s P_s} f_1 \left(\frac{P_e}{P_{aq}} \right) \tag{16.26}$$

When C_{de}, C_{ds}, and port widths w_e and w_s are known, Eq. 16.26 may be solved for P_{aq} by trial and error by use of Fig. 16.5. The value of C_{de}/C_{ds} is usually nearly unity. The value of port-width ratio w_e/w_s is a significant design parameter because it largely determines the value of the quiescent pressure ratio P_{aq}/P_s in the ram chamber.

In a high-pressure pneumatic servomotor, $P_e \ll P_s$ or P_a and $f_1(P_e/P_{aq}) = 1$. For the special case when $C_{de} w_e / C_{ds} w_s = 1$, the value of P_{aq}/P_s is the same as the value of $f_1(P_{aq}/P_s)$. From Fig. 16.5, P_{aq}/P_s is approximately 0.82 for this special case, and increasing w_e/w_s results in decreasing P_{aq}/P_s.

Once P_{aq}/P_s is known, curves of W_a/W_q versus P_a/P_s can be plotted for various values of X/U. This can be done by systematically selecting values of P_a/P_s and X/U and solving for W_a/W_q by use of Fig. 16.5 or the *Gas Tables*.[9] This has been done for the special case when $C_{de} w_e / C_{ds} w_s = 1$ and when $P_e \ll P_s$ or P_a, and the results are shown in Fig. 16.6. Dashed curves represent values of X greater than U. As the

[9] J. H. Keenan and J. Kaye, *Gas Tables*, John Wiley and Sons, New York, 1948, pp. 139–142.

value of the valve underlap, U, approaches zero, X/U is greater than unity for even small valve displacements from center, and the solid curves tend to become an insignificant part of the valve characteristics.

Although valves can be manufactured with zero underlap or even overlap, it is impossible to manufacture valves with zero clearance be-

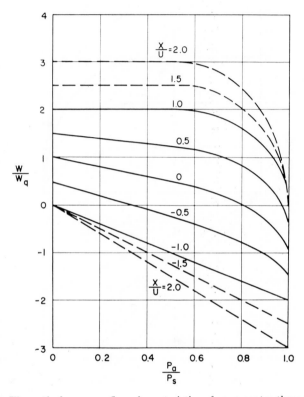

Fig. 16.6. Theoretical pressure-flow characteristics of open-center three-way valve.

$[C_{de}w_e/C_{ds}w_s = 1, \ P_e \ll P_s \ \text{or} \ P_a$

$$W_q = (1.08C_{ds}w_s UP_s/\sqrt{T_s})f_1(P_{aq}/P_s)]$$

tween the moving parts of the valve. Effects of leakage through the clearance spaces cannot be neglected in a nominally closed-center valve operating with air. Clearance spaces at the orifices and rounding of the orifice corners in themselves do not affect seriously the value of the quiescent ram pressure since these effects are very nearly the same as valve underlap when the valve is centered. Side leakage to atmosphere, however, does tend to lower P_{aq}. Although side leakage has not proved

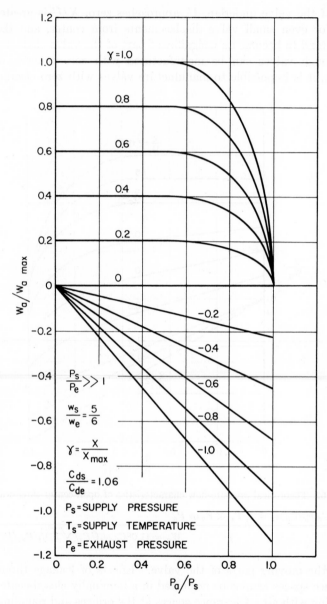

Fig. 16.7. Computed dimensionless pressure-flow curves of ideal sliding-plate valve (see Fig. 16.4).

$$[W_{a\ max} = 0.53(P_s/\sqrt{T_s})(0.89w_sX_{max})]$$

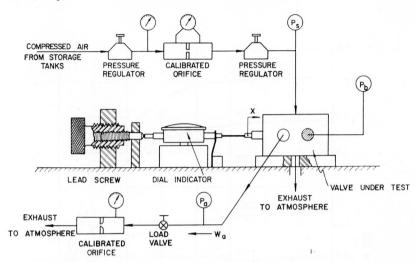

Fig. 16.8. Schematic diagram of experimental apparatus used to measure valve characteristics.

to be significant in hydraulic valves because of high resistance to flow of viscous liquids in capillary clearance spaces, it is very important in pneumatic valves because of the very low viscosity of air compared with that of hydraulic oil. The flow of air in small clearance spaces has been investigated by Grinnell [10] and Egli.[11] Their work shows that unless the length of the leakage path is several thousand times its thickness in a typical valve with a clearance space of 0.0001 in., the resistance to flow is due mostly to momentum effects, with the result that the leakage approaches the amount that would occur through an orifice having an area equal to the cross-sectional area of the clearance spaces (that is, the area normal to the leakage flow).

Figure 16.7 shows the computed dimensionless pressure-flow curves of a closed-center four-way sliding-plate valve having a port-width ratio of $W_e/W_s = 1.2$, and where $W_{a\,max}$ is the maximum flow to the ram when $X = X_{max}$. The values of C_{ds} and C_{de} used in the computations were 0.89 and 0.84, respectively (the valve has to be tested at large openings to find these values of C_{ds} and C_{de}).

To measure the characteristics of this valve, the experimental setup shown in Fig. 16.8 was used. The valve stroke was adjusted by a dif-

[10] S. K. Grinnell, "A Study of Pressurized Air Bearing Design, Steady Loading—No Rotation," S.M. Thesis, Department of Mechanical Engineering, Massachusetts Institute of Technology, Cambridge, Mass., 1954.

[11] A. Egli, "The Leakage of Gases Through Narrow Channels," *J. Appl. Mechanics*, Vol. 59 (June 1937), pp. A-63–A-67.

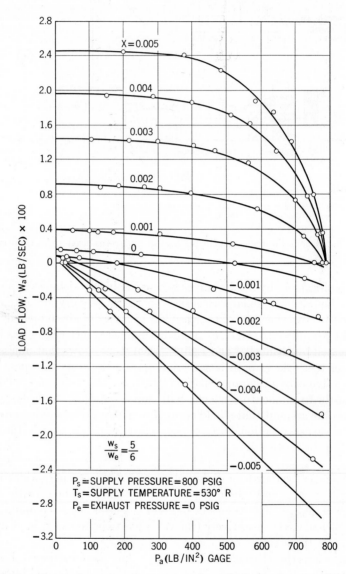

Fig. 16.9. Measured pressure-flow curves of nominal closed-center sliding-plate valve (see Fig. 16.4).

ferential lead screw and measured by a precision dial indicator. Pressures were measured with Bourdon type gages, and flows were measured with calibrated orifices.

The characteristics for negative values of W_a were obtained by employing a differential-pressure gage with orifices of various sizes connected to supply pressure rather than exhausting to atmosphere. Figure 16.9 gives the measured pressure-flow characteristics for values of valve opening from -0.005 in. to $+0.005$ in. Zero valve opening is the position where the quiescent pressure in ram chamber a is the same as the quiescent pressure in ram chamber b when the ram is motionless (that is, when the valve is in its center or neutral position). For valve openings between -0.001 in. and $+0.001$ in., the characteristics resemble those of an open-center valve, yet this valve has a slight overlap condition because the sleeves were made slightly larger than the metering holes in order to establish a firm press fit for them in the valve blocks. The open-center effect is due to the effects of clearances and rounding of the corners at the metering orifices. The measured quiescent pressure, P_{aq}, is $0.66P_s$.

For larger valve openings, the characteristics approach those shown in Fig. 16.6 for valve positions larger than the valve underlap. The gentle slope of the left-hand portions of the measured curves for positive values of X arises largely from side-leakage flow to atmosphere through the clearance spaces. The nearly constant slope indicates that this leakage flow is very nearly proportional to P_a. The value of C_{ds} was calculated by dividing the change in W_a caused by opening the valve from $X = 0.002$ in. to $X = 0.005$ in. with $P_a = 0$ by the change that would occur if the discharge coefficient for the supply orifice were 1.00. The value of C_{de} was calculated by dividing the change in W_a caused by opening the valve from $X = -0.002$ in. to $X = -0.005$ in. with $P_a = 800$ psig by the change that would occur if the discharge coefficient for the exhaust orifice were 1.00. The values of C_{ds} and C_{de} were found to be 0.89 and 0.84, respectively, within the range of Stenning's measurements (fn. 6). The difference between C_{ds} and C_{de} may be attributed to a stronger three-dimensional contraction effect in the exhaust orifice.

The measured curves for negative values of W_a were obtained with the temperature of the flow W_a equal to the supply temperature, T_s. When the temperature of this flow is appreciably different from T_s, the flow can be expected to vary inversely as the square root of this temperature. This effect partly cancels the effect noted previously when the gas is leaving the ram chamber. There, an error in the term $T_{la}W_a$ would be incurred whenever T_a was appreciably different from T_{la} as W_a changed direction from positive to negative. Since $W_a \sim \sqrt{1/T_a}$ when W_a is negative, $T_{la}W_a \sim \sqrt{T_a}$, and there is further justification for assuming that the temperature T_{la} is the same for both positive and negative values of W_a. Because the flow through the valve is very nearly adiabatic, T_{la} can be considered equal to T_s for all but very exceptional cases.

Since the b end of the four-way valve is identical with the a end, characteristic curves for the b end of the valve are the same as those for the a end except that curves for positive values of X of the a end correspond to curves for negative values of X of the b end of the valve. Otherwise, P_a and W_a may be replaced by P_b and W_b, respectively, and the same curves may be used for both ends of the valve. In general

$$W_a = f_v(X, P_a) \tag{16.27}$$

and

$$W_b = f_v(-X, P_b) \tag{16.28}$$

where f_v denotes the functional relationship graphed in Fig. 16.9.

When all variables undergo small changes from a point where $X = X_i$, $P_a = P_{ai}$, and $W_a = W_{ai}$,

$$\Delta W_a = \frac{\partial W_a}{\partial X}(\Delta X) + \frac{\partial W_a}{\partial P_a}(\Delta P_a) \tag{16.29}$$

$$\Delta W_b = \frac{\partial W_b}{\partial X}(\Delta X) + \frac{\partial W_b}{\partial P_b}(\Delta P_b) \tag{16.30}$$

or

$$\Delta W_a = k_{1a}(\Delta X) + k_{2a}(\Delta P_a) \tag{16.31}$$

$$\Delta W_b = k_{1b}(-\Delta X) + k_{2b}(\Delta P_b) \tag{16.32}$$

Figure 16.10 illustrates graphically how the values of k_{1a} and k_{2a} may be measured from a family of pressure-flow curves for a valve.

The variation of W_a with X when $P_a = P_{aq}$ is of particular interest because this shows the flow sensitivity of the valve when the ram pressure does not vary appreciably from its quiescent value. Figure 16.11 is a plot of W_a versus X when $P_a = P_{aq} = 525$ psig. The slope of this curve is k_{1a}.

Everything that has been discussed about k_{1a} and k_{2a} can be repeated for k_{1b} and k_{2b}, with care being taken to change the sign of X_i throughout; and a plot of $-W_b$ versus X when $P_b = P_{bq} = 540$ psig would be the same as the curve of W_a versus X. The curve of W_a versus X consists of two straight-line portions of nearly the same slope with a small S-curved section near the origin, where the minimum slope is less than one third the slope of the straight portions. This region of low valve sensitivity is due in part at least to a slight overlap condition, which has

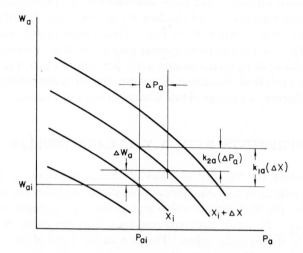

Fig. 16.10. Graphical determination of k_{1a} and k_{2a}.

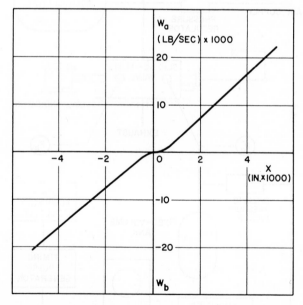

Fig. 16.11. Plot of W_a and W_b versus X when $P_a = P_b = P_{aq} = 525$ psig.

been discussed already. For large variations of X, the low-sensitivity region becomes negligible, and the flow W_a is very nearly linearly related to valve opening X when $P_a = P_{aq}$. This means that a subsequent linear analysis of the complete system may allow large variations of X if the variations of P_a do not exceed a few per cent of P_{aq}. A more comprehensive analytical discussion of pneumatic valve characteristics appears in Chapter 8, Pressure-Flow Characteristics of Pneumatic Valves.

16.4. EXPERIMENTS WITH A VALVE-CONTROLLED TANK

To ascertain the validity of some of the assumptions made in the dynamic analysis of ram chambers, some simple experiments were conducted with the sliding-plate valve and ½-gal tank arranged as shown in Fig. 16.12. The tank was connected by a short line to the port at the a end of the sliding-plate valve. The port at the b end of the valve was

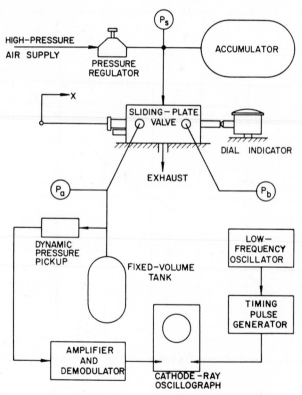

Fig. 16.12. Schematic diagram of valve-controlled tank experiment.

plugged.   Calibrated Bourdon type pressure gages were installed to make
steady measurements of P_s, P_a, and P_b.   A 1-gal hydraulic accumulator
with all of the oil removed was employed in conjunction with a pressure-
regulating valve to maintain a constant pressure supply to the sliding-
plate valve.

Dynamic measurement of P_a was accomplished with a bonded-strain-
gage-type pressure pickup developed at DACL by S.-Y. Lee.   The pressure signal was amplified, demodulated,
and displayed on a cathode-ray oscillograph.   The cathode ray was driven
horizontally at low speed by the sweep
generator of the oscillograph, and accurate timing marks were superimposed
on the display of P_a by modulating the
intensity of the cathode ray with pulses
from a pulse generator driven by a low-
frequency oscillator.

Before each test the valve was centered, and the pressure and temperature
in the fixed-volume tank were allowed to
reach steady values.   Then the valve
plate was suddenly displaced to a new

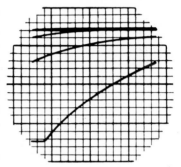

Fig. 16.13.   Photographic record
of tank pressure as a function of
time, charging of tank.

position and held there for the duration of the test.   The display of P_a on
the face of the cathode-ray oscillograph was recorded photographically,
and the initial and final values of P_a were read from the Bourdon type
gages to provide a means of calibrating the photographic records.
Valve-plate displacement was measured with a dial indicator.   Both
charging and discharging tests were made.   Figure 16.13 reproduces
one of the photographic records made during charging of the tank.

To evaluate the analytical work in the dynamic analysis of ram
chambers, enlargements were made of the photographic records and
compared with the results computed by performing a step-by-step
numerical integration of Eq. 16.13 together with the measured pressure-
flow characteristics, Fig. 16.9, of the valve.   Figures 16.14 and 16.15
graph the measured and calculated responses of P_a as a function of time
for charging and discharging of the tank.   The excellent agreement between measurements and calculations during the early part of the
responses lends strong support to the assumptions made in the analysis
for adiabatic changes.   After 3 sec, the divergence of the measured response from the calculated response begins to become noticeable, presumably because of the effects of heat transfer.   Yet, the correspondence
between measured and calculated results during the later periods is still

good enough for most engineering work. According to the results of the work by Skinner and Wagner (fn. 3) the effects of heat transfer up to this time should be negligible.

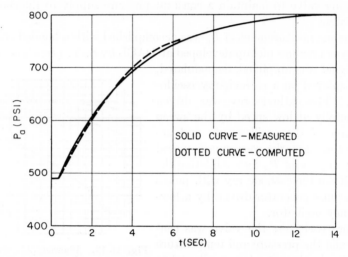

Fig. 16.14. Results of valve-controlled tank experiment, charging of tank with air.

On the other hand, the results of the discharging tests also lend support to the assumption that the temperature of the gas leaving the tank after tank pressure has fallen may be taken as equal to the initial temperature

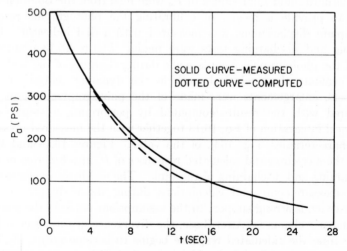

Fig. 16.15. Results of valve-controlled tank experiment, discharging of tank with air.

in the tank because the calculations were made on this basis. This apparent paradox exists because the time rate of change of temperature in the tank is of much greater dynamic significance than the change of flow through the valve due to the effects of changes in the value of this temperature. The situation is analogous to that existing in the pressurization of liquids where the rate of change of fluid density is of much greater significance than the change in the value of density itself (fn. 4).

No attempt was made to make dynamic measurements of temperature in the tank because of practical considerations. If careful measurements were to be made of gas temperature in the tank, measurements at many places in the tank would be required during charging because there is no direct evidence that perfect mixing occurs. The known methods of making sufficiently rapid measurement of gas temperature are very elaborate, and the need for knowing temperatures was not sufficient to justify the effort required to measure temperatures dynamically.

The main differences between pressurization of a chamber filled with a gas and pressurization of the same chamber filled with a liquid may be summarized as follows: rate of change of chamber temperature is a significant factor in the dynamic behavior of the gas system but the variation of temperature itself is a minor factor. In a liquid system, pressurization of the liquid results in negligible rate of change of temperature and negligible change in temperature itself.

The experiments with a valve-controlled tank show that the basic equations which were derived to describe the pneumatic processes in the servomotor are in good agreement with experimental results.

The complete servomotor system driving a mass load may now be analyzed for its response to small changes of valve opening and to changes of external load force.

16.5. ANALYSIS OF VALVE, RAM, AND LOAD SYSTEM

To perform a simple mathematical analysis, a number of simplifying assumptions may be made about the valve, ram, and load system. Justification for or refutation of the following assumptions can come only from experience with real systems or detailed analyses of systems of this kind:

1. Supply pressure, P_s, is constant.
2. Supply temperature, T_s, is constant.
3. Heat transfer between working gas and its environment is negligible.

4. The working gas obeys the perfect-gas law: $p = \rho R T$.

5. Temperature of the gas flowing between the valve and the ram is at all times equal to supply temperature, T_s.

6. The ram moves only small distances from its center position.

7. Ram pressures P_a and P_b vary by only small amounts from an initial steady value of P_i.

8. Control-valve constants k_{1a} and k_{1b} are equal and do not vary with valve opening.

9. Control-valve constants k_{2a} and k_{2b} are equal and do not vary with valve opening.

10. Valve opening does not exceed its maximum design value.

11. Passages connecting valve and ram are very short and offer negligible resistance to flow.

12. Friction forces on the ram are viscous (no dry friction).

13. External load force, F_l, is sufficiently small to make assumption 7 possible.

A small change is defined as one that is small enough to provide a given degree of accuracy in the equation where this change may appear. Equations 16.15, 16.16, and 16.17, which were developed for the case when the ram moves only small distances from its center position, may be used to illustrate this definition of "small." When Eqs. 16.15, 16.16, and 16.17 are repeated for convenience, with T_s substituted for T_l, T_{la}, and T_{lb},

$$T_s\,(\Delta W_a) - \frac{gP_{ai}}{c_p}\frac{d(\Delta V_a)}{dt} = \frac{g}{kR}\left[P_{ai}\frac{d(\Delta V_a)}{dt} + V_{ai}\frac{d(\Delta P_a)}{dt}\right] \quad (16.33)$$

$$T_s\,(\Delta W_b) - \frac{gP_{bi}}{c_p}\frac{d(\Delta V_b)}{dt} = \frac{g}{kR}\left[P_{bi}\frac{d(\Delta V_b)}{dt} + V_{bi}\frac{d(\Delta P_b)}{dt}\right] \quad (16.34)$$

and with $P_{ai} = P_{bi} = P_i$,

$$(\Delta W_a - \Delta W_b) = \frac{g}{RT_s}\left[2P_iA\frac{d(\Delta Y)}{dt} + \frac{V_i}{k}\frac{d}{dt}(\Delta P_a - \Delta P_b)\right] \quad (16.35)$$

The quantities V_{ai}, P_{ai}, V_{bi}, P_{bi}, V_i, and P_i may be employed as constant coefficients only as long as V_a, P_a, V_b, and P_b do not vary appreciably from their initial steady-state values. Much mathematical simplification results from use of the initial values of these variables as constant coefficients; that is, using the initial values facilitates combination of Eqs. 16.33 and 16.34 into a single equation. The error involved in using the initial value of V_a, for example, in Eq. 16.33 is proportional to the change in V_a from its initial value, with the result that if V_a changes by 5 per cent of its initial value, the coefficient V_a will be in error by approximately 5 per cent after the change of V_a has occurred, and the computation probably will be in error by no more than 2.5 per cent. The same reasoning holds for variation of P_a because the coefficient gP_a will be in error by the same amount that P_a departs from P_{ai}. Since most of the engineering measurements possibly may be in error by at least 2 per cent, changes of variables from initial values can be considered to be small if the computations based on coefficients using initial values are not in error by more than 2 or 3 per cent.

Although a given system may undergo large changes during operation, usually it may be expected to reach a steady-state condition at some time. The way in which it finally reaches steady state is usually of primary concern in the design of control systems. As a system finally reaches a steady-state condition, its variables change by very small amounts from their final values. Thus, when these final values are taken as the initial values of an analysis based on small changes of all variables, the analysis will give a reasonably accurate description of the performance of the system as it reaches steady state.

To have a completely symmetrical system, and thus to be able to combine Eqs. 16.33 and 16.34 into Eq. 16.35, the initial steady values of P_a and P_b must be equal, a condition possible only when the initial value of load force F_l is zero. The analysis will hold for small changes of F_l consistent with the allowable changes of P_a and P_b. At the same time, initially steady equal values of P_a and P_b are possible only when the control valve is centered ($X = 0$). Then each of its three-way valves is operating with identical pressure, flow, and opening conditions. If, in the pressure-flow curves, the changes in P_a, P_b, and X are sufficiently small, the constants k_{1a} and k_{1b} are equal, and the constants k_{2a} and k_{2b} are equal. Here a small change of X must be a change that is a small percentage of its maximum value rather than of its initial value. The flows W_a and W_b may be expressed in terms of small changes as

$$\Delta W_a = C_1 (\Delta X) - C_2 (\Delta P_a) \qquad (16.36)$$

$$\Delta W_b = -C_1 (\Delta X) - C_2 (\Delta P_b) \qquad (16.37)$$

where C_1 is the value of k_{1a} and k_{1b}, and C_2 is the value of $-k_{2a}$ and $-k_{2b}$. Equation 16.37 may be subtracted from 16.36 to give

$$\Delta W_a - \Delta W_b = 2C_1 (\Delta X) - C_2(\Delta P_a - \Delta P_b) \qquad (16.38)$$

Equation 16.35 gives

$$\Delta W_a - \Delta W_b = \frac{gV_i}{kRT_s} \frac{d}{dt} (\Delta P_a - \Delta P_b) + \frac{2gP_iA}{RT_s} \frac{d}{dt} (\Delta Y) \qquad (16.39)$$

Subtracting Eq. 16.39 from Eq. 16.38 yields

$$\left(\frac{gV_i}{kRT_s} D + C_2\right)(\Delta P_a - \Delta P_b) = 2C_1\,(\Delta X) - \frac{2gP_iA}{RT_s} D(\Delta Y) \quad (16.40)$$

where D denotes the derivative with respect to time. Equation 16.40 may be rearranged to give

$$AD(\Delta Y) = k_1\,(\Delta X) - (k_2 + k_3 D)(\Delta P_a - \Delta P_b) \quad (16.41)$$

where

$$k_1 = \frac{C_1 RT_s}{gP_i} \cdots \frac{\text{sq in.}}{\text{sec}} \cdots \text{(no-load valve-flow sensitivity)} \quad (16.42)$$

$$k_2 = \frac{C_2 RT_s}{2gP_i} \cdots \frac{\text{in.}^5}{\text{lb sec}} \cdots \text{(loss of valve flow per unit load pressure)} \quad (16.43)$$

$$k_3 = \frac{V_i}{2kP_i} \cdots \frac{\text{in.}^5}{\text{lb}} \cdots \text{(fluid compliance)} \quad (16.44)$$

Equation 16.41 is identical in form to the equation used in describing the behavior of a valve-controlled hydraulic servomotor (fn. 4). A pneumatic system undergoing small changes behaves much like the corresponding hydraulic system except that the effective bulk modulus of compressed gas is equal to k times its pressure, whereas the bulk modulus of hydraulic fluid is very nearly independent of its pressure.

An analysis of the forces acting on the mechanical part of the system, consisting of the ram and the mass plus viscous damping load, yields

$$A(\Delta P_a - \Delta P_b) = mD^2\,(\Delta Y) + bD(\Delta Y) + \Delta F_l \quad (16.45)$$

Combining Eqs. 16.41 and 16.45 produces the second-order differential equation

$$\left[\left(\frac{k_3 m}{k_2 b + A^2}\right)D^2 + \left(\frac{k_2 m + k_3 b}{k_2 b + A^2}\right)D + 1\right]D(\Delta Y)$$
$$= \frac{k_1 A\,(\Delta X) - (k_2 + k_3 D)\,(\Delta F_l)}{k_2 b + A^2} \quad (16.46)$$

Equation 16.46 is identical with the differential equation that may be derived for a similar hydraulic system, and is discussed in some detail in an earlier paper (fn. 4). The degree of system damping is determined by the value of the viscous-damping coefficient b and/or the coefficient k_2.

The steady-state load sensitivity of a system of this kind is an important factor in many applications. The steady-state equation relating ram velocity, $D(\Delta Y)$, to valve stroke, ΔX, and external load, ΔF_l, is

$$[D(\Delta Y)]_{ss} = \frac{k_1 A (\Delta X)_{ss} - k_2 (\Delta F_l)_{ss}}{k_2 b + A^2} \qquad (16.47)$$

Only when k_2 is zero, is zero steady-state load sensitivity possible. Otherwise, in holding a steady velocity, a readjustment in $(\Delta X)_{ss}$ must be made to counteract load changes.

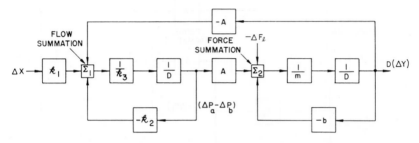

Fig. 16.16. Block-diagram representation of the differential equations derived from a linearized analysis of valve, ram, and mass-load system.

Since large steady loads may be imposed on such a system, its sensitivity to steady loads should be minimized. Thus, quiescent power-drain limitations and load-sensitivity requirements often necessitate the use of a closed-center valve. Allowing leakage to by-pass the ram, in order to gain damping, also leads to steady-state load sensitivity and to dissipation of power during periods of steady loading because the same pressure difference across the ram, which is required to hold the load, causes a steady flow of gas through the by-pass resistance.

The presence of viscous damping does not make the system sensitive to steady external load but does tend to reduce the velocity attainable per unit valve stroke, as demonstrated in Eq. 16.47. Viscous damping on the ram also may dissipate a great deal of power when the ram is moving at high velocity.

The block diagram of Fig. 16.16 may be constructed by performing the summations Σ_1 and Σ_2 required to satisfy Eqs. 16.45 and 16.46. Since the differential operator, D, denotes differentiation with respect to time, $1/D$ is used to denote integration with respect to time.

As shown previously (fn. 4), the degree of damping of a system with given values of ram area, A, fluid compliance, k_3, and mass load, m, is determined by the values of k_2 and b present in the system. Figure 16.17

is an analogue block diagram of the servomotor system. The coefficients k_2 and b serve as the feedback paths of two small minor loops of the system. Thus, system stability seems to be related directly to the degree with which these respective minor loops are controlled by negative feedback. This suggests that other, perhaps artificial, means may be employed to close these loops to gain system stability. For instance, it might be possible to measure $(\Delta P_a - \Delta P_b)$ and to provide a feedback control system that would tend to decrease valve displacement by an amount proportional to $(\Delta P_a - \Delta P_b)$. In addition to practical diffi-

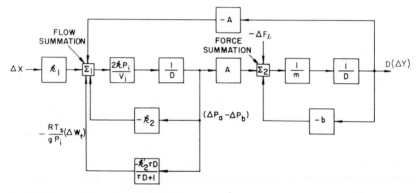

Fig. 16.17. Analogue block diagram based on linearized analysis and showing effect of transient-pressure feedback.

culties involved in providing this pressure feedback, the load sensitivity of the system would increase with increasing pressure feedback just as it does with increasing k_2. Similarly, an artificial feedback scheme might be employed to simulate the effects of b but would result in excessive dissipation of power at high ram velocities.

Another possibility, investigated by Dinerstein [12] is to measure ram velocity and employ feedback-compensation techniques around the whole system in order to stabilize it. Practical limitations due to the complexity and expense of this scheme tend to nullify some of the theoretical advantages. Nevertheless, sufficient stability can be gained by this means without increasing the steady-state load sensitivity or power dissipation at high ram velocity.

During an electronic-analogue study of the valve, ram, and load system, the decision was made to investigate the possibilities of using a transient-pressure-feedback effect, such as the block diagram of Fig.

[12] J. Dinerstein, "The Development of a Valve-Controlled Pneumatic Servomechanism," S.M. Thesis, Department of Electrical Engineering, Massachusetts Institute of Technology, Cambridge, Mass., 1954.

16.17, to supplement the effects of k_2. A transient flow, W_t, must be provided to and from the ram chambers a and b which satisfies a differential equation of the following form

$$\frac{RT_s}{gP_i} \Delta W_t = -k_2' \left(\frac{\tau D}{\tau D + 1}\right)(\Delta P_a - \Delta P_b) \qquad (16.48)$$

The transfer characteristic $-k_2'\tau D/(\tau D + 1)$ represents the dynamic behavior of many physical systems, and it is easily instrumented with the arrangement of analogue-computer components shown in Fig. 16.18.

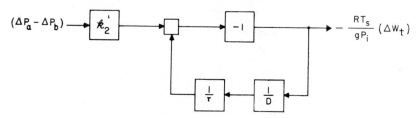

Fig. 16.18. Arrangement of analogue-computer components to simulate:

$$\frac{RT_s}{gP_i} W_t = -k_2'[\tau D/(\tau D + 1)](\Delta P_a - \Delta P_b)$$

This feedback effect occurs only when $(\Delta P_a - \Delta P_b)$ is time variant. When $(\Delta P_a - \Delta P_b)$ reaches a steady value, $W_t = 0$. Therefore, it may be called a transient-pressure-feedback effect.

A preliminary study made on a Philbrick electronic-analogue computer indicated that this scheme would ensure system stability if sufficiently large values of the time constant τ could be provided. Obviously, when $\tau = \infty$, simple pressure feedback exists. The analogue study indicated, however, that extremely large values of τ would not be required and that the value of k_2' necessary to attain a given degree of stability would be nearly the same as the value of k_2 that would be required otherwise.

At this point, some thought was given to the means of accomplishing the transient-pressure-feedback scheme in a simple, effective manner. Several arrangements requiring elaborate control and instrumentation techniques were discarded when the desired effect seemed obtainable by employing auxiliary tanks and flow resistances, as shown in Fig. 16.19. Such a scheme looks promising because flows to and from tanks a and b do occur through the flow resistances during periods when P_a and P_b are changing, yet the tank pressures P_{ta} and P_{tb} must approach P_a and P_b when P_a and P_b reach steady values, and the flows W_{ta} and W_{tb} approach zero as $(P_a - P_{ta})$ and $(P_b - P_{tb})$ approach zero. In order to ascertain the values of resistances and tank sizes required, the system must be analyzed with this added complexity. As in earlier portions of this section, greatest emphasis will be given to the linearized analysis.

Capillary flow resistances have been chosen because their characteristics are more nearly linear than those of orifices. Momentum effects of fluid flow in the capillary passages and heat transfer to the fluid passing through the resistances are assumed to be negligible. The

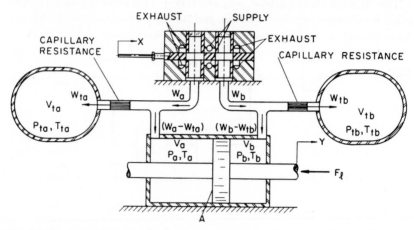

Fig. 16.19. Schematic diagram showing arrangement of tanks and flow resistances to provide system damping.

equation for the flow of a compressible fluid in capillary passages * may be applied to each of the two flow resistances.

$$W_{ta} = \frac{g w_c t_c^3}{24 \mu R T_s L_c} (P_a^2 - P_{ta}^2) \tag{16.49}$$

and

$$W_{tb} = \frac{g w_c t_c^3}{24 \mu R T_s L_c} (P_b^2 - P_{tb}^2) \tag{16.50}$$

where w_c = capillary-passage width, in.
$\quad t_c$ = capillary-passage thickness, in.
$\quad L_c$ = capillary-passage length, in.

For simplicity, a capillary flow coefficient is defined as

$$C_c = \frac{w_c t_c^3}{12 \mu L_c}$$

---

* See Chapter 3, Fundamentals of Fluid Flow.

hence

$$W_{ta} = \frac{gC_c}{2RT_s} (P_a{}^2 - P_{ta}{}^2) \qquad (16.51)$$

and

$$W_{tb} = \frac{gC_c}{2RT_s} (P_b{}^2 - P_{tb}{}^2) \qquad (16.52)$$

And if only small changes occur from the initially steady conditions existing when $P_a = P_b = P_{ta} = P_{tb} = P_i$,

$$\Delta W_{ta} = \frac{gC_cP_i}{RT_s} (\Delta P_a - \Delta P_{ta}) \qquad (16.53)$$

and

$$\Delta W_{tb} = \frac{gC_cP_i}{RT_s} (\Delta P_b - \Delta P_{tb}) \qquad (16.54)$$

The energy equations for identical tanks may be written

$$W_{ta} = \frac{gV_t}{kRT_s} \frac{dP_{ta}}{dt} \qquad (16.55)$$

and

$$W_{tb} = \frac{gV_t}{kRT_s} \frac{dP_{tb}}{dt} \qquad (16.56)$$

or in terms of small changes

$$\Delta W_{ta} = \frac{gV_t}{kRT_s} \frac{d}{dt} (\Delta P_{ta}) \qquad (16.57)$$

$$\Delta W_{tb} = \frac{gV_t}{kRT_s} \frac{d}{dt} (\Delta P_{tb}) \qquad (16.58)$$

Combining Eq. 16.53 with Eq. 16.57 and Eq. 16.54 with Eq. 16.58 gives

$$\Delta W_{ta} = \frac{gV_t}{kRT_s} \frac{d}{dt} \left[\Delta P_a - \frac{RT_s}{gC_cP_i} (\Delta W_{ta}) \right] \qquad (16.59)$$

and

$$\Delta W_{tb} = \frac{gV_t}{kRT_s} \frac{d}{dt} \left[\Delta P_b - \frac{RT_s}{gC_cP_i} (\Delta W_{tb}) \right] \qquad (16.60)$$

Collecting terms yields

$$\frac{V_t}{kC_cP_i}\frac{d(\Delta W_{ta})}{dt} + \Delta W_{ta} = \frac{gV_t}{kRT_s}\frac{d}{dt}(\Delta P_a) \qquad (16.61)$$

and

$$\frac{V_t}{kC_cP_i}\frac{d(\Delta W_{tb})}{dt} + \Delta W_{tb} = \frac{gV_t}{kRT_s}\frac{d}{dt}(\Delta P_b) \qquad (16.62)$$

Thus Eqs. 16.61 and 16.62 are of the desired form given by Eq. 16.48.

The net flows to chambers a and b are $(\Delta W_a - \Delta W_{ta})$ and $(\Delta W_b - \Delta W_{tb})$, respectively. The energy equations may be applied again to chambers a and b.

$$\Delta W_a - \Delta W_{ta} - \frac{gP_iA}{RT_s}\frac{d}{dt}(\Delta Y) = \frac{gV_i}{kRT_s}\frac{d}{dt}(\Delta P_a) \qquad (16.63)$$

and

$$\Delta W_b - \Delta W_{tb} + \frac{gP_iA}{RT_s}\frac{d}{dt}(\Delta Y) = \frac{gV_i}{kRT_s}\frac{d}{dt}(\Delta P_b) \qquad (16.64)$$

Solution of Eqs. 16.61 and 16.62 for ΔW_{ta} and ΔW_{tb} gives

$$\Delta W_{ta} = \frac{(gV_t/kRT_s)D(\Delta P_a)}{(V_t/kC_cP_i)D + 1} \qquad (16.65)$$

and

$$\Delta W_{tb} = \frac{(gV_t/kRT_s)D(\Delta P_b)}{(V_t/kC_cP_i)D + 1} \qquad (16.66)$$

Combination of Eqs. 16.63 and 16.65 and Eqs. 16.64 and 16.66 gives

$$\Delta W_a - \frac{(gV_t/kRT_s)D(\Delta P_a)}{(V_t/kC_cP_i)D + 1} - \frac{gP_iA}{RT_s}D(\Delta Y) = \frac{gV_i}{kRT_s}D(\Delta P_a) \quad (16.67)$$

$$\Delta W_b - \frac{(gV_t/kRT_s)D(\Delta P_b)}{(V_t/kC_cP_i)D + 1} + \frac{gP_iA}{RT_s}D(\Delta Y) = \frac{gV_i}{kRT_s}D(\Delta P_b) \quad (16.68)$$

Subtracting Eq. 16.68 from Eq. 16.67 gives

$$(\Delta W_a - \Delta W_b) - \frac{(gV_t/kRT_s)D(\Delta P_a - \Delta P_b)}{(V_t/kC_cP_i)D + 1}$$

$$- \frac{2gP_iA}{RT_s}D(\Delta Y) = \frac{gV_i}{kRT_s}D(\Delta P_a - \Delta P_b) \quad (16.69)$$

Since from Eq. 16.38

$$(\Delta W_a - \Delta W_b) = 2C_1 (\Delta X) - C_2(\Delta P_a - \Delta P_b)$$

substitution for $(\Delta W_a - \Delta W_b)$ and division through by $2gP_i/RT_s$ gives

$$k_1 (\Delta X) - k_2(\Delta P_a - \Delta P_b) - k_2' \frac{\tau D(\Delta P_a - \Delta P_b)}{\tau D + 1}$$

$$- AD(\Delta Y) = k_3D(\Delta P_a - \Delta P_b) \quad (16.70)$$

where $k_2' = \dfrac{C_c}{2}$, in.^{5}/lb sec $\qquad\qquad\qquad (16.71)$

$$\tau = \frac{V_t}{kC_cP_i}, \text{ sec} \qquad\qquad\qquad (16.72)$$

Therefore, the desired transient-pressure-feedback effect is provided by the flow resistances and tanks.

Since the stabilization really is caused by a flow that is diverted from, or added to, the valve flow during transient operation, the term transient-flow stabilization will be employed to describe this means of providing system damping. Addition of the transient-flow stabilization to the system requires a third-order differential equation to describe the system. This equation is obtained by combining Eqs. 16.70 and 16.45.

$$k_1(\Delta X) - \left(k_2 + k_2' \frac{\tau D}{\tau D + 1} + k_3D \right)$$

$$\left[\left(\frac{m}{A} D + \frac{b}{A} \right) D(\Delta Y) + \frac{\Delta F_l}{A} \right] = AD(\Delta Y) \quad (16.73)$$

Since a closed-center valve is employed in this investigation and no by-pass leakage is expected past the ram, k_2 is negligible. At the same time, minimizing the viscous-damping coefficient, b, to avoid loss of power at high ram velocities would be advantageous. Setting k_2 and b equal to zero gives a somewhat simpler equation relating $D(\Delta Y)$ to ΔX and ΔF_l.

$$\left\{ [k_2'\tau D + k_3D(\tau D + 1)] \frac{mD}{A} + A(\tau D + 1) \right\} D(\Delta Y)$$

$$= k_1(\tau D + 1) (\Delta X) - [k_2'\tau D + k_3D(\tau D + 1)] \frac{(\Delta F_l)}{A} \quad (16.74)$$

Equation 16.74 may be written more simply if the following parameters are used

$$\pi_v = \frac{V_i}{V_t} \text{ (dimensionless)} \tag{16.75}$$

$$\tau_i = \pi_v \tau = \frac{V_i}{kC_cP_i} \text{ (sec)} \tag{16.76}$$

$$\pi_s = \frac{A^2V_i^2\tau^2}{k_3mV_t^2} = \frac{2A^2V_i}{kmC_c^2P_i} \text{ (dimensionless)} \tag{16.77}$$

The resulting relationship is

$$[(\tau_iD)^3 + (1 + \pi_v)(\tau_iD)^2 + \pi_s\tau_iD + \pi_s\pi_v]D(\Delta Y)$$

$$= \frac{k_1}{A} \pi_s\pi_v \left(\frac{\tau_i}{\pi_v}D + 1\right)(\Delta X) - \frac{\tau_i^2}{m(\pi_v + 1)}\left(\frac{\tau_iD}{\pi_v + 1} + 1\right)D(\Delta F_l)$$

$$\tag{16.78}$$

For convenience, Eq. 16.42 is repeated in computing k_1.

$$k_1 = \frac{C_1RT_s}{gP_i}, \text{ sq in./sec} \tag{16.79}$$

Equation 16.78 is a useful equation for this system because it is in dimensionless form and can be applied readily to design work. Before application, however, the relationship of the roots of the characteristic equation to the dimensionless parameters of the system must be found. The characteristic equation is

$$[(\tau_iD)^3 + (1 + \pi_v)(\tau_iD)^2 + \pi_s\tau_iD + \pi_s\pi_v]D(\Delta Y) = 0 \tag{16.80}$$

It will have three roots, one of which is real and the other two of which may be real or complex. When the system is underdamped, the other two roots will be complex, and the following identity may be used

$$(\tau_iD)^3 + (1 + \pi_v)(\tau_iD)^2 + \pi_s\tau_iD + \pi_s\pi_v$$

$$\equiv \pi_s\pi_v(\tau_1D + 1)\left[\left(\frac{1}{\omega_{ns}}\right)^2 D^2 + \frac{2\zeta_s}{\omega_{ns}}D + 1\right] \tag{16.81}$$

The use of dimensional analysis is helpful in determining how the quantities τ_1, ω_{ns}, and ζ_s are related to the quantities τ_i, π_v, and π_s. In general

$$\tau_1 = f_a(\tau_i, \pi_v, \pi_s) \tag{16.82}$$

or

$$\frac{\tau_1}{\tau_i} = f_\tau(\pi_v, \pi_s) \tag{16.83}$$

and

$$\omega_{ns} = f_b(\tau_i, \pi_v, \pi_s) \tag{16.84}$$

or

$$\omega_{ns}\tau_i = f_\omega(\pi_v, \pi_s) \tag{16.85}$$

and

$$\zeta_s = f_\zeta(\pi_v, \pi_s) \tag{16.86}$$

The functional relationships shown symbolically by Eqs. 16.83, 16.84, and 16.85 may be obtained mathematically by systematically solving for the roots of the characteristic equation in terms of π_v, π_s, and τ_i. The results of such a procedure are given in graphical form in Figs. 16.20, 16.21, and 16.22, respectively. These graphs, together with Eq. 16.78, are the key to gaining an understanding of the dynamic characteristics of the valve, ram, and load system with transient-flow stabilization.

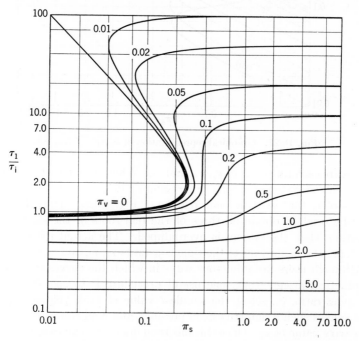

Fig. 16.20. Graphical representation of $\tau_1/\tau_i = f_\tau(\pi_v, \pi_s)$.

When the system is overdamped, the characteristic equation has three real roots, and the following identity may be used:

$$(\tau_i D)^3 + (1 + \pi_v)(\tau_i D)^2 + \pi_s \tau_i D + \pi_s \pi_v$$

$$\equiv \pi_s \pi_v (\tau_1 D + 1)(\tau_2 D + 1)(\tau_3 D + 1) \quad (16.87)$$

It can be shown by dimensional analysis that the time constants τ_1, τ_2,

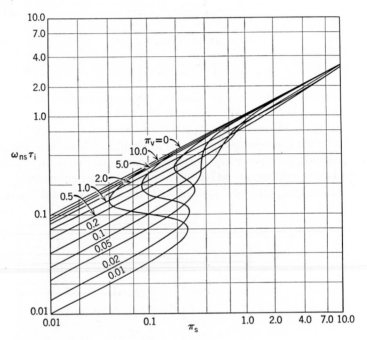

Fig. 16.21. Graphical representation of $\omega_{ns}\tau_i = f_\omega(\pi_v, \pi_s)$.

and τ_3 are each functions of π_s, π_v, and τ_1. When the characteristic equation is solved systematically for τ_1, τ_2, and τ_3, they coincide with the three values of τ_1 that exist in some regions of Fig. 16.20. Thus, Fig. 16.20 may be used to obtain all three of the real roots of the characteristic equation when the system is overdamped. The actual roots, incidentally, are the reciprocals of the time constants τ_1, τ_2, and τ_3.

Although there are numerous ways in which the curves may be used in practical design work, one useful procedure is to specify a desired damping ratio, ζ_s, select the smallest value of tank volume ratio, π_v, that makes such a damping ratio possible, and read from Fig. 16.22 the necessary value of π_s. Once this has been done, the values of the various physical parameters making up π_s often can be juggled to evolve a satisfactory design. In many cases, it is desirable to obtain the fastest response consistent with space and power limitations. The values of all of the system time constants are closely related to the values of τ_i and π_s. The dynamic terms on the right-hand side of Eq. 16.78 must be included when assessing the speed of response of the system. As in most

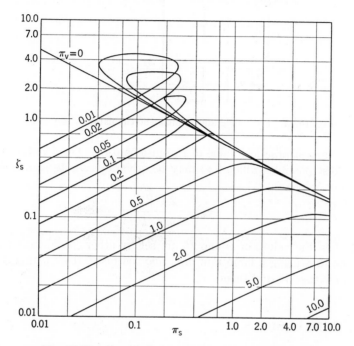

Fig. 16.22. Graphical representation of $\zeta_s = f_\zeta(\pi_v, \pi_s)$.

design problems, trial-and-error techniques may be combined with the analytical results presented here to work out a system design.

Since this system is potentially useful for the continuous control of position, velocity, or acceleration, there are many possible considerations which may enter into its practical application.

The electronic analogue computer has proved to be a valuable tool in the design process. The rather simple block diagram for the simulation of Eq. 16.78 is shown in Fig. 16.23. This diagram serves to emphasize the pervading influence of τ_i on system speed of response. Since

$$\tau_i = \frac{V_i}{(kC_cP_i)}$$

the volume of fluid under compression in the ram, V_i, must be kept small,

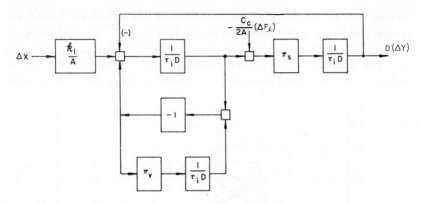

Fig. 16.23. Block diagram for analogue simulation of Eq. 16.78.

and the largest practical value of P_i must be used. On the other hand, the largest possible value of π_s may be advantageous to gain speed of response. Since

$$\pi_s = \frac{2A^2 V_i}{kmC_c^2 P_i}$$

then when the values of m and P_i are set by other considerations, it is possible that using a large ram area may be justified even though this would increase V_i and thence τ_i.

16.6. CONCLUSIONS

A fundamental study of the pneumatic process in a system used for the continuous control of motion with compressed air shows how linearized equations may be used to analyze the system. The use of analogue block diagrams and an electronic analogue computer made possible a simple means of attaining system damping with very little steady-state load sensitivity and viscous-friction power loss.

NOMENCLATURE

Symbol	Definition	Unit
A	ram area	in.^{2}
A_e	area of exhaust-orifice opening	in.^{2}
A_o	orifice area	in.^{2}
A_s	area of supply-orifice opening	in.^{2}

Symbol	Definition	Unit
b	viscous-damping coefficient	lb sec/in.
C	compressible-flow coefficient	°R/sec
C_1	partial derivative of weight rate of flow with respect to valve opening	lb/sec in.
C_2	partial derivative of weight rate of flow with respect to load pressure	in.²/sec
C_c	capillary-resistance coefficient	in.⁵/lb sec
C_d	orifice-discharge coefficient	
d	diameter	in.
D	denotes derivative with respect to time, d/dt	1/sec
e	subscript denoting exhaust	
f	subscript to designate final condition	
f	to indicate a functional relationship	
f_1	function of pressure ratio	
f_a	functional relationship between τ_1, τ_i, π_s, and π_v	
f_b	functional relationship between ω_{ns}, τ_i, π_s, and π_v	
f_v	functional relationship between valve flow, W, valve opening, X, and load pressure, P	
f_ζ	functional relationship for ζ_s	
f_τ	functional relationship for τ_1/τ_i	
f_ω	functional relationship for $\omega_{ns}\tau_i$	
F_l	total load force	lb
g	acceleration due to gravity	386 in./sec²
i	subscript to designate initial condition	
k	ratio of specific heats, c_p/c_v, for air: 1.4	
k_1	coefficient denoting no-load flow sensitivity	in.²/sec
k_2	coefficient denoting loss of flow per unit load pressure	in.⁵/lb sec
k_2'	transient-pressure-feedback coefficient	in.⁵/lb sec
k_3	coefficient denoting fluid compliance	in.⁵/lb
k_{1a}	same as C_1	
k_{1b}	same as C_1	
k_{2a}	same as C_2	
k_{2b}	same as C_2	
L	external load force	lb
L_c	capillary-passage length	in.
m	load mass	lb sec²/in.
P_a	pressure in chamber a	psi
P_b	pressure in chamber b	psi
P_{aq}	quiescent value of P_a ($W_a = 0$ and $x = 0$)	psi
P_{bq}	quiescent value of P_b ($W_b = 0$ and $x = 0$)	psi
P_d	downstream pressure	psi
P_e	exhaust pressure	psi
P_i	initial pressure	psi

Symbol	*Definition*	*Unit*
P_q	quiescent value of P_a and P_b when they are equal	psi
P_s	pressure of gas supply	psi
P_{ta}	pressure in tank at a end of ram	psi
P_{tb}	pressure in tank at b end of ram	psi
P_u	upstream pressure	psi
Q_h	heat transferred to system	in. lb
		(1 Btu = 9336 in. lb)
R	gas constant (for air: 2.47×10^5)	in.^{2}/sec^2 °R
s	subscript denoting supply	
ss	subscript denoting steady state	
t	time	sec
t_c	capillary-passage thickness	in.
T_l	line temperature	°R
T_{la}	stagnation temperature of gas in the line entering chamber a	°R
T_{lb}	stagnation temperature of gas in the line entering chamber b	°R
T_a	temperature in chamber a	°R
T_b	temperature in chamber b	°R
T_e	stagnation temperature upstream of exhaust orifice	°R
T_i	initial temperature	°R
T_s	temperature of supply gas	°R
T_u	upstream stagnation temperature	°R
U	valve underlap	in.
V_a	volume of chamber a	in.^{3}
V_b	volume of chamber b	in.^{3}
V_i	initial volume at each end of ram (including lines to valve)	in.^{3}
V_t	tank volume (when $V_{ta} = V_{tb}$)	in.^{3}
V_{ta}	tank volume at a end of ram	in.^{3}
V_{tb}	tank volume at b end of ram	in.^{3}
w_c	capillary-passage width	in.
w_e	width of exhaust port	in.
w_s	width of supply port	in.
W	weight rate of flow	lb/sec
W_a	weight rate of flow from a end of valve	lb/sec
W_b	weight rate of flow from b end of valve	lb/sec
W_e	weight rate of flow through A_e	lb/sec
W_q	quiescent-leakage flow	lb/sec
W_s	weight rate of flow through A_s	lb/sec
W_t	transient flow	lb/sec
W_{ta}	weight rate of flow to tank at a end of ram	lb/sec
W_{tb}	weight rate of flow to tank at b end of ram	lb/sec

Symbol	Definition	Unit
X	valve displacement measured from center position	in.
Y	ram displacement measured from center position	in.
Δ	designation of small change	
ζ_s	servomotor damping ratio	
μ	absolute viscosity	lb sec/in.^{2}
	(for air at 530°R, $\mu = 2.65 \times 10^{-9}$ lb sec/in.^{2})	
π_s	dimensionless parameter $2A^2V_i/(kmC_c^2P_i)$	
π_v	tank volume ratio, V_i/V_t	
ρ_a	density of fluid in chamber a	lb sec^2/in.^{4}
ρ_b	density of fluid in chamber b	lb sec^2/in.^{4}
Σ_1	flow summation	
Σ_2	force summation	
τ	transient-pressure-feedback time constant	sec
τ_1	servomotor time constant	sec
τ_2	servomotor time constant	sec
τ_3	servomotor time constant	sec
τ_i	integrating time constant, $V_i/(kC_cP_i)$	sec
ω_{ns}	servomotor undamped natural frequency	radians/sec

17

Gerhard Reethof
J. L. Shearer

≡ Closed-Loop Systems

17.1. INTRODUCTION

The evolution of automatic control systems has been in great measure responsible for the developments that have taken place in the field of fluid-power control. The need to know how to specify component performance is tightly linked with the analysis of closed-loop systems. It seems appropriate to include a chapter on closed-loop systems which illustrates, by example, some of the typical problems that have been encountered in this field. This chapter is not intended as a complete coverage of closed-loop-system design—it is rather intended as a glimpse at some relatively simple practical problems and possible ways of solving them. The material in this chapter has been published in various places and it has been brought together here with the permission of the copyright owners.

17.2. PROPORTIONAL CONTROL OF RATE-TYPE SERVOMOTORS [1]

When rate-type servomotors are put under proportional control, the resulting closed-loop systems may become unstable when the amount of

[1] J. L. Shearer, "Proportional Control of Rate-Type Servomotors," *Trans. ASME*, Vol. 76 (August 1954), pp. 889–903.

proportional control is increased excessively. A rate-type servomotor is defined here as a servomotor with a rate of change of output proportional to the input during steady-state operation. It is usually advantageous to introduce as much proportional control as possible in order to obtain the fastest speed of response. More elaborate means of control may be employed to improve speed of response without sacrificing system stability, but it is important to know first what performance may be attained with simple proportional control. There are many instances in which it is necessary to understand thoroughly the performance of rate-type servomotors having second-order lag effects when under proportional control. A high-speed electronic analogue computer proved to be a satisfactory means of gaining this understanding for various types of second-order lag effects.

17.21. Characteristics of Rate-Type Servomotors

The power required to manipulate the input to a process under automatic control, such as that shown schematically in Fig. 17.1, often makes necessary the use of a means of power amplification which is

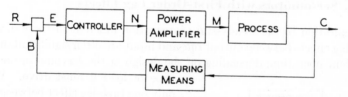

Fig. 17.1. Typical automatic control system.

capable of manipulating a variable M so that it is proportional to the controller output N.

Instantaneous response of M to changes in N would be highly desirable, but such a response, if truly instantaneous, would require the expenditure of large amounts of power for rapidly changing values of N. The need for fast response in each instance must be reconciled with the availability of power for the amplification to be attained.

The power level required in many mechanical systems is attainable only by using a rate-type servomotor with a *rate of change* of output proportional, in the steady state, to its input. In other words, the servomotor steady-state output is proportional to the integral of its input. The valve-controlled fluid motor and the resistance-controlled electric motor, shown schematically in Fig. 17.2, are examples of rate-type servomotors when motor shaft *position* is the output. Unless such a steady-state integration is required specifically in the automatic control system, some means must be employed to convert this steady-state integrating action into the desired steady-state proportional action.

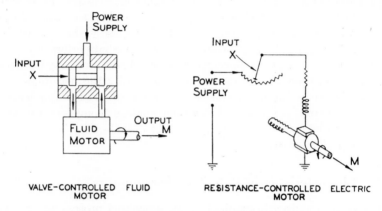

Fig. 17.2.   Examples of rate-type servomotors.

17.22. Servomotors with First-Order Lag Effects

In addition to the steady-state integration inherent in these devices, time-lag effects between output *rate* and input are often significant during transient operation, depending on the design of the servomotor and the

Fig. 17.3.   Block-diagram representation of rate-type servomotor having single time lag between output rate and input.

type of load it must drive. When only one time-lag effect between output rate and input is significant, the servomotor may be represented symbolically by a functional block diagram, as shown in Fig. 17.3. A servomotor of this type may be employed in a system such as that shown symbolically in Fig. 17.4 in order to obtain an output M which is proportional, in the steady state, to an input N. This is basically a closed-loop control system in which the servomotor output is measured and fed back in such a way that it is compared with the input N. If the measuring means is sufficiently fast (negligible time lags), the relationship between the output M and the input N may be expressed by the following second-order differential equation

$$(\tau D^2 + D + k_s k_f)M = k_s N \qquad (17.1)$$

where τ = time lag of uncontrolled servomotor, sec

D = the differential operator, $\dfrac{d}{dt}$

k_s = servomotor gain

k_f = measuring-means gain

M = manipulated variable

N = controller output

and in the steady state

$$M_{ss} = \left(\frac{1}{k_f}\right) N_{ss} \tag{17.2}$$

The dynamic characteristics of such a system are ascertained readily from the coefficients of Eq. 17.1, and its transient and frequency-response

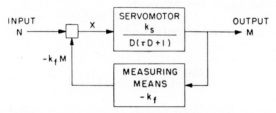

Fig. 17.4. Proportional control of rate-type servomotor having single time-lag effect.

characteristics are discussed thoroughly elsewhere.[2-5] It is worth while noting here that the undamped natural frequency, ω_n, and the damping ratio, ζ, of this system are determined as

$$\omega_n = \sqrt{K_l/\tau} \tag{17.3}$$

$$\zeta = \frac{1}{2\sqrt{K_l\tau}} \tag{17.4}$$

where $K_l = k_s k_f$.

[2] G. S. Brown and D. P. Campbell, *Principles of Servomechanisms*, John Wiley and Sons, New York, 1948, pp. 48–56.

[3] W. R. Ahrendt and J. F. Taplin, *Automatic Feedback Control*, McGraw-Hill Book Co., New York, 1951, pp. 53–63.

[4] H. Chestnut and R. W. Mayer, *Servomechanism and Regulating System Design*, Vol. 1, John Wiley and Sons, New York, 1951, pp. 59–63.

[5] C. S. Draper, S. Lees, and W. McKay, "Methods for Associating Mathematical Solutions with Common Forms," *Instrument Engineering*, Vol. 2, McGraw-Hill Book Co., New York, 1953, Chapter 19, pp. 184–357.

Thus, increasing the loop gain, K_l, increases the speed of response and decreases the degree of stability of this system. Fortunately, many servomotors have only one significant time-lag effect when driving certain types of loads, and their performance under various operating conditions can be surveyed readily with Eqs. 17.3 and 17.4. In instances when τ is negligible, the system is stable for all values of K_l.

17.23. Servomotors with Second-Order Lag Effects

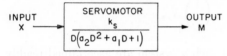

Fig. 17.5. Block-diagram representation of rate-type servomotor having second-order lag effect between output rate and input.

Experience has shown that a second-order lag effect is significant in some servomotors. When this condition exists, the servomotor may be represented symbolically as shown in Fig. 17.5.

The relationship between output rate, DM, and X is expressed mathematically as

$$DM = \frac{k_s X}{a_2 D^2 + a_1 D + 1} \tag{17.5}$$

where X = servomotor input and a_1 and a_2 are servomotor constants.

If this system is overdamped, the denominator of Eq. 17.5 may be expressed as

$$a_2 D^2 + a_1 D + 1 = (\tau_1 D + 1)(\tau_2 D + 1) \tag{17.6}$$

where

$$\tau_1, \tau_2 = \frac{a_1}{2}\left(1 \pm \sqrt{1 - \frac{4a_2}{a_1{}^2}}\right) \tag{17.7}$$

The time constants, τ_1 and τ_2, which are positive real numbers, denote two distinct time lags. If this system is underdamped, the denominator of Eq. 17.5 may be written as

$$a_2 D^2 + a_1 D + 1 = \left(\frac{1}{\omega_{ns}}\right)^2 D^2 + \left(\frac{2\zeta_s}{\omega_{ns}}\right) D + 1 \tag{17.8}$$

$$\omega_{ns} = \sqrt{1/a_2} \tag{17.9}$$

$$\zeta_s = \frac{a_1}{2\sqrt{a_2}} \tag{17.10}$$

and the subscript s refers to the uncontrolled servomotor.

When a servomotor with a second-order lag effect is put under proportional control as shown in Fig. 17.6, the relationship between output M and input N may be expressed by the third-order differential equation

$$(a_2D^3 + a_1D^2 + D + k_sk_f)M = k_sN \tag{17.11}$$

and in the steady state

$$M_{ss} = \frac{1}{k_f}N_{ss} \tag{17.12}$$

The dynamic characteristics of this system cannot be revealed by a simple analytical treatment similar to that employed for a system with a

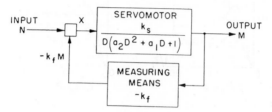

Fig. 17.6. Proportional control of rate-type servomotor having second-order lag effect.

second-order differential equation. It is possible to find the roots of the characteristic equation

$$a_2D^3 + a_1D^2 + D + k_sk_f = 0 \tag{17.13}$$

But this process usually proves cumbersome if a comprehensive survey of all system parameters is to be made in order to achieve an optimum system design.

17.24. Electronic Analogue Study

An electronic analogue computer was found to be very useful in gaining a thorough understanding of the dynamic performance of this type of third-order system under all possible conditions of operation. A dimensional analysis may be employed as in Sec. 17.25 to show that the loop gain, $K_l = k_sk_f$, of the system in Fig. 17.6 is related to the second-order lag effect, as shown in the following equations:

$$K_l\tau_1 = f_1\left(\frac{\tau_1}{\tau_2}\right) \qquad \text{(overdamped servomotor)} \tag{17.14}$$

$$\frac{K_l}{\omega_{ns}} = f_2(\zeta_s) \qquad \text{(underdamped servomotor)} \tag{17.15}$$

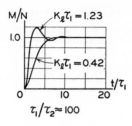

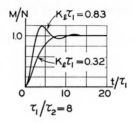

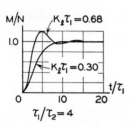

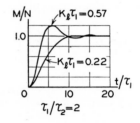

(a)

(a) Servomotor overdamped.

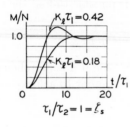

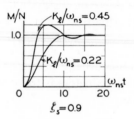

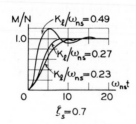

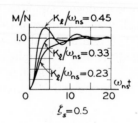

(b)

(b) Servomotor underdamped.

Fig. 17.7.   Transient-response curves obtained from electronic analogue study.

if a given type of response of M is to be obtained when a step change in N is made. Here $\omega_{ns} =$ undamped natural frequency, and $\zeta_s =$ damping ratio of uncontrolled servomotor.

The types of response of most interest are (1) the fastest response with no overshoot and (2) the response with 20 per cent overshoot. Figures 17.7a and 17.7b show the actual transient-response curves that were obtained during an electronic analogue study. When this study was made,

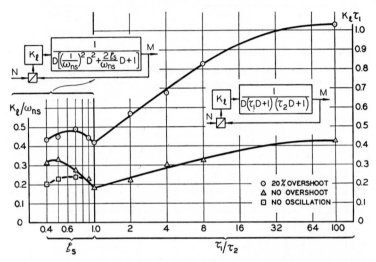

Fig. 17.8. Dimensionless loop gain versus ζ_s and τ_1/τ_2 for typical transient responses.

unity feedback was employed and the effects of k_f and k_s were combined into K_l as shown in the upper parts of Fig. 17.8 in order to simplify presentation of the results. It is interesting to note that the rate of decay of the 20 per cent overshoot response becomes small and that the fastest response with no overshoot is still oscillatory when the uncontrolled servomotor is well underdamped. This shows clearly that care must be taken in specifying the transient-response performance of such systems. In cases where this effect occurred, the fastest nonoscillatory reponse also was found. Figure 17.8 shows graphically the functional relationships of Eqs. 17.14 and 17.15 plotted from the data given in Fig. 17.7.

Furthermore, it can be shown by dimensional analysis that the response times T_1, T_2, and T_3, as defined in Fig. 17.9, are related to the second-order lag characteristics as follows:

$$\frac{T}{\tau_1} = f_3\left(\frac{\tau_1}{\tau_2}\right) \tag{17.16}$$

$$\omega_{ns}T = f_4(\zeta_s) \tag{17.17}$$

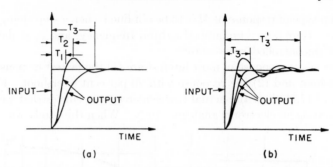

Fig. 17.9. Typical transient responses showing characteristic response times.

(a) Proportional control of *overdamped* rate-type servomotor.
(b) Proportional control for *underdamped* rate-type servomotor.

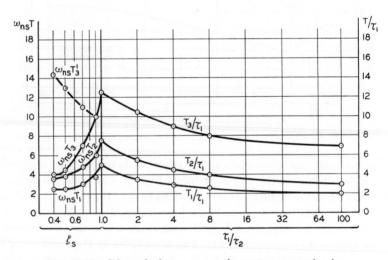

Fig. 17.10. Dimensionless response times versus ζ_s and τ_1/τ_2.

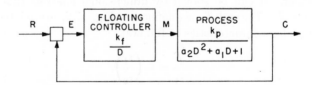

Fig. 17.11. Floating control of second-order process.

These functional relationships of Eqs. 17.16 and 17.17 are shown in Fig. 17.10.

This section has been limited to the treatment of proportional control of rate-type servomotors because this type of control is the simplest way to eliminate the steady-state integration inherent in a rate-type servomotor, and it was believed that it would be useful to survey comprehensively the capabilities of this type of control before investigating more elaborate types.

The application of this work to a typical problem is demonstrated in Sec. 17.26.

The results of this investigation also are applicable to the reset control (with no proportioning action) of second-order processes, as shown in Fig. 17.11, because this is dynamically the same type of problem.

17.25. Dimensional Analysis

Consider the case when the uncontrolled servomotor is overdamped. Let M_{max} be the maximum value that M attains on the first overshoot and M_{ss} be the final steady-state value of M after a step change in N. In general, M_{max} will be a function of all of the other variables and we may write

$$M_{max} = f_a(M_{ss}, K_l, \tau_1, \tau_2)$$

where M_{max} and M_{ss} have the same dimensions, K_l has the dimension $(1/t)$ and τ_1 and τ_2 have the dimension (t). Forming dimensionless π's, we find

$$\frac{M_{max}}{M_{ss}} = \pi_1$$

$$K_l\tau_1 = \pi_2$$

$$\frac{\tau_1}{\tau_2} = \pi_3$$

Then we may write

$$\pi_2 = f_b(\pi_1, \pi_3)$$

or

$$K_l\tau_1 = f\left(\frac{\tau_1}{\tau_2}\right) \qquad \text{(for a given value of } \pi_1\text{)}$$

17.26. Solution of a Typical Control Problem

The results of this work may be employed in the solution of a typical control problem in order to illustrate how they may be used in engineering practice.

Consider the problem of controlling the position of a moving member of a machine. In order to meet requirements such as maximum velocity, acceleration, load forces, and speed of response, a rate-type hydraulic servomotor is employed as shown schematically in Fig. 17.12. The load consists of the member to be positioned and effects associated with its motion. The input to this system is the position, X, of the servomotor-stroking mechanism and the output is the position, M, of the load. In order to determine how M is related to X, the dynamic characteristics of the servomotor and its load must be ascertained. In reality, the load must be considered as part of the servomotor system, and henceforth

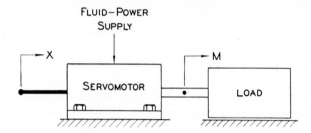

Fig. 17.12. Rate-type hydraulic servomotor and load.

the term "servomotor system" shall mean the servomotor and its load.

Suppose that a dynamic analysis of the servomotor system yields the following relationship between M and X:

$$(3 \times 10^{-4}D^2 + 10^{-2}D + 1)DM = 100X \qquad (17.18)$$

where the damping ratio of the servomotor system is found to be

$$\zeta_s = \frac{10^{-2}}{2\sqrt{3 \times 10^{-4}}} = \frac{1}{2\sqrt{3}} = 0.285 \qquad (17.19)$$

Since the analogue study indicates that proportional control is not very satisfactory with a rate-type servomotor system having a damping ratio less than 0.7, it is necessary first to improve the servomotor-system characteristics. Suppose that this can be done and that the system now yields the following relationship between M and X:

$$(4 \times 10^{-4}D^2 + 5 \times 10^{-2}D + 1)DM = 100X \qquad (17.20)$$

Now the damping ratio is

$$\zeta_s = \frac{5 \times 10^{-2}}{2\sqrt{4 \times 10^{-4}}} = \frac{5}{4} = 1.25 \qquad (17.21)$$

and the time constants τ_1 and τ_2 are found to be

$$\tau_1, \tau_2 = \frac{5 \times 10^{-2}}{2}\left(1 \pm \sqrt{1 - \frac{16 \times 10^{-4}}{25 \times 10^{-4}}}\right) = 2.5(1 \pm 0.6) \times 10^{-2} \text{ sec}$$

(17.22)

or

$$\tau_1 = 0.04 \text{ sec}$$

(17.23)

and

$$\tau_2 = 0.01 \text{ sec}$$

(17.24)

In order to establish control of the output position, M, a simple feedback linkage and summation lever may be employed as shown in Fig.

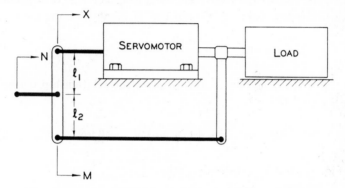

Fig. 17.13. Proportional control with mechanical feedback.

17.13. An analysis of small motions of this linkage yields the following equation:

$$X = \left(\frac{l_1 + l_2}{l_2}\right)N - \left(\frac{l_1}{l_2}\right)M$$

(17.25)

The complete system now may be represented by the functional block diagram shown in Fig. 17.14. The loop gain is given by

$$K_l = 100\frac{l_1}{l_2}$$

(17.26)

and using the value of τ_1 found in Eq. 17.23 gives

$$K_l\tau_1 = 4\frac{l_1}{l_2}$$

(17.27)

In order to attain highest speed of response, a transient response with 20 per cent overshoot is considered to be acceptable, and, from Fig. 17.8,

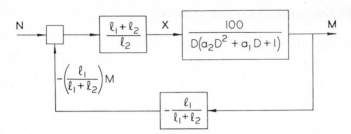

Fig. 17.14. Block-diagram representation of complete system.

it is seen that when the value of $K_l \tau_1$ is equal to 0.69, this system will respond with 20 per cent overshoot when a step change in N occurs.

Substituting this value of $K_l \tau_1$ in Eq. 17.27 and solving for l_1/l_2, we find

$$\frac{l_1}{l_2} = \frac{0.69}{4} = 0.17 \tag{17.28}$$

When Eq. 17.12 is employed, the steady-state relationship between M and N is found to be

$$M_{ss} = \frac{N_{ss}}{k_f} = \left(\frac{l_1 + l_2}{l_1}\right) N_{ss} = 6.9 N_{ss} \tag{17.29}$$

From Fig. 17.10 it is seen that a period of

$$T_2 = 4.6 \tau_1 = 0.18 \text{ sec} \tag{17.30}$$

must elapse before the value of M can reach its first peak.

17.27. Conclusions

The results of this work show clearly that a rate-type servomotor with second-order lag effects may be controlled satisfactorily by simple proportional feedback if its damping ratio without control is greater than 0.7. The ratio of the loop gain, K_l, to the servomotor natural frequency, ω_{ns}, required to produce a given type of response does not vary appreciably with servomotor damping ratio, ζ_s, if the servomotor is underdamped. By expressing the second-order lag effect of an overdamped rate-type servomotor in terms of its real time constants, τ_1 and τ_2, where τ_1 is the larger time constant, it may be seen that the value of $K_l \tau_1$ required to produce a given type of response is roughly doubled as the value of τ_1/τ_2 varies from 1 to 10. As the value of τ_1/τ_2 increases above 10, little variation of $K_l \tau_1$ occurs, indicating that the effect of τ_2 is negligible when it is less than $\tau_1/10$.

It may be seen that the characteristic response times with either under-damped or overdamped servomotors are affected most by the values of ω_{ns} and τ_1, respectively, and they never vary by more than a factor of 2 as the degree of damping in the uncontrolled servomotor is changed.

17.3. DYNAMIC STUDY OF PRESSURE-CONTROLLED HYDRAULIC SYSTEMS *

17.31. Introduction

In the design of complex hydraulic systems the questions of speed of response and stability are of paramount importance. The heavy landing gear of an aircraft in flight and the cutting tool in a machine should move quickly and without hydraulically induced vibrations. Furthermore, if such systems are disturbed during their operation, as by wind gusts for the case of the landing gear or by load changes on the cutting tool, they should respond rapidly without excessive oscillation. A typical hydraulic system may consist of a positive-displacement pump driven by a prime mover, the combination representing essentially a constant-flow source, a pressure-limiting or pressure-regulating device, a flow-control valve, an energy-storage element such as an accumulator, and the load to be moved. The load may be located at some distance from the pressure source, necessitating the motion of a long fluid column between the pressure source and the load.

Problems in the design of such a system include the proper selection of commercially available components, the development of new components as needed, and the choice of optimum system characteristics to ensure the desired dynamic operation of the over-all system. Much insight into the problem can be gained by an analytical study supplemented by such modern tools as the electronic analogue computer. The computer method is, by and large, an experimental procedure of great flexibility where the mechanical components are replaced by their electronic equivalents. The value of this method lies primarily in the facility with which the physical constants of the system such as the mass, size, spring stiffness, damping, and so on, can be changed by the mere turning of a calibrated dial and the ease with which the effects of such changes on the behavior of the system can be observed. A complete understanding of the physical relationships that characterize each component and the interrelation of these characteristics in the system is

* The work here reported was supported by the Pantex Manufacturing Corp., Pawtucket, R. I., and is presented with their permission.

essential to successful simulation, particularly when it becomes necessary, as it will in many instances, to idealize the actual physical problem.

An analytical study based on certain idealizations can augment the effectiveness of computer studies by giving a better insight into the broadest phases of the problem. Experimentation on the computer will, for example, bring out certain peculiarities of hydraulic systems only after extended variations of the many adjustable parameters of the system. The combination of several of these parameters into dimensionally homogeneous groups as part of a stability study will bring such peculiarities quickly to light.

The general problem of the stability of hydraulic systems has been investigated in many different ways.[6-11] The complexity of the problem even for the simple groupings of components to be discussed here makes it advisable to separate as many factors as possible and to study a number of smaller and more manageable problems individually. A great deal of insight can be gained by subdividing the problem.

17.32. Derivation of the Fundamental Physical Relationships

The system to be studied is illustrated in Fig. 17.15.

For the purpose of this discussion the flow source will be considered to be immune to system-pressure demand changes since the response of the prime-mover–flow-source combination is slow, compared with the system dynamics under study.

The load system can be either a ram controlling the velocity of an inertia, Fig. 17.16, with certain acceleration or force demands, or a positive-displacement fluid motor of the piston or vane type controlling the angular velocity of an inertia with certain angular acceleration or torque requirements.

The relief or pressure-control valve, in the most general case, consists of a pressure-sensing means, a device which compares the system pres-

[6] G. Reethof, *On the Dynamics of Pressure-Controlled Hydraulic Systems.* Paper No. 54-SA-7, presented at an ASME meeting, Pittsburgh, June 1954.

[7] G. Friedensohn, *Stability of Control Valves.* S.M. Thesis, Department of Mechanical Engineering, Massachusetts Institute of Technology, Cambridge, Mass., 1954.

[8] F. D. Ezekiel, *Effect of a Hydraulic Conduit with Distributed Parameters on Control-Valve Stability.* Sc.D. Thesis, Department of Mechanical Engineering, Massachusetts Institute of Technology, Cambridge, Mass., 1955.

[9] S.-Y. Lee and J. F. Blackburn, "Steady-State Axial Forces on Control-Valve Pistons," *Trans. ASME*, Vol. 74 (August 1952), pp. 1005–1011.

[10] F. W. Ainsworth, "The Effect of Oil-Column Acoustic Resonance on Hydraulic Valve 'Squeal,'" *Trans. ASME*, Vol. 78 (May 1956), pp. 773–778.

[11] S.-Y. Lee and J. F. Blackburn, "Transient Flow Forces and Valve Instability," *Trans. ASME*, Vol. 74 (August 1952), pp. 1013–1016.

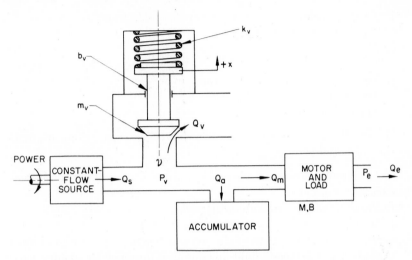

Fig. 17.15. Schematic diagram of a pressure-controlled hydraulic system.

sure with a desired pressure, and a device for employing this difference to control the actual system pressure. In its simplest form such a pressure-control valve is illustrated in Fig. 17.15.

The single-stage valve consists of a conical-seat poppet of mass m_v (lb sec^2/in.) loaded by a coil spring of stiffness k_v (lb/in.). The viscous and other friction forces acting on the poppet are lumped into a damping

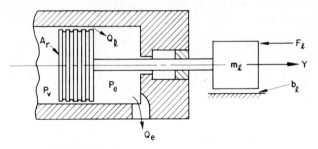

Fig. 17.16. Load system, schematic.

constant b_v (lb sec/in.). The pressure-sensitive area of the poppet valve is denoted by A_s (in.^{2}).

Accumulators are used for several purposes, such as:

1. To store hydraulic power to reduce intermittent-peak-demand pump requirement.

2. To filter out the effects of pulsations in pump flow.

3. To absorb shocks due to suddenly changing loads.

4. To attain dynamic stability.

Three types of accumulators are considered:

1. The added-volume accumulator.

2. The spring-loaded accumulator.

3. The air-loaded accumulator.

17.321. Basic Assumptions. The following assumptions are made in the derivation:

1. All damping forces are viscous in nature, Coulomb friction in seals having been neglected.

2. Viscous leakage flows are neglected.

3. Head losses in fluid passages are negligible.

4. Only small flow variations through the poppet valve are considered.

5. Steady and unsteady fluid-flow-induced forces on the poppet valve are included in the effective spring and damping constants.[9-11]

6. The inertia of the fluid in the passages is neglected.

7. All of the walls of the fluid passages and chambers are rigid.

8. Viscous-friction effects are negligible in the flow through the poppet.

The summation of forces on the poppet valve results in Eq. 17.31.

$$m_v D^2 x + b_v Dx + k_v x = P_v A_s - \delta \qquad (17.31)$$

where x = the regulator valve displacement, in.

D = the differential operator, d/dt, sec^{-1}

P_v = the pressure drop across the poppet valve, psi

δ = the external disturbance force on the poppet, lb

If a pressure drop, P_v, exists across the poppet-valve orifice of area wx, the flow of fluid, Q_v, will be expressed by Eq. 17.32.

$$Q_v = C_o wx \sqrt{P_v} \qquad (17.32)$$

where w = the port width of the valve, in.

$C_o = C_d \sqrt{2/\rho}$, the orifice coefficient for the regulator valve, in.^{2}/ sec $\sqrt{\text{lb}}$

C_d = a discharge coefficient

ρ = the density of the fluid, lb $\text{sec}^2/\text{in.}^4$

The fluid, although commonly treated as incompressible, does undergo a small change in density with change in pressure. Thus, a change in flow to the load will not necessarily result in an instantaneous change in load motion. The phenomenon of compressibility can be expressed by a bulk modulus, β, as defined in Eq. 17.33.

$$\beta = \rho \, \frac{dP}{d\rho} \qquad (17.33)$$

where ρ = density of the fluid, lb sec^2/in.^{4}

P = pressure in the system, psi

The variation in density of a typical hydraulic fluid for the commonly encountered pressure changes is small and will be neglected, but the rate of change of density may be quite large compared to the other quantities and cannot be omitted. Thus, Eq. 17.33 can be written as follows:

$$\frac{dP}{dt} = \frac{\beta}{V} \frac{dV}{dt} \qquad (17.34)$$

where V is the volume of system hydraulic fluid under compression, in.^{3} It is important to note that β can be greatly affected by the presence of small amounts of air in the hydraulic fluid.

The leakage flow past the ram is given by

$$Q_l = C_l P_m \qquad (17.35)$$

where C_l = a leakage flow coefficient, in.^{5}/lb sec

P_m = the pressure difference across the ram, psi

The flow to the fluid motor, Q_m, therefore consists of three parts, the load velocity flow, the compressibility flow, and the leakage flow.

$$Q_m = A_r \, DY + \frac{V}{\beta} DP_v + C_l P_m \qquad (17.36)$$

where A_r = the motor effective area, in.^{2}

Y = the small change in load displacement, in.

Equating the external forces acting on the load inertia to the inertia force gives

$$m_l \, D^2 Y = P_m A_r - b_l \, DY - F_l \qquad (17.37)$$

where m_l = the load-motor mass, lb sec^2/in.

b_l = the load-motor damping constant, lb sec/in.

F_l = the load force, lb.

Three types of accumulators will be discussed here. Although the constructions of the several types differ, their basic functions are identical, namely to store pressure energy.

A. THE ADDED-VOLUME ACCUMULATOR. Adding a large volume of fluid to the volume under compression is sometimes used to lower system natural frequency. This type of accumulator, illustrated in Fig. 17.17, merely provides a supplemental volume to increase the volume of fluid, V, under compression between pump and motor. The resistance to flow into the added volume must be negligible if the total volume is to be

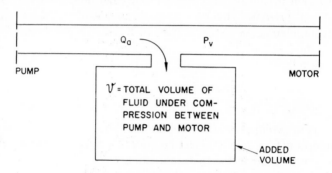

Fig. 17.17.   Added-volume type of accumulator.

lumped in this way. It is also important to check the compliance of the walls of the fluid passages and chambers to see if the line elastic coefficient * should be included in the compliance of the system.

B. THE SPRING-LOADED PISTON ACCUMULATOR. As shown in Fig. 17.18, the spring-loaded accumulator is a commonly used means of storing fluid energy.

The steady-state force equilibrium equation for the piston is given by Eq. 17.38.

$$(P_v - P_{atm})A_a = k_a(l_0 - l) \qquad (17.38)$$

where P_{atm} = atmospheric pressure, psi
A_a = the accumulator piston area, in.²
l_0 = the unstretched length of accumulator spring, in.
l = the instantaneous compressed length of accumulator spring, in.
k_a = the accumulator spring stiffness, lb/in.

Dynamic parameters such as mass and damping of the accumulator piston are neglected as a first approximation.

* See Chapter 3 of this volume, Sec. 3.351.

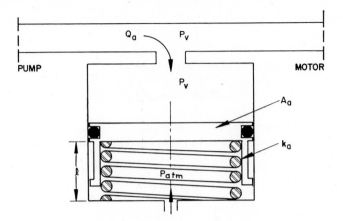

Fig. 17.18.   Spring-loaded accumulator.

Differentiating Eq. 17.38 with respect to time gives

$$DP_v = \frac{k_a}{A_a} Dl \tag{17.39}$$

The rate of flow of oil into the accumulator is given by

$$Q_a = -A_a \, Dl \tag{17.40}$$

Substituting Eq. 17.39 into Eq. 17.40 gives

$$Q_a = \frac{A_a{}^2}{k_a} DP_v \tag{17.41}$$

When a spring-loaded piston accumulator is used, the effects of the compressibility of the hydraulic fluid itself are usually negligible.

C. Air-Loaded Piston Accumulator.   Let the spring above the piston of Case B be replaced by compressed air at pressure P_a.   Force equilibrium requires that

$$A_a P_a = A_a P_v \tag{17.42}$$

Assuming isothermal compression or expansion of the air and using the equation for a perfect gas, we obtain

$$P_a V_a = m_a R T_a = \text{constant} \tag{17.43}$$

where V_a = the volume of the air in the accumulator, in.^{3}
  m_a = the mass of the air in the accumulator, lb sec^2/in.
  T_a = the temperature of the air in the accumulator, °R
  R = the perfect-gas constant, in.^{2}/sec^2 °R

Furthermore

$$V_a = lA_a \tag{17.44}$$

where $l =$ (now) the length of the air-column "spring" and

$$P_a l A_a = m_a R T_a \tag{17.45}$$

For isothermal conditions, T_a is constant.

Differentiating with respect to time, combining with Eq. 17.42, gives

$$DP_a = DP_v = - \frac{m_a R T_a}{l^2 A_a} Dl = - \frac{P_a A_a}{V_a} (Dl) \tag{17.46}$$

The flow into the accumulator is then given by

$$Q_a = A_a \, Dl = \left(\frac{V_a}{P_a}\right) DP_v \tag{17.47}$$

The compressibility of the hydraulic fluid itself is again usually negligible.

Equations 17.41 and 17.47 relate the flow into the accumulator for the two types considered. In the spring-loaded type, the flow is directly proportional to the rate of change of pressure. In the air-loaded type, the flow can be taken as proportional to the rate of change of pressure only when small changes are considered because the value of (V_a/P_a) changes appreciably when large piston displacements occur.

In the previous three cases it has been assumed that there is no resistance to flow into the accumulator. In some installations a considerable amount of resistance to flow may be encountered in the fittings if the designer is not careful to provide adequate flow passages between the line and the accumulator.

17.33. Transformation of the Equations into Linearized Incremental Form

The fundamental equations derived above completely describe the system, but they contain certain nonlinearities even though a number of idealizations were made in deriving them. Notably, Eq. 17.32 contains the variable $P_v^{1/2}$, which in turn is multiplied by the variable X. Since this study is concerned with the incremental stability of the system at various operating conditions, the variables can be divided into a steady-state value and a small change about the steady-state value.

This change in variables is indicated as follows:

$$X = X_i + \Delta X \qquad (17.48)$$

$$P_v = P_{vi} + \Delta P_v \qquad (17.49)$$

$$Y = Y_i + \Delta Y \qquad (17.50)$$

$$DY = (DY)_i + \Delta(DY) \qquad (17.51)$$

where subscript i denotes the initial steady operating value of the variable and Δ denotes a small deviation from the steady-state operating value.

For an initially steady-state set of conditions, $(DY)_i = 0$, and there we may write

$$DY = \Delta(DY) = D(\Delta Y) \qquad (17.52)$$

Thus, only small oscillations of each variable about the steady value can be studied. The relative ease with which the linear differential equations may be studied usually more than offsets the disadvantage of being limited to small variations of the system variables.

The substitution of Eqs. 17.48, 17.49, 17.50, and 17.51 into the previously derived equations results in a set of linearized differential equations. From Eq. 17.31

$$m_v D^2(\Delta X) + b_v D(\Delta X) + k_v (\Delta X) = A_s (\Delta P_v) + \delta \qquad (17.53)$$

From Eq. 17.32 for the flow through the poppet valve we may write

$$\Delta Q_v = \left.\frac{\partial Q_v}{\partial X}\right|_{P_v=\text{const}} (\Delta X) + \left.\frac{\partial Q_v}{\partial P_v}\right|_{X=\text{const}} (\Delta P_v) \qquad (17.54)$$

where

$$\left.\frac{\partial Q_v}{\partial X}\right|_{P_v=\text{const}} = C_o w \sqrt{P_v} = \frac{Q_{vi}}{X} \qquad (17.55)$$

and

$$\frac{\partial Q_v}{\partial P_v} = \frac{C_o w X_i}{2\sqrt{P_{vi}}} = \frac{Q_{vi}}{2P_{vi}} \qquad (17.56)$$

Hence

$$\Delta Q_v = \frac{Q_{vi}}{X_i} (\Delta X) + \frac{Q_{vi}}{2P_{vi}} (\Delta P_v) \qquad (17.57)$$

The incremental load rate of flow is obtained from Eq. 17.36.

$$\Delta Q_m = A_r D(\Delta Y) \qquad (17.58)$$

The incremental rates of flow into the various types of accumulators are given by Eqs. 17.59, 17.60, and 17.61.

For the added-volume accumulator

$$\Delta Q_a = \frac{V}{\beta} D(\Delta P_v) \qquad (17.59)$$

For the spring accumulator, from Eq. 17.41,

$$\Delta Q_a = \frac{A_r^2}{k_a} D(\Delta P_v) \qquad (17.60)$$

For the air accumulator, from Eq. 17.47,

$$\Delta Q_a = \frac{V_{ai}}{P_{ai}} D(\Delta P_v) \qquad (17.61)$$

where the subscript i denotes the initial value of each variable.

For the air-loaded and spring-loaded accumulators, the compressibility of the oil itself has been neglected.

The terms (V/β), (A_r^2/k_a), and (V_{ai}/P_{ai}) in Eqs. 17.59, 17.60, and 17.61 can be considered to be system capacitances, and a general case may be introduced with the capacitance coefficient, C_c, as follows

$$\Delta Q_a = C_c D(\Delta P_v) \qquad (17.62)$$

Applying the continuity equation to the fluid system between the pump and motor yields for the case of negligible leakage flow

$$\Delta Q_m + \Delta Q_v + \Delta Q_a = 0 \qquad (17.63)$$

It is important to note that the displacement of the poppet-spring combination in the relief valve creates an additional "accumulator" effect which should be included in the capacitance coefficient, C_c, whenever this effect is great enough to be important.

The force equation for the load in the absence of spring coupling to ground is obtained from Eq. 17.37.

$$m_l D^2(\Delta Y) = A_r(\Delta P_v) - b_l D(\Delta Y) - \Delta F_l \qquad (17.64)$$

17.34. Dynamic Stability Based on Linearized Equations

Equations 17.53, 17.57, 17.58, and 17.62 through 17.64 may be combined into the fourth-order differential equation

$$(D^4 + a_3 D^3 + a_2 D^2 + a_1 D + a_0)(\Delta P_v)$$

$$= \text{(summation of input disturbance effects)} \qquad (17.65)$$

where the coefficients a_0, a_1, a_2, and a_3 are given by

$$a_3 = 2\zeta_l\omega_{nl} + 2\zeta_v\omega_{nv} + C_b \tag{17.66}$$

$$a_2 = \omega_{nl}^2 + \omega_{nv}^2 + 4\zeta_l\zeta_v\omega_{nl}\omega_{nv} + C_b(2\zeta_l\omega_{nl} + 2\zeta_v\omega_{nv}) \tag{17.67}$$

$$a_1 = 2\zeta_l\omega_{nl}\omega_{nv}^2 + 2\zeta_v\omega_{nv}\omega_{nl}^2 + C_a + C_b(\omega_{nv}^2 + 4\zeta_l\zeta_v\omega_{nl}\omega_{nv}) \tag{17.68}$$

$$a_0 = \omega_{nl}^2\omega_{nv}^2 + 2\zeta_l\omega_{nl}C_a + 2\zeta_l\omega_{nl}\omega_{nv}^2C_b \tag{17.69}$$

and the physical characteristics of the system are related to the parameters C_a, ω_{nl}, ζ_l, ω_{nv}, ζ_v, and C_b, which are given in Table 17.1.

It can be shown that the most critical stability condition occurs for very small openings of the poppet valve, that is, when $C_b \approx 0$. Equation 17.65 then becomes

$$(D^4 + a_3'D^3 + a_2'D^2 + a_1'D + a_0')\,(\Delta P_v)$$
$$= \text{(summation of input disturbance effects)} \tag{17.70}$$

where

$$a_3' = 2\zeta_l\omega_{nl} + 2\zeta_v\omega_{nv} \tag{17.71}$$

$$a_2' = \omega_{nl}^2 + \omega_{nv}^2 + 4\zeta_l\zeta_v\omega_{nl}\omega_{nv} \tag{17.72}$$

$$a_1' = 2\zeta_l\omega_{nl}\omega_{nv}^2 + 2\zeta_v\omega_{nv}\omega_{nl}^2 + C_a \tag{17.73}$$

and

$$a_0' = 2\omega_{nl}\omega_{nv}^2 + 2\zeta_l\omega_{nv}C_a \tag{17.74}$$

The characteristic equation of the system is obtained when all the inputs are zero. For the system to be dynamically stable none of the roots of this equation may contain a positive real part. Finding the roots of an equation of a degree higher than the third usually requires an inordinate amount of labor, and one useful method of ascertaining whether or not the system is stable (without, however, finding the margin of stability) is the use of Routh's stability criterion, [12] as described in Sec. 14.191.

For a fourth-degree equation, the criteria of stability are:

$$a_3a_2 > a_1 \tag{17.75}$$

and

$$a_1(a_3a_2 - a_1) > a_3^2a_0 \tag{17.76}$$

where all the coefficients are assumed to be positive. Since in the present case Inequality 17.75 is always true, the necessary and sufficient condition for stability is Inequality 17.76. When it becomes an equation,

$$a_1(a_3a_2 - a_1) = a_3^2a_0 \tag{17.77}$$

[12] M. F. Gardner and J. L. Barnes, *Transients in Linear Systems*, Vol. 1, John Wiley and Sons, New York, 1942, pp. 198–200.

Table 17.1

	General Case	Spring-Loaded	Air-Loaded	Added-Volume
C_a	$\left(\dfrac{A_s}{m_v}\right)\left(\dfrac{1}{C_c}\right) C_o w \sqrt{P_{vi}}$	$\left(\dfrac{A_s}{m_v}\right)\left(\dfrac{k_a}{A_a^2}\right) C_o w \sqrt{P_{vi}}$	$\left(\dfrac{A_s}{m_v}\right)\left(\dfrac{P_{vi}}{V_{ai}}\right) C_o w \sqrt{P_{vi}}$	$\left(\dfrac{A}{m_v}\right)\left(\dfrac{\beta}{V}\right) C_o w \sqrt{P_{vi}}$
ω_{nl}^2	$\dfrac{A_r^2}{m_l}\cdot\dfrac{1}{C_c}$	$\dfrac{A_r^2}{m_l}\cdot\dfrac{k_a}{A_a^2}$	$\dfrac{A_r^2}{m_l}\cdot\dfrac{P_{vi}}{V_a}$	$\dfrac{A_r^2}{m_l}\cdot\dfrac{\beta}{V}$
ζ_l	$\dfrac{b_l}{2A_r}\sqrt{C_c/m_l}$	$\dfrac{b_l}{2A_r}\sqrt{A_a^2/k_a m_l}$	$\dfrac{b_l}{2A_r}\sqrt{V_{ai}/P_{vi}m_l}$	$\dfrac{b_l}{2A_r}\sqrt{V/\beta m_l}$
ω_{nv}^2	$\dfrac{k_v}{m_v}$	$\dfrac{k_v}{m_v}$	$\dfrac{k_v}{m_v}$	$\dfrac{k_v}{m_v}$
ζ_v	$\dfrac{b_v}{2\sqrt{k_v m_v}}$	$\dfrac{b_v}{2\sqrt{k_v m_v}}$	$\dfrac{b_v}{2k_v m_v}$	$\dfrac{b_v}{2k_v m_v}$
C_b	$\dfrac{Q_{vi}}{2P_{vi}}\cdot\dfrac{1}{C_c}$	$\dfrac{Q_{vi}}{2P_{vi}}\cdot\dfrac{k_a}{A_a^2}$	$\dfrac{Q_{vi}}{2V_{ai}}$	$\dfrac{Q_{vi}}{2P_{vi}}\cdot\dfrac{\beta}{V}$

the system is said to be *marginally stable;* it will respond to an input disturbance with an oscillation that persists indefinitely at constant amplitude.

An important limitation of Routh's criterion is that it gives no indication of *degree* of stability of a stable system as evidenced by the rate of decay of the oscillation of the system following a sudden input disturbance. If quantitative data on the response to a particular input

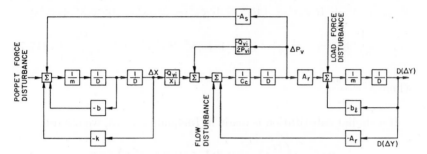

Fig. 17.19. Functional block diagram of system.

disturbance are desired, it is necessary to employ other more laborious techniques.

The system of equations that was combined into Eq. 17.65 can also be represented by the functional block diagram of Fig. 17.19. The complete system under study is essentially a single-loop system with a number of internal subsidiary loops, and the primary closed-loop effect is that of the pressure acting on the area, A_s, of the poppet valve. All

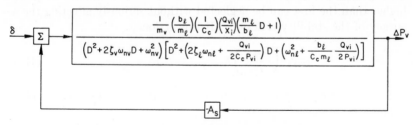

Fig. 17.20. Simplified block diagram of system.

of the minor loops may be lumped into a single block as shown in Fig. 17.20, provided all input disturbances except δ are zero.

Workers in the field of automatic control have developed many different techniques which may be applied to the analysis of closed-loop systems such as the one under study. Although these techniques make it possible to study the characteristic equation in detail and to determine the degree of system stability, the inequalities given by the application of Routh's method often afford a much greater insight into a complex problem than do the servosynthesis techniques, and with considerably less labor.

Substitution of the complete expressions for each of the coefficients in Inequality 17.76 gives Routh's stability criterion in terms of the system parameters of Table 17.1.

$$2\zeta_v\omega_{nv}{}^3[\zeta_v\lambda^5 + 4\zeta_l\zeta_v\lambda^4 + (4\zeta_l{}^3 + 4\zeta_l\zeta_v{}^2 - 2\zeta_l)\lambda^3 + 4\zeta_l{}^2\zeta_v\lambda^2 + \zeta_l{}^2\lambda]$$

$$> C_a\left[(4\zeta_l{}^3 + \zeta_l)\lambda^3 + (4\zeta_l\zeta_v + \zeta_v)\lambda^2 + \zeta_v\lambda - \zeta_v + \frac{C_a}{2\omega_{nv}{}^3}\right] \quad (17.78)$$

where $\lambda = \dfrac{\omega_{nl}}{\omega_{nv}}$

If a similar substitution is made in Inequality 17.75, we obtain

$$2\zeta_v\omega_{nv}{}^3\left(\frac{\zeta_l}{\zeta_v}\lambda^3 + 4\zeta_l{}^2\lambda^2 + 4\zeta_l\zeta_v\lambda + 1\right) > C_a \quad (17.79)$$

17.35. Special Cases

17.351. Small λ. It often occurs in practice that the complicated Inequality 17.78 can be considerably simplified. For example, for both air-loaded and spring-loaded accumulators the quantity λ, the ratio of load undamped natural frequency, ω_{nl}, to control-valve natural frequency, ω_{nv}, is usually $1/10$ or less so that terms containing λ to powers higher than the first can be neglected without seriously affecting the accuracy of the stability criterion. If this is done,

$$\zeta_v\zeta_l{}^2\left(\frac{2\omega_{nv}{}^3}{C_a}\right)\lambda > \zeta_v(\lambda - 1) + \frac{C_a}{2\omega_{nv}{}^3} \quad (17.80)$$

Several interesting conclusions can be drawn from this simplified expression:

1. Without load damping ($\zeta_l = 0$) the system may or may not be stable, depending upon the relative magnitudes of ζ_v and $C_a/2\omega_{nv}{}^3$.

2. Without adequate regulator damping, the system will be unstable. A numerical example is given below.

3. For the cases of the stable air-loaded and spring-loaded accumulators, increasing accumulator volume will: (*a*) increase the margin of stability, (*b*) increase the damping coefficient, (*c*) decrease load-system natural frequency.

17.352. Large λ. The opposite situation $(\lambda \gg 1)$ also occurs at times, particularly when a sensitive low-spring-rate large-area regulator is used with a pure volume accumulator. In such cases Inequality 17.78 may be approximated by

$$\frac{\zeta_v^2}{\lambda}\left(\frac{2\omega_{nl}^3}{C_a}\right)^2 > \zeta_l(4\zeta_l^2 - 1)\left(\frac{2\omega_{nl}^3}{C_a}\right) + 1 \qquad (17.81)$$

In this case the conditions of stability are quite different from those of the last section. Variation of the load damping ratio, ζ_l, with all other factors held constant presents an interesting paradox. The system will be most stable for $\zeta_l = 0.288$, and either a decrease or an increase of the load damping from this optimum value will decrease the stability of the system. Increasing the value of V, the total volume of liquid under compression, may or may not improve system stability, depending upon the value of ζ_l.

17.36. Numerical Examples

Some numerical examples may help to clarify the statements of the preceding sections

(a) Suppose we have a system for which

$\omega_{nl} = 250$ rad/sec $A_s = 0.1$ in.^{2}
$\omega_{nv} = 2500$ rad/sec $m_v = 1.76 \times 10^{-4}$ lb sec^2/in.
$\zeta_v = 0.2$ $P_{vi} = 2000$ psi
$\zeta_l = 0.1$ $1/C_c = 2000$ lb/in.^{5}
$C_a = 56.81 \times 10^8$ 1/sec^3 $w = 1.12$ in.

Here $\lambda \gg 1$ and Expression 17.80 applies. When the numbers are substituted into the inequality, it becomes $0.0011 \not> 0.002$, and the system is unstable.

If we increase the generalized system capacitance so that $1/C_c = 1000$ lb/in.^{5}, the inequality becomes $0.0022 > -0.88$, and the system is stable.

(b) Assume a second system for which

$\omega_{nl} = 600\pi$ rad/sec $A_s = 0.3$ in.^{2}
$\omega_{nv} = 120\pi$ rad/sec $m_v = 8.26 \times 10^{-5}$ lb sec^2/in.
$\zeta_v = 0.2$ $P_{vi} = 200$ psi
$\zeta_l = 0.2$ $1/C_c = 2000$ lb/in.^{5}
$C_a = 200 \times 10^8$ 1/sec^3 $w = 1.945$ in.

Since $\lambda \gg 1$, Inequality 17.81 applies. Substituting the numbers, we get $0.0036 \ngtr 0.9865$, and the system is unstable.

If we change the natural frequencies of both the load and the regulator, while preserving their ratio, we get $\omega_{nl} = 1200\pi$ radians/sec and $\omega_{nv} = 240\pi$ radians/sec, and the inequality becomes $0.235 > 0.094$. The system is now stable.

17.37. Summary

A linearized analysis, based upon a number of simplifying assumptions about a given type of pressure-control system, has revealed a number of possible causes of system instability. The conclusions drawn from this analysis are subject to the following limitations:

1. An analysis which ignores distributed-parameter effects in the fluid lines and which neglects the effects of fluid mass throughout the system except near the throat of the poppet-valve orifice.

2. The assumption that the viscosity of the fluid has only a negligible effect upon the flow through the poppet valve and upon the forces exerted by the fluid on the poppet.

3. The assumption that no static friction exists in the system.

4. The assumption that all system variables undergo only "small" changes.

5. The assumption that pressure variations downstream of the poppet valve are negligibly small.

There are many other facets in the general picture of hydraulic system stability. Some of these are discussed elsewhere in this book; others are not even recognized as definable problems at the present time. Three investigations of current interest in this field are those of Ezekiel [8] and of Ainsworth [10] on control-valve stability and that of Stone [13] on poppet-valve flow and force characteristics.

[13] J. A. Stone, *Design and Development of an Apparatus to Study the Flow-Induced Forces in a Poppet-Type Flow Valve.* S.B. Thesis, Department of Mechanical Engineering, Massachusetts Institute of Technology, Cambridge, Mass., 1955; and J. A. Stone, *An Investigation of Discharge Coefficients and Steady State Flow Forces for Poppet Type Valves,* S.M. Thesis, Department of Mechanical Engineering, Massachusetts Institute of Technology, Cambridge, Mass., 1957.

17.4. ANALOGUE COMPUTER STUDY OF A VELOCITY-CONTROL HYDRAULIC SYSTEM [14]

17.41. Introduction

A phenomenal growth of the application of the analogue computer to the study of control-system problems has occurred since the war. In many of these studies, an investigation is made of the performance of an assemblage of commercially available components. Nevertheless, a much neglected and very promising field for the analogue computer lies in its use as a design tool in the conceptual stages of a complex development.

The purpose of this section is to demonstrate an approach to a hydraulic design problem. The method utilizes simple mathematical techniques first to form a good physical understanding of the system. This phase then is followed by an exhaustive "designing with the analogue computer." The first stage of familiarization is of the greatest importance to the proper utilization of the analogue computer potentials. It can be said without minimizing the value of the analogue approach that unless a good qualitative-performance prediction can be made from a basic physical understanding of the problem, the use of the computer is unjustified. This first phase then sets the stage for an effective use of the analogue computer as a design tool for the determination of optimum system parameters.

17.42. Description of Hydraulic System

The hydraulic system to be discussed controls the velocity of a large mass. A schematic diagram of the system under study is shown in Fig. 17.21. The flow from the downstream side of the fluid motor is to remain constant, thereby maintaining a constant ram velocity. A constant pressure drop across the fixed-area downstream metering orifice assures a constant ram velocity unless there is leakage flow from seals.

The control mechanism in the form of a variable-area control orifice on the upstream side attempts to maintain this constant pressure drop across the metering orifice on the downstream side by by-passing the excess amount of fluid supplied by the pump. The fixed-displacement

[14] Taken from Gerhard Reethof, "The Analog Computer as a Design Tool in the Study of a Velocity-Control Hydraulic System," *Proceedings of the Eleventh National Conference on Industrial Hydraulics*, Chicago, 1955.

pump is driven by an electric motor. In the steady state, the pressure force acting on the right-hand side of the by-pass control valve (see Fig. 17.21) is in balance with the spring force and the flow-induced valve-closing force. A load velocity in excess of the desired value results in an increased pressure drop across the metering orifice; consequently, an excessively high downstream motor pressure, by acting on the spool valve, causes a larger portion of the pump flow to be by-passed through the larger control-orifice opening. As the flow from the supply pump

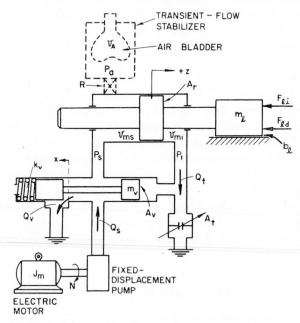

Fig. 17.21. Schematic diagram of velocity-control hydraulic system.

remains essentially constant, the fluid motor slows down, bringing the downstream pressure back to the desired value. The upstream motor pressure adjusts itself to accommodate load requirements. As the load pressure drop increases, the by-pass orifice area required to permit a certain flow decreases. The flow valve assumes a new position at a lower spring force. The required force equilibrium calls for a lower downstream motor pressure that results in a lower load velocity for the higher load pressure. A certain degree of compensation of this steady-state error in control is obtained at the orifice from the flow-induced force. This force tends to close the valve, thereby augmenting the reduced spring force.[9]

17.43. Derivation of Fundamental Equations

A preliminary fundamental analysis of the system just described should serve to give a better insight into the quantitative aspects of the problem. Its purpose is twofold:

1. To set up the algebraic and differential relationships best describing the system and to present these equations in a form suitable for analogue computation.

2. To manipulate these relationships before the analogue study, in order to understand the speeds of response of the various components of the system. Thus, certain portions of the system may respond so fast to variations in the system variables as to appear to act instantaneously relative to some of the slower portions. The justifiable omission of the dynamics of this faster but stable system may result in a simpler analogue setup. An earlier study of the fast system alone must have proved its own stability.

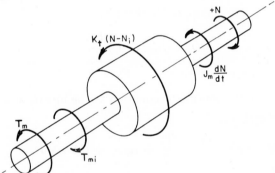

Fig. 17.22. Torques on armature.

The summation of torques acting on the electric-motor armature as illustrated in Fig. 17.22 gives

$$T_m = T_{mi} - K_t(N - N_i) - J_m \frac{dN}{dt} \tag{17.82}$$

where T_m = electric-motor torque at speed N, in. lb (torque on the motor armature is positive if tending to slow it down)

T_{mi} = initial value of electric-motor torque at initial speed, N_i

K_t = torque constant of electric motor, in. lb sec/radian

N = electric-motor speed, radians/sec

N_i = initial value of N, radians/sec

J_m = moment of inertia of electric-motor-pump assembly, in. lb sec^2/radian

The summation of the torques acting on the pump gives [15]

$$T_p = P_s D_p (1 + C_f) + C_{dv} D_p \mu N \qquad (17.83)$$

where T_p = torque on pump shaft, in. lb (positive if tending to speed
up pump)

P_s = pressure drop across pump, psi

D_p = pump displacement, in.^{3}/radian

C_f = pressure-drag coefficient of pump

C_{dv} = viscous-drag coefficient of pump, 1/radian

μ = absolute viscosity of fluid, lb sec/in.^{2}

From the definition of the torques,

$$T_p = T_m \qquad (17.84)$$

and therefore

$$T_{mi} - K_t(N - N_i) - J_m \frac{dN}{dt} = P_s D_p (1 + C_f) + C_{dv} D_p \mu N \quad (17.85)$$

The flow from the pump, Q_s, is given by

$$Q_s = D_p N - C_s \frac{D_p P_s}{\mu} \qquad (17.86)$$

where C_s = leakage coefficient.

Application of the continuity equation to the upstream ram chamber
and passages gives

$$Q_s = \frac{V_{ms}}{\beta} \cdot \frac{dP_s}{dt} + A_r \frac{dz}{dt} + C_q w x \sqrt{P_s} \qquad (17.87)$$

where V_{ms} = volume of fluid under compression on upstream side of
hydraulic motor, in.^{3}

β = bulk modulus of fluid, psi

A_r = area of ram piston, in.^{2}

w = peripheral width of rectangular valve port, in.

x = axial displacement of valve spool from cutoff, in.

and C_q is the orifice-flow coefficient in in.^{2}/sec $\sqrt{\text{lb}}$. It is defined by the
equation

$$Q = C_q A_o \sqrt{P_o} \qquad (17.88)$$

where Q = rate of flow through orifice, in.^{3}/sec

A_o = area of orifice, in.^{2}

P_o = pressure drop across that orifice, lb/in.^{2}

[15] W. E. Wilson, "Hydraulic Pumps and Motors," *Machine Design*, Penton Publishing Company, Cleveland, 1949, pp. 133–138.

Also

$$Q = C_d A_o \sqrt{2P_o/\rho} \tag{17.89}$$

where ρ = mass density of fluid in lb sec^2/in. If $\rho = 8 \times 10^{-5}$ and the discharge coefficient $C_d = 0.61$, $C_q = 96$ in.^{2}/sec $\sqrt{\text{lb}}$.

Combining Eqs. 17.86 and 17.87 gives

$$D_p N - C_s \frac{D_p P_s}{\mu} = \frac{V_{ms}}{\beta} \frac{dP_s}{dt} + A_r \frac{dz}{dt} + C_q wx \sqrt{P_s} \tag{17.90}$$

Equating the forces which act on the valve spool gives the equation

$$A_v P_1 = m_v \frac{d^2 x}{dt^2} + b_v \frac{dx}{dt} + k_v(x + x_s) + C_v P_s wx \tag{17.91}$$

where A_v = pressure-sensitive area of control valve, in.^{2}
m_v = mass of valve spool, lb sec^2/in.
b_v = damping constant of valve, lb sec/in.
k_v = valve spring stiffness, lb/in.
x_s = initial deflection of valve spring for zero opening, in.

and $C_v = 0.465$ = flow-force coefficient. It is defined by the equation

$$F_v = C_v wx P_s \tag{17.92}$$

where F_v is the flow force on the valve spool in pounds. F_v is defined by

$$F_v = 2C_d wx P_s \cos 69° \tag{17.93}$$

for the usual "square-land" valve.

The continuity equation for the downstream ram chamber and passages can be written

$$A_r \frac{dz}{dt} = C_q A_t \sqrt{P_1} + A_v \frac{dx}{dt} + \frac{V_{m1}}{\beta} \frac{dP_1}{dt} \tag{17.94}$$

where z = load displacement, in.
and V_{m1} = volume of fluid under compression on downstream side of hydraulic motor, in.^{3}

Finally, the forces acting on the ram piston rod can be expressed by

$$(P_s - P_1)A_r = m_l \frac{d^2 z}{dt^2} + b_l \frac{dz}{dt} + F_{li} + F_{ld} \tag{17.95}$$

where m_l = mass of load, lb sec^2/in.
b_l = load damping constant, lb sec/in.
F_{li} = initial load force, lb
F_{ld} = load disturbance force, lb

For convenience, the five equations describing the system are collected Table 17.2 and are presented as a complete system block diagram in Fig. 17.23.

Table 17.2. SYSTEM EQUATIONS

$$T_{mi} - K_t(N - N_i) - J_m \frac{dN}{dt} = P_s D_p(1 + C_f) + C_{dv}D_p\mu N \quad (17.85)$$

$$D_p N - C_s \frac{D_p P_s}{\mu} = \frac{V_{ms}}{\beta} \frac{dP_s}{dt} + A_r \frac{dz}{dt} + C_q wx\sqrt{P_s} \quad (17.90)$$

$$A_v P_1 = m_v \frac{d^2 x}{dt^2} + b_v \frac{dx}{dt} + k_v(x + x_s) + C_v P_s wx \quad (17.91)$$

$$A_r \frac{dz}{dt} = A_v \frac{dx}{dt} + C_q A_t\sqrt{P_1} + \frac{V_{m1}}{\beta} \frac{dP_1}{dt} \quad (17.94)$$

$$(P_s - P_1)A_r = m_l \frac{d^2 z}{dt^2} + b_l \frac{dz}{dt} + F_{li} + F_{ld} \quad (17.95)$$

The following assumptions were made in the derivation of the system equations:

1. Shaft connecting electric motor and pump was assumed very stiff.
2. Fluid inertia in passages and ram chamber was neglected.
3. Fluid friction in lines and orifices was neglected.
4. Transmission delays in signal lines were neglected.
5. Acoustic effects in other passages were neglected.
6. Static friction was neglected throughout the system.
7. Walls of all fluid passages were assumed very stiff.

In view of the other large system inertias, the neglect of fluid-inertia effects is justified. The neglect of transmission-line delays, particularly in the transmission of the downstream-motor pressure signal to the pressure-sensitive area of the control valve, may not be justified in some cases.[8]

17.431. Linearization of Problem. The equations of Table 17.2 may now be expressed in terms of incremental quantities of each variable.

$$N = N_i + \Delta N \quad (17.96)$$

$$P_s = P_{si} + \Delta P_s \quad (17.97)$$

$$\frac{dz}{dt} = \left(\frac{dz}{dt}\right)_i + \frac{d}{dt}(\Delta z) \quad (17.98)$$

$$x = x_i + \Delta x \quad (17.99)$$

$$P_1 = P_{1i} + \Delta P_1 \quad (17.100)$$

where the subscript i denotes the initial (steady-state) value of each variable.

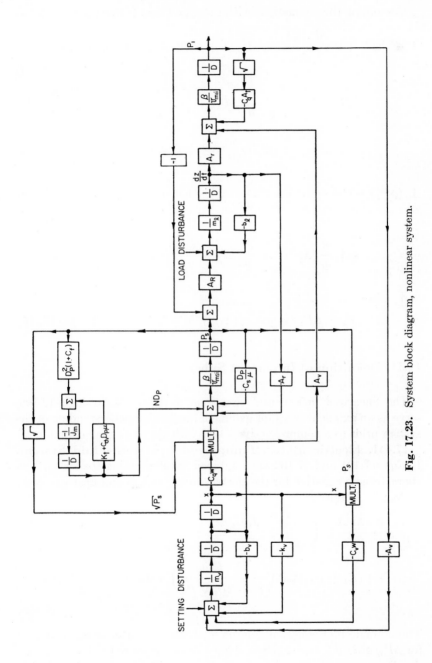

Fig. 17.23. System block diagram, nonlinear system.

Combining these equations with those of Table 17.2, we get

$$+ |K_t| \Delta N + J_m \frac{d(\Delta N)}{dt} = -D_p (1 + C_f) (\Delta P_s) - C_{dv} D_p \mu (\Delta N) \tag{17.101}$$

$$D_p (\Delta N) = \frac{V_{ms}}{\beta} \frac{d}{dt} (\Delta P_s) + \left(C_s \frac{D_p}{\mu} + C_q w x_i \frac{1}{2\sqrt{P_{si}}} \right) (\Delta P_s)$$

$$+ A_r \frac{d}{dt} (\Delta z) + C_q w \sqrt{P_{si}} (\Delta x) \tag{17.102}$$

$$A_v (\Delta P_1) - C_v w x (\Delta P_s) = m_v \frac{d^2}{dt^2} (\Delta x) + b_v \frac{d}{dt} (\Delta x) + (k_v + C_v w P_{si}) (\Delta x) \tag{17.103}$$

$$A_r \frac{d}{dt} (\Delta z) = A_v \frac{d}{dt} (\Delta x) + \frac{C_q A_t}{2 P_{1i}} (\Delta P_1) + \frac{V_{m1}}{\beta} \frac{d}{dt} (\Delta P_1) \tag{17.104}$$

and

$$A_r(\Delta P_s - \Delta P_1) = m_l \frac{d^2}{dt^2} (\Delta z) + b_l \frac{d}{dt} (\Delta z) + F_{ld} \tag{17.105}$$

17.44. Definition of Important Subsystems

The linearized incremental equations, Eqs. 17.101 through 17.105, serve as an excellent basis for gaining a qualitative understanding of the system prior to a comprehensive study with the analogue computer.

17.441. Electric Motor, Pump, and Upstream-Fluid System. If Eqs. 17.101 and 17.102 are combined by eliminating the speed, N, a characteristic equation for the electric motor, pump, and upstream-fluid system is obtained.

$$\frac{J_m V_{ms}}{D_p} \frac{d^2(\Delta P_s)}{dt^2} + \left(\frac{J_m C_s}{\mu} + \frac{J_m C_q w x_i}{2 D_p \sqrt{P_{si}}} + \frac{V_{ms} K_t}{D_p \beta} + \frac{V_{ms} C_{dv} D_p \mu}{D_p \beta} \right) \frac{d(\Delta P_s)}{dt}$$

$$+ \left[(K_t + C_{dv} D_p \mu) \left(\frac{C_s}{\mu} + \frac{C_q w x_i}{2 D_p \sqrt{P_{si}}} \right) + D_p(1 + C_f) \right] (\Delta P_s) = 0 \tag{17.106}$$

This second-order equation appears in the denominator of an expression for $\Delta P_s / \Delta(dz/dt)$, the response of the system pressure to changes in hydraulic motor velocity. Thus a system consisting only of the electric motor, pump, and compliant volume, V_{ms}, is considered. The undamped natural frequency, ω_{np}, of this system is closely approximated by

$$(\omega_{np})^2 = \frac{[(K_t C_s/\mu) + D_p(1 + C_f)]D_p\beta}{J_m V_{msi}} \tag{17.107}$$

The various factors determining the damping coefficient are contained in the terms associated with $d(\Delta P_s)dt$. Thus, hydraulic-pump leakage, throttle-valve flow, electric-motor torque characteristics, and pump drag enter prominently.

17.442. Downstream-Fluid, Motor, and Load System. The downstream side of the fluid motor with the pressure P_s held constant represents a second-order system with the load inertia, m_l, acting against a hydraulic spring. If load damping is ignored, combining Eqs. 17.104 and 17.105 results in

$$\frac{V_{m1}m_l}{\beta A_r} \cdot \frac{d^3z}{dt^3} + \frac{C_q A_t m_l}{2\sqrt{P_{1i}}\,A_r} \cdot \frac{d^2z}{dt^2} + A_r\frac{dz}{dt} = A_v\frac{dx}{dt} \tag{17.108}$$

$$(\omega_{nm})^2 = \frac{A_r{}^2\beta}{V_{m1i}m_l} \tag{17.109}$$

17.443. Spool-Valve and Spring System. The third important subsystem is the spool valve and its spring. From Eq. 17.103, the undamped natural frequency, ω_{nv}, is given by

$$(\omega_{nv})^2 = \frac{(k_v + C_v w P_{si})}{m_v} \tag{17.110}$$

Equations 17.101 through 17.105 are represented in operational block-diagram form as shown in Fig. 17.24 by applying conventional techniques to the system of equations.[16]

17.45. Steady-State Considerations

17.451. Specifications. The stage is now set for the study of a specific system. A machine-tool application with the following specifications is used as an example:

Feed rate (max)	50 ft/min
Cutting force	14,000 lb
Supply pressure (max)	1,000 psi
Weight of table plus work	20,000 lb

[16] *A Palimpsest on the Electronic Analog Art*, H. M. Paynter, ed., G. A. Philbrick Researches, Inc., Boston, 1955.

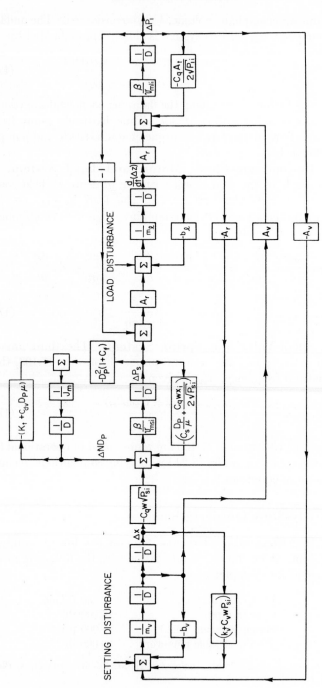

Fig. 17.24. Complete block diagram, linearized system.

17.452. Maximum Steady Values of System Variables.   If the maximum pressure drop across the fluid motor is to be 750 psi and the downstream motor pressure is to be 250 psi, the following system parameters result:

$$A_r = 18.7 \text{ in.}^2$$

$$Q_{m(\max)} = 48.5 \text{ gpm}$$

$$Q_s = 55 \text{ gpm at 1000 psi (commercially available vane pump)}$$

$$D_v = \tfrac{1}{2} \text{ in. (valve-spool diameter)}$$

$$A_v = 0.196 \text{ in.}^2 \text{ (pressure-sensitive area of control valve)}$$

$$Q_v = 6.5 \text{ gpm at 1000 psi}$$

$$x = 0.0052 \text{ in.}$$

$$k = 1960 \text{ lb/in. (if 10 psi causes 0.001-in. motion)}$$

$$A_{t(\max)} = 0.123 \text{ in.}^2$$

$$D_{t(\max)} = 0.397 \text{ in.}$$

$$D_p = 1.84 \text{ in.}^3/\text{radian}$$

17.453. Steady-State Load Sensitivity.   The steady-state load sensitivity of the control system is the deviation of the steady load velocity from a desired value as the load force is changed to a new steady value.   The steady-state-system equations are as follows.

For the downstream side of the motor

$$\frac{dz}{dt} A_r = C_q A_t \sqrt{P_1} \tag{17.111}$$

For the upstream side of the motor

$$\left(Q_{si} - C_s \frac{D_p P_s}{\mu} \right) - C_q w x \sqrt{P_s} = \frac{dz}{dt} A_r \tag{17.112}$$

The forces on the valve spool are

$$P_1 A_v = k_v(x + x_s) + C_v P_s x w \tag{17.113}$$

Since the flow-induced-force spring stiffness is in this instance small compared to the mechanical spring stiffness, it is neglected.   Thus,

$$P_1 A_v = k_v(x + x_s) \tag{17.114}$$

Eliminating P_1 and x yields

$$A_r \frac{dz}{dt} = \left(\frac{C_q A_t^2 k_v}{2w\sqrt{P_s}\, A_v} \right)$$

$$\times \left[\sqrt{1 + \frac{4w\sqrt{P_s}\, A_v(Q_{si})}{C_q A_t^2 k_v} - \left(\frac{4wP_s^{3/2} A_v}{C_q A_t^2 k_v} \right) \left(\frac{C_s D_p}{\mu} \right) + \frac{4w^2 P_s A_v x_s}{A_t^2 k_v}} - 1 \right]$$

$$(17.115)$$

The substitution of appropriate values results in the values shown in Table 17.3.

Table 17.3. STEADY-STATE REGULATION OF SYSTEM

Pump Pressure (psi)	High Feed Rate ($A_t = 0.123$ in.²) (ft/min)	Low Feed Rate ($A_t = 0.025$ in.²) (ft/min)
1000	50.7	14.7
500	51.5	16.0
300	52.1	17.6

Adjustments in the feed rate are obtained by changing the size of the metering orifice as shown. Since the system attempts to maintain a constant downstream pressure, a decrease in the area of the metering orifice causes a decrease in feed rate, with a larger portion of the pump flow being by-passed through the control valve.

17.46. Analogue Computer Study of Linearized System

The complete block diagram of the linearized system is given in Fig. 17.24. The characteristic frequencies as previously derived, Eqs. 17.107, 17.109, and 17.110, are next calculated to give an order of magnitude of the speed of response of the various subsystems. For the maximum pressure, P_1, the flow-force spring stiffness becomes

$$C_v w P_{s\ \max} = 182.5 \text{ lb/in.}$$

which is only 10 per cent of the mechanical spring stiffness.

The volume under compression on each side of the ram in midposition is here assumed to be

$$V_{ms} = V_{m1} = \frac{V_t}{2} = 1350 \text{ in.}^3$$

corresponding to 72 in. of travel from center.

The commercially available electric motor is found to have a torque constant of

$$K_t = 59 \text{ in. lb sec}$$

with a rotating inertia of

$$J_m = 10.0 \text{ in. lb sec}^2$$

The pump inertia is negligible by comparison. The mass to be moved, a 20,000-lb weight, becomes

$$m_l = 51.8 \text{ lb sec}^2/\text{in.}$$

The hydraulic fluid in the system is taken as D.T.E. light hydraulic oil at 100°F. Its absolute viscosity at 100°F is

$$\mu = 4.3 \times 10^{-6} \text{ lb sec/in.}^2$$

The vane pump has a displacement of

$$D_p = 1.84 \text{ in.}^3/\text{radian}$$

and performance coefficients (fn. 15) of

$$C_s = 0.5 \times 10^{-7}$$

$$C_f = 0.212$$

$$C_{dv} = 7.3 \times 10^4$$

The electric-motor-pump-hydraulic-fluid system has an undamped natural frequency given by the following approximation to Eq. 17.107:

$$\omega_{np} \approx \sqrt{D_p{}^2\beta/J_m V_{msi}} \qquad (17.116)$$

After substitution of values for the ram motor near its center position, we get

$$2\pi\omega_{np} \approx 1.26 \text{ cps}$$

The load inertia acting against the fluid spring has an undamped natural frequency of

$$\omega_{nm} = \sqrt{A_r{}^2\beta/V_{mli} m_l} \qquad (17.117)$$

Substitution of values gives

$$\omega_{nm}/2\pi = 5.63 \text{ cps}$$

The control-valve system has an undamped natural frequency of

$$\omega_{nv} = \sqrt{(k_v + C_v w P_{si})/m_v} \qquad (17.118)$$

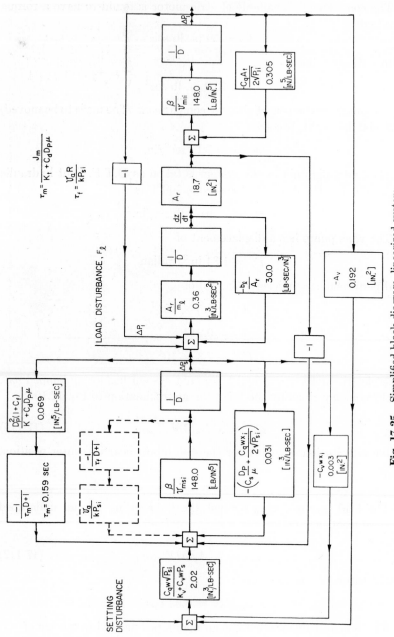

Fig. 17.25. Simplified block diagram, linearized system.

(*a*) Step change of cutting force (times 1).

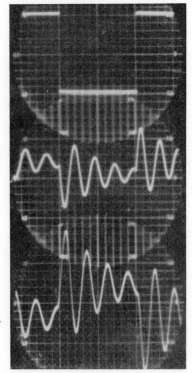

(*b*) Velocity response of load (times 1).

(*c*) Spool-valve position response (times 10).

Fig. 17.26. Step response of system. $V_{msi} = V_{m1i} = 0.5V_t$.

The following system characteristics are used throughout in the analogue studies except as noted:

$$P_{si} = 1000 \text{ psi}$$
$$F_{li} = 14{,}000 \text{ lb}$$
$$\left(\frac{dz}{dt}\right)_i = 10 \text{ in./sec} = 50 \text{ ft/min}$$
$$V_{msi} = V_{m1i} = 1350 \text{ in.}^3$$
$$W = 20{,}000 \text{ lb}; \quad m_l = 51.8 \text{ lb sec}^2/\text{in.}$$
$$A_r = 18.7 \text{ in.}^2$$
$$A_v = 0.192 \text{ in.}^2$$
$$\frac{C_s D_p}{\mu} = 0.032 \text{ in.}^5/\text{lb sec}$$
$$\tau_m = 0.159 \text{ sec}$$

Time scale for Figs. 17.26 through 17.36: Computing interval (duration of step) corresponds to 0.478 sec.

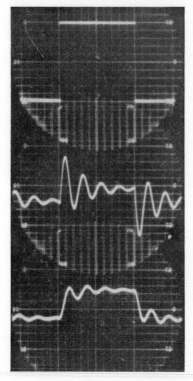

(a) Step change in cutting force (times 1).

(b) Velocity response (times 1).

(c) Spool-valve position response (times 10).

Fig. 17.27. Step response of system. $V_{msi} = 0.9V_t$; $V_{m1i} = 0.1V_t$.

Taking $m = 4.4 \times 10^{-4}$ lb sec^2/in. and substituting values gives

$$\omega_{nv}/2\pi = 336 \text{ cps}$$

Because the speed of response of the control-valve subsystem is so fast compared with the other subsystems, the simplified block diagram of Fig. 17.25 may be used instead of Fig. 17.24. Figure 17.25 includes only the steady-state characteristics of the control-valve subsystem.

The final aim of the analogue computer study is to design a dynamically stable system for all operating conditions. The equipment used in this study was manufactured and marketed by George A. Philbrick Researches, Inc., of Boston, Massachusetts. The equipment was of the high-speed repetitive display type with an integrator time constant of 0.7 millisecond. Although nonlinear computing elements are available to handle the type of block diagram in Fig. 17.23, it was decided to use the linearized approach. In contrast to the nonlinear problem, the linearized problem presents far fewer technical difficulties that might

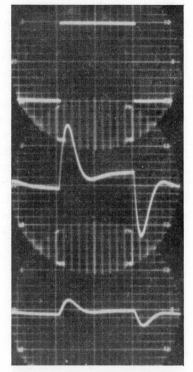

(a) Step change in cutting force (times 1).

(b) Velocity response (times 1).

(c) Spool-valve position response (times 1).

Fig. 17.28.   Step response of system. $V_{msi} = 0.1V_t$; $V_{mli} = 0.9V_t$.

discourage experimentation on the analogue computer.   However, the steady-state part of the problem, whether it be linear or nonlinear, can be handled more accurately by purely analytical means.   The nonlinearities encountered are sufficiently moderate to put confidence in the results from a linearized analysis.   Such drastic nonlinearities as backlash, deadtime, and large static-friction forces cannot be treated adequately by linearization, and recourse to a nonlinear analogue study is usually necessary.

The system operating at maximum feed rate, maximum load, and a maximum cutting force is now studied.   The load-velocity response to a sudden, small decrease in cutting force is shown in Fig. 17.26b.   Figure 17.26a shows the step change in cutting force; Fig. 17.26c, the change in spool-valve position.   The response is seen to be markedly oscillatory but stable.   Before an examination of design changes leading to a less oscillatory response, some other operating conditions of this system will be studied to give added familiarity with the problem.

Next, operation at either end of the motor travel is investigated. Since $V_t = V_{m1i} + V_{msi}$ is constant and since some volume remains at the end of each stroke, the case when $V_{m1i} = 0.1V_t$ and $V_{msi} = 0.9V_t$ is illustrated in Fig. 17.27. The system is more stable for this condition. Similarly, in Fig. 17.28 the ram is at the other end of its travel, with $V_{m1i} = 0.9V_t$ and $V_{msi} = 0.1V_t$, all other conditions remaining unchanged. The high degree of stability of these conditions indicates that the critical case occurs with the ram in the central position. The large oscillations of this system are undesirable, and some changes leading to a more stable system are called for. A reduction of 50 per cent in the pressure-sensitive area of the spool valve results in the performance indicated in Fig. 17.29. The steady-state regulation of this new system is, of course, not so good as that of the previous system because the pressure sensitivity of the control valve is reduced. An indication of this poorer regulation is demonstrated by Table 17.4.

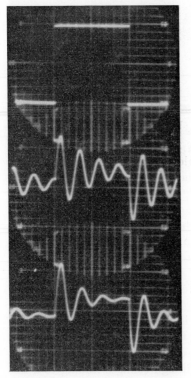

(a) Step change in cutting force.

(b) Velocity response: $A_v = 0.192$ in.^{2}

(c) Velocity response: $A_v = 0.096$ in.^{2}

Fig. 17.29. Step response of system. $V_{msi} = V_{m1i} = 0.5V_t$ and A_v reduced by 50 per cent.

Table 17.4

$(dz/dt)_{ss}$ ft/min

P_s, psi	Spool Valve Area$_v$ = 0.196 in.^{2}	Spool Valve Area$_v$ = 0.098 in.^{2}
1000	50.7	51.4
300	52.1	52.8

The sensitivity of dynamic response of this system to pump leakage is indicated in Fig. 17.30. A pump with 90 per cent volumetric efficiency is shown in Fig. 17.30a; Fig. 17.30b is for 80 per cent volumetric efficiency; Fig. 17.30c is for a volumetric efficiency of 70 per cent. The remainder of the studies are with a pump of volumetric efficiency of 90 per cent. Some of the other operating variables now can be investigated.

(a) $C_s D_p/\mu = 0.032$.

(b) $C_s D_p/\mu = 0.064$.

(c) $C_s D_p/\mu = 0.096$.

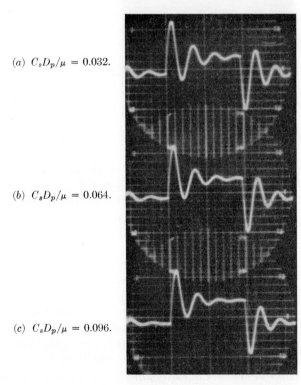

Fig. 17.30. Effect of pump leakage on velocity response to a cutting-force step change.

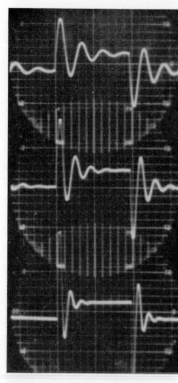

(a) $W = 20{,}000$ lb.

(b) $W = 10{,}000$ lb.

(c) $W = 5000$ lb.

Fig. 17.31. Effect of table weight on velocity response to a cutting-force step change.

Figure 17.31 shows the effect of changing the table weight. In Fig. 17.31a, the weight is 20,000 lb; in Fig. 17.31b, it is 10,000 lb; and in Fig. 17.31c, it is 5000 lb. Figure 17.32 shows the effect of reducing the initial magnitude of the cutting force. In Fig. 17.32a, the initial cutting force is 14,000 lb; in Fig. 17.32b, it is 7000 lb; in Fig. 17.32c, it is 3500 lb.

Changes in the feed rate are studied in Fig. 17.33. Figure 17.33a represents the response to a small step change in cutting force with an initial feed rate of 50 ft/min, Fig. 17.33b with 25 ft/min, and Fig. 17.33c with 12.5 ft/min. The oscillatory character of a condition of low feed rate presents a new problem, and some means must be found to stabilize the system further without undue loss in feed-rate regulation.

The assumed load damping of 500 lb sec/in. is open to question. The effect of changes in load damping on the velocity response to a step disturbance in cutting force is shown in Fig. 17.34. Since the lowest load

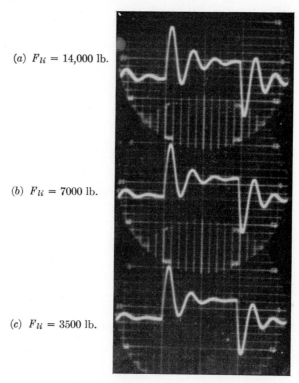

(a) $F_{li} = 14,000$ lb.

(b) $F_{li} = 7000$ lb.

(c) $F_{li} = 3500$ lb.

Fig. 17.32. Effect of initial magnitude of cutting force on the velocity response to a cutting-force step change.

damping of 166 lb sec/in. is entirely possible, the oscillatory nature of the response calls for additional stabilizing means.

17.47. Transient-Flow Stabilization

The transient-flow damper [17] consists of an air-oil accumulator which is connected to the upstream side of the motor through a capillary or orifice-type resistance. The function of the transient-flow stabilizer, shown dotted in Fig. 17.21, is to reduce large dynamic-pressure peaks without changing the steady-state performance of the feed system.

[17] J. L. Shearer, *Continuous Control of Motion with Compressed Air*, Sc.D. Thesis, Department of Mechanical Engineering, Massachusetts Institute of Technology, Cambridge, Mass., 1954.
Also see Chapter 16 of this book.

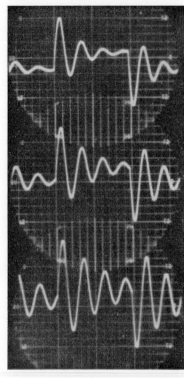

(a) 50 ft/min.

(b) 25 ft/min.

(c) 12.5 ft/min.

Fig. 17.33. Effect of initial magnitude of feed rate on the velocity response to a cutting-force step change.

The flow across the capillary orifice is given by

$$Q_a = (P_s - P_a) \frac{1}{R_c} \qquad (17.119)$$

where Q_a = flow into transient-flow stabilizer, in.^{3}/sec
P_a = pressure in transient-flow stabilizer, psi
R_c = capillary resistance of transient-flow stabilizer, lb sec/in.^{5}

The rate of pressure rise in the stabilizer chamber is expressed by

$$Q_a = \frac{V_{ai}}{kP_{si}} \cdot \frac{dP_a}{dt} \qquad (17.120)$$

where V_{ai} = initial volume in air bladder of transient-flow stabilizer, in.^{3}

(a) $b_l = 500$ lb sec/in.

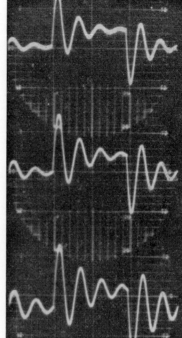

(b) $b_l = 330$ lb sec/in.

(c) $b_l = 166$ lb sec/in.

Fig. 17.34. Effect of load damping on the velocity response to a cutting-force step change.

The continuity equation, Eq. 17.87, is now changed into the form

$$Q_s = Q_a + C_q wx\sqrt{P_s} + \frac{V_{ms}}{\beta}\frac{dP_s}{dt} + A_r\frac{dz}{dt} \qquad (17.121)$$

Combining Eqs. 17.119 and 17.120, solving for Q_a as a function of P_s and t, and then substituting into Eq. 17.121 results in

$$\Delta Q_s = \left(\frac{\dfrac{V_{ai}}{kP_{si}}}{\dfrac{V_{ai}R_c}{kP_{si}}D+1} + \frac{V_{ms}}{\beta}\right)(\Delta P_s) + C_q wx\sqrt{P_s}\,(\Delta x) + A_r\,D(\Delta z) \qquad (17.122)$$

where D denotes derivative with respect to time. Since the volume of the air bladder is inversely proportional to the pressure, under quiescent

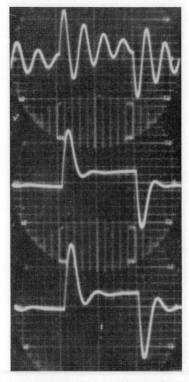

(a) Velocity response to a cutting-force step input of original high-gain system.

(b) Velocity response with transient-flow stabilizer introduced. High load damping ($b_l = 500$ lb sec/in.)

(c) Velocity response with transient-flow stabilizer. Low load damping ($b_l = 166$ lb sec/in.)

Fig. 17.35. Effect of introducing the transient-flow stabilizer into the upstream side of the motor.

conditions, Eq. 17.122 can be written in the form

$$\Delta Q_s = \left(\frac{\dfrac{C}{kP_{si}^2}}{\dfrac{CR_c}{kP_{si}^2}D + 1} + \frac{V_{ms}}{\beta} \right)(\Delta P_s) + C_q wx\sqrt{P_s}\,(\Delta x) + A_r\,D(\Delta z)$$

$$(17.123)$$

where $C = V_{ai}P_{si}$. The addition of the transient-flow stabilizer to the system is represented in dotted lines on the block diagram of Fig. 17.25. An additional feedback loop from the rate-of-change-of-pressure signal to the flow summing point consists of a coefficient and a simple lag.

A system is now studied which has the original pump leakage (C_sD_p/μ), load damping b_l, and the highest cutting force of 14,000 lb at a high velocity of 50 ft/min with a weight of 20,000 lb. Its response was origi-

(a) 50 ft/min.

(b) 25 ft/min.

(c) 12.5 ft/min.

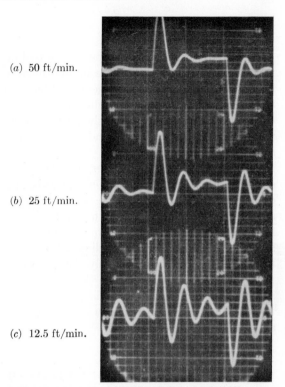

Fig. 17.36. Effect of feed rate on the velocity response to a cutting-force step input with the transient-flow stabilizer. This system has low load damping ($b_l = 166$ lb sec/in.).

nally shown in Fig. 17.26b and is now repeated in Fig. 17.35a. A transient-flow stabilizer, consisting of an air bladder of 5.2 in.^{3} at a supply pressure of 1000 psi and a capillary resistance which consists of a tube with 6.8-in. length and 0.068-in. diameter, will give the improved transient response shown in Fig. 17.35b. If the load damping is reduced from a high of 500 lb sec/in. to a low of 166 lb sec/in., the response shown in Fig. 17.35c is still acceptable.

In Fig. 17.33, the lowering of the feed rate resulted in a less stable system. Figure 17.36 shows the effect of lowered feed rate on the stabilized system. At the very low feed rates, the system again becomes oscillatory. However, a doubling of the volume of the transient-flow stabilizer would stabilize the system markedly. Therefore, the simultaneous adjustment of accumulator volume and feed rate could be used to attain optimum stability.

17.48. Conclusions

The use of the analogue computer as a tool in the design and development of hydraulic equipment is unquestionably established. The analogue computer ought to become as familiar a piece of equipment as the slide rule. The design and development engineer should have free access to a computer. Moreover, he should be charged with the setting of the problem into the computer and the subsequent "designing with the analogue."

A word of warning is, however, in order. The analogue computer will solve only the problem that the engineer visualizes. If certain important distributed or lumped effects are neglected, the answers from an analogue study can be embarrassingly misleading. If, however, a reasonable facsimile of the physical system is studied by the analogue approach, an excellent physical understanding of the apparatus will be gained. The results will reduce immeasurably the experimental time needed for the prototype-development phase. An important aspect of the use of the analogue computer is the challenge to experiment analytically, to attempt to outguess the computer, and thereby to arrive at a solution to an otherwise difficult problem.

17.5. NONLINEAR ANALOGUE STUDY OF A HIGH-PRESSURE PNEUMATIC SERVOMECHANISM [18]

17.51. Introduction

Hydraulic fluids have been used widely as the working medium in systems developed to control the motion of mass loads. Recent development of pneumatic control systems [19, 20] has demonstrated that compressed gases can be used effectively for the continuous control of motion of mass loads.

[18] J. L. Shearer, "Nonlinear Analog Study of a High-Pressure Servomechanism," *Trans. ASME*, Vol. 79 (April 1957), pp. 465–472.

[19] J. L. Shearer, "Study of Pneumatic Processes in the Continuous Control of Motion with Compressed Air," Parts I and II, *Trans. ASME*, Vol. 78 (February 1956), pp. 233–249 (this article is presented in somewhat condensed form in Chapter 16); and J. L. Shearer, *Continuous Control of Motion with Compressed Air*, Sc.D. Thesis, Department of Mechanical Engineering, Massachusetts Institute of Technology, Cambridge, Mass., 1954.

[20] H. Levenstein "Tie Simplicity to Power with Pneumatic Servomechanisms," *Control Engineering*, Vol. 2, No. 6 (June 1955), pp. 65–70.

The schematic diagram, Fig. 17.37, shows a positioning servomechanism employing a valve-controlled pneumatic servomotor to drive a mass load and external load force with a simple feedback mechanism to sense output position, Y, and to stroke the valve in such a manner that the

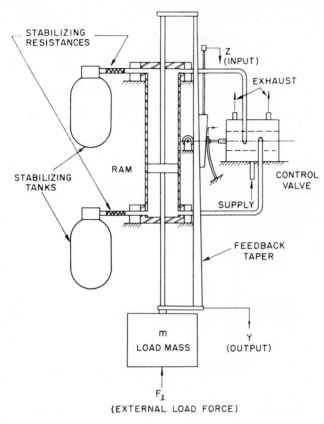

Fig. 17.37. Schematic diagram of pneumatic positioning servomechanism.

pneumatic servomotor will attempt to provide changes in Y that are proportional to a low-energy-level input motion, Z. Chapter 16 discussed the pneumatic processes involved in the operation of a system like that of Fig. 17.37, and a linearized analysis was employed to demonstrate how stabilizing resistances and stabilizing tanks could be used effectively to provide damping in the servomotor.

When the complete servomechanism was studied experimentally in the laboratory by means of frequency-response measurements,[17] the output motion was not sinusoidal at all frequencies and the measured amplitude and phase characteristics were somewhat different from those computed from a linearized analysis. In particular, the ram came to rest when its direction of motion changed in the way shown qualitatively in Fig. 17.38a when the input was varied sinusoidally. This dwell in

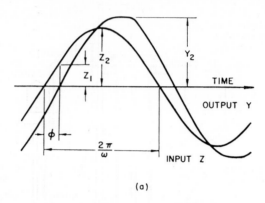

(a)

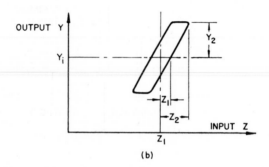

(b)

Fig. 17.38. Sketches of input and output wave forms.

(a) Input, Z, and output, Y, as functions of time.
(b) Output, Y, as function of sinusoidally varying input, Z.

output motion, which was observed with a fairly wide range of input frequencies, appears more distinctly when the output is plotted against the input sinusoid, as shown in Fig. 17.38b. A small amount of Coulomb friction was found from static measurements of the ram, and the dwell in output motion tentatively was attributed to this Coulomb friction. It was apparent, however, that either an exhaustive experimental study or a detailed analogue study would be required to show quantitatively how the various known nonlinearities in the system contributed to system performance. Estimates indicated that a thorough experimental study, in which many important system parameters were changed, would be much more costly than an analogue study. Although a digital computer was employed in a check solution intended to ensure the validity of the analogue results, the long computing time and limited information-storage facilities then available made a thorough digital-computer study inadvisable. All of the required analogue computer components were available in the Generalized Computer of the Dynamic Analysis and Control Laboratory (DACL) of the Massachusetts Institute of Technology.

17.52. Basic Equations and Functional Relationships

17.521. Valve. The characteristics of the control valve are discussed in detail in Chapter 7. Figure 17.39 is a graphical representation of the relationship between the weight rate of flow toward one end of the ram, W_a, valve position, X, and ram-chamber pressure, P_a. The flow toward the other end of the ram, W_b, is related to valve position, X, and the other ram-chamber pressure, P_b, by an identical set of curves with only the sign of the valve position changed. In other words,

$$W_a = f_v(X, P_a) \tag{17.124}$$

$$W_b = f_v(-X, P_b) \tag{17.125}$$

where f_v denotes the functional relationship given by the curves.

17.522. Ram Chambers. Applying the energy equation to the ram chambers, as shown in Fig. 17.40, gives

$$(W_a - W_{ta})T_s - \frac{gP_a}{C_p}\frac{dV_a}{dt} = \frac{g}{kR}\frac{d}{dt}(P_a V_a) \tag{17.126}$$

$$(W_b - W_{tb})T_s - \frac{gP_b}{C_p}\frac{dV_b}{dt} = \frac{g}{kR}\frac{d}{dt}(P_b V_b) \tag{17.127}$$

where W_{ta} = weight rate of flow into tank at a end, lb/sec
   W_{tb} = weight rate of flow into tank at b end, lb/sec
   T_s = temperature of gas supply, °R
   g = acceleration due to gravity, 386 in./sec^2
   C_p = specific heat of gas at constant pressure, in.^{2}/sec^2 °R
   V_a = volume of chamber in a end of ram, in.^{3}
   V_b = volume of chamber in b end of ram, in.^{3}
   t = time, sec
   k = ratio of specific heat at constant pressure to specific heat at constant volume, for air = 1.4
   R = gas constant for air = 2.47 × 10^5 in.^{2}/sec^2 °R

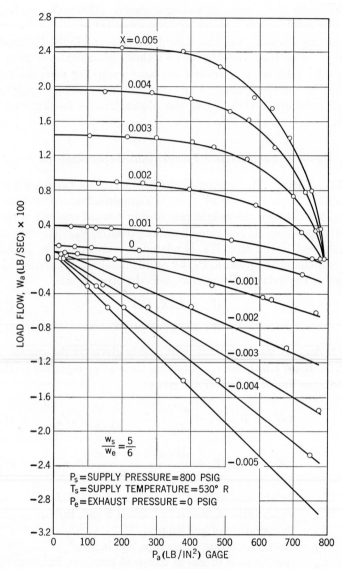

Fig. 17.39. Measured pressure-flow characteristics of nominal closed-center sliding-plate valve.

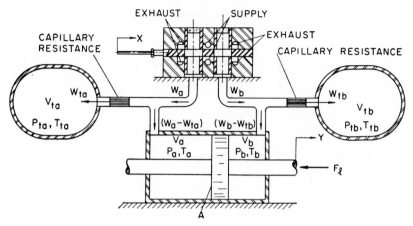

Fig. 17.40.   Schematic diagram showing important variables in pneumatic system.

17.523. Stabilizing Resistances. The flow through the stabilizing resistances is obtained by using the expression for flow of a gas through a capillary resistance.

$$W_{ta} = \frac{gC_c}{2RT_s} (P_a{}^2 - P_{ta}{}^2) \qquad (17.128)$$

$$W_{tb} = \frac{gC_c}{2RT_s} (P_b{}^2 - P_{tb}{}^2) \qquad (17.129)$$

where C_c = capillary-resistance coefficient, in.^{5}/lb sec
P_{ta} = pressure in tank at a end, psi
P_{tb} = pressure in tank at b end, psi

Reasonably good approximations for Eqs. 17.128 and 17.129 are given by

$$W_{ta} = \frac{gC_cP_a}{RT_s} (P_a - P_{ta}) \qquad (17.130)$$

$$W_{tb} = \frac{gC_cP_b}{RT_s} (P_b - P_{tb}) \qquad (17.131)$$

17.524. Stabilizing Tanks. Applying the energy equation to the stabilizing tanks gives

$$W_{ta}T_s = \frac{gV_{ta}}{kR} \frac{d}{dt} (P_{ta}) \qquad (17.132)$$

$$W_{tb}T_s = \frac{gV_{tb}}{kR} \frac{d}{dt} (P_{tb}) \qquad (17.133)$$

17.525. Ram and Mass Load. Application of Newton's second law to the ram and mass load yields

$$(P_a - P_b)A = m\frac{d^2Y}{dt^2} + F_l + F_c \tag{17.134}$$

where A = ram area, in.^{2}
m = load mass, lb sec^2/in.
Y = ram position, in., zero when ram is in center of cylinder
F_l = external load force, lb
F_c = Coulomb-friction force, lb

17.526. Ram Volumes. The ram volumes V_a and V_b are related to Y by

$$V_a = V_i + AY \tag{17.135}$$

$$V_b = V_i - AY \tag{17.136}$$

where Y is measured from the point where $V_a = V_b = V_i$.

17.527. Feedback. The valve motion, X, is related to the input motion, Z, and the output motion, Y, by

$$X = k_i Z - k_f Y \tag{17.137}$$

where k_i = input cam slope, in./in.
Z = position of input cam, in.
k_f = feedback cam slope, in./in.

17.528. Coulomb Friction. The Coulomb-friction force is constant (independent of ram velocity) when the ram is moving, and when the ram is motionless, this friction force is equal to the sum of all other forces acting on the ram.

$$F_c = \frac{dY/dt}{|dY/dt|} F_{cl} \cdots \left[\frac{dY}{dt} > 0\right] \tag{17.138a}$$

$$F_c = (P_a - P_b)A - F_l \cdots \left[\frac{dY}{dt} = 0\right] \tag{17.138b}$$

where F_{cl} = limiting (maximum) value of F_c, lb.

17.53. Description of Analogue System

17.531. Block Diagram of Complete System. The block diagram shown in Fig. 17.41 represents the complete set of equations describing the system and gives an over-all picture of the operation of the system.

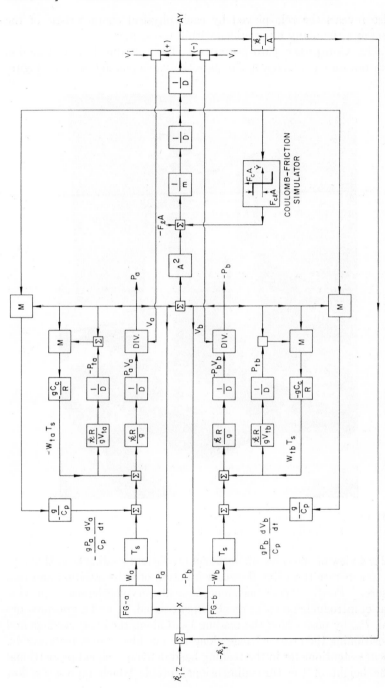

Fig. 17.41. Block diagram of complete pneumatic servomechanism. (M denotes multiplier; DIV denotes divider; FG denotes function generator; 1/D denotes integration with respect to time; Σ denotes summation.)

It also reveals the role played by each physical characteristic of the system and the many interactions within the system.

17.532. Computer Components. Two of the DACL function generators were employed to simulate the valve characteristics. Figure

Fig. 17.42. View of function-generator setup to simulate a end of control valve and to perform the multiplications $P_a \cdot dV_a/dt$ and $P_a \cdot (P_a - P_{ta})$.

17.42 is a view of one three-dimensional profile and reading head that was used to generate the valve flow as a function of valve position and ram pressure. Position-type electromechanical servomechanisms provide means of introducing a valve-displacement signal, X, and a ram-pressure signal, P_a, by positioning the reading head along the cross carriage and by positioning the cross carriage along the machine frame, respectively. A linear potentiometer in the reading head delivers a signal proportional to the height of the three-dimensional profile which represents flow

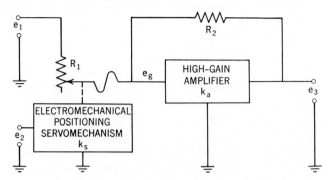

Fig. 17.43. Schematic diagram of electromechanical system used to divide e_1 by e_2.
$e_3 = -(R_1/k_s)(e_1/e_2)$ when $(R_2 + R_1)/k_a = R_1$.

Fig. 17.44. Over-all view of analogue computer equipment.

toward the ram. Also shown is a constant-slope profile and accompanying reading head clamped to the cross carriage. This reading head contains two potentiometers; one is excited with a voltage representing dV_a/dt, and the other is excited with a voltage representing $(P_a - P_{ta})$. Thus the voltages picked up by the two potentiometer wipers represent P_a times dV_a/dt and P_a times $(P_a - P_{ta})$, respectively. A similar arrangement simulates the situation at the other end of the ram and valve.

Each of the quotients $P_a V_a/V_a$ and $-P_b P_b/V_b$ is obtained with an electromechanical dividing system like that shown in Fig. 17.43.

Conventional summing, coefficient, and integrating circuits are used with chopper-stabilized d-c amplifiers to perform these necessary operations. Coulomb friction was simulated with a high-gain amplifier and precision limiter. Figure 17.44 is a view of the complete analogue setup. Each system variable is represented by an electric signal having a voltage proportional to the value of the variable. The time scale used in the greater part of this study was 8 to 1; in other words, the integrators were 8 times slower than real-time integration.

17.54. Results of Dynamic System Studies

The system studied in greatest detail with the analogue was the one which had been measured in the laboratory. The control-valve characteristics are shown in Fig. 17.39; supply pressure, P_s, was 800 psig; supply temperature, T_s, was 530°R; ram area was 4.34 in.²; half ram volume, V_i, was 30 in.³; stabilizing-tank volume, V_t, was 130 in.³; stabilizing-resistance coefficient, C_c, was 4.46 in.⁵/lb sec; maximum Coulomb-friction force, F_{cl}, was 13.0 lb; load mass, m, weighed 70 lb;

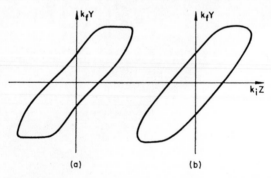

(a)   (b)

Fig. 17.45. Comparison of frequency-response test of experimental system with simulated test on analogue.

    (a) Y versus Z of analogue at 1.0 cps.
    (b) Y versus Z of experimental system at 1.0 cps.

feedback taper, k_f, was 0.022 in./in.; and input taper, k_i, was 0.25 in./in. The result of an analogue simulation is compared in Fig. 17.45 to the measured response of the experimental system when the input, Z, was varying sinusoidally with an amplitude of ±0.050 in.

Several series of steady-state frequency-response tests were simulated on the analogue in order to determine the differences between the analogue system and the real system, and to observe the effects of varying the input amplitude, the Coulomb friction in the ram, and the operating

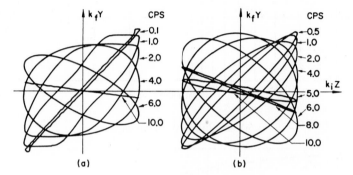

Fig. 17.46.   Two series of tests simulated with analogue computer.

(*a*) With Coulomb friction.
(*b*) Without Coulomb friction.

position of the ram.   Figure 17.46 illustrates two of the series of tests that were simulated on the analogue with the ram operating near its center position.   The series of Fig. 17.46*a* was made with the measured Coulomb friction set into the analogue, and the series of Fig. 17.46*b* was made with zero Coulomb friction set into the analogue.   This shows clearly that the dwell in output motion, which was discussed earlier, is caused by Coulomb friction.   Each series required only a few minutes to run off.   Other tests that were simulated on the analogue included frequency-response measurements with the Coulomb friction doubled and with the input amplitude reduced from ±0.050 in. to ±0.025 in. The results of all of these tests appear in Fig. 17.47, where the response computed from a linearized analysis [19] is given also for purposes of comparison.   The results from tests of the experimental system are shown in Fig. 17.48.   The greatest discrepancy between the analogue simulation and the experimental system itself appears in the frequency range of 3 to 5 cps.

The results of series of frequency-response tests that were simulated for the ram operating near positions 2 in., 4 in., and 6 in. from its center

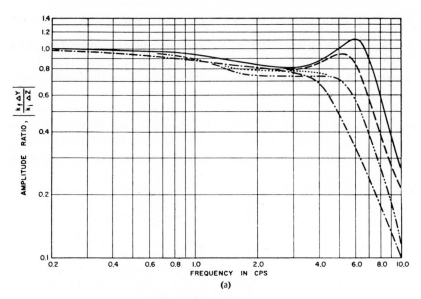

(a) Amplitude versus frequency.

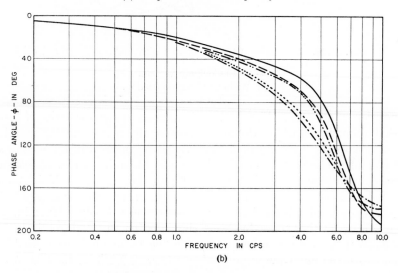

(b) Phase angle versus frequency.

Fig. 17.47. Results of several series of steady-state frequency-response tests simulated with analogue computer (broken lines) compared to response computed from linearized analysis (solid lines).

———————————Computed from linearized analysis.
———— ————Simulator, no friction, ±0.050-in. input.
················Simulator, twice-measured Coulomb friction, ±0.050-in. input.
—··—··—··—··—Simulator, measured Coulomb friction, ±0.050-in. input.
—·—·—·—·—··Simulator, measured Coulomb friction, ±0.025-in. input.

628

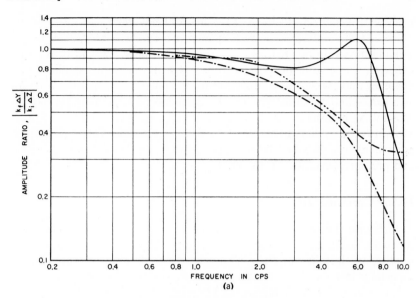

(a) Amplitude versus frequency.

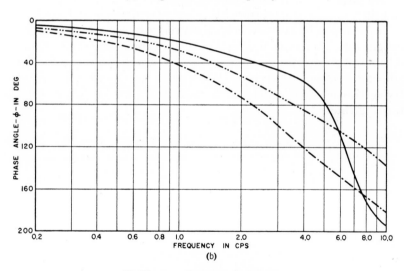

(b) Phase angle versus frequency.

Fig. 17.48. Results of two series of steady-state frequency-response tests on experimental system (broken lines) compared to response computed from linearized analysis (solid lines).

———————————Computed from linearized analysis.
—··—··—··—··—·Measured with ±0.050-in. input.
—·—·—·—·—·—·—·Measured with ±0.025-in. input.

position (maximum stroke = ±6.5 in.) did not vary perceptibly from those of tests made for the ram operating near the center position. It was observed, however, that during operation near one end of the ram cylinder, the major pressure variation occurred in the smallest ram chamber and attached stabilizing tank, thus demonstrating the need for two tanks and resistances. The magnitude of the pressure variations was always less than 100 psi, with the result that the control valve operated in a nearly linear fashion throughout the tests. Reasonably good agreement was obtained between pressure variations in the real system and pressure variations in the analogue.

17.55. Conclusions

Although the linearized analysis does not predict actual system performance accurately, it does provide an excellent means of gaining a qualitative insight into the dynamic characteristics of the system, especially in regard to the effectiveness of the stabilizing tanks and resistances. A fact predicted by the linearized analysis is that this system is very unstable when the stabilizers are shut off. Both the real system and the analogue demonstrated this fact very dramatically.

The analogue proved to be a highly effective tool in attaining optimum designs for a system of this kind having different sets of requirements. A number of such designs were completed in the space of a few hours after the analogue was set up. The cost in time, money, and effort of trying various modifications of the real system would have been many times greater.

18

T. E. Hoffman

Power Steering

18.1. INTRODUCTION

Vehicular power-assisted steering, known commonly as "power steering," has been in evidence for a considerable period of time, having been applied mainly to heavy trucks and earth-moving machinery until recent years. With the introduction of the low-pressure tire on modern passenger automobiles, turning resistance, especially during parking maneuvers, became prohibitively high unless steering ratios were increased proportionately. However, increasing steering ratios also increased the response time of steering to the danger point. Power steering is one solution to this problem because it provides the extra force needed for ease of handling and at the same time permits a low steering ratio as a safety measure.

The following features may be considered as most desirable when designing an automotive power steering system:

1. The system must have fast enough response and high enough output force to ensure safe and adequate operation in all normal driving and parking situations.

2. The system must be reversible; that is, the road reaction forces must be "felt" by the driver at the steering wheel.

3. The system must be simple and cheap to produce, but foolproof, and also must provide manual steering in the event of system failure.

4. The system must have a relatively low stand-by power drain.

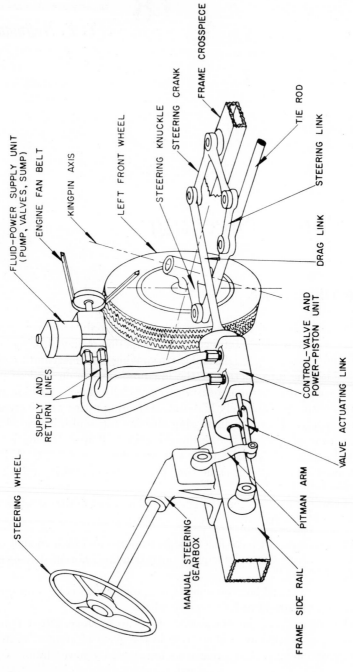

Fig. 18.1. Pictorial diagram of automotive power-assisted steering installation of the drag-link type.

Many types of systems have been proposed which comply with all or most of these requirements, but it will be sufficient for our present purpose to choose as an example a simple one of the type that replaces the drag link in the conventional manual steering system. A pictorial diagram of the entire steering system is shown in Fig. 18.1. As can be seen, the power steering unit consists of two main components: the fluid-power supply unit and the control-valve and power-piston unit. Since these two components are physically separated and are connected only by fluid supply and return lines, they may be designed independently provided their characteristics are suitably matched.

18.2. THE FLUID-POWER SOURCE

We shall consider first the design of the fluid-power source, enumerating and discussing the various factors which influence the design or selection of the individual components. A fluid-power source is usually composed of a pump with appropriate control valves plus a fluid sump. Presumably, the pump will be driven continuously by the automobile engine at some set ratio to the engine speed. Thus it must operate at speeds proportional to the range of 400 to 4000 rpm, the approximate limits of modern automobile engine speeds. Pump drives have been proposed which operate the pump over a proportionately smaller speed range or at constant speed, but they have usually been prohibitively expensive. Therefore, we shall assume that the pump will be operating continuously while the engine is running, at a speed proportional to the engine speed, although the demand for fluid power will be variable and/or intermittent. This demand will be greatest in the low engine-speed range since the steering-system force output must be highest during parking, because of the high twisting friction of stationary or slowly moving tires on the road surface when the engine is running at an idling speed, and also since the greatest output velocity will be required at low road speeds and thus at low engine speeds.

The input conditions and output requirements for the fluid-power source having thus been established, the choice of the pump and control valves will depend upon whether the system will operate at constant flow or at constant pressure. These factors and the desirable feature of low stand-by power drain indicate that a constant-flow system using a constant-displacement pump and a flow-control valve would be preferable to a constant-pressure system. The latter would require either a variable-displacement pump and a stroking control, which would be expensive, or a constant-displacement pump using a pressure-regulating valve, which would cause excessive power drain at engine speeds above idling. Several other advantages are obtained from constant-flow operation: the constantly circulating fluid quickly brings about temperature equilibrium in the several portions of the system; the open-center control valve required to pass the constant flow with small pressure drop also provides considerable load damping; and open-center valves have fewer close dimensional tolerances and, therefore, are less expensive.

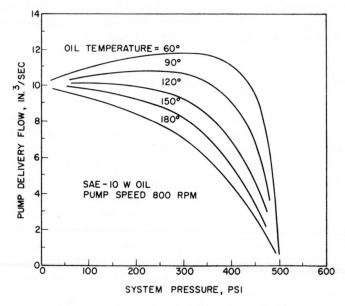

Fig. 18.2. Typical pump output characteristics.

The required value of the constant flow is determined by the area of the power piston and the velocity of the drag link during the maximum expected rate of turning of the front wheels. This maximum velocity is approximately 3 in./sec for a typical automotive steering system. Using a constant-displacement pump in a constant-flow system requires a flow-control valve which will bypass the excess portion of the pumped fluid when the engine speed is above idling. This flow-control valve, consisting of a spring-loaded plunger partially opening a return port with a calibrated bleed passage across the plunger, can be inexpensively and simply made. The analysis and detailed design of such a valve will not be covered here.

Any fluid-power source must also contain a relief valve to limit the maximum pressure of the system as a safety measure, a filter to prevent dirt and chips from circulating with the oil, and an adequate sump. The first two items are selected by routine consideration, while the selection of the last item, the sump, must take into account several factors. The important consideration in the specification of an adequate sump is that it have sufficient cooling capacity to keep the oil temperature below some specified value. If the oil temperature becomes too high, the oil viscosity will become so low that the internal leakage of the pump unit will limit the turning rate. The output characteristic curves of a typical fluid-power source of a 10 in.^{3}/sec flow with the relief valve set at 500 psi are given in Fig. 18.2. These curves indicate clearly the adverse effect of temperature on the delivery of usable fluid power. The factors to consider in specifying the sump are thus the volume of oil (the quantity of oil needed in a typical system is about 2 quarts), the viscosity and viscosity index of the oil, the cooling area available, the ambient temperature, which may be high in the engine compartment of an automobile, and the demand characteristics of the steering unit. The detailed design of the sump involving these factors will not be discussed further. The usual over-all packaging scheme is to combine all the components of the fluid-power source into one physical package with the flow-control valve and relief valve placed within the pump-body casting and with the sump, containing the filter, attached externally to the pump body, thus allowing internally manifolded flow paths and providing a neat compact package.

18.3. THE ACTUATOR UNIT

18.31. Preliminary Design

The design of the control-valve power-output unit depends mainly on the steering characteristics of the particular automobile in which the system is to be installed. Any automobile, by virtue of its weight distribution, kingpin inclination, and tire characteristics, will require a certain minimum torque about the axis of the kingpin to turn the front wheels in the static parking situation. As this condition determines the maximum output force required, the power-piston area can be specified for a given equivalent steering-arm radius and a given maximum system pressure. A typical power-piston area is 3 in.^{2} for a maximum system pressure of 500 psi and a 7.5-in. steering-arm radius. Here it should be stressed again that the high internal leakage caused by high oil temperature must be considered to ensure that the fluid-power supply will provide ample flow at the high pressure required for static turning to allow the maximum expected standstill turning rate. The maximum flow rate that the supply system must furnish will be required during

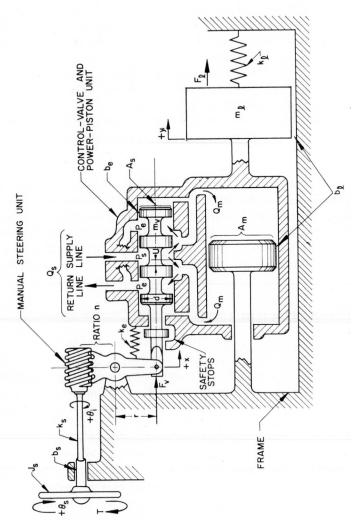

Fig. 18.3.   Schematic diagram of control-valve-power-piston unit.

maneuvers at low road speeds, as when turning corners, but these low-speed turns require much lower pressures so the effect of internal leakage in the supply unit is small.

After the required output ram size has been established, the control valve is considered. As previously mentioned, the use of a constant-flow fluid-power source requires an open-center valve. The flow resistance of this valve in the neutral position plus the flow resistance of the remaining parts of the system establishes the stand-by drain from the power source. The type of open-center valve chosen depends on the valve characteristics and the required dynamic performance as well as on the cost of manufacture. Because it does not require a bulky differential-type ram and has a larger linear operating region, a four-way valve is used in preference to a three-way valve. This four-way valve will be of the type which has all four orifices variable in order to utilize the full available pressure. The characteristics of such a valve are shown in Fig. 7.16. To arrive at design values for the remaining valve and system parameters, we must analyze the entire steering system on a dynamic basis, assuming an idealized constant-flow fluid-power source. The pictorial diagram of Fig. 18.1 is further simplified to the schematic diagram of Fig. 18.3 to establish nomenclature and geometrical relationships. The small passages connecting each of the two output ram chambers to the chambers on either side of the control-valve spool are included to produce "road feel." This is accomplished by the action of the output differential pressure on the valve-spool area, which offers a resistance to valve-spool motion (road feel) proportional to the force output of the system. The amount of proportional road feel is thus equal to the ratio of valve-spool area to power-piston area. This ratio must be kept small if the over-all force amplification (or gain) of the steering system is to be in a desirable range. The usual ratio is approximately 0.1.

18.32. The Basic Equations

In analyzing the system the following equations can be written for its several parts. From the characteristic equation of an underlapped four-way valve operating about the neutral position we get

$$Q_m = k_1(x - y) - k_2 P_m \qquad (18.1)$$

where Q_m = flow to power piston, in.^{3}/sec
  x = displacement from neutral of valve spool, in.
  y = displacement from neutral of power piston, in.
  P_m = pressure drop across power piston, psi

and the k's are given by

$$k_1 = \frac{Q_s}{U} = \text{flow sensitivity, in.}^2/\text{sec} \qquad (18.2)$$

where Q_s = constant supply flow, in.^{3}/sec
$\quad U$ = valve underlap, in.

and by

$$k_2 = \frac{Q_s}{2P_{s0}} = \frac{(C_0 A_v)^2}{Q_s}, \text{ in.}^5/\text{lb sec} \qquad (18.3)$$

where P_{s0} = supply pressure with valve centered, psi
$\quad C_0 = C_d\sqrt{2/\rho}$ = control-valve flow coefficient, in.^{2}/sec$\sqrt{\text{lb}}$
$\quad C_d$ = valve-orifice discharge coefficient
$\quad \rho$ = density of fluid, lb sec^2/in.^{4}
$\quad A_v$ = area of each valve metering orifice when $x = 0$, in.^{2}

We also define k_3 as

$$k_3 = \frac{k_1}{k_2} = \frac{2P_{s0}}{U} = \frac{Q_s{}^2}{U(C_0 A_v)^2} = \text{pressure sensitivity, lb/in.}^3 \quad (18.4)$$

By equating torques on the steering wheel, using the Newtonian dot notation for derivative with respect to time, we obtain

$$T = J_s\ddot{\theta}_s + b_s\dot{\theta}_s + k_s(\theta_s - \theta_i) \qquad (18.5)$$

where T = external torque applied to steering wheel, in. lb
$\quad J_s$ = moment of inertia of steering wheel, in. lb sec^2/radian
$\quad \theta_s$ = rotation angle of steering wheel, radians
$\quad \theta_i$ = rotation angle of gearbox input shaft, radians
$\quad b_s$ = torsional damping coefficient of steering wheel, in. lb sec/radian
$\quad k_s$ = torsional spring constant of steering-wheel shaft, in. lb/radian

By equating forces on the valve and linkage, we obtain

$$F_v = m_v\ddot{x} + b_v\dot{x} + b_e(\dot{x} - \dot{y}) + k_e(x - y) + A_s P_m \qquad (18.6)$$

where F_v = force applied to valve spool by pitman arm of steering gear, lb
$\quad m_v$ = mass of valve spool plus associated linkages (referred to spool), lb sec^2/in.
$\quad b_v$ = damping coefficient of valve spool and associated linkages, lb sec/in.

b_e = equivalent damping coefficient between valve spool and load, lb sec/in.

k_e = equivalent spring constant between valve spool and load, lb/in.

A_s = cross-sectional area of valve spool, in.^{2}

By equating forces on the power piston and load, we get

$$F_l = m_l\ddot{y} + b_l\dot{y} - b_e(\dot{x} - \dot{y}) + k_ly - k_e(x - y) - A_mp_m \quad (18.7)$$

where F_l = external force applied to power piston and load, lb

m_l = mass of load and linkages referred to power piston, lb sec^2/in.

b_l = damping coefficient of load, lb sec/in.

k_l = spring constant of load, lb/in.

A_m = effective area of power piston, in.^{2}

By the equation of continuity, assuming that the oil is incompressible,

$$Q_m = A_m\dot{y} + A_s(\dot{y} - \dot{x}) \quad (18.8)$$

From the geometry of the pitman arm,

$$F_v = \left(\frac{n}{r}\right)k_s(\theta_s - \theta_i) \quad (18.9)$$

and

$$\theta_i = \left(\frac{n}{r}\right)x \quad (18.10)$$

where n = gear ratio from steering-wheel shaft to pitman-arm shaft

r = radius of pitman arm, in.

All the foregoing equations are based on the assumptions that the automobile frame and all linkages not assigned a spring constant are infinitely stiff, linkage joints and gearing are free from backlash, compressibility effects and flow losses of the oil are negligible, and power-piston leakage is negligible.

18.33. Dynamic Analysis

In analyzing the system, two modes of operation are apparent. First, in steering the automobile, the system input is mainly a steering-wheel position, θ_s, while the output is a power-piston (or wheel) position, y. Thus the dynamics of the steering wheel are excluded while those of the load must be introduced for the forward direction of operation of the system. The second mode occurs during recovery from a turn or while driving over a rough roadway where the wheels tend to position the power piston and thus to operate the system in reverse. For this mode, the input is piston position, y, and the output is the steering-wheel position, θ_s. For this case, it is assumed that the dynamics of piston and load may be excluded and the steering-wheel dynamics must be included in the analysis.

18.331. Forward Operation. In analyzing the system for the forward direction of operation, several assumptions may be made that will considerably simplify the mathematics. At any road speeds other than those encountered in parking, the physical damping on the load and output members (b_l and b_e) will be negligibly small compared to the load damping derived from the valve. Since the parking situation does provide further damping of the load, the condition just mentioned is considered to be the worst case and is the proper basis for the forward-direction analysis. The spring force on the power piston due to k_e will be negligibly small when compared to the spring force due to k_l. Thus Eq. 18.7 becomes

$$F_l = m_l \ddot{y} + k_l y - A_m P_m \qquad (18.11)$$

In considering Eq. 18.6, it is desirable to obtain the major portion of the valve damping by utilizing the inherent load-damping characteristics of the valve output applied through the road-feel passages connecting to the valve end chambers. We must choose our valve parameters to promote this condition and thus we can optimistically assume that the actual physical damping will be negligible in comparison. Furthermore, since road feel is obtained by output-pressure feedback, no physical spring between the valve spool and valve body will be needed to obtain this feel, and the spring indicated by k_e will include only the effect of flow forces. Since these flow forces will be small compared to the pressure and inertia forces, we may neglect k_e. Thus Eq. 18.6 reduces to

$$F_v = m_v \ddot{x} + A_s P_m \qquad (18.12)$$

Combining Eqs. 18.1, 18.4, and 18.8 through 18.12 to eliminate the variables x and θ_i and employing the operational notation $D = d/dt$, we obtain a solution for y as a function of θ_s and F_l. This equation for the forward direction is of the form

$$y = \frac{C_1 \theta_s + C_2 F_l}{D^4 + G_3 D^3 + G_2 D^2 + G_1 D + G_0} \qquad (18.13)$$

where the C's and G's are various combinations of the system physical parameters and are given as follows:

$$C_1 = \left(\frac{n}{r}\right)\frac{k_s}{m_l m_v}\left(\frac{A_m A_s}{k_2}D + A_m k_3\right) \tag{18.14}$$

$$C_2 = \frac{1}{m_l m_v}\left\{m_v D^2 + \frac{A_s{}^2}{k_2}D + \left[A_s k_3 + \left(\frac{n}{r}\right)^2 k_s\right]\right\} \tag{18.15}$$

$$G_3 = \frac{1}{k_2}\left[\frac{A_s{}^2}{m_v} + \frac{A_m(A_m + A_s)}{m_l}\right] \tag{18.16}$$

$$G_2 = \left[\frac{\left(\frac{n}{r}\right)^2 k_s + A_s k_3}{m_v} + \frac{k_l + A_m k_3}{m_l}\right] \tag{18.17}$$

$$G_1 = \frac{1}{m_l m_v k_2}\left[\left(\frac{n}{r}\right)^2 k_s A_m(A_m + A_s) + k_l A_s{}^2\right] \tag{18.18}$$

and

$$G_0 = \frac{1}{m_l m_v}\left[\left(\frac{n}{r}\right)^2 k_s(k_l + A_m k_3) + k_l A_s k_3\right] \tag{18.19}$$

Equation 18.13 expresses the dynamic behavior of the system in the forward direction with the given basic assumptions. From Eq. 18.13 it can be seen that the output ram position, y, is dependent not only on the steering-wheel position, θ_s, but also on the external reaction force applied to the ram by the wheels. This latter dependence is the system load sensitivity and is inherent in any system employing an open-center valve, as well as in all systems which use a control valve having internal leakage. The system dynamics in the forward direction depend only on the denominator of Eq. 18.13; the characteristic equation of the system is

$$D^4 + G_3 D^3 + G_2 D^2 + G_1 D + G_0 = 0 \tag{18.20}$$

Solving Eq. 18.20 and inserting numerical values will give the response of the system to a steering command. This response must be neither too oscillatory nor too sluggish. Since we are more or less free to choose them, we can vary the valve dimensions until the response of the system is of the proper character. We must also assign numerical values to all the other parameters, and these values and their justifications are as follows: Let

$A_m = 3$ in.^{2}, since we need approximately a 1500-lb maximum output force with a maximum pressure of 500 psi

$A_s = 0.3$ in.^{2}, since we have chosen approximately 10 per cent "road feel." The spool diameter, d, will therefore be $\frac{5}{8}$ in.

$Q_s = 10$ in.^{3}/sec, since a maximum drag-link velocity of 3 in./sec is required

and

$C_0 = 100$ in.^{2}/sec $\sqrt{\text{lb}}$, by calculation

The following approximate values were measured or calculated on the basis of an actual automobile in the medium-weight class:

$$n = 16$$
$$r = 6 \text{ in.}$$
$$m_l = 0.6 \text{ lb sec}^2/\text{in. (neglecting gyroscopic forces)}$$
$$m_v = 7.5 \times 10^{-3} \text{ lb sec}^2/\text{in.}$$
$$k_l = 10 \text{ lb/in.}$$
$$k_s = 1.6 \times 10^4 \text{ in. lb/radian}$$
$$J_s = 0.7 \text{ lb in. sec}^2/\text{radian}$$

In choosing the valve parameters, by specifying the neutral-valve supply pressure, P_{s0}, the remaining values are established by the relations of Eqs. 18.2, 18.3, and 18.4. By trial and error we have found that the value $P_{s0} = 50$ lb/in.^{2} gives a desirable balance between system performance and stand-by power loss. The stand-by power loss associated with the valve for $P_{s0} = 50$ psi is 0.9 hp. Additional losses in the lines will increase the total power loss slightly but it will still be acceptable. Having thus chosen P_{s0}, we calculate the valve parameters to be

$$U = 5.0 \times 10^{-3} \text{ in.}$$

$$A_v = 10^{-2} \text{ in.}^2$$

so that

$$k_1 = 2 \times 10^3 \text{ in.}^2/\text{sec}$$

$$k_2 = 10^{-1} \text{ in.}^5/\text{lb sec}$$

and

$$k_3 = \frac{k_1}{k_2} = 2.0 \times 10^4 \text{ lb/in.}^3$$

These parameter values result in the following numerical version of the characteristic equation:

$$D^4 + 285D^3 + (1.61 \times 10^7)D^2 + (2.5 \times 10^9)D + 1.52 \times 10^{12} = 0$$

$$(18.21)$$

which has the two quadratic factors

$$(D^2 + 155.6D + 9.5 \times 10^4)(D^2 + 129.4D + 1.6 \times 10^7) = 0 \quad (18.22)$$

This numerical solution of Eq. 18.20 indicates that the theoretical response to an input will be composed primarily of a low-frequency (49-cps) quadratic lag, which has a 0.25 damping ratio, and a superimposed lightly damped, high-frequency (640-cps) vibration. This response is too oscillatory, but because several known sources of physical damping were neglected in the analysis, the primary response will actually be damped further. It would be desirable to achieve a damping ratio of 0.6 to 0.8. The high-frequency nature of the superimposed vibration will introduce appreciable physical damping from many sources and the vibration will thus die out before it can be effective. It will also be possible to increase the damping ratio by altering one or more of the actual parameter values pertaining to the automobile. Some of these parameters can be more easily changed physically than others, so the process of increasing the damping ratio by this method becomes an involved trial-and-error process combined with engineering judgment. Actual tests on a system that had a low calculated damping ratio showed considerably more damping than that indicated by the analysis.

18.332. Reverse Operation. By varying the valve parameters, we have obtained a predicted response which is satisfactory, although not the most desirable, with respect to the forward direction of operation. We must now check the reverse direction to ensure satisfaction in that respect also. In this direction the actual response is not of so much interest as is the question of stability. Since we wish to determine only if the system is stable or unstable, we will apply Routh's stability criterion.[1] In order to use this criterion, we must again combine the differential equations for the various parts of the system involved. Since the input will be considered as a displacement of the power piston, y, Eq. 18.7 will not be used; however, Eq. 18.5 must now be included to introduce the dynamics of the steering wheel. Combining Eqs. 18.1, 18.4 through 18.6, and 18.8 through 18.10 to eliminate the variables x and θ_i results in the following solution for θ_s as a function of y and T:

$$\theta_s = \frac{E_1 y + E_2 T}{D^4 + H_3 D^3 + H_2 D^2 + H_1 D + H_0} \qquad (18.23)$$

[1] M. F. Gardner and J. L. Barnes, *Transients in Linear Systems*, Vol. I, John Wiley and Sons, New York, 1942, pp. 197–201.

where the E's and H's are

$$E_1 = \left(\frac{n}{r}\right)\frac{k_s}{J_s m_v}\left\{\left[b_e + \frac{A_s(A_s + A_m)}{k_2}\right]D + (k_e + A_s k_3)\right\} \qquad (18.24)$$

$$E_2 = \frac{1}{J_s m_v}\left\{m_v D^2 + \left(b_e + b_v + \frac{A_s{}^2}{k_2}\right)D + \left[k_e + A_s k_3 + \left(\frac{n}{r}\right)^2 k_s\right]\right\}$$

$$\qquad (18.25)$$

$$H_3 = \left[\frac{b_s}{J_s} + \frac{b_e + b_v + \left(\frac{A_s{}^2}{k_2}\right)}{m_v}\right] \qquad (18.26)$$

$$H_2 = \left[\frac{k_s}{J_s} + \frac{k_e + A_s k_3 + \left(\frac{n}{r}\right)^2 k_s}{m_v} + \frac{b_s\left(b_e + b_v + \frac{A_s{}^2}{k_2}\right)}{J_s m_v}\right] \qquad (18.27)$$

$$H_1 = \left[\frac{k_s\left(b_e + b_v + \frac{A_s{}^2}{k_2}\right) + b_s\left[k_e + A_s k_3 + \left(\frac{n}{r}\right)^2 k_s\right]}{J_s m_v}\right] \qquad (18.28)$$

and

$$H_0 = \frac{k_s(k_e + A_s k_3)}{J_s m_v} \qquad (18.29)$$

The denominator of Eq. 18.23,

$$D^4 + H_3 D^3 + H_2 D^2 + H_1 D + H_0, \qquad (18.30)$$

is characteristic of the dynamic behavior of the reverse operation of the system and is thus the basis to which Routh's stability criterion must be applied. The symbolic statements of Routh's stability criterion for a fourth-order characteristic equation are

$$H_3, H_2, H_1, \text{ and } H_0 > 0 \qquad (18.31)$$

and

$$H_1(H_3 H_2 - H_1) - H_3{}^2 H_0 > 0 \qquad (18.32)$$

The first condition, Eq. 18.31, is satisfied since all the individual quantities are positive. The second condition, Eq. 18.32, requires considerable algebraic manipulation and results in the final statement

$$H_1(H_3H_2 - H_1) - H_3{}^2H_0 = \left(\frac{n}{r}\right)^2 k_s{}^2 \left[\frac{b_s}{J_s} + \frac{b_e + b_v + \left(\frac{A_s{}^2}{k_2}\right)}{m_v}\right]^2$$

$$+ \frac{b_s\left(b_e + b_v + \frac{A_s{}^2}{k_2}\right)}{J_s m_v} \left\{k_s\left(b_e + b_v + \frac{A_s{}^2}{k_2}\right)\right.$$

$$\left. + b_s\left[k_e + A_s k_3 + \left(\frac{n}{r}\right)^2 k_s\right]\right\} \left[\frac{b_s}{J_s} + \frac{b_e + b_v + \left(\frac{A_s{}^2}{k_2}\right)}{m_v}\right]$$

$$+ b_s\left(b_e + b_v + \frac{A_s{}^2}{k_2}\right)\left[\frac{k_s}{J_s} - \frac{k_e + A_s k_3 + \left(\frac{n}{r}\right)^2 k_s}{m_v}\right]^2 > 0 \quad (18.33)$$

This condition is also satisfied, for the same reason as given for Eq. 18.31. Thus for the reverse direction of operation the system satisfies Routh's stability criterion and is stable. This criterion does not indicate the margin of stability, but it is a very useful tool for theoretical system analyses.

18.34. Experimental Results

By appropriate methods, we have obtained an acceptable response for the forward direction of operation and have shown that the system is stable for the reverse direction. This completes the system analysis. A more detailed analysis would be of dubious value because assumptions such as linear valve characteristics and lumped parameters would become less valid and the mathematical manipulation would become extremely involved. The usual procedure at this point is to complete a mechanical design of the system, to manufacture a prototype, and then to conduct thorough tests on the individual components as well as on the system as a whole. This procedure was followed by the author in designing and building an actual power-steering system. Some details were different, but the actual system and the system just described were essentially similar. Comparison of the actual system performance with the theoretically predicted performance showed that, as frequently happens, considerably more damping appeared in the real system than was accounted for on paper. The reverse situation may also occur, but this can usually be attributed to poor judgment on the part of the designer. The response of the system actually built showed a damping ratio close to unity in the forward direction compared with a calculated ratio of 0.25, while the measured natural frequency was close to the predicted value. Also, as previously mentioned, the predicted lightly damped high-frequency oscillation was not observable in the actual system. When installed in an automobile, the system designed and built by the author performed in a completely satisfactory manner and confirmed the results of the laboratory tests.

The design procedure based on a detailed dynamic analysis such as that just presented is at best useful only as a basis for initial design, as a guide with which to compare test results, and as a means of gaining an insight into the factors which influence system performance.

19

J. L. Shearer

═ Comparison of Hydraulic and Pneumatic Servomechanisms

19.1. INTRODUCTION

In order to gain a better perspective of the roles that hydraulic and pneumatic drives play in the field of automatic control, it is useful to compare them on the basis of a given configuration and physical application. Consider the two modes of control for a system such as that shown in Fig. 19.1, namely a valve-controlled ram with inertia load operating from a constant-pressure supply.

19.2. QUALITATIVE COMPARISON

Many of the qualitative factors that have proved to be important are considered in Table 19.1, which summarizes the relative merits of hydraulic and pneumatic control for this configuration.[1]

---

[1] This table, together with some of the other material in this chapter, was published in J. L. Shearer and S.-Y. Lee, "Selecting Power Control Valves," Parts I and II, *Control Engineering*, Vol. 3, No. 3 (March 1956), pp. 72–78, No. 4 (April 1956), pp. 73–76.

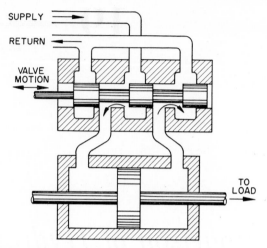

Fig. 19.1. A valve-controlled-ram-type servomotor.

19.3. QUANTITATIVE COMPARISON

To demonstrate and compare the design of hydraulic and pneumatic servomotors, let a valve-controlled ram type of servomotor be employed to position a 500-lb mass within a maximum stroke of ±3 in. in a closed-loop system (servomechanism), described schematically by the block diagram in Fig. 19.2. Assume that the measuring means and the amplifier have negligible lags (often true with simple linkages as feedback members, but generally not true when electric transducers are employed).

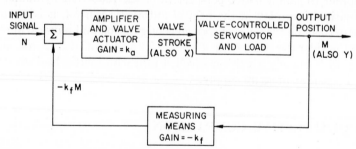

Fig. 19.2. Schematic block diagram of servomechanism using a valve-controlled servomotor.

19.31. System Specifications

The system specifications are: a maximum load acceleration of 500 in./sec² at zero velocity with a 500-lb external opposing load; a maxi-

Table 19.1. HYDRAULICS VERSUS PNEUMATICS
FOR VALVE-CONTROLLED DRIVE

Factor	Hydraulic Fluid	Compressed Gas
1. Over-all system complexity	Pump, pressure control, sump, filter, and heat exchanger.	Multistage compressor with interstage cooling, pressure control, filter, dryer (no return line).
2. Working fluid	High-quality mineral-base oils with additives, water-base solutions, synthetic liquids (all expensive—some flammable).	Air, nitrogen, products of combustion.
3. Efficiency	Seldom over 60 per cent.	Seldom over 30 per cent.
4. Over-all size	Approximately the same for a given supply pressure.	
5. Ease of energy storage	Air- or spring-loaded accumulator.	Simple tank.
6. Lubrication of moving parts	Usually good—sticking valves usually due to poor design or dirty fluid.	Small amounts of grease help seals to last—air has negligible lubricating qualities.
7. Susceptibility to moisture	Will not attack corrosion-resistant materials—small amounts of water will eventually dissolve traces of acids or salts in mineral-base oils and attack corrodible materials. Regular change of working fluid generally recommended at fixed intervals of time.	Corrosion-resistant materials recommended throughout pneumatic systems because of combination of small amounts of moisture and other contaminants in atmosphere. Small amounts of water can freeze valves and render them inoperable below 32°F.
8. Susceptibility to contamination (various kinds of dirt)	Closely fitted valves (0.0002-in. clearance) stick very easily except when best dirt filters are used. Oil has "washing" action that dislodges particles that would adhere to walls of flow passages leading to valve orifices.	Less trouble than with oil. Foreign particles seem to drop out of air before it gets to valve. Little "washing" action apparent with air. Dirty combustion products can be troublesome.

Table 19.1.   HYDRAULICS VERSUS PNEUMATICS
FOR VALVE-CONTROLLED DRIVE.   (*Continued*)

Factor	Hydraulic Fluid	Compressed Gas
9. Ease of manufacture	Valves usually require great care. Precision machining usually required throughout.	Same as hydraulic, but valves need more care and tight seals are difficult to attain.
10. Safety of operation	Leakage of flammable fluids a fire hazard. High-velocity jet can pierce skin and cause blood poisoning. Fluid can cause serious eye inflammations. Not explosive.	Flying debris from a rupture can be very dangerous. Explosions possible when small amounts of volatile fuel such as oil are present in air—not with pure nitrogen.
11. Temperature sensitivity	Most hydraulic-fluid viscosities change greatly with temperature changes. Unequal expansion of dissimilar materials can give serious trouble. (Some "orifices" behave as orifices only when fluid viscosity is low.) If temperature is too high, fluid vaporizes—if too low, it solidifies.	Since viscosity of most gases is much less than that of hydraulic fluids, relatively little trouble can be expected from viscosity changes. Same trouble as with hydraulic systems when dissimilar materials work together. Neither liquefaction nor freezing is common.
12. Valve stroking forces	Both steady and dynamic flow forces are often significant. Stiction due to dirt or lateral forces (pressure unbalance) is common in control valves.	Neither steady nor are dynamic flow forces often significant because of low fluid density. Steady-flow force can result from high velocity (1000 fps) at throat of control valve. Lateral forces very troublesome because of lack of lubrication. Stiction due to dirt seems less troublesome.
13. Pressure range	50–5000 psig	5–3000 psig
14. Relative speed of response	Usually better than any other means of control at same output power level.	Not so good as same system with hydraulic fluid; better than electric drive at same power level.

Table 19.1. HYDRAULICS VERSUS PNEUMATICS
FOR VALVE-CONTROLLED DRIVE. (*Continued*)

Factor	Hydraulic Fluid	Compressed Gas
15. Dry friction acting on output shaft	Can cause excessive steady errors in a position servomechanism, especially when an open-center valve is used.	Steady errors in position servomechanism more serious than in some hydraulic drives. Can also cause large low-frequency phase shift.
16. Quiescent power drain	Usually less than 10 per cent of maximum output power with a closed-center valve. Up to 100 per cent with open-center valve.	Up to 20 per cent of maximum output power with closed-center valve. Up to 100 per cent of maximum output power with open-center valve.
17. Rotary motors for continuous control	Many types commercially available. Some work at pressures up to 5000 psig.	Few available—none at pressures above about 100 psig.

mum steady velocity of 3 in./sec with a steady 500-lb external opposing load; a fluid supply pressure of 800 psi; a stand-by power loss (load motionless, external load zero) not to exceed 0.5 hp. The mass-loaded servomotor should have a damping ratio of at least 0.5 to make possible the best closed-loop performance [2] (also see Chapter 17).

19.32. Ram-Area Calculation

If the valve is displaced far enough from its neutral position, the full supply pressure is available to move the ram when its velocity is low. On the basis of acceleration and external load specifications, summing forces (Newton's second law) gives

$$P_s A = m \frac{d^2 Y}{dt^2} + b \frac{dY}{dt} + F_l \tag{19.1}$$

or, at zero velocity,

$$A = \frac{(1.3)(500) + 0 + 500}{800} = 1.44 \text{ in.}^2$$

A commercially available equal-area double-acting cylinder with a net

[2] J. L. Shearer, "Proportional Control of Rate-Type Servomotors," *Trans. ASME*, Vol. 76 (August 1954), pp. 889–894.

working area of 1.50 in.^{2} and a total stroke of 6.5 in. has a measured viscous-damping coefficient of 1.5 lb sec/in. Load damping measures 15.9 lb sec/in. Thus, the pressure across the ram, P_m, that is needed for a 3 in./sec maximum steady velocity with a 500-lb external opposing load is

$$P_m A = b \left(\frac{dY}{dt}\right)_{\max} + F_{l\,\max} \qquad (19.2)$$

or

$$P_m = \frac{(17.4)(3) + 500}{1.50} = 368 \text{ psi}$$

The dynamic characteristics of the valve-controlled ram for "small" changes of all variables are [3]

$$\left(\frac{k_3 m}{k_2 b + A^2}\right)\frac{d^2[d(\Delta Y)/dt]}{dt^2} + \left(\frac{k_2 m + k_3 b}{k_2 b + A^2}\right)\frac{d[d(\Delta Y)/dt]}{dt} + \frac{d(\Delta Y)}{dt}$$

$$= \frac{k_1 A(\Delta X) - k_3[d(\Delta F_l)/dt] - k_2(\Delta F_l)}{k_2 b + A^2} \qquad (19.3)$$

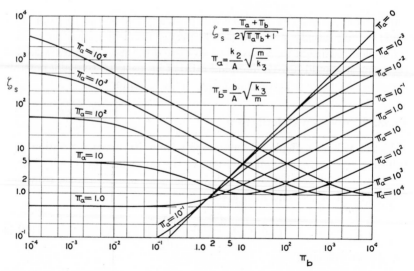

Fig. 19.3. Graph showing damping ratio, ζ_s, as a function of the dimensionless parameters π_a and π_b.

[3] J. L. Shearer, "Dynamic Characteristics of Valve Controlled Hydraulic Servomotors," *Trans. ASME*, Vol. 76 (August 1954), pp. 895–903. Also see Chapters 15 and 16.

19.33. Damping Ratio

The damping ratio, ζ_s, of the servomotor is given by the following equation, or by the graph in Fig. 19.3:

$$\zeta_s = \frac{k_2 m + k_3 b}{2\sqrt{k_3 m (k_2 b + A^2)}}$$

$$= \frac{\dfrac{k_2}{A}\sqrt{\dfrac{m}{k_3}} + \dfrac{b}{A}\sqrt{\dfrac{k_3}{m}}}{2\sqrt{\left(\dfrac{k_2}{A}\sqrt{\dfrac{m}{k_3}}\right)\left(\dfrac{b}{A}\sqrt{\dfrac{k_3}{m}}\right) + 1}} \tag{19.4}$$

19.34. Closed-Loop Gain

The closed-loop differential equation relating change of output position, ΔN, to change of input signal, ΔM (see Fig. 19.2), is

$$\left(\frac{k_3 m}{k_2 b + A^2}\right)\frac{d^3(\Delta M)}{dt^3} + \left(\frac{k_2 m + k_3 b}{k_2 b + A^2}\right)\frac{d^2(\Delta M)}{dt^2}$$

$$+ \frac{d(\Delta M)}{dt} + \frac{k_1 A k_a k_f}{k_2 b + A^2}\,(\Delta M)$$

$$= \frac{k_1 A k_a\,(\Delta N) - k_3\dfrac{d(\Delta F_l)}{dt} - k_2\,(\Delta F_l)}{k_2 b A^2} \tag{19.5}$$

and in the steady state

$$(\Delta M)_{ss} = \frac{(\Delta N)_{ss}}{k_f} - \frac{k_2(\Delta F_l)_{ss}}{A k_1 k_a k_f} \tag{19.6}$$

Although k_f can be adjusted to meet most requirements, $k_1 k_a k_f$ can cause system instability (fn. 2) (also see Chapter 17). Large values of $k_1 k_a k_f$ reduce load sensitivity, but

$$K_l = \frac{k_1 k_a k_f A}{k_2 b + A^2} \tag{19.7}$$

should not exceed $0.45\omega_{ns}$, where ω_{ns} is given by

$$\omega_{ns} = \sqrt{(k_2 b + A^2)/k_3 m} \tag{19.8}$$

19.35. Hydraulic

For oil, the fluid compliance is (fn. 3) (also see Chapters 6 and 15)

$$k_3 = \frac{V_i}{2\beta} \qquad (19.9)$$

where β = bulk modulus of pure hydraulic fluid with no entrained air, which is 2.5×10^5 lb/in.^{2} Therefore,

$$\frac{b}{A}\sqrt{\frac{k_3}{m}} = \frac{17.4}{1.50}\sqrt{\frac{(3.25)(1.50)}{2(2.5 \times 10^5)(1.30)}}$$

$$= 3.13 \times 10^{-2} \qquad (19.10)$$

assuming negligible volume in passages connecting valve to ram. Using the graph of Fig. 19.3, for a damping coefficient, ζ_s, of 0.5, we obtain

$$\frac{k_2}{A}\sqrt{m/k_3} = 1.0$$

from which

$$k_2 = A\sqrt{k_3/m} = (1.50)(2.7 \times 10^{-3})$$

$$= 4.05 \times 10^{-3} \text{ in.}^5/\text{lb sec} \quad (19.11)$$

As defined, $k_2 = C_1 + C_2$. For the ram motionless at the center with no load, the valve must be centered.

Assuming no leakage past the ram,

$$C_2 = 0 \qquad (19.12)$$

$$C_1 \cong C_d U w \sqrt{1/(P_s\rho)} \qquad (19.13)$$

$$U w = \frac{C_1}{C_d}\sqrt{P_s\rho}$$

$$= \frac{4.05 \times 10^{-3}}{0.65}\sqrt{800 \times 8 \times 10^{-5}}$$

$$U w = 1.58 \times 10^{-3} \text{ in.}^2 \qquad (19.14)$$

So, each of the four orifices would have an area of 0.00158 in.^{2} with the valve centered. The quiescent power loss would be

19.36. Pneumatic

For air, the fluid compliance is [4] (also see Chapter 16)

$$k_3 = \frac{V_i}{2kP_i} \qquad (19.26)$$

where k = ratio of specific heats, 1.4 for air; and P_i = initial value of ram pressure, 539 psi abs. Then

$$\frac{b}{A}\sqrt{\frac{k_3}{m}} = \frac{17.4}{1.50}\sqrt{\frac{(3.25)(1.50)}{2(1.4)(539)(1.30)}}$$

$$= 0.577 \qquad (19.27)$$

assuming negligible volume in passages connecting valve to ram. Using the graph of Fig. 19.3, for a damping coefficient, ζ_s, of 0.5, we obtain

$$\frac{k_2}{A}\sqrt{m/k_3} \cong 0.5$$

from which

$$k_2 = (0.5)A\sqrt{k_3/m}$$

$$= (0.5)(1.50)(5.0 \times 10^{-2})$$

$$= 3.75 \times 10^{-2} \text{ in.}^5/\text{lb sec} \quad (19.28)$$

As defined, $k_2 = C_1 + C_2$. For the ram motionless at the center with no load, the valve must be centered.

$$C_1 = \frac{RT_s}{2gP_i}\frac{\partial W_a}{\partial P_a} \qquad (19.29)$$

$$= \frac{(2.47 \times 10^5)(530)(5.7 \times 10^{-6})}{(2)(386)(540)}$$

$$C_1 = 1.79 \times 10^{-3} \text{ in.}^5/\text{lb sec} \quad (19.30)$$

The above data are based on measured characteristics shown in Fig. 16.9.

For a measured quiescent leakage flow of 0.001 lb/sec for a complete four-way valve, the stand-by power to compress this flow rate isothermally from 15 psia to 815 psia is

---

[4] J. L. Shearer, "Continuous Control of Motion with Compressed Air," Sc.D. Thesis, Department of Mechanical Engineering, Massachusetts Institute of Technology, Cambridge, Mass., 1954.

Hydraulic

$$\frac{P_sQ_q}{6600} = \frac{Q_q}{8.25} \qquad (19.15)$$

$$Q_q = 2C_dUw\sqrt{P_s/\rho} \text{ in.}^3/\text{sec} \quad (19.16)$$

$$\frac{P_sQ_q}{6600} = \frac{(2)(0.65)(1.58 \times 10^{-3})}{8.25}$$

$$\times \sqrt{800/(8 \times 10^{-5})}$$

$$= 0.79 \text{ hp} \qquad (19.17)$$

which exceeds the specified 0.5 hp. Hence Uw must be reduced by about one-third to 1.0×10^{-3} in.^{2}

To satisfy Uw, C_1 must equal $\frac{2}{3}k_2$; therefore, C_2 must equal $\frac{1}{3}k_2$, or $C_2 = 1.35 \times 10^{-3}$ in.^{5}/lb sec. A bypass capillary can be designed to provide C_2 because it is easier to control in manufacture than leakage past the ram and permits adjustment.

Maximum Hydraulic Power. The maximum hydraulic power which must be supplied to the valve is

$$\frac{P_sQ_{max}}{6600} = \frac{P_sC_dA_{o\ max}\sqrt{P_s/\rho}}{6600} \quad (19.18)$$

where

$$A_{o\ max} = \frac{A\left(\dfrac{dY}{dt}\right)_{max} + C_2P_{m\ max}}{C_d}$$

$$\times \sqrt{\frac{\rho}{(P_s - P_{m\ max})}}$$

$$= 3.3 \times 10^{-3} \text{ in.}^2 \qquad (19.19)$$

The maximum power thus calculated is 0.83 hp (compare the high stand-by power of 0.5 hp), but it is well to provide at least 100 per cent extra valve capacity ($A_{o\ max}$) to provide for transient demands.

Pneumatic

$$\frac{W_qRT_s}{6600g}(\ln 815 - \text{n } 15)$$

$$= \frac{(0.001)(2.47 \times 10^5)(530)(6.71 - 2.71)}{(6600)(386)}$$

$$= 0.206 \text{ hp} \qquad (19.31)$$

Although the allowable power drain is 0.5 hp, C_1 falls too far short of k_2 to think of making the valve intentionally open-centered. Besides, some production valves might have twice as much leakage flow as the valve measured.

Damping, therefore, is needed, and can be provided by a capillary passage connecting the two ends of the ram. The average velocity in the capillary must be low enough to avoid nonlinear "saturation" due to momentum effects.[5] A good rule is to make its cross-sectional area equal to that of the valve-to-ram passages. Matrix-type capillaries can be adjusted to the resistance needed.

Maximum Pneumatic Power. The maximum pneumatic power which must be supplied to the valve may be calculated from the maximum flow, W_a, needed for a ram velocity of 3 in./sec when P_a is 650 psi (that is, when load force is 500 lb). Note that this does not include leakage flow past the ram.

$$W_a = \frac{P_agA}{RT_s}\left(\frac{dY}{dt}\right)$$

$$= \frac{(650)(386)(1.5)(3.0)}{(2.47 \times 10^5)(530)}$$

$$= 8.6 \times 10^{-3} \text{ lb/sec} \qquad (19.32)$$

And the power which would be required to compress air isothermally to 815 psia at this rate would be

$$\text{hp}_{max} = \frac{W_{a\ max}RT_s}{6600g}(\ln 815 - \ln 15)$$

$$= 0.95 \text{ hp} \qquad (19.33)$$

[5] S. K. Grinnell, "Flow of a Compressible Fluid in a Thin Passage," *Trans. ASME,* Vol. 78 (May 1956), pp. 765–771.

Hydraulic

Pneumatic

It is seldom necessary to compress air at this peak rate because of the tremendous storage capacity of even relatively small storage tanks.

Response. The servomotor's natural frequency is

$$\omega_{ns} = \sqrt{\frac{4.05 \times 10^{-3}(17.4) + (1.5)^2}{9.75 \times 10^{-6}(1.3)}}$$

$$= 428 \text{ radians/sec} \qquad (19.20)$$

so that

$$K_l = 0.45(428) = 192 \text{ sec}^{-1} \qquad (19.21)$$

and

$$k_1 k_a k_f = \frac{k_2 b + A^2}{A}(K_l)$$

$$= \frac{4.05 \times 10^{-3}(17.4) + (1.5)^2}{1.5}(192)$$

$$= 294 \text{ in.}^2/\text{sec} \qquad (19.22)$$

The value of k_1 is (fn. 3) (also see Chapter 15)

$$k_1 = 2C_d w \sqrt{P_s/\rho} \qquad (19.23)$$

for the zero-flow zero-load condition and depends on the port width, w, which can be chosen for reasonable values of k_1, k_a, and k_f. The port width often determines the method of valve fabrication.

The steady-state load sensitivity is

$$\left(\frac{\Delta M}{\Delta F_l}\right)_{ss} = -\frac{k_2}{k_1 k_a k_f A}$$

$$= \frac{-4.05 \times 10^{-3}}{(294)(1.5)}$$

$$= -9.2 \times 10^{-6} \text{ in./lb} \qquad (19.24)$$

The response of the output shaft to a step change in input signal will overshoot by about 20 per cent, and the time to first crossover will be (fn. 2)

$$T_1 = \frac{2.5}{\omega_{ns}} = \frac{2.5}{428} = 5.85 \times 10^{-3} \text{ sec} \qquad (19.25)$$

Response. The servomotor's natural frequency is

$$\omega_{ns} = \sqrt{\frac{3.75 \times 10^{-2}(17.4) + (1.5)^2}{3.22 \times 10^{-3}(1.3)}}$$

$$= 8.3 \text{ radians/sec} \qquad (19.34)$$

so that

$$K_l = 0.45(8.3) = 3.74 \text{ sec}^{-1} \qquad (19.35)$$

and

$$k_1 k_a k_f = \frac{k_2 b + A^2}{A}(K_l)$$

$$= \frac{3.75 \times 10^{-2}(17.4) + (1.5)^2(3.74)}{1.5}$$

$$= 7.25 \text{ in.}^2/\text{sec} \qquad (19.36)$$

The value of k_1 is (fn. 4) (also see Chapter 16)

$$k_1 = \frac{RT_s}{gP_i}\frac{\partial W_a}{\partial X} \qquad (19.37)$$

for the zero-flow zero-load condition and depends on the port width, w, which can be chosen for reasonable values of k_1, k_a, and k_f. The port width often affects the simplicity of valve fabrication.

The steady-state load sensitivity is

$$\left(\frac{\Delta M}{\Delta F_l}\right)_{ss} = -\frac{k_2}{k_1 k_a k_f A}$$

$$= \frac{-3.75 \times 10^{-2}}{(7.25)(1.5)}$$

$$= -3.44 \times 10^{-3} \text{ in./lb} \qquad (19.38)$$

The response of the output shaft to a step change in input signal will overshoot by about 20 per cent, and the time to first crossover will be (fn. 2)

$$T_1 = \frac{2.5}{\omega_{ns}} = \frac{2.5}{8.3} = 0.30 \text{ sec} \qquad (19.39)$$

Thus, the hydraulic system is about 50 times as fast as the pneumatic for the same load mass and supply pressure.

19.37. Increasing Pneumatic Sensitivity

As calculated, the hydraulic system turns out to be almost 400 times as "stiff," $\Delta M/\Delta F_l$, as the pneumatic system for a steady load, F_l. Instead of the bypass capillary, the resistance and tank scheme devised by Levenstein [6] can be used for damping. This so-called "transient-flow stabilizer" was tested experimentally (fn. 4), and with its use only the effects of C_1 appear in the load-sensitivity equation so that the steady-state pneumatic stiffness can be increased by a factor of 40.

NOMENCLATURE

Symbol	Definition	Unit
A	area	in.^{2}
b	viscous damping in ram and load	lb sec/in.
C_1	valve characteristic—partial derivative with respect to pressure, at operating point	in.^{5}/lb sec
C_2	laminar leakage-flow coefficient	in.^{5}/lb sec
C_d	discharge coefficient, 0.65, dimensionless	
g	acceleration due to gravity	386 in./sec^2
K_l	loop gain	1/sec
k_1	valve characteristic (flow sensitivity)—partial derivative of flow with respect to valve position	in.^{2}/sec
k_2	equals $C_1 + C_2$	
k_3	fluid compliance	in.^{5}/lb
k_a	amplifier and valve-actuator gain	
k_f	feedback gain	
F_l	external load force	lb
m	load mass, $500/386 = 1.30$ lb sec^2/in.	
P	pressure	psi
P_a	pressure at a end of ram	psia
P_i	initial pressure on both sides of ram	psia
P_s	supply pressure	psia
Q	volume rate of flow	in.^{3}
R	gas constant, 2.47×10^5 in.^{2}/sec^2 °F for air	
t	time	sec
T	absolute temperature	°R
T_s	supply temperature	°R
U	underlap	in.
V_i	volume in one ram chamber (ram centered) plus one passage to valve	in.^{3}

[6] H. Levenstein, "Tie Simplicity to Power with Pneumatic Servomechanisms," *Control Engineering*, Vol. 2, No. 6 (June 1955), pp. 65–70.

Symbol	Definition	Unit
w	port width of valve	in.
W	weight rate of flow	lb/sec
W_a	weight rate of flow to a end of ram	lb/sec
X	valve position (stroke from center)	in.
Y or M	ram position	in.
Δ	small change from initial steady condition	
ζ_s	servomotor damping ratio	
ρ	density of hydraulic fluid, 8×10^{-5} lb sec^2/in.^{4} for mineral-base hydraulic oils	
π_a	dimensionless parameter containing k_2	
π_b	dimensionless parameter containing b	
ω_{ns}	natural frequency of uncontrolled servomotor	1/sec

Subscripts

Symbol	Definition
i	denotes initial condition
m	motor
q	denotes quiescent flow

20

Gerhard Reethof

Analysis and Design
of Servomotors Operating
on High-Pressure Hot Gas[1]

20.1. INTRODUCTION

The development of aircraft and guided missiles which fly at supersonic and hypersonic speeds has resulted in extremely high temperature requirements for component operation. Thus at a 20,000-ft altitude the temperature created by ram compression on the skin surface of an airplane flying at 1800 mph is of the order of 600°F. The temperature rise in valve-controlled hydraulic systems from the power dissipated in the control orifices may raise this temperature another 100 to 150°F. Temperatures of this magnitude are considerably beyond the capabilities of present-day hydraulic fluids. The process industry for several decades has utilized relatively-low-pressure pneumatic devices. The air-supply pressure used in these applications seldom exceeds 150 psig, and the speeds of response required are orders of magnitude lower than those called for in control systems for aircraft, and especially for small guided

---

[1] Some of the material in this chapter is taken from the author's ASME paper "Analysis and Design of a Servomotor Operating on High-Pressure Compressed Gas," *Trans. ASME*, Vol. 79 (May 1957), pp. 875–879.

missiles. These low speeds of response of the industrial process controls have led to a mistaken belief that pneumatic controls cannot be designed to meet the high-speed requirements of aircraft, missile, or machine-tool controls. Nevertheless, current and projected temperature requirements as well as other considerations are causing a renewed interest in pneumatic servomotors. Recent work with high-pressure pneumatic systems has demonstrated that pneumatic controls can be designed to meet some of the high-speed requirements.

There has been strong interest, in recent years, in jet vane controls of rocket motors, exhaust nozzle controls for jet engines and other control mechanisms which operate in the high-temperature environment of airborne propulsion devices. At the present time oil hydraulic actuation means are contemplated for these actuators, yet simple hot-gas controls offer very definite advantages.

The control of the flight path of guided missiles requires the accurate positioning of aerodynamic control surfaces. Oil hydraulic systems are almost exclusively used for this purpose. Again the high-temperature environment and large-temperature-range requirement have presented difficult design problems. The short duration of flights of the guided missile makes the use of a hot gas particularly attractive, since heat sinks and adequate insulation can be designed to limit the temperatures of such devices as electromagnetic actuator and feedback means.

A third field which opens broad possibilities is the control of nuclear reactors. At the present time electric and hydraulic controls are used. Yet in some cases, such as that of a steam-type reactor, a readily available source of high-pressure, high-temperature gas is available.

20.2. SYSTEM CONSIDERATIONS

A servomotor system consists of four basic elements:

1. Power generation.
2. Power control.
3. Power transmission.
4. Power utilization.

The choice of the means to implement the four basic elements depends largely on:

1. Duration of operation.
2. Continuous or intermittent power requirements (duty cycle).
3. Size considerations.

4. Weight considerations.
5. Environmental conditions.

The great care which must be exercised in making these initial decisions will be evident if working fluid temperatures lie in the range of 2000°F and higher. Needless to say, heat-transfer considerations then take on a major importance.

20.3. PNEUMATIC-POWER GENERATION

As the present treatment is concerned with relatively high speed of response applications, such low pressures as may be derived from ram effects on aircraft and missiles will be excluded. The following possible sources of power may be considered:

1. Compressor-accumulator power packs.
2. Bleed air from compressor stage of jet engines (750–1100°F).
3. Liquid-monopropellant gas generators (900–2000°F).
4. Superheated steam or hot gas from nuclear reactors (1200°F).
5. Bleed gas from turbine-entrance stage of jet engines (1500–2400°F).
6. Liquid-bipropellant gas generators (2000–3400°F).
7. Solid-propellant generators (2000–3600°F).
8. Gaseous-propellant generators (2000–3600°F).

The gas generators are arranged roughly in order of increasing gas temperatures. Even the lowest temperatures are high by normal control standards; furthermore, most of the available energy rests in the high temperature so that heat losses should be minimized. As a corollary, the control, transmission, and utilization of fluid power should occur at the highest possible energy, unless intercoolers can be used with a minimum of power loss, as for heating liquid propellants prior to combustion.

The nature of the gaseous products of the various gas generators is of the utmost importance. Only the compressor-accumulator air and jet-engine compressor bleed air may be clean and relatively free of solid particles. The compressor-accumulator air will usually contain traces of lubricant from the compressor which have been found to be very effective in lubricating control valves. However, the products of combustion of solid, liquid, and gaseous propellants contain solid particles of larger than micronic size, which will require special consideration in the design of the control and utilization devices. Although the solid, hot particles are primarily soft carbonaceous matter, some of the solid propellants produce very hard and highly erosive particles.

20.31. Solid Propellants

The choice of a particular solid propellant as a gas generator fuel depends, furthermore, on such factors as:

1. Linear burning rate.
2. Effect of ambient temperature on linear burning rate.
3. Effect of variations in pressure on the linear burning rate at the operating point.
4. Ease and reliability of ignition.
5. Burning temperature.

The linear burning rate of some propellants varies exponentially with the combustion-chamber pressure, whereas some recently developed propellants exhibit virtually no change in burning rate over a large range of pressures about the operating point. Therefore, a system consisting of such a gas generator and an orifice type of control element is almost self-regulatory.

The ambient temperature (or soaking temperature) of the propellant prior to combustion will markedly change the burning rate of most solid propellants. Thus for a commercial propellant the burning rate will vary from 0.08 in./sec at $-65°F$ soaking temperature to 0.11 in./sec at $+160°F$ soaking temperature. This variation is representative of most solid propellants.

Pressure of the combustion can be controlled either by matching the propellant burning characteristics with the nozzle flow-pressure relationship or by the use of dump-type pressure regulators.

The high linear burning rates of most solid propellants limit their use to applications having durations of not over 5 minutes. The high burning temperatures of these propellants also dictate short-duration applications.

20.32. Liquid Propellants

Either liquid monopropellants or liquid bipropellants require fairly elaborate fuel-flow control systems to maintain reasonable chamber-pressure control. The monopropellants such as hydrazine, propyl nitrate hydrogen peroxide, and ethylene oxide are capable of autodecomposition, whereas bipropellant systems consist of such fluids as kerosene or other hydrocarbon fuels and an oxidizer such as liquid oxygen or fuming nitric acid.

20.4. POWER CONTROL

The gas generators discussed in the previous section can supply high-pressure, high-temperature gas with some entrained solid particles. The control of the flow of such a medium presents a series of difficult problems.

1. The fluid has no lubricity so that sliding of closely fitting parts, which are common in hydraulic controls, must be minimized to reduce wear and friction. Note should also be taken of the fact that length-to-clearance ratios of better than 1000:1 are required to assure laminar flow of gases. Seating-type valves are therefore advised in preference to sliding-type valves in the power-control system.

2. The leakage of a hot gas past clearance spaces will result in the rapid heating of adjacent parts.

3. Since almost all leakage flows will be of the orifice type, the greatest possible temperature drop from the expansion is given by an absolute-temperature ratio of about 0.88. Thus if the high-pressure gas is at 2000°F abs (1540°F), the expanded gas will be at 1760°F abs (1300°F) and will still be at a high enough temperature to cause damage to most materials.

4. The solid particles which exist in the products of combustion of most of the gas generators may be highly erosive and damaging to the seats or orifice edges of flow-control valves. The high temperatures in addition may weaken the valve materials.

20.5. POWER TRANSMISSION AND CONVERSION

The transmission of fluid power from the gas generator to the power-control unit and then to the power-utilization unit should occur with the minimum of heat loss. Thus the passages should be made as large as possible to reduce mass velocities and in turn reduce surface coefficients of heat transfer in the conduits.

20.51. Turbo-Clutch System

One means of transmitting and connecting the hot-gas power is by means of the turbo-clutch system. Here the energy contained in the hot gases is transduced in a high-speed–low-torque gas turbine. The required high torques are obtained by a high-ratio gear-reduction unit.

If an angular-position control system is required, the low-speed–high-torque power can be made available in two counter-rotating discs which, through a clutch-control system, control the angular position of a light clutch plate as shown in Fig. 20.1. The attractive feature of this system concept is the lack of hot-gas control and transmission problems. Products of combustion of a solid or liquid propellant drive the turbine at essentially constant speed with a minimum of control. Such a system, however, is inefficient, requires a complex clutch-control system, and

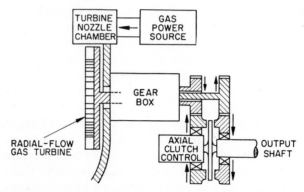

Fig. 20.1. Schematic diagram of turbo-clutch control system.

presents a difficult gear-reduction problem. For these reasons the turbo-clutch servo, although often proposed, has not found extensive use.

20.52. Valve-Controlled Motor System

The useful fluid power which is made available by the flow-control valve is transduced into mechanical power by the fluid motor. If small and limited linear or angular motion is considered, a vane or ram type of motor appears most suitable.

VANE MOTOR. The vane motor deserves consideration because of its simplicity of construction. For angular output motions this arrangement would appear ideal. However, sealing between vane chambers and on the output shaft presents difficult design problems. As will be shown in a later section, pneumatic systems are very sensitive to non-viscous friction. A vane motor, from first considerations, appears to be limited by unavoidably high friction, yet good and ingenious design may make this mode of power utilization feasible.

RAM MOTORS. The conventional double-acting or single-acting ram motors offer excellent possibilities. Again nonviscous friction phenomena should be minimized by good design to ease the problems of stabilizing the high-performance system. If a split-ram rocker-arm arrangement is used, leakage past the pistons must be avoided to reduce the dangers of overheating. The single-acting system, consisting, for example, of a

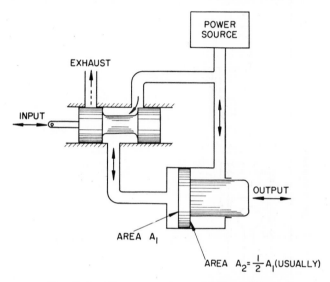

Fig. 20.2. Three-way valve and differential ram.

three-way valve which modulates the pressure on the larger area of the motor (see Fig. 20.2) with constant supply pressure on the smaller area, results in a minimum number of sealing elements.

For the present discussion a split-ram rocker-arm arrangement has been chosen because of the simplicity of construction and the ease with which such variables as leakage, areas, and loads could be monitored and controlled. (See Fig. 20.3).

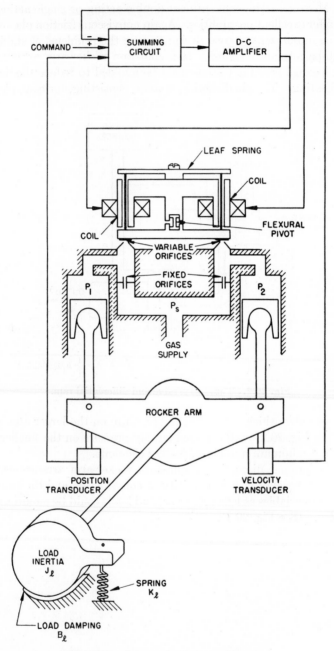

Fig. 20.3. Schematic diagram of system.

20.6. DESCRIPTION OF THE SYSTEM

The system is illustrated schematically in Fig. 20.3. The flow valve is supplied with a high-pressure gas from a constant-pressure gas power source. The power-control valve consists of two upstream orifices of fixed area and two downstream orifices with differentially variable areas. The control pressures in the chambers between the upstream and downstream orifices are piped to each side of the fluid motor. In this instance, the fluid motor consists of two single-acting, self-lubricating pistons which, through two spherical-ended connecting rods, apply a torque to the rocker arm. The rocker arm can be connected to the load by means of a splined shaft. The position of the output member is measured by a linear differential transformer which delivers an a-c signal. This signal is rectified in a keyed demodulator and fed back to the adding circuit of the forward-loop d-c amplifier. The push-pull output stage of the amplifier supplies the polarizing currents and differentially varying signal currents to the two coils of the E-style electromagnetic valve actuator. The small motion of the plugs on the rotor of the valve actuator controls the flow and pressure conditions in the ram-type motor. An I-beam flexure-pivot mounting between rotor and stator ensures frictionless motion of the rotor and maintains correct spacing of the air gaps. Since the magnetic fields produce a statically unstable valve actuator, a leaf spring is mounted on the stator and connected to the rotor by wire links. A constant tension in the wire links prevents backlash in the motion of the rotor.

The desired high speed of response of the complete control system is obtained by feeding back a signal proportional to the velocity of the output member through an amplifier with an adjustable gain.

20.7. COMPONENT ANALYSIS

20.71. Flow Valve

The control of the flow of fluid power from the gas generator to the motor-load system is accomplished by a four-way valve consisting of two three-way valves operating in a push-pull arrangement. The analysis of the weight flow through the orifices is based on the assumption of frictionless adiabatic flow of a perfect gas.[2] The topic of valve

---

[2] A. H. Shapiro, *The Dynamics and Thermodynamics of Compressible Fluid Flow*, Vol. 1, Ronald Press Co., New York, 1953.

characteristics is discussed in detail in Chapter 8, and the topic of flow through nozzles and orifices is discussed in Sec. 3.332 of Chapter 3.

20.711. Flow through a Single Orifice. For a specific gas, the weight rate of flow through an orifice is given by

$$W = C_d C_2 A \frac{P_u}{\sqrt{T_u}} f_1 \left(\frac{P_d}{P_u}\right) \tag{20.1}$$

where W = weight rate of flow of gas, lb/sec
C_d = discharge coefficient of orifice
A = area of orifice, in.^{2}
f_1 = a function of $\dfrac{P_d}{P_u}$
P_d = downstream stagnation pressure, psia
P_u = upstream stagnation pressure, psia
T_u = upstream stagnation temperature, °R

and

$$C_2 = g \sqrt{\frac{k}{R\left(\dfrac{k+1}{2}\right)^{(k+1)/(k-1)}}} \tag{20.2}$$

where g = gravitational acceleration, in./sec^2
R = gas constant, in.^{2}/sec^2 °R
k = ratio of specific heats

The function of (P_d/P_u) is equal to unity except when $P_d/P_u > \left[\dfrac{2}{(k+1)}\right]^{k/k-1}$, and then it is given by

$$f_1(P_d/P_u) = \sqrt{\left(\frac{2}{k-1}\right)\left(\frac{k+1}{2}\right)^{(k+1)/(k-1)}} \left(\frac{P_d}{P_u}\right)^{1/k} \sqrt{1 - \left(\frac{P_d}{P_u}\right)^{(k-1)/k}} \tag{20.3}$$

20.712. Flow through a Three-Way Valve. From geometric considerations, the area of the variable orifice, A_{v1}, of the left-hand three-way valve is given by

$$A_{v1} = \pi D_n (x_0 - x) \tag{20.4}$$

where A_{v1} = variable downstream orifice area of left-hand three-way valve, in.^{2}
D_n = nozzle diameter, in.
x_0 = mid-position opening of each variable orifice, in.
x = upward displacement of left end of valve-actuator arm, from mid-position, in.

With the valve at mid-position ($x = 0$), the orifice area is

$$A_{v0} = \pi D_n x_0 \tag{20.5}$$

Division of Eq. 20.4 by Eq. 20.5 results in the dimensionless equation

$$\frac{A_{v1}}{A_{v0}} = 1 - \frac{x}{x_0} \tag{20.6}$$

The weight flow through the downstream variable orifice for the expected critical-flow conditions is given by

$$W_{e1} = C_d C_2 A_{v0} \left(1 - \frac{x}{x_0}\right) \frac{P_1}{\sqrt{T_s}} \tag{20.7}$$

The weight flow through the fixed upstream area orifice is given by

$$W_{s1} = C_d C_2 A_f \frac{P_s}{\sqrt{T_s}} f_1 \left(\frac{P_1}{P_s}\right) \tag{20.8}$$

From continuity considerations, the weight flow to the motor is

$$W_{m1} = W_{s1} - W_{e1} \tag{20.9}$$

The appropriate substitutions in Eq. 20.9 result in an expression for the volume flow to the motor from the left-hand three-way valve

$$Q_{m1} = C_d C_2 A_{v0} \frac{R\sqrt{T_s}}{g} \left[\frac{1}{\alpha} \frac{P_s}{P_1} f_1 \left(\frac{P_1}{P_s}\right) - (1 - \gamma)\right] \tag{20.10}$$

where

$$\alpha = \frac{A_{v0}}{A_f} \tag{20.11a}$$

and

$$\gamma = \frac{x}{x_0} \tag{20.11b}$$

A convenient reference flow rate is defined as

$$Q_{mr} = C_d C_2 A_{v0} \frac{R\sqrt{T_s}}{g} \tag{20.12}$$

Division of Eq. 20.10 by Eq. 20.12 results in the dimensionless volume rate of flow to the motor:

$$q_{m1} = \frac{1}{\alpha} \frac{P_s}{P_1} f_1 \left(\frac{P_1}{P_s}\right) - (1 - \gamma) \tag{20.13}$$

Similarly, the expression for the volume flow to the motor from the right-hand three-way valve is

$$q_{m2} = \frac{1}{\alpha} \frac{P_s}{P_2} f_1\left(\frac{P_2}{P_s}\right) - (1 + \gamma) \qquad (20.14)$$

Owing to the complexity of the analytical relationships, Eqs. 20.13 and 20.14 were solved by graphical methods with data from the *Gas Tables*.[3] The assumptions made in the derivation of Eqs. 20.13 and 20.14 were tested by an experimental investigation of the weight-flow versus pressure relationships of this type of three-way valve. The results of these tests agree very well with the predicted valve characteristics. Similar agreement for a closed-center three-way valve is reported by Shearer.[4] (Also see Chapters 8 and 16.)

20.713. Four-Way Valve. The steady-state load-flow versus load-pressure characteristics of the four-way valve are obtained by graphically

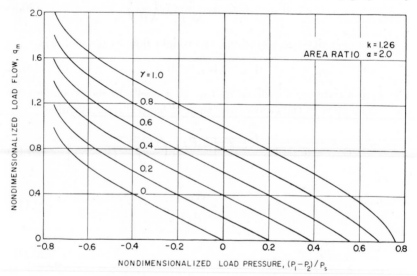

Fig. 20.4. Theoretical load-flow versus load-pressure characteristics of a four-way valve.

combining Eqs. 20.13 and 20.14. This operation is accomplished by noting that during steady state

$$q_{m1} = -q_{m2} = q_m \qquad (20.15)$$

[3] J. H. Keenan and J. Kaye, *Gas Tables*, John Wiley and Sons, New York, 1948.

[4] J. L. Shearer, "Study of Pneumatic Processes in the Continuous Control of Motion with Compressed Air—I," *Trans. ASME*, Vol. 78 (February 1956), pp. 233–242.

The configuration of the four-way valve requires equal values of the variable γ for both three-way valves. The load pressure, P_l, is

$$P_l = P_1 - P_2 \tag{20.16}$$

The plot of load flow versus load pressure for a four-way valve with an area ratio $\alpha = 2.0$ is given in Fig. 20.4.

The pressure sensitivity of a flow valve, defined by

$$\left. \frac{\partial P_l}{\partial \gamma} \right|_{q_m=0}$$

expresses the ability of the flow valve to furnish a certain load pressure per unit valve displacement with no flow to the motor. The greatest pressure sensitivity of a flow valve would exist if full supply pressure

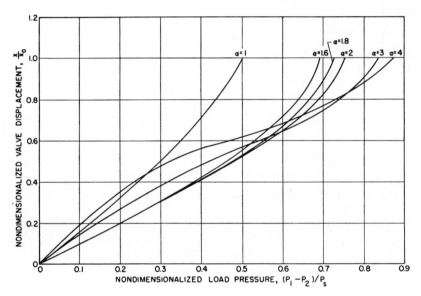

Fig. 20.5. Plots of load pressure versus valve displacement for several values of α with motor stationary.

could be achieved across the load for an infinitesimal motion of the valve from its center position. The fixed upstream-orifice construction results in a much lower pressure sensitivity than this because one side of the fluid motor remains at an appreciable pressure when the other side reaches supply pressure by the closing of its downstream orifice. The pressure sensitivity for valves of different ratios, α, is given by the inverse slope of the plots in Fig. 20.5.

The total weight flow to the valve is given by the sum of the weight flows through the upstream orifices.

$$W_{s1} + W_{s2} = C_d C_2 \frac{P_s}{\sqrt{T_s}} A_f \left[f_1 \left(\frac{P_1}{P_s} \right) + f_1 \left(\frac{P_2}{P_s} \right) \right] \quad (20.17)$$

Plots of the dimensionless weight flow to the valve, $(W_{s1} + W_{s2})/C_d C_2 (P_s/\sqrt{T_s}) A_f$, versus load pressure for several values of α are

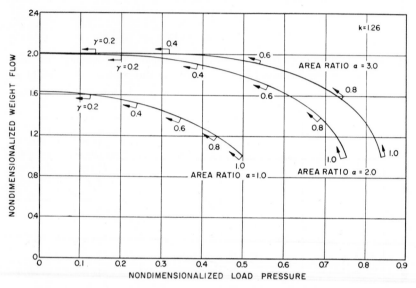

Fig. 20.6. Plots of weight flow versus load pressure for several values of α.

given in Fig. 20.6. Of interest are the facts that: (1) for a fixed value of α, one curve describes the weight flow for all conditions and (2) the weight-flow demand remains remarkably constant over a wide range of operating conditions.

The steady-state output power of the flow valve may be defined as the product of load pressure and load-volume flow:

$$\text{IP} = (P_1 - P_2)Q_m \quad (20.18)$$

Use of a convenient reference power defined by

$$\text{IP}_r = P_s Q_{mr} \quad (20.19)$$

results in the nondimensional expression for the power

$$\frac{\text{IP}}{\text{IP}_r} = \frac{P_1 - P_2}{P_s} q_m \quad (20.20)$$

Figures 20.7*a*, *b*, and *c* are plots of nondimensional power as expressed in Eq. 20.20 for various values of α. A plot of the peaks of the power curves as a function of α for several values of γ is given in Fig. 20.8.

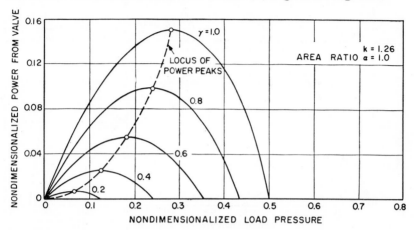

Fig. 20.7(*a*). Plots of nondimensionalized fluid power to the load versus load pressure for $\alpha = 1.0$.

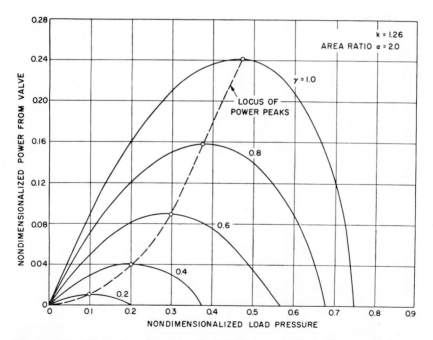

Fig. 20.7(*b*). Plots of nondimensionalized fluid power to the load versus load pressure for $\alpha = 2.0$.

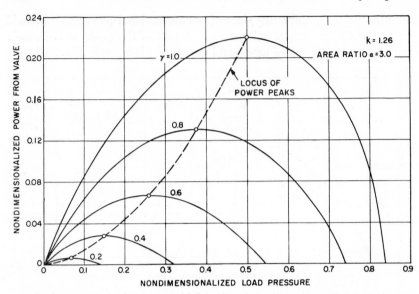

Fig. 20.7(c). Plots of nondimensionalized fluid power to the load versus load pressure for $\alpha = 3.0$.

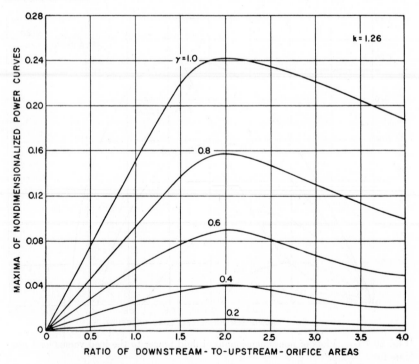

Fig. 20.8. Plot of maximum power versus α for several values of γ.

A value of $\alpha = 2$ was chosen for the valve on the basis of:

1. Linearity of load-flow versus load-pressure characteristics.
2. Optimum pressure sensitivity.
3. Peak power ratio.
4. Compressibility considerations.

The weight-flow curves, Fig. 20.6, show that the higher values of α result in less variation in weight rate of flow to the valve due to load-pressure changes. However, a large value of α will be shown to result in a large compressibility coefficient for the motor.

The choice of α is, therefore, a compromise between compressibility considerations, the nature of the gas generator, and the valve characteristics.

20.72. Gas-Motor Analysis

The analysis of motor performance is based on the following assumptions:

1. Perfect gas.
2. No resistance to the flow into the chambers.
3. Adiabatic conditions throughout the charging and discharging processes.
4. Temperature of gas entering the chamber equal to temperature of gases in the chamber.
5. Uniform chamber pressure at all times.

In Fig. 20.9, the control volume represents an artificial boundary around the chosen volume of gas, across which energy and mass flow.

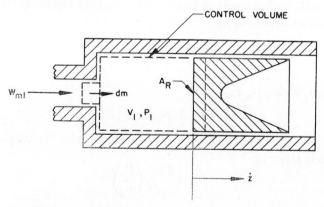

Fig. 20.9. Diagram of control volume in fluid motor.

The first law of thermodynamics, when applied to the control volume of Fig. 20.9, is

$$\left(u_1 + \frac{P_1}{\rho_1}\right) dm - P_1\, d\mathcal{V}_1 = d(\rho_1 \mathcal{V}_1 u_1) \tag{20.21}$$

where u_1 = specific internal energy of gas, in.^{2}/sec^2
ρ_1 = mass density of gas, lb sec^2/in.^{4}
dm = incremental mass of gas entering chamber, lb sec^2/in.
$\mathcal{V}$ = volume of chamber, in.^{3}

The first term represents the energy added to and the work done on the system by an incremental mass of gas, dm, crossing the system boundaries. The second term is the work being done by the system on the piston. The third term is the change in internal energy of the gas within the control volume. Equation 20.21 can be modified for use in the gas-motor analysis by substituting the perfect-gas relationships and noting that $d\mathcal{V}/dt = \dot{z}A_r$. Equation 20.21 then becomes

$$\frac{dP_1}{dt} + \frac{kP_1}{\mathcal{V}_1}(A_r\dot{z} - Q_{m1}) = 0 \tag{20.22}$$

where A_r = ram area, in.^{2}
$\dot{z}$ = ram velocity, in./sec

Equation 20.22 expresses the action of one side of the fluid motor. Since a positive load flow on one side results in a negative load flow for the other side, the action of the second side of the fluid motor can be expressed by

$$\frac{dP_2}{dt} - \frac{kP_2}{\mathcal{V}_2}(A_r\dot{z} + Q_{m2}) = 0 \tag{20.23}$$

For small load-pressure changes near a no-load condition, the pressures may be expressed as

$$P_1 = P_0 + \Delta P_1 \tag{20.24}$$

$$P_2 = P_0 + \Delta P_2 \tag{20.25}$$

and if $\Delta P_1 \approx -\Delta P_2$, then

$$P_1 + P_2 = 2P_0 \tag{20.26}$$

where the constant no-load cylinder pressure, P_0, with the valve in mid-position is given by

$$P_0 = \frac{P_s}{\alpha} f_1\left(\frac{P_0}{P_s}\right) \tag{20.27}$$

Furthermore, operation about the mid-point of the motor travel is assumed, and therefore, volume $\mathcal{U}_1$ remains almost equal to volume $\mathcal{U}_2$:

$$\mathcal{U}_1 \approx \mathcal{U}_2 = \mathcal{U}_i \tag{20.28}$$

Motor action about the mid-position has been shown by an extensive non-linear analogue computer study [5] to result in the least stable operation.

20.73. Combination of Valve and Motor Equations

Substitution of Eqs. 20.13, 20.14, 20.26, 20.27, and 20.28 into the difference of Eqs. 20.22 and 20.23 results in

$$\frac{\mathcal{U}_i}{kP_s}\left[\frac{\alpha}{2f_1(P_0/P_s)}\right]\frac{dP_l}{dt} + A_r\dot{z}$$

$$= Q_{mr}\left(\gamma - \frac{\alpha}{2f_1(P_0/P_s)}\left\{\frac{P_l}{P_s} - \frac{2}{\alpha}\left[f_1\left(\frac{P_1}{P_s}\right) - f_1\left(\frac{P_2}{P_s}\right)\right]\right\}\right) \tag{20.29}$$

Equation 20.29 represents the operation of the flow-valve and motor system. The first term represents the compressibility flow necessary for a rate of change in load pressure. The second term is the flow necessary to maintain a steady ram velocity. The right side of the equation represents the flow from the four-way valve and, therefore, is an analytical expression of the valve characteristics. The first term in the outer parentheses, γ, represents the spacing, and the second term represents the slope of the curves in Fig. 20.4.

For the case of $\alpha > 1.8$, $f_1(P_0/P_s) = 1$. Furthermore, for the assumed small changes in P_1 and P_2, $f_1(P_1/P_s) = f_1(P_2/P_s) = 1$. With these assumptions, Eq. 20.29 becomes

$$\frac{\alpha}{2}\frac{\mathcal{U}_i}{kP_2}\frac{dP_l}{dt} + A_r\dot{z} = Q_{mr}\left(\gamma - \frac{\alpha}{2}\frac{P_l}{P_s}\right) \tag{20.30}$$

The right-hand side of Eq. 20.30 is a linearized expression for steady-state characteristics of the four-way valve. The validity of this expression is substantiated by the excellent linearity that exists within the expected operating region of the valve characteristics.

20.74. Electromechanical Valve Actuator

The electromechanical transducer was treated as a position and force source having negligible dynamic lag. The force equation of the simpli-

---

[5] J. L. Shearer, "Continuous Control of Motion with Compressed Air," Sc.D. Thesis, Department of Mechanical Engineering, Massachusetts Institute of Technology, Cambridge, Mass., 1954.

fied actuator and its electronic amplifier is given by (see Chapter 11)

$$F_m = K_1 K_a e - K_3 x \tag{20.31}$$

where F_m = force developed by valve actuator, lb
    K_1 = current sensitivity of valve actuator, lb/amp
    K_a = gain of d-c amplifier, amp/volt
    e = error-signal voltage, volts
    K_3 = effective stiffness of valve actuator, lb/in.

Since all dynamic effects are neglected, the only force acting on the rotor of the actuator results from the load pressure,

$$F_m = P_l A_p \tag{20.32}$$

where A_p = pressure-sensitive area of flapper, in.^{2}

20.75. Gas-Motor Load System

The fluid motor applies a force to the mass, m, in the presence of a load spring. All damping effects are assumed viscous. The force equation for the load is

$$m_l \ddot{z} + b_l \dot{z} + k_l z = P_l A_r - F_l \tag{20.33}$$

20.76. Gas Generator

The gas generator is considered a constant-pressure gas source which is unaffected by any system variables.

20.8. SYSTEM ANALYSIS

20.81. Open-Loop Characteristics

The open-loop characteristics of the simple system are obtained by combining Eqs. 20.30, 20.31, 20.32, and 20.33 with the load-spring stiffness $k_l = 0$. The resulting differential equation expressing the open-loop system response to a time-varying input, $e(t)$, in the absence of a load disturbance is given by

$$\left\{ (m_l D + b_l) \left[\frac{\alpha \mho_i}{2kP_s} D + \left(\frac{Q_{mr}}{P_s} + \frac{Q_{mr} A_p}{x_0 K_3} \right) \right] + A_r{}^2 \right\} Dz$$
$$= \left(\frac{Q_{mr} K_1 A_r K_a}{x_0 K_3} \right) e \tag{20.34}$$

where D represent derivative with respect to time, d/dt. The characteristics of the second-order term in the form of the undamped natural frequency, ω_n, with the damping ratio, ζ, are instructive in predicting closed-loop behavior with position feedback. The undamped natural frequency is given by

$$\omega_n{}^2 = \frac{2\left[A_r{}^2 kP_s + kb_l Q_{mr}\left(1 + \dfrac{A_p P_s}{x_0 K_3}\right)\right]}{\alpha m_l \mathcal{V}_i} \tag{20.35}$$

The damping ratio is given by

$$\zeta = \frac{\dfrac{b_l \mathcal{V}_i \alpha}{2kP_s} + m_l \dfrac{Q_{mr}}{P_s} + \dfrac{Q_{mr} A_p}{x_0 K_3}}{2\sqrt{\left(A_r{}^2 + \dfrac{b_l Q_{m0}}{P_s} + \dfrac{b_l Q_{mr} A_p}{x_0 K_3}\right)\left(\dfrac{\alpha m_l \mathcal{V}_i}{2kP_s}\right)}} \tag{20.36}$$

Therefore, the system damping consists of three independent phenomena:

1. Viscous damping acting on the load.
2. Load-pressure sensitivity of the flow.
3. Load-pressure feedback through the flapper.

20.82. Steady-State Closed-Loop Characteristics

In order for the servomechanism to operate as a position-control system, the position of the output member must be compared with the desired position, which is in the form of a voltage signal, e^*. The load position, z, therefore, must be transduced into a voltage, e_l. The error voltage is given by

$$e = e^* - e_l \tag{20.37}$$

The position-transducing constant, K_f, which is the position-feedback gain, is defined by

$$e_l = K_f z \tag{20.38}$$

The steady-state gain of the closed-loop system in the presence of a load spring is given by

$$\frac{\Delta z}{\Delta e^*} = \frac{K_a K_1 P_s A_r}{k_l(A_p P_s + K_3 x_0) + K_f K_a K_1 P_s A_r} \tag{20.39}$$

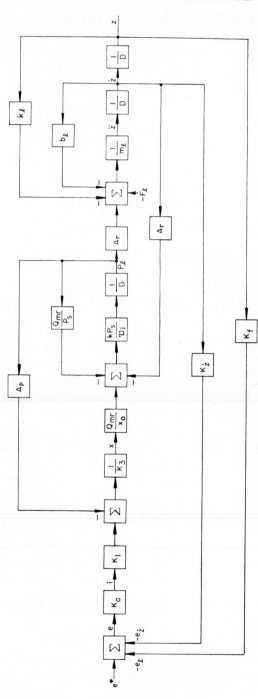

Fig. 20.10.   Block diagram of the closed-loop system.

In the absence of the load spring, Eq. 20.39 simplifies to

$$\frac{\Delta z}{\Delta e^*} = \frac{1}{K_f} \tag{20.40}$$

An important steady-state characteristic of the system is the stiffness of the output member to changes in the externally applied forces, F. The load stiffness is given by

$$\frac{\Delta F_l}{\Delta z} = k_l + \frac{K_1 K_f A_r P_s}{K_3 x_0 + A_p P_s} \tag{20.41}$$

The second term represents the internal stiffness of the control system.

20.83. Dynamic Closed-Loop Study

The system equations, Eqs. 20.30, 20.31, 20.32, 20.33, 20.37, and 20.38, can be presented in block-diagram form. A diagram of this third-order closed-loop system is shown in Fig. 20.10; both velocity feedback and position feedback are indicated. The aim of this study is to select values of the system parameters which will result in a satisfactory dynamic response of the system. The well-known graphical techniques of control-system synthesis, when applied to higher than second-order systems, lend themselves best to the study of system performance when a single system variable such as position-feedback gain is changed. Thus, every change in any other system constant requires the redrawing of either the root locus or the normalized gain-phase diagrams. The correct feedback gains for a specific response requirement can be determined for each case. The large amount of labor that the graphical techniques entail discourages graphical and analytical experimentation. A further limitation of these methods is the complexity that results if nonlinearities are included. In this study, none of the system constants is fixed. The recourse to machine computing techniques to aid in the final selection of the system constants was, therefore, a natural consequence.

The following example illustrates the design process.

20.9. DESIGN EXAMPLE

The performance specifications for a small pneumatic control system having angular output motion are tabulated in Table 20.1 along with equivalent linear output characteristics.

Table 20.1. PERFORMANCE SPECIFICATIONS

Steady-State Response

Motor torque at maximum angular velocity	396 in. lb
External load force at maximum ram velocity	330 lb
Maximum angular velocity	10 radians/sec
Maximum ram velocity	12 in./sec
Inertia of load, J_l	0.008 in. lb sec^2
Equivalent load mass, m_l	0.0056 lb sec^2/in.
External-load spring stiffness (angular), K_l	1344 in. lb/radian
Equivalent-load spring stiffness (linear), k_l	1120 lb/in.

Transient Response

Rise time to a step input (time final amplitude is first reached)	8 milliseconds
Maximum overshoot	50 per cent

20.91. Steady-State Performance Calculations

The maximum power output of the system from the data of Table 1 is $\text{HP}_{max} = 3960$ in. lb/sec. The nondimensionalized power curves, Fig. 20.7, for $\alpha = 2$ give a maximum power ratio of $(\text{HP}/\text{HP}_r)_{max} = 0.24$. The maximum power ratio occurs at a load-pressure ratio of $P_l/P_s = 0.47$. The downstream-orifice area with the valve centered is calculated from the expressions for the nondimensional power, Eqs. 20.12, 20.19, and 20.20:

$$A_{v0} = \frac{\text{HP}_{max}g}{P_s C_1 R \sqrt{T_s} \,(\text{HP}/\text{HP}_r)_{max}} \qquad (20.42)$$

For air at a temperature of 520°F abs and a pressure of 2000 psi, the mid-position downstream-orifice area is $A_{v0} = 0.00107$ sq in. In order to prevent a loss of control at the downstream orifice, the maximum area of the downstream orifice should be no greater than approximately one-third of the approach area; therefore, the mid-position flow area is one-sixth of the approach area, with the result that $D_n = 24x_0$. An assumed orifice opening of 0.003 in. requires a nozzle diameter of $D_n = 0.072$ in. An area ratio of $\alpha = 2$ results in the following valve design:

$$D_n = 0.113 \text{ in.}$$

$$x_0 = 0.003 \text{ in.}$$

$$A_n = 0.010 \text{ in.}^2$$

$$D_f = 0.026 \text{ in.}$$

If there is an extensive flat on the nozzle, then the value of the pressure-sensitive area, A_p, will be larger than that of the approach area, A_n. In this case $A_n = A_p$.

For a load-pressure ratio of 0.47, the nondimensionalized motor flow $q_m = 0.52$ (Fig. 20.4). With appropriate substitutions in Eqs. 20.12 and 20.13, the ram flow at maximum motor power is given by

$$Q_m = q_m C_1 \frac{R}{q} \sqrt{T_s}\, A_{v0} \tag{20.43}$$

with the result that $Q_m = 4.20$ in.^{3}/sec and $Q_{mr} = 8.27$ in.^{3}/sec. The load pressure for the maximum-power condition is 940 psi. If the radius of the rocker arm is assumed to be 1.20 in., the required ram area is $A_r = 0.421$ in.^{2} The volume under compression for one side of the fluid motor including all passages is assumed to be 0.30 in.^{3}, which permits an angular deflection of $\pm 25°$.

20.92. Open-Loop Performance Calculations

The validity of the choice of the system parameters based on the static requirements is examined with the characteristics of the open-loop system. The open-loop undamped natural frequency from Eq. 20.35 is

$$\omega_n |_{b_l=0} = 86.5 \text{ cps}$$

The open-loop damping ratio with no load damping and an estimated stiffness of $K_3 = 4000$ lb/in. for the electromechanical actuator is

$$\zeta |_{b_l=0} = 0.10$$

The undamped natural frequency has a sufficiently high value, but the open-loop damping ratio is too low.[6] An appreciable amount of open-loop damping is desirable in order to reduce the effect of variations in the load damping on the stability of the closed-loop system. The system can be stabilized effectively with open-loop damping ratios as low as 0.10, but since only a low value of position-feedback gain can be used in this case, the system becomes very load-sensitive.

A study of Eq. 20.36 reveals that several courses of action may be taken to increase the open-loop damping:

1. The pressure-sensitive area, A_p, can be increased. If this increase is made and no other system constants are changed, a higher force loading of the valve actuator and a small mid-position orifice opening, x_0,

---

[6] J. L. Shearer, "Dynamic Characteristics of Valve-Controlled Hydraulic Servomotor," *Trans. ASME*, Vol. 76 (August 1954), pp. 895–903.

result. However, any drastic reduction of x_0 below the chosen value of 0.003 in. is undesirable.

2. The quiescent flow through the valve can be doubled by doubling Q_{mr}. If x_0 remains constant, a doubling of Q_{mr} will quadruple A_p. In this manner, the damping ratio is increased to 0.55. However, the power drain on the supply source is doubled.

3. The electromagnetic-actuator net spring stiffness, K_3, could be reduced by a factor of 2. However, since a very fast actuator response was assumed in the system analysis, this reduction in the speed of response of the actuator no longer may be neglected.

A compromise solution whereby the upstream-orifice diameter is increased to 0.031 in. and the electromagnetic-actuator stiffness is decreased to 2000 lb/in. yields the following results for the same value of x_0:

$$Q_{mr} = 12.0 \text{ in.}^3/\text{sec}$$

$$D_n = 0.165 \text{ in.}$$

$$A_n = 0.0211 \text{ in.}^2$$

$$\zeta = 0.413$$

The revised design parameters are not final and are subject to changes during the closed-loop dynamic-response study on the analogue computer.

20.93. Closed-Loop System Design

The block diagram of the closed-loop system with tachometric feedback is shown in Fig. 20.10. A high-speed electronic analogue computer made by George A. Philbrick Researches, Inc., of Boston, Mass., was used in this design study. The investigation was limited to a study of a linear approximation to the system. The following assumptions form the basis for the use of linear differential equations:

1. Small perturbations are considered in order to simulate linear valve operation.

2. The ram motor is operated near its mid-position; as a result, the volumes under compression remain essentially equal.

3. The electromagnetic-actuator operation is linear.

4. All friction phenomena are limited to viscous-type drags.

5. Backlash in the system is negligible.

$Q_{mr} = 8.27$ in.^{3}/sec
$D_f = 0.026$ in., $A_f = 0.00064$ in.^{2}
$D_n = 0.113$ in., $A_n = 0.10$ in.^{2}
$K_3 = 4000$ lb/in.
$A_r = 0.421$ in.^{2}
$J_l = 0.008$ in. lb sec^2
$m_l = 0.0056$ in. lb sec^2
$\mathcal{V}_i = 0.30$ in.^{3}

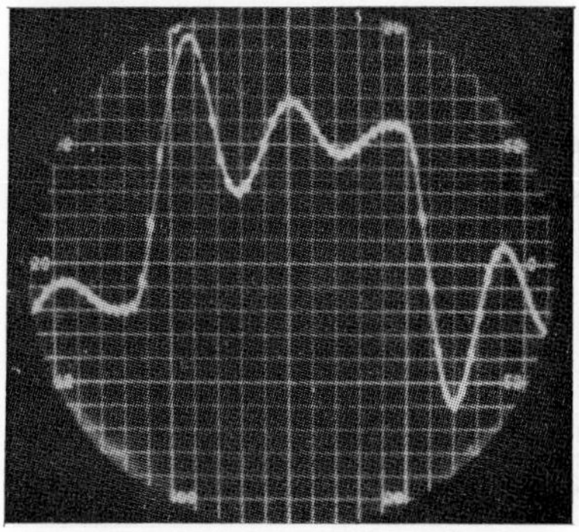

Fig. 20.11. Open-loop transient response of statically designed system.

The accuracy of the analogue computer results was checked by the open-loop velocity response to a step input signal with zero load spring stiffness. Thus, in Fig. 20.11, the predicted undamped natural frequency of 86.5 cps and the damping ratio of 0.10 are verified for the unmodified system. An increase in the quiescent flow through the valve and a decrease in the valve-actuator stiffness result in an open-loop velocity response with better damping. The effect of introducing position feedback to the modified system is shown in Fig. 20.12. The system is oscillatory. The closed-loop position response of the system with a small amount of load damping was effectively stabilized by the introduction of some velocity feedback as shown in Fig. 20.13. An optimum amount of

$K_1 K_a K_f = 0.40$
$K_1 K_a K_{\dot{z}} = 0$

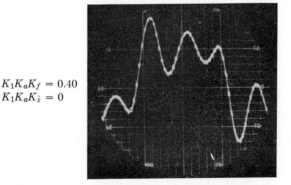

Fig. 20.12. Closed-loop transient position response of modified system.

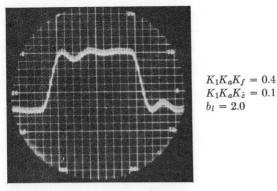

$$K_1 K_a K_f = 0.4$$
$$K_1 K_a K_{\dot{z}} = 0.1$$
$$b_l = 2.0$$

Fig. 20.13. Effect of introducing velocity feedback.

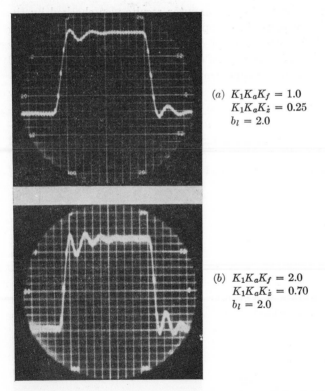

(a) $K_1 K_a K_f = 1.0$
$\quad K_1 K_a K_{\dot{z}} = 0.25$
$\quad b_l = 2.0$

(b) $K_1 K_a K_f = 2.0$
$\quad K_1 K_a K_{\dot{z}} = 0.70$
$\quad b_l = 2.0$

Fig. 20.14. Effect of simultaneously increasing velocity-feedback and position-feedback gains.

(a) $b_l = 4.0$ lb sec/in.

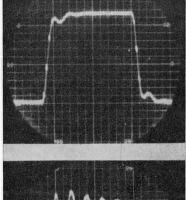

(b) $b_l = 1.0$ lb sec/in.

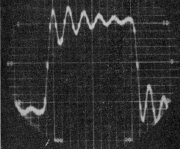

Fig. 20.15. Effect of varying load damping.

velocity feedback, based on the least oscillatory system, was found to exist for each value of position-feedback gain and load damping. Furthermore, the adjustment of the velocity-feedback gain for least oscillatory response became increasingly critical with reduced open-loop damping.

The effect of a simultaneous increase in both position-feedback and velocity-feedback gains to attain a faster yet adequately damped system is demonstrated in Figs. 20.14a and b. An attempt was made to increase the feedback gains in such a way that the first peak of the transient response would occur at the final amplitude. The rise time in Fig. 20.12 is 7 milliseconds; in Fig. 20.14a this has been decreased to 6 milliseconds, and in Fig. 20.14b, to 4 milliseconds.

The sensitivity of the transient response of the optimized system to variations in load damping is shown in Figs. 20.15a and b. The effects of halving and doubling the load mass are shown in Figs. 20.16a and b.

The results of the analogue computer study predict that a pneumatic control system with an acceptable response can be designed.

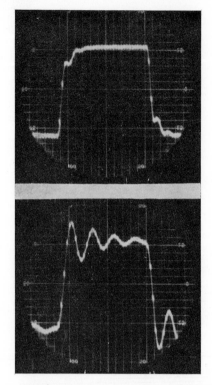

(a) $m_l = 0.0048$ lb sec^2/in.

(b) $m_l = 0.0192$ lb sec^2/in.

Fig. 20.16. Effect of varying load mass.

20.94. Test Result with Experimental Model

An experimental model utilizing the dimensions that resulted from the analogue computer study has been built. The design details of the mechanism are illustrated in Fig. 20.17. The steady-state pressure-sensitivity tests for the flow valve and the load-flow versus load-pressure characteristics showed reasonable agreement with the predicted results. The transient response of the system for small travel of the motor about its mid-position is shown in Fig. 20.18. The 2000-cps carrier frequency of the position signal indicated that the servomotor has a rise time of about 4 milliseconds, which is in good agreement with the predicted response. The sinusoidal nature of the overshoot of the position response confirmed the assumptions of linear system characteristics, in particular, the viscous load damping resulting from the grease-extruding pistons. Frequency-response tests in the form of Lissajous figures provided further proof that the assumptions of linearity were justified for small motor travel.

Several types of piston seal were experimented with before the grease-extruding type of piston was devised. Because the analysis and the analogue study apply only to viscous friction of the motor, the detrimental effect of stiction could not be predicted. The stabilization of the system with nonviscous friction is far more difficult, and the resulting motion of the output member occurs in small steps. Owing to the reduced compressibility, the performance of an oil hydraulic control system was found to be far less sensitive to such nonviscous phenomena than the performance of the equivalent pneumatic system. Rubber and Teflon O-rings, Graphalloy piston rings, and cast-iron piston rings were tested with various boundary lubricants. With the exception of the grease-extruding pistons, semicircular rubber rings placed in undersized grooves with the circular section facing the cylinder wall produced the best results.

20.95. Conclusions

A pneumatic servomotor was developed to meet exacting requirements. The assumptions used in the linear analysis of the components were checked experimentally. The design of the system was based on a thorough study of the system on an analogue computer. Because the effects of variations in system parameters were demonstrated by the analogue computer study, a great deal of time was saved during the process of refining the experimental model. The advisability of designing a pneumatic motor with a minimum of nonviscous friction is an important conclusion that can be drawn from these tests.

To the writer, the most gratifying experience of this project was derived from the fact that the system performance would be predicted quantitatively on an analytical basis and with a minimum of cut-and-try fumblings.

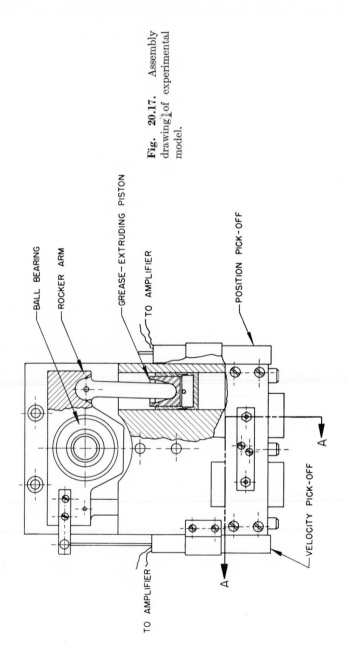

Fig. 20.17. Assembly drawing of experimental model.

BALL BEARING

ROCKER ARM

GREASE-EXTRUDING PISTON

TO AMPLIFIER

POSITION PICK-OFF

VELOCITY PICK-OFF

TO AMPLIFIER

A

A

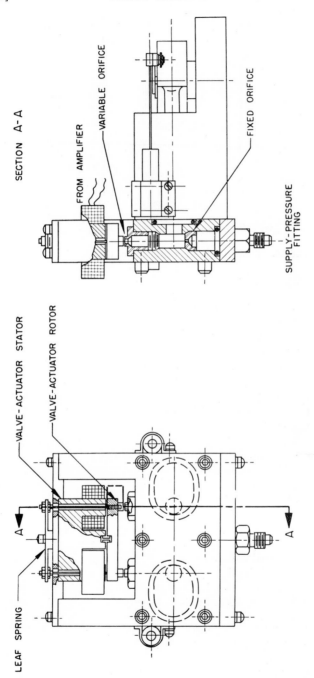

SECTION A-A

VARIABLE ORIFICE

FROM AMPLIFIER

FIXED ORIFICE

SUPPLY-PRESSURE
FITTING

VALVE-ACTUATOR STATOR

VALVE-ACTUATOR ROTOR

A

A

LEAF SPRING

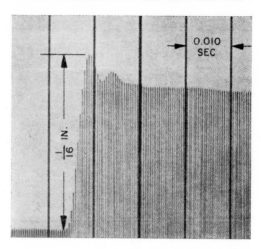

Fig. 20.18. Transient position response of the experimental system.

NOMENCLATURE

Symbol	Definition	Unit
A	minimum area of flow	in.^{2}
A_f	fixed upstream-orifice area	in.^{2}
A_n	area of approach passage to downstream orifice (nozzle area 0)	in.^{2}
A_p	flapper pressure-sensitive area	in.^{2}
A_r	ram area	in.^{2}
A_v	flow area of variable downstream orifice	in.^{2}
A_{v0}	downstream-orifice mid-position flow area	in.^{2}
b_l	load-damping coefficient	lb sec/in.
C_2	weight-flow coefficient	$(\degree\mathrm{R})^{\frac{1}{2}}/\mathrm{sec}$
C_d	discharge coefficient of orifice	
dm	incremental weight flow of gas (not a rate of flow)	lb
D_n	diameter of approach passage to downstream orifice (nozzle diameter)	in.
e	error-signal voltage	volts
e^*	command signal	volts
e_l	voltage proportional to load position	volts
e_z	voltage proportional to load velocity	volts
F_l	load disturbance	lb
F_m	force developed by valve actuator	lb
g	gravitational acceleration	in./sec^2
i	valve actuator current	amp
J_l	load inertia	in. lb sec^2
k	ratio of specific heats	

Symbol	Definition	Unit
k_l	load spring (linear equivalent) stiffness	lb/in.
K_1	current sensitivity of valve actuator	lb/amp
K_3	effective stiffness of valve actuator	lb/in.
K_a	gain of d-c amplifier	amp/volt
K_f	position-feedback gain	volts/in.
K_l	load spring stiffness	in. lb/radian
$K_{\dot{z}}$	velocity-feedback gain	volt sec/in.
m_l	load mass	lb sec²/in.
P_0	zero-load cylinder pressure with valve in mid-position	psia
P_1	pressure on ram 1	psia
P_2	pressure on ram 2	psia
P_d	downstream stagnation pressure	psia
P_l	load pressure, $P_1 - P_2$	psia
P_s	supply-source pressure	psia
P_u	upstream stagnation pressure	psia
HP	power	lb in./sec
HP_r	nondimensionalizing power	lb in./sec
q_m	nondimensional volume flow to motor	
Q_m	volume flow to motor	in.³/sec
Q_{mr}	reference flow rate to motor	in.³/sec
R	gas constant	in.²/sec² °R
T	temperature	°R (°F abs)
T_1, T_2	gas temperature in chambers 1 and 2	°R
T_a	upstream stagnation temperature	°R
T_s	supply-gas temperature	°R
u	specific internal energy of gas	in.²/sec²
$\mathcal{V}_1, \mathcal{V}_2$	volumes of motor chambers 1 and 2	in.³
$\mathcal{V}_i$	initial volume under compression with motor at mid-position	in.³
W	weight rate of flow of gas	lb/sec
W_e	exhaust flow through downstream orifice	lb/sec
W_m	motor flow	lb/sec
W_s	supply flow through upstream orifice	lb/sec
x	upward displacement of left end of valve actuator from mid-position	in.
x_0	mid-position opening of each variable orifice	in.
z	ram position	in.
$\dot{z}$	ram velocity	in./sec
$\ddot{z}$	ram acceleration	in./sec²
α	orifice-area ratio, A_{v0}/A_f	
γ	nondimensionalized valve displacement, x/x_0	
Δ	incremental change in quantity	
ζ	damping ratio	
ρ	density of gas	lb/sec²/in.⁴
ω_n	undamped natural frequency of open-loop servomotor	radians/sec

═ *Index*

695

THE UNIVERSITY OF STRATHCLYDE
ANDERSONIAN
LIBRARY